Grazing in Temperate Ecosystems: Large Herbivores and the Ecology of the New Forest

GRAZING IN TEMPERATE ECOSYSTEMS
LARGE HERBIVORES AND THE ECOLOGY OF THE NEW FOREST

R. J. PUTMAN
Department of Biology,
University of Southampton

CROOM HELM
London & Sydney

TIMBER PRESS
Portland, Oregon

Croom Helm Ltd, Provident House, Burrell Row,
Beckenham, Kent BR3 1AT

Croom Helm Australia Pty Ltd, Suite 4, 6th Floor,
64-76 Kippax Street, Surry Hills, NSW 2010, Australia

British Library Cataloguing in Publication Data

Putman, Roderick J.
Grazing in temperate ecosystems: large herbivores and the ecology of the New Forest.
1. Ecology—England—New Forest
2. Herbivora—England—New Forest
I. Title
574.5′264 QH138.N4

ISBN 0-7099-4036-X

First published in the USA 1986 by
Timber Press
9999 S.W. Wilshire
Portland OR 97225
USA

ISBN 0-88192-071-1

Typeset in Plantin Light by Leaper & Gard Ltd., Bristol, England

**Printed and bound in Great Britain by
Biddles Ltd, Guildford and King's Lynn**

Contents

Preface

The New Forest in southern England is an area of mixed vegetation set aside as a Royal Hunting Forest in the eleventh century and since that time subjected to heavy grazing pressure from large herbivores. The entire structure of the Forest and its various communities has been developed under this continued history of heavy grazing, with the establishment of a series of vegetational systems unique within the whole of Europe. The effects of large herbivores in the structuring of this ecosystem in the past, and the pressure of grazing continuing to this day, have in turn a profound influence, indeed the dominating influence, on the whole ecological functioning of the Forest system. Because of its assemblage of unique vegetation types, the area is clearly of tremendous ecological interest in its own right. In addition, its long history of heavy grazing and the continued intense herbivore pressure make the New Forest an ideal study-site for evaluation of both short-term and long-term effects of grazing upon temperate ecosystems.

The New Forest (some 37,500 ha in total area) currently supports a population of approximately 2,500 wild deer (red, roe, sika and fallow); in addition 3,500 ponies and 2,000 domestic cattle are pastured on the Forest under Common Rights. From 1977, I have, together with a number of associates, undertaken a series of research studies on the ecology and behaviour of the large herbivores within the Forest, examining the various ways in which the different species use their Forest environment and considering their combined influence and impact upon the Forest vegetation. Slowly we are beginning to piece together some understanding of the complex functioning of this multi-species system, and to appreciate the influence of the heavy grazing pressure by the Forest herbivores on vegetational processes. Nor are the effects of grazing restricted to vegetational change: it is clear that, through its dominating effect upon the vegetation, the intense grazing pressure imposed by the larger herbivores has repercussions *throughout* the system — 'knock-on' effects upon other organisms reliant on this shared vegetational environment — until it affects in practice the whole ecological shape and functioning of the New Forest system.

In this book I have attempted to draw together the results of this work to present some kind of synthesis. It may be read as a series of separate studies within the New Forest of the ecology and behaviour of a number of different species of large ungulate: presenting a current review of our knowledge to date of the autecology of these species. I hope that it may also be read as a whole: as an investigation of the effects of heavy grazing on the dynamics and functioning of a temperate ecosystem.

A synthesis of this sort necessarily draws upon the work of many besides myself. It is a pleasure to acknowledge the debt I owe to all those whose work is quoted here, and to whom belongs the full credit for the tremendous contribution each has made to our understanding of the New Forest and its ecology. Particularly I would honour my research assistants Bob Pratt and Rue Ekins, who undertook most of the monumental amount of fieldwork that went into our studies of the Forest ponies and cattle, my various research colleagues and research students: Andy Parfitt (working on fallow deer), Chris Mann (sika deer), Steve Hill and Graham Hirons (whose work unravelled the complex effects of grazing upon the Forest rodent populations and their dependent predators) and Elaine Gill. I owe in addition an immense debt to my long-standing friend and colleague Peter Edwards, who has advised and assisted with supervision of the more botanical elements of our work throughout. Thanks are also due to various others who have worked within the Forest and have generously allowed me to quote their work: Norman Rand, John Jackson and Stephanie Tyler.

While I owe a tremendous debt to all these scientists, my thanks must also go to the Nature Conservancy Council, the Forestry Commission, the Verderers and all the New Forest Commoners who individually or collectively have supported our studies. Colin Tubbs of the Nature Conservancy Council first focused our attention on the New Forest system and persuaded the NCC to finance our initial studies of the cattle and ponies. His local knowledge of the Forest and its ecology is unrivalled and it has been a pleasure to have had his close interest and support in our work ever since. We owe a debt also to the various officers and staff of the New Forest Forestry Commission, who have over the years most patiently tolerated our tireless but frequently tiresome interest in the Forest. Far from merely putting up with us, they too have been active in support of our studies; we have made many friends in the

District Office and among the keepers and foresters. To all of these go my grateful thanks: I hope that this book, as the culmination of all those years of study, will contain something of use to them in the future management of the Forest.

Finally, I would thank Dawn Trenchard for coping so nobly with the horrendous task of creating an ordered typescript out of my scrawled chaos, Barry Lockyer and Raymond Cornick for help with the diagrams and other colleagues known and unknown who have commented on various bits of the manuscript itself.

Chapter 1

Introduction

Much attention has been devoted in recent years to the role played by large herbivores in shaping and maintaining vegetational systems. It is perhaps self-evident that in any system where grazing animals, whether large *or* small, occur in any density they are bound to exercise a significant influence upon that system: altering vegetational structure, diversity and productivity.

Although, typically, herbivores remove perhaps only in the region of 10% of the annual green-matter production in any community, in certain instances their impact may be far in excess of this. Wiegert and Evans (1967) estimate that ungulates may remove between 30% and 60% of primary production of East African grasslands, and Sinclair and Norton–Griffiths (1979) calculate that herbivores (both vertebrate and invertebrate) in the Serengeti National Park in Tanzania are removing up to 40% of the annual primary production. Even smaller herbivores can have a dramatic effect if they occur in sufficient density: in the tundra, arctic lemmings may remove up to 90% of available primary production (Schultz 1969), while even tiny invertebrates, massed together, can account for losses of between 10% and 30% of annual production of natural grasslands (Andrzejewska 1967; Schuster *et al.* 1971) or field crops (e.g. Bullen 1970, Clements 1978). Nor is the impact accurately measured by the amount of material actually ingested. Animal feeding is often wasteful: in studies of grasshoppers feeding on the black needle rush (*Juncus roemerianus*), Parsons and de la Cruz (1980) found that the insects consumed only 0.33% of annual net production; yet, as a result of the insects' habit of feeding in mid leaf, loss of production was three times that figure because of material clipped off and discarded. Red and roe deer browsing in deciduous forests in southern Poland show a comparable wastage: the deer remove some 46 kg (dry weight) of browse material per hectare; of this they actually consume only 19 kg (Bobek *et al.* 1979), the balance being destroyed during feeding.

Further, it is clear that grazing may have a *far* greater impact on the community than is suggested by mere consideration of the absolute quantities of plant material removed: a herbivore ingesting perhaps 10% of a plant's production is going to have a far more significant effect upon the plant if that 10% is made up of primordia, destined for future growth, than if it merely results in a loss of 10% in the form of mature leaves. In the study of Bobek *et al.*, potential browse production of an unbrowsed forest was estimated as 172 kg/ha; actual production in practice (including the 46 kg later removed by deer) totalled only 160 kg/ha. Damage by deer *suppressed* productivity by 12 kg/ha, in addition to its effect in merely removing 46 kg of that production (Bobek *et al.* 1979). In another, similar study browsing by moose in a pine/mountain ash forest in Russia reduced forage biomass from 181 kg/ha to 109 kg/ha; only 3.5 kg of this loss was directly related to moose feeding, the remaining 68.5 kg being due to the reduced growth rate of the damaged trees (Dinesman 1967).

In fact, we may note a whole variety of these more subtle effects of grazers upon vegetation, where quite minimal *absolute* consumption by herbivores may have far-reaching significance.

(i) We may see a change in productivity. We have noted here a suppression of plant production due to herbivore pressure, but under other circumstances grazing may equally result in an *increase* in productivity.
(ii) Selective grazing by herbivores in a multi-species system may result in a change in species composition within the plant community as particularly graze-sensitive species are eliminated, and resistant, or tolerant, plant species may increase in dominance.
(iii) Grazing or browsing may alter natural patterns of nutrient cycling within the system.
(iv) Grazing may cause a change in the physical structure of the vegetation, altering the physical habitat and microclimate and making it more or less suitable for other plant and animal species, causing further changes in species composition — both within the plant assemblage and within the associated animal community (review by Putman 1985a).

The effects of herbivores upon *primary production* may be direct — through defoliation in feeding — or indirect: the trampling effects of hoofed mammals may lead to soil compaction and thus affect plant growth, while return of dung and urine to the system, resulting in local changes in soil-nutrient status, may also affect productivity. Direct removal of plant tissue by herbivores can directly affect rate of photosynthesis, respiration rate, location of nutrient storage, growth rates and phenology of the affected plant. While, as we have noted, heavy grazing and browsing may reduce production, moderate levels of grazing may actually increase plant productivity, through stimulating some compensatory growth. Growth will be inhibited — as in our example of deer browsing in Polish forests, where herbivores damage the growth primordia of the plant, or where excessive defoliation reduces the effective leaf area of the plant below a minimum threshold for efficient photosynthesis — but there are numerous examples in the literature where lighter grazing pressure can be shown to increase productivity. Productivity of natural meadowland in Russia was greatest at vole densities of 100 per hectare (Coupland 1979), when about 20% of plant growth was being harvested; in grazed sheep pastures in Australia, net production of a *Phalaris tuberosa/Trifolium repens* sward was greatest at a stocking density of 10 sheep per hectare (Vickery 1972). Similar increases in productivity in response to herbivory can be demonstrated for woody species. Various studies (e.g. Ellison 1960; Grant and Hunter 1966; Krefting *et al.* 1966; Wolff 1978) have established that trees and shrubs which are regularly browsed by ungulates frequently show increases in productivity under *light* herbivore pressure.

Such increase in production may be due to a number of factors. McNaughton (1979) lists nine possible mechanisms which may compensate for plant-tissue loss from herbivory and may result in increased production following grazing or browsing.

1. Increased photosynthetic rates in residual tissue.
2. Re-allocation of substrates from elsewhere in the plant.
3. Mechanical removal of older tissues functioning at less than a maximum photosynthetic level.
4. Consequent increased light intensities upon potentially more active underlying tissues.
5. Reduction of the rate of leaf senescence, thus prolonging the active photosynthetic period of residual tissue.
6. Hormonal redistributions promoting cell division and elonga-

tion and activation of remaining meristems, thus resulting in more rapid leaf growth and promotion of tillering.

7. Enhanced conservation of soil moisture by reduction of the transpiration surface and reduction of mesophyll resistance relative to stomatal resistance.

8. Nutrient recycling from dung and urine.

9. Finally, it has even been suggested that one potential stimulatory effect of grazing ruminants upon productivity may arise from plant-growth-promoting agents which have been found in ruminant saliva (Vittoria and Rendina 1960; Reardon *et al.* 1972, 1974). Direct growth stimulations up to 50% above control levels have been recorded following addition of ungulate saliva to surfaces of manually clipped leaves.

Note, however, that each plant species reacts differently to different grazing pressure — and to different grazers. While we suggested that maximum production of Australian sheep pasture was realised at a stocking density of 10 sheep per hectare, rabbits grazing on such pastures at densities as low as 40 per hectare can *reduce* pasture growth by up to 25%, merely because they feed in a different way on different parts of the plants; among woody species, too, treatments which best increase production of one shrub species may have different results for another species (review by Gessaman and MacMahon 1984). Defoliation by herbivores can thus stimulate *or depress* productivity of individual plants. Clearly, the actual result observed in any one instance will depend both on the degree of defoliation and also on the *timing* at which damage occurs in relation to the growth stage and growth characteristics of the particular plant. Different plant species will have markedly different responses depending on structure and growth pattern: the level of offtake at which productivity of woody species (with terminal growth points) begins to decline is far lower than that at which production of many grasses — which grow from the base — would be suppressed.

Where defoliation is sufficient to depress productivity, continued grazing may ultimately result in eradication of particular plant species from the community. Thus, through its effect upon the individual plant, grazing can alter the entire species composition of the community, with continued eradication of species sensitive to grazing and an increase in abundance of those species which, through chemical or physical defence, or because of their growth form, may have greater abilities to resist, tolerate or escape defoliation. Even without such deletion of species, however, defoliation may still have a profound effect on community composition. By reducing the leaf area of preferred forage species, the herbivore may reduce competition for light and space experienced by other plants, which may therefore be able to colonise or increase in abundance within communities from which they would normally be out-competed.

On upland grasslands in Britain, sheep show a distinct selection for palatable grasses such as *Agrostis* and *Festuca* species (the bents and fescues): these are grazed heavily in preference to the unpalatable *Nardus stricta* (mat-grass). Each year, therefore *Nardus* tends to increase at the expense of the other species (Chadwick 1960; Nicholson *et al.* 1970). Even the initial *Agrostis/Festuca* sward is itself a result of grazing. When sheep are excluded from such areas, the *Agrostis/Festuca* grasslands develop a patchy sward dominated by *Deschampsia cespitosa* (tufted hair-grass) and *Holcus mollis* (creeping soft-grass), and herbs such as lady's-mantle (*Alchemilla glabra*), sorrel (*Rumex acetosa*), *Galium* spp. and nettles (*Urtica dioica*) (Rawes 1981). (Interestingly, once *Nardus* has been allowed to become dominant within the sward, the change in species composition brought about by grazing is not so easily reversed. Long-term sheep exclosures on *Nardus* grasslands in the Pennines have shown little change in species composition in 24 years: Rawes 1981.)

Grazing of saltmarshes by sheep leads to the replacement of broadleaved species such as sea-purslane (*Halimione portulacoides*) and sea aster (*Aster tripolium*) by the saltmarsh-grass *Puccinellia maritima* (Gray and Scott 1977), and numerous other examples of changes in species composition in response to grazing may be cited. Thus, on semi-arid grasslands in the Serengeti National Park in Tanzania, McNaughton (1979b) records marked differences in species composition of grazed areas from that observed within plots protected by fences since 1963. In one comparison he noted that the grasses *Andropogon greenwayi* and *Sporobolus marginatus* made up 56% and 20% respectively of biomass outside the exclosure, but were not recorded in samples taken inside; in contrast, *Pennisetum stramineum* and *P. mezianum* made up only 5% and 3% of the biomass in grazed areas, while contributing to the biomass of ungrazed plots by 72% and 26% respectively. Perhaps the oldest and most quoted example of how grazers can affect the species composition of whole vegetational communities is in the development of the specific plant assembly we now associate with chalk grassland in Britain, under the influence of first sheep, and later rabbit grazing (Tansley 1922;

Tansley and Adamson 1925; Hope Simpson 1940).

Grazing herbivores not only affect productivity and species composition of the vegetational communities on which they graze: they may also have considerable impact on nutrient cycling (which may in itself have yet further repercussions for species composition and productivity of the vegetation: McNaughton 1979, Crawley 1983). By feeding on plant materials, animal lock up within their own tissues essential nutrients, making them unavailable to the next 'generation' of plants. In systems where nutrients are relatively abundant, and there exists in the soil a relatively large pool of 'free' nutrients, this has but little effect upon plant growth; but in other, nutrient-poor systems, the effects of having significant quantities of a limited nutrient supply bound up in animal tissue may well be quite marked. In the tundra zone of Alaska, for example, lemmings, in years of peak abundance, may consume 40% of the available nitrogen and 50% of the phosphorus. Although ultimately all of this is returned to the soil-nutrient pool in the form of animal wastes (in faeces or as carcases), there is a considerable time delay, and for some period the vegetation becomes markedly nutrient-impoverished. The situation is complicated in this example by the cyclic nature of lemming population abundance. When lemming populations are low, few of the available nutrients are tied up in animal tissue, so plants have adequate nutrients available to them and grow well. Offered such abundant and nutrient-rich vegetation, lemming populations grow fast and reproduce rapidly. As lemming numbers increase, so fewer and fewer nutrients are left available for plant growth. Productivity and quality of the vegetation falls, which leads in turn to a decline in the reproductive performance of the lemmings themselves. At this stage in the cycle, natural mortality outstrips reproductive recruitment: reproductive rates are low, and those individuals born into the population in the last boom are reaching the end of their allotted span. Lemming numbers fall; decomposers release the nutrients of their body tissues, which once more become available to the vegetation — and the cycle begins again (Schultz 1964, 1969).

While such effects are really important only in nutrient-poor systems, *patterns* of nutrient cycling may be affected by grazing in any system. Animals that feed over a wide area but defecate in a small area can have a substantial effect on local nutrient distribution. Sheep, for example, graze widely over pasture during daylight, but congregate in camps at night or for shade; in consequence, 35% of their faeces are deposited on less than 5% of the grazing area — resulting in a gradual impoverishment of the wider grazing range, but continued enrichment of small areas within it (nutrient '*dislocation*': Spedding 1971). Other animals may show different habitat preferences for grazing and for elimination, so that nutrients may be removed from some habitats and returned to others; this *translocation* of nutrients, too, has a profound effect on the nutrient dynamics of any community. While these are effects of elimination more than grazing, pure defoliation itself may alter the water balance of the community, reducing the area of transpiring leaves, and exposing areas of the soil surface to the drying influence of sun and wind (Crawley 1983).

Finally, as we have noted, although most work has concentrated on the *direct* effects of grazers upon the vegetation itself, it is clear that the effects of the herbivores are not restricted to this level, but have a whole series of 'knock-on' effects. All the various effects of grazing and elimination result in modification of the habitat itself — and also of the environment it offers to others. A changing microclimate, through structural modification of the primary vegetation, will have an effect on the secondary plant species, and animals, which may colonise the 'new' environment. Changing species composition, and species dominance, will affect relative availability of food to other smaller herbivores. In short, through the changes that the grazing process causes within the vegetation — in structure, species composition and productivity — it has at once additional effects on the rest of the community dependent on that vegetation.

In conclusion, the potential effects of heavy grazing upon vegetation may be summarised as: increase or decrease in primary production; alteration to nutrient cycling and gross distribution of nutrients; changes in species composition (due in turn to the selective elimination of sensitive species, to changes in the competitive equilibrium of the plant community under grazing pressure, and to modification of nutrient flows within the system); and changes in the actual physical structure of the vegetation itself, affecting microhabitats offered to dependent plant and animal species.

There is yet one further whole category of animal influences on plants, where the consumer affects *other consumers* which in turn affect the plants. Gessaman and MacMahon (1984) cite the example of a predatory sea-otter (*Enhydra lutris*) feeding on invertebrate herbivores such as abalones and sea urchins (Simenstad *et al.* 1978). The removal of the invertebrate grazers increases the density and biomass of macro-algae. But predation is by no means the only biotic interaction between consumers that may alter their impact upon vegetation: competition between the herbivores themselves, in multi-grazer systems, may influence their individual effects upon the vegetation. Illustration of each of these effects may be drawn from examination of the multi-species grazing system found in the New Forest in Hampshire — a system which also provides us with a unique opportunity to investigate the overall combined effect when all these separate, potential influences act together within a single ecological system.

The New Forest

The New Forest in southern England is an area of mixed vegetation set aside as a Royal Hunting Forest in the eleventh century and since that time subjected to heavy grazing pressure from a variety of large herbivores. The entire structure of the Forest and its various communities has been developed under this continued history of 900 years of heavy grazing. Many of the Forest's vegetational communities owe their very existence to past and present patterns of grazing, and the area boasts a number of particular vegetational formations now unique within Europe. Such a system offers an unparalleled opportunity to study in detail both long-term and short-term effects of intensive grazing within a single system, while the current ecology of the area itself can only be interpreted in relation to the various effects of past and present grazing. Any attempt to describe the ecology of the Forest — in accounting for the curious lack of diversity of many vegetational systems, low numbers and diversity of small mammals, curious behaviour of birds of prey and other predators, unusual patterns of habitat use of the larger herbivores themselves — any attempt to *explain* rather than just describe, forces the attention back to the dominating effect of grazing in the shaping of this ecosystem. An account of the 'local' ecology of the New Forest, and an examination in more general terms of the ecological effects of heavy grazing, thus prove quite inseparable.

Set in the heart of the Hampshire Basin and covering an area of some 37,500 ha between the Solent and the Avon (Figure 1.1), the New Forest is more easy to define as an administrative area than as an entity. It is a diffuse region of Common land, dissected by roads and railways, perforated by towns and quite sizable agricultural holdings which have grown up within its boundaries on the better soils.

Figure 1.1: Location of the New Forest in the South of England

Even the term Forest is somewhat misleading, suggesting as it does an area densely wooded, for in fact only some 10,000 ha are actually covered with trees. In common with most royal forests, the word 'Forest' is in this context used in its older sense of an area set aside for hunting game (the modern association of a forest with trees is a purely secondary association), and in practice much of the area is open heathland or grassland.

The diversity of vegetation types within the Forest is in fact far greater than might be expected for the area, and this may be attributed to considerable variations in edaphic factors over the area. Heavily leached and base-poor plateau gravels and sands are widespread, particularly in the north, and support a *Calluna*-dominated *dry-heath* community. At lower altitude, and where the plateau gravel has been eroded, more fertile clays and loams support a mixed *deciduous woodland.* These are predominantly of beech and oak, with an understorey of holly; the common bent (*Agrostis tenuis*) colonises the woodland floors, in openings and glades. (Many of the more fertile woodland sites have been enclosed over the last 100 years and now support commercial plantations.) Also common on the more fertile soils are a range of *acid-grassland* communities, dominated by the coarse grass *Agrostis setacea* and to a lesser extent the purple moorgrass (*Molinia caerulea*), usually colonised to a greater or lesser degree by bracken and, particularly in the south, also by gorse (*Ulex europaeus*).

Where drainage is impeded in the valley bottoms, domination of the heathland by *Calluna* is diminished, and the species diversity of the whole heath community increases. A gradation is observed from the dry-heath community through *humid* and *wet heath*, with increasing abundance of cross-leaved heath (*Erica tetralix*) and *Molinia* and the appearance of true wetland plants such as bog asphodel (*Narthecium ossifragum*) and *Juncus* species. This progression frequently ends in bog communities. The *valley bogs* offer some of the richest vegetation types in the New Forest in terms of plant diversity, and are one of the formations unique to this area. The species composition varies considerably in relation to how eutrophic the water supply is, and several distinct communities can be recognised. Perhaps the most widespread in base-poor water is that dominated by *Molinia* tussocks with common cottongrass (*Eriophorum angustifolium*) and *Sphagnum* mosses abundant between the tussocks. In many heathland catchments, *carr woodland* develops in the valley bottom where drainage waters have a definite axis of flow. These carr woodlands are composed of *Salix atrocinerea*, alder buckthorn (*Frangula alnus*), alder (*Alnus glutinosa*) and other tree species, and have a diverse herb layer including the greater tussock-sedge (*Carex paniculata*).

Where the area is well drained by one of the many small streams which transect the Forest, the bogs are replaced and the heathland progression terminates abruptly at the edge of alluvial strips bordering the streams. These alluvial deposits are covered by grassland, often dominated by *Agrostis canina*, interrupted with patches of riverine woodland. These *streamside lawns* are particularly nutrient-rich because of regular annual flooding from the rivers which they border, which carry base-rich compounds from north of the Forest.

Very little of the New Forest vegetation can be considered natural, and most areas have at various times been subjected to management by man. This has resulted in the creation of a number of new community types, the most obvious of which are the *re-seeded lawns*, areas of acid grassland which were fenced during the Second World War, ploughed, fertilised, and cropped for potatoes and oats. At the end of the war, these were re-seeded with a ley and after the grassland was established the fences were removed. Many of the sown species have since disappeared from the sward, which has been recolonised by natural grassland species, but the areas still comprise a very distinct vegetation type. In the late 1960s and early 1970s, other attempts to improve the grazing and grassland were made by swiping bracken from certain areas and liming the grassland. These improved areas once again form a distinct and characteristic vegetational community. Finally, the heathland communities are subjected to continued management, being cut, or burnt on a rotational basis, so that any extensive area of heath contains a patchwork of sub-communities from 0 to 12 years of maturity.

Of the total administrative area of the Forest, some 9,000 ha are occupied by towns, villages or agricultural land; 8,300 ha have been enclosed for commercial plantation; while only some 20,000 ha remain as 'wasteland' or 'Open Forest' whose vegetation we have just described. This 20,000 ha, with its mixture of vegetational communities, supports a big population of large herbivores. Some 2,500 wild deer (red, roe, sika and fallow) have access to the entire area, while the open, unenclosed Forest supports in addition approximately 3,500 ponies and 2,000 cattle, pastured on the Forest under ancient Rights of Common.

Of the four deer species present on the Forest today, red deer (*Cervus elaphus*) and sika (*Cervus nippon*) are essentially local in distribution: populations are restricted to limited areas of the Forest. Roe (*Capreolus capreolus*) are everywhere uncommon, and only fallow deer (*Dama dama*) are both widespread and abundant (numbers estimated at about 2,000 in 1976: Strange 1976). (It has been suggested that muntjac deer, *Muntiacus reevesi*, may also now be resident within the Forest woodlands, but reports are unconfirmed and numbers are in any case low.) Red deer and roe are both native to Great Britain. Fallow deer were introduced by the Romans or Normans and may be considered an old-established British species; sika have been introduced more recently. Within the New Forest, populations of both fallow and roe are natural, and must be presumed to have existed since the Forest's creation in 1069. Red deer were also almost certainly native to the area, but present-day populations derive from two separate introductions in different parts of the Forest (page 112).

Domestic animals are also grazed upon the 20,000 ha of the Open Forest. One of the concessions granted to the local populace when the area was declared a Royal Hunting Forest was the right of common grazing. On the right payment of an appropriate 'marking fee' (presently set at £10 an animal), local cottagers and farmers could turn out cattle and horses to exploit the rough grazing of the forest lands. These rights are still honoured, and in 1980 marking fees were paid for some 3,400 ponies and 2,000 cattle (although this of course does not mean that all these animals were necessarily on the Forest all through the year).

In times past, the Forest deer were probably by far the most significant grazing pressure upon the New Forest vegetation: the area was, after all, set aside for the preservation of game and, until the 1850s, deer populations over the area numbered between 8,000 and 9,000. At the end of the nineteenth century, however, the area was 'disafforested'; deer populations were decimated and have only recently recovered to their current status. Their present, lowered, density now casts the free-ranging domestic stock as the major herbivores. Cattle and ponies have been depastured on the Forest alongside the deer ever since its designation as a Royal Hunting Preserve in the eleventh century, and probably considerably before that time, but numbers were probably lower than at present and in addition a far larger area of land was unenclosed and available for common grazing. As a result, the impact of the common stock upon the Forest was probably secondary to that of the deer. With reduction in the numbers of deer and simultaneous increased effective density of domestic stock at the end of the nineteenth century, however, cattle and ponies emerged as the major grazing influence, and have continued so to this day.

The New Forest area has, then, always sustained a tremendous grazing pressure from large herbivores. At present, 20,000 ha of some of the poorest possible grazing (current land-use surveys class the majority of the area as grade 5, or non-agricultural land) support some 8,000 head — a total biomass in excess of 2,500 tonnes — and it is clear that equivalent grazing pressure must have existed over the centuries. Yet the system has maintained itself in some kind of balance, however uneasy, for more than 900 years, this without any formal attempt at long-term management. Indeed, the system has survived *despite* dramatic and drastic changes in management and land-use (Chapter 2). This history of continued grazing has, however, stamped its mark upon the Forest vegetation, and provided a major influence on the shaping of the Forest and on its present-day ecology.

Over the last 40-50 years the Forest has come under increasing pressure, from commercial interests, tourism and recreation. Rising beef and dairy prices and the establishment of a ready market for horse flesh have produced a steady increase in the numbers of domestic stock commoned on the Forest since the war; commercial forestry interests in the area continue, and the Forest has also become more widely used as a recreational centre, for camping, walking etc. In order to maintain the Forest for the future, to conserve all its varied resources for its many users, and to conserve its unique vegetational structure, some deliberate policy of management must now be adopted. To be of value, this must itself be based on a firm understanding of the ecological functioning of the New Forest system, past and present. To this end, the Nature Conservancy Council commissioned a study in 1977 of the current interrelationship between the grazing herbivores of the New Forest and the Forest vegetation. Initially restricted to a study of the ecology of the Common animals — cattle and ponies — as the most significant of the herbivores, its aims were to discover the pattern of use by the cattle and ponies of the different vegetation types available to them within the Forest and also to estimate the impact of that pattern of habitat use and feeding activity upon the vegetational community in calculating the effects of grazing pressure on vegetational productivity and diversity.

The deeper we delved into the ecology of the Forest animals, the more fascinating became the complexity of interaction between animals and vegetation — and animal and animal. Over the years our studies ranged wider, to include consideration of the current role of the Forest deer in this complex ecological community, and to consider the effects of such heavy grazing on other parts of the ecosystem. Grazing proved to have such a dominating effect on vegetational structure and species composition that it became clear that the effects of the grazers on the vegetation must also have considerable repercussions on the ecology of other organisms dependent on the same vegetation; we therefore extended our studies to consider the effects of large herbivore pressure on the smaller herbivorous mammals of the Forest, and, through them, on their various predators. This book aims to bring together the results of all these various studies, offering a comprehensive review of our accumulating knowledge of the autecology of the various animal species of the Forest. Through our apprecation of the interactions between these species and between them and the Forest vegetation, it attempts then to draw the separate threads together into an integrated whole, to produce a new insight into the actual ecological processes underpinning the Forest we observe. In this sense the book builds naturally on Colin Tubbs' classic 'ecological history' of the New Forest (Tubbs 1968), in updating our information in the light of recent work and in presenting a picture of the ecological functioning of the Forest as it is today.

The ecological functioning of the New Forest is so dominated by the effects of heavy grazing that the book has in addition a secondary interest: as a detailed analysis of the effects of continued heavy grazing on a temperate ecosystem. Although the focus of the book remains an analysis of the ecological functioning of one complex, multi-species ecosystem, our studies in the New Forest have important implications in terms of the effects of grazing on vegetation in general. We shall hope therefore to provide, through this 'case study', insight into the more general effects of grazing in the dynamics of such temperate ecosystems, in terms of those same, potential effects outlined in the first few pages of this chapter.

Chapter 2
The History of the Forest

Despite the fact that it may claim to be the largest single unit of 'unsown' vegetation remaining in lowland Britain (Tubbs 1968), the New Forest is none the less a man-made system. The shaping of the Forest owes as much to its past history of human management as to its current grazers. The two are, of course, not entirely independent. Man's activities clearly affect relative numbers of wild herbivores within the Forest area today, as in the past. Man's domestic animals, too, are major contributors to overall levels of grazing within the Forest. But other human activities have also affected the Forest and its development, and continue to do so. The whole development of the Forest may be seen as a chronicle of human activity and human pattern of land-use — and Nature's continued response to the different pressures of each generation. The current ecological functioning of the Forest system is influenced by this history and by current methods of land-use; these modern activities are themselves restricted by the laws and traditions of the Forest, and by an administrative system which has evolved over 900 years. The Forest today is a complex product of *all* such factors, human and biological. Fully to appreciate the ecology of this curious system it is therefore critical to understand its social context, both historical and present-day.

The Early History of the Forest

Regrettably, very little is known of the history of the New Forest area before its annexation as a Royal Forest in the eleventh century. Using what little evidence there is and a great deal of deduction, it is, however, possible to piece together a picture of land-use. Mesolithic sites on heathland are widespread, and are the first good evidence for extensive use of the area. Mesolithic peoples, however, left few obvious traces on the ground: they did not cultivate the land and lived essentially as hunters. Their major impact on the Forest system was probably in the clearance of trees. The rapid regeneration of ground vegetation and shrubs in the secondary succession of forest clearings provides, in the short term, a greatly increased supply of rich forage for wild game. Forest clearance in provision of this increased grazing has always been a part of the human hunter-gatherer economy, and Mesolithic man was no exception. Large-scale clearance of forest and forest edge, usually by fire, was quite common in this period; other areas may have been accidentally cleared by fires lit to drive or flush game.

While Neolithic and Mesolithic peoples left few traces on the landscape, the succeeding Bronze Age cultures left more tangible evidence of their occupation in the form of burial mounds or round barrows. In both the Middle and Late Bronze Age, there is therefore abundant evidence for human occupation of heathlands within the Forest. Pollen analysis of soils buried beneath some of the barrows suggests that, at least by the Middle Bronze Age, some form of actual cultivation of the heathlands was being practised. This cultivation was short-lived, however, and apparently did not persist into the Iron Age; indeed, evidence for Iron Age occupation of the Forest heathlands is in any case very sparse. Tubbs (1968) suggests that this scant evidence of Iron Age occupation, by comparison with the abundant evidence for Bronze Age settlement, reflects not just an archaeological vacuum, but actual desertion of the Forest heathlands because their fertility had declined to a point where cultivation was no longer possible.

The forest clearance of Neolithic and Mesolithic times probably continued through the Bronze Age, though its purpose may have shifted towards the creation of clearings for agriculture. In the short term forest clearance may encourage rapid growth of lush grazing for wild or domestic stock; it may provide relatively fertile areas for cultivation. But this apparent productivity is short-lived: on poor soils like those of the New Forest, clearance of forest ultimately leads to a dramatic deterioration in soil fertility as changed rainfall and run-off patterns lead to leaching of the limited soil nutrients. Thus Bronze Age peoples probably settled in areas which were already beginning to decline in fertility; by the Iron Age, the soils were in general no longer suitable for cultivation by the techniques then available.

After the end of the Bronze Age therefore, and certainly by the end of the Iron Age, the open heathland areas of the Forest were exhausted of nutrients, and usable only for rough grazing. Instead of their previous widespread distribution across all the Forest lands, settlements became restricted only to those areas of better soil which could still support cultivation. A pattern of settlement became established which persisted through Saxon and Norman times, and which indeed has largely persisted until the present day, with communities focused upon the limited number of areas where soils are rich enough to support arable agriculture, and the greater part of the Open Forest deserted except for exploitation of its limited grazing.

The Royal Forest

Undoubtedly, the most significant event in the history of the Forest was the appropriation of the area in 1069 by the Crown as a Royal Forest. This dramatically affected its ecological development, in markedly influencing subsequent patterns of land-use. It has also, to an extent, safeguarded the Forest and its peculiar mixture of communities. Although by the middle of the eighteenth century agricultural expertise had developed to a point where reclamation of even the most infertile soils for agriculture was a practicality, the Forest and its open wastes, as Crown land, were safe from reclamation.

As noted in the previous chapter, designation of an area as a Royal Forest implied establishment of a hunting preserve for the Crown. It did not necessarily imply that an area was wooded, nor indeed did it necessarily imply Crown ownership of land. Afforestation was based on the prerogative enjoyed by the Sovereign that all wild animals were in his possession. The later Saxon kings claimed as part of this prerogative the right to reserve to themselves the chase, at least of deer, over any part of their kingdom they might choose to define; under the Norman kings, the restrictions of the use of these royal hunting grounds became even more severe and were established in statute. The King had, in effect, the right to subject any land, whether of his ownership or owned by others, to Forest Law: a forest, while it has a secondarily assumed implication of woodland cover, was thus defined in fact simply as an area coming under Forest Law as opposed to Common Law. While imposition by the Crown of Forest Law did not imply any legal change in ownership, it none the less brought with it severe restrictions: lands under Forest Law could be neither enclosed nor cultivated; the rights to take game were reserved for the Crown. The restrictions were such that the private owner of such land gained little from that ownership; in 'compensation' he was allowed to retain the rights of free-grazing of stock over the area. Clearly, it would have been unrealistic to prohibit fencing and at the same time to enforce restrictions on the roaming of stock from private land on to Crown land. Thus the right of free range of stock over the entire area seems to have gained mutual acceptance (and would appear to be the origin of many later rights of Common grazing on Crown lands, although these were not legally granted on *any* Crown land until very much later, in 1598).

The New Forest was designated a Royal Forest by William I, and of all the many Royal Forests created during the eleventh, twelfth and thirteenth centuries it alone persists today in near entirety. The form of management imposed by Forest Law created a pattern of land-use which shaped the whole development of the area, and which remains much the same to the present day. Although subsequent reforms progressively sweetened the old Forest Laws, they did not fundamentally change their import. (As a result of growing disaffection among landowners for the restrictions of Forest Law, the *Charta de Foresta* of 1217 conceded extensive disafforestation and apparent relaxation of restrictions. Before the *Charta* no land under Forest Law could be enclosed or cultivated, nor could timber be felled or game taken. Under the *Charta* these practices were permitted on private land, but only under special licence from the Crown. In effect, the old Forest Laws remained in force right up until the New Forest Deer Removal Act of 1851, when the Crown relinquished its right to keep deer in the New Forest.) But, while the administration of the Forest changed little over the next few centuries, its boundaries did. Increasing demands on land, and increasing land values, put tremendous pressure on the Crown's hold of the Forest preserves. Neighbouring estates encroached on the outer boundaries. Richer lands within the perambulation were enclosed for cultivation, either under licence from the Crown or in flagrant disregard of the Forest Law. The initial area of lands afforested by the Conqueror had diminished to some 67,000 acres (27,110 ha) by 1279, when a firm perambulation or legal boundary was established. But, while these same pressures resulted in the gradual fragmentation or total dissolution of other Royal Forests, the New Forest somehow survived virtually intact: the perambulations of 1279 remained more or less unchanged until recast by the New Forest Act of 1964.

Patterns of land-use also changed, even on the open Forest. While the original purpose of the Forest was the preservation of deer, the value of the area for timber production became more and more important as England emerged as a seafaring nation, and as the country's total timber resources began to dwindle as they were rapidly consumed in the shipyards. From the seventeenth century onwards, the Crown's interests in the New Forest were focused more and more on the production of timber. Tree production on a commercial scale, however, required the enclosure of land against browsing and grazing by deer and commonable stock. More and more land within the Forest was enclosed — both by the Crown and by private landowners — for silviculture, culminating in the Enclosure Acts of 1698 and 1808. This new emphasis on timber production and the resultant enclosures began to change the character of the Forest and was the next major stage in its development. As more and more land was enclosed for silviculture, less remained available for Common grazing. Tubbs (1968) notes that the documentary evidence of the seventeenth, eighteenth and nineteenth centuries shows that a vigorous pastoral economy based on the use of these Common grazings had developed in the Forest. While numbers of animals depastured on the Forest doubtless decreased as land was enclosed, actual densities of animals on the remaining acres of open ground must thus have increased tremendously, leading to extremely heavy grazing pressure on these areas — and another major 'event' in the shaping of the Open Forest's characteristic vegetation patterns.

This same conflict of interests between those of the Crown (in enclosing land for silviculture) and those of the Commoners (in retention of the open grazings) culminated in the New Forest Deer Removal Act of 1851, by which, as noted earlier, the Crown relinquished its right to keep deer in the New Forest and released the area from many of the restrictions of Forest Law. Under the same law, the Crown took for itself instead powers for silvicultural enclosures on an enormous scale — well beyond the total 6,000 acres (2,430 ha) provided for by the Enclosure Acts of 1698 and 1808. Twenty-five years later, in the New Forest Act of 1877, the powers of the Crown to enclose land were, however, severely curtailed and the Commoners were assigned a statutory area of 45,000 acres (18,210 ha) over which they might exercise their rights in perpetuity. But, from this time on, forestry — in its sense of timber management — has been an important part of the New Forest economy, and an important influence on the vegetational structure of the Forest. There are currently some 20,000 acres (8,300 ha) of statutory silvicultural enclosures within the Forest, although only 17,600 acres (7,100 ha) may be behind fences at any one time.

Management of the Crown-owned lands of the New Forest passed to the Forestry Commission in 1923 — with responsibility for timber production within the Statutory Inclosures, and maintenance of Common grazing. At the same time the Forest has come under increased pressure for recreation. Management of the area, both Inclosures and Open Forest, must thus be framed to preserve in addition the amenity value of the Forest; indeed, the Commission now has a statutory requirement so to do. In our century, then, as in all the years before, the Forest must serve a multitude of purposes. Today its management is for timber production, for maintenance of Common grazing and Common agriculture, for recreation and for conservation. These management aims will play as important a part in shaping its future as its past history has had in shaping its present.

The Forest Administration

During the thirteenth century, Forest Law was administered by two main courts. The lower court, the Court of Swainmote or Attachment, was presided over by appointed Verderers of each Forest. Its function was essentially to hold preliminary hearings, binding over those prosecuted by Forest Officers for breach of the law to the higher court, namely the Justice in Eyre. The Court of Swainmote, or the Verderers' Court, had no authority itself to impose fines or punishments. The Justice in Eyre was supposed to hold court in each Forest every three years, but by the end of the thirteenth century it was held only irregularly; in the seventeenth century only two courts were held within the New Forest, the second, in 1670, being the last court ever held. This gradual demise of the Justice in Eyre was doubtless in large part due to the progressive relaxation of Forest Law, softened from the original toughness of Norman times by successive Charters. But without a Supreme Court to which to refer, the Court of Swainmote and Attachment became to a large extent powerless in any legal sense, and began to function more as a manorial court, merely administering privileges and rights of both Crown and Commoner within the Forest itself.

The executive officer in charge of a Forest was the Warden or

Lord Warden, a post which could be hereditary or by Royal Appointment; the Verderers of the Forest were elected by the County. Day-to-day management and policing of the Forest were carried out by Foresters and other minor officials. Agisters were employed by the Verderers to look after stock on the Forest and to enforce the by-laws, and the Regarders (originally 12 knights of the county) were responsible for examining the Forest perambulations and reporting on the general state of the Forest.

In the New Forest, the Verderers' Court still remains; the new role as protector of Common Rights, adopted by the Court of Swainmote and Attachment after the collapse of the Justice in Eyre, was confirmed and given statutory definition by the New Forest Act of 1877. This Act redefined the Verderers as a statutory body responsible not for safeguarding the interests of the Crown (for which the Forest Courts had originally been intended), but for safeguarding instead the interests and rights of the Commoners. The New Forest Act of 1949 reconstituted the Verderers for a second time, increased their number from six to ten and extended their powers. The Court now consists of the Official Verderer, nominated by the Crown; five Verderers elected by persons occupying an acre or more of ground to which Common Rights attach; and four Verderers (added by the 1949 Act) appointed respectively by the Ministry of Agriculture, the Forestry Commission, the Local Planning Authority and the Council for the Preservation of Rural England. The Verderers' responsibilities are to make by-laws for the control of stock on the Forest and for the benefit of the health of the Commonners' animals; to employ officers as required for the control and management of commonable stock (the Agisters); to defray costs by levying dues ('marking fees') for each animal turned out on the Forest; and to maintain the Register of Common Rights.

In brief, the responsibilities of the Verderers remain those of defending and safeguarding the interests of the Commoners. Common pasturage, however, is only one facet of the current economy of the New Forest. As we have noted, it is still a Royal Forest where the Crown retains silvicultural interests. In effect, the ancient and perpetual conflict of interests between Crown and Commoner which has for so long been a part of Forest history continues to the present day, with the modern-day interests of the Commoners represented by the Verderers and the interests of the Crown represented by the Forestry Commission (to whom exercise of the Crown's interests has been devolved). While the conflict between the two *bodies* has to a large extent been overcome in recent years (the Forestry Commission is represented on the Court of Verderers, while the Verderers are invited to join in discussions about Commission management policy), the conflict of *interest* between silviculture and Common pasturage is a real one.

The *administration* of the Forest is thus the responsibility of two statutory bodies (Commission and Verderers), in opposition or in collaboration. The actual *management* of the Forest is equally torn by conflict of interests: to the interests of Commoners in grazing, and of Crown in timber, are added statutory requirements for conservation (whose interests are represented in the New Forest by the local officer of a third body, the Nature Conservancy Council) and for provision of recreational amenities. The New Forest Act of 1964 adds to the responsibilities of the Forestry Commission and Verderers the obligation to 'have regard for the desirability of conserving flora, fauna and geological and physiographic features of special interest' within the Forest; the interests of amenity are safeguarded in earlier Acts, particularly in regard to the preservation of the ancient woodland of the Open Forest (the so-called Ancient and Ornamental Woodlands of the Open Forest were secured under the 1949 Act).

The current administration of the Forest must thus try to satisfy four major aims, often in conflict; and that administration is the responsibility of *two* statutory bodies, in consultation with other interested parties such as the Nature Conservancy Council and the Countryside Commission.

Rights of Common

While the history of changing patterns of land-use within the Forest has clearly had a profound influence on its ecological development — and while its current complex administration determines the direction of overall management policy in the future — these are long-term and indirect influences. On a day-to-day basis the ecology of the Forest was, and is, affected *directly* by the exercise of various Rights of Common, and it seems appropriate to end this section with a brief review of these Forest rights.

Before afforestation in 1069 the open wastes of the Forest would have been freely grazed. Pigs would have been turned out to feed on acorn and beech mast in the autumn; timber, turves or peat taken for fuel; and bracken cut for bedding and litter. After the imposition of

Forest Law, these practices were allowed to continue, but under close control and regulation — eventually, within the New Forest, being defined in law in the Acts of 1689. These rights may be summarised as Rights of Common of pasture for commonable beasts (cattle, ponies, donkeys); Right of Common of mast (the right to turn out pigs during the mast season, or 'pannage'); Right of Common of turbary (the right to take turf fuel); Right of Estovers (the right to take fuel wood); and Right of Common of marl (the right to take marl from recognised pits in the Open Forest). To these may be added rights to cut 'fern' (bracken) for bedding and litter, and sheep rights claimed by the adjoining manors of Beaulieu and Cadland.

Common Rights could be claimed by any cottager or landowner holding more than an acre of land within the Forest perambulation. While most claims to pasturage conceded the old rules under Forest Law of *levancy* and *couchancy* (limitation of the number of stock turned out in spring and summer to that number which could be maintained on the holding over winter), this was clearly not a rigid requirement (nor is it today) and many claims are for the right of pasturage in all months.

Both cattle pasturage and pannage of pigs were, however, subject to certain restrictions under Forest Law. Medieval Forest Law provided for the removal of cattle from the Royal Forests during the midsummer 'fence month' (20 June to 20 July) when the deer were calving; and during that part of the year when keep was short (22 November to 4 May) — the winter 'heyning'. It would appear that the fence month was observed for the most part within the New Forest, but it is clear from the various Registers of Claims recorded from 1635 onwards that the winter heyning was largely neglected. Under Forest Law, the period during which pigs might be turned out to mast was restricted to about two months in the autumn (the precise dates were set each year, according to the mast-fall). Yet, once again, while presumably rigorously observed in the early days when the area first came under Forest Law, these restrictions were but weakly enforced by the sixteenth and seventeenth centuries and in the first detailed Register of Claims compiled by the Regarders in 1635 many claims for Right of Common of mast are not confined to the true pannage season: many are claimed for all times of year.

Rights of pannage and Common pasturage were undoubtedly the most significant and important rights which could be claimed by Forest Commoners in terms of maintaining a livelihood; it is clear that many of the holdings were of such a size that they would not have been viable without these rights upon the Forest. In evidence given before Government Select Committees in 1868 and 1875, it was stated that the Right of Common of grazing enabled a Commoner to maintain at least three times as many cattle as he would have been able to maintain without that right. But almost all claims registered in 1635 are for every Right of Common, and there is no doubt that the rights to take bracken for bedding and litter and, more particularly, the rights to take turf and fuel wood must have contributed significantly to the cottage economy.

While Registers of Claims to Common Rights on the New Forest have been compiled at regular intervals since the seventeenth century, such records cannot show the extent to which these rights were exercised. Thus it is difficult to assess the impact that the exercise of the different Rights of Common might have had upon the Forest at different times. Precise figures for the number of stock depastured on the Forest are available for the years 1875, 1884-93 and from 1910 onwards. A census of stock carried out in 1895 gave a figure of 2,903 ponies, 2,220 cattle and 438 sheep. The number of stock for which marking fees have been paid has fluctuated considerably since that time — and presumably did before it — depending largely upon the outside economic situation. By 1910 the number of ponies on the Forest had fallen to 1,500; cattle numbers remained between 2,000 and 2,500. In 1920 numbers rose again to a peak of 4,550 stock (mostly cattle) before the depression led to a slump in Common practice, and on the outbreak of war in 1949 there were only 1,757 domestic animals on the Forest. Economic ups and downs are sensitively reflected by ups and downs in the numbers of large stock turned out to Common: numbers have fluctuated between 1,500 and 5,500 over the last 100 years (Table 2.1). These same differences in the degree of exercise of Common Rights have had profound effects upon patterns of vegetational change within the Forest.

The commonable stock and other large herbivores have grazed upon the Forest vegetation for centuries, and have shaped and modified the structure and function of the entire Forest community. Whatever effects they may have upon the Forest's ecology they have had for hundreds of years, yet remarkably little is known of the animals' day-to-day use of the Forest in which they live and which they have helped to shape. Their alleged role in the dynamics of the Forest system past and present is largely supposition, based on anecdote and circumstantial evidence. It is of course not now possible to

Table 2.1: Commoning activity in five periods of time since 1789

Historical period	Number of Commoners	Number practising		Number of ponies	Number of cattle
1789-1858	2,314	800 to 1,000	MAX	3,000	6,000
			MIN	1,800	5,400
1858-1900	2,000[a]	400 to 800	MAX	2,903	3,450
			MIN	2,250	2,220
1900-1930	c.1,300	250 to 350	MAX	2,068	2,482
			MIN	not recorded	
1930-1943	c.1,300	100 to 250	MAX	not recorded	
			MIN	416	908
1943-1980	c.1,300	250 to 400	MAX	3,219	3,682
			MIN	1,416	1,139

Note: [a] 1,200 approved claims included many 'block' claims by large estates, including up to 150 tenancies.
Source: Countryside Commission Report 1984.

do more than speculate about the interactions between herbivores and vegetation in the past, but the *current* relationship between animals and vegetation can be more objectively assessed. The next four chapters will thus review what is known of the ecology and behaviour of the main herbivore species within the New Forest today, and their impact upon their environment. From such study and the understanding it gives us of the interrelationships between the herbivores and their vegetation, we can hope to design effective management for the future.

We must emphasise that what we shall attempt is a description of the current functioning of the New Forest ecosystem and the interrelationships of its various herbivores, both with each other and with the vegetation. In this chapter we have considered the history of the Forest and the influence of herbivores in the past. We now turn to a consideration of the ecology of the Forest in the present day, and it must be stressed that what we shall present in the succeeding pages is a description only of the *current* situation. From the studies to be described, restricted as they are to the ecology of the Forest as it is now, we can draw conclusions primarily about the present influence of the larger herbivores within the system; we may, however, from the understanding we gain of the underlying principles of the relationships between herbivores and vegetation, also make intelligent deductions about the effects of past grazing pressure (Chapters 7 and 8).

Chapter 3

The Grazers: Ecology and Behaviour of the Common Stock

The most abundant of the animals pastured under Common Rights today on the New Forest are the cattle and ponies. While the cattle are for the most part common everyday beef breeds, such as Hereford cross and other crossbreeds, turned out for the rough grazing the Forest affords, the New Forest pony has more of a tradition attached to it. Indeed, no-one knows precisely how long ponies have run upon the Forest. Certainly they are mentioned in Domesday Book — though it is not clear whether even by this stage these were domestic animals turned out to graze by private owners, much as they are today, or whether the Forest area may have supported truly wild stock — and it seems probable that the Forest has supported its own distinctive breed of pony for many centuries. Valerie Russell writes:

> In the days when the moors and forests of Southern England stretched practically unbroken from Southampton to Dartmoor, and even possibly to the fringes of Exmoor, wild ponies are believed to have wandered freely over the area. With the advance of civilisation sections of the Forest and moors were cultivated so that the ponies were restricted much more to the areas from which the present-day breeds take their name. (Russell 1976)

The New Forest Pony

The New Forest breed has suffered many attempts at 'improvement' over the centuries, so that it is now difficult to describe a distinctive type, but the typical Forest pony is a rather small, stocky animal, with an unfashionably large and coarse head; usually bay and somewhat shaggy in appearance, these are hardy, sturdy animals, shaped by the rigours of the environment in which they live as much as by the breeding of those keen to 'improve' the stock. Market pressure has seen many such attempts to alter the characteristics of the breed over the years: the chief aim of the improvers was to introduce some quality into the New Forest ponies and particularly to increase their size. To this end, at various times Arab and thoroughbred stallions were used on Forest mares — and even stallions of hackney and carthorse blood. Almost without exception such attempts at improvement have led to a larger, but less hardy stock, ill-adapted to withstand the pressures of year-long subsistence on the Forest. Although such cross-breeding has left its mark on the Forest stock — and it is possible even now to identify animals of distinctly Arab type or Welsh type on the Forest — the effects were less marked than might be anticipated. In the 1930s came a ban on further infusions of outside blood, and such is the power still of *natural* selection upon the Forest that the New Forest pony has assimilated all these outside influences but still remains today a sturdy, cobby, shaggy bay not unlike the unimproved Exmoor (Plate 1). In recent years, there has again been increased selection for more 'sophisticated' stock; better-quality stallions are selected for release on the Forest — and are run only from May till September, so there is no need for the hardiness required if an animal is dependent upon the Forest all year. In result, the Forest pony once more is becoming more variable, more mixed in breeding, and it is again rather difficult to define a clear distinctive 'type'.

Most of the attempts to improve the Forest ponies have been prompted by market requirements. There is no doubt that the ponies grazed upon the Forest initially were the Commoner's own stock for use on their small holdings. As time progressed, however, the breed was developed for sale in various directions as outside markets came and went. Before the general use of wheeled transport, they found a ready market as pack animals; when wheeled vehicles became more widespread, the Commoners had another market for their ponies and Russell (1976) considers that it was chiefly as draught and harness animals that they were used until the early days of this century. With the coming of the Industrial Revolution, many of the native breeds were in great demand as pit ponies in the coal workings; the Forest ponies, however, were in general too large for such work and, while some of the smaller, stockier individuals went to the mines, the majority remained more popular as harness ponies. More recently, the demand for riding ponies, particularly for children, has opened up another lucrative market: the current attempts to upgrade Forest stock are directed chiefly at modifying the 'Forester' better to suit this new demand.

Allegations are still made from time to time that many of the Forest ponies are sold to the meat trade. Certainly, during the war years, many *were* bought for human consumption, and, when prices are low enough, large numbers are doubtless sold for the pet-food trade. Whether or not there is at present an export market to the Continent as has been suggested is, however, more doubtful. Russell notes:

> While it cannot be stated categorically that this has never happened, a certain amount of sleuthing has been done, and so far no proof has been forthcoming. In 1965 the council of the pony society asked for any evidence of such a trade, and the National Pony Society offered a substantial reward; but nothing even remotely conclusive was produced. With transport costs at their present level, it does not seem an economic proposition for foreign buyers to deal in New Forest ponies which, for the most part, do not carry much meat.

Social Organisation and Behaviour

Studies of the social organisation of most populations of wild or semi-wild horses (e.g. Feist and McCullough 1976; Berger 1977; Welsh 1975; Wells and von Goldschmidt-Rothschild 1979) have shown that the normal social 'unit' is a stallion-maintained harem of mares and sub-adults: a harem which is maintained throughout the year. Stallions are aggressively territorial and defend not only their group of mares, but also an exclusive home range (Feist and McCullough 1976).

The ponies of the New Forest, while equivalent in many respects to these feral populations, are more intensively managed. Groupings are often artificial, resulting as fixed associations of animals belonging to a single holding which were put out on the Forest together and retain their association. In addition, many owners do not leave their animals out all year but take them off the Forest over the winter, returning them to free range the following spring; such a practice reinforces each year the social bondings of the animals of a given holding. Not all owners, however, do take their stock in over the winter period, and many mares and foals — perhaps the majority — remain on the Forest throughout the year. None the less, 'natural' groupings are difficult to maintain: while many owners may be prepared to leave their mares and foals on the Forest all winter, few indeed allow stallions to overwinter. With a few exceptions, stallions are run on the Forest only from April to September. As a result, natural harem groups are rare. Females may form permanent associations (maintained throughout the year), and a stallion may join up with such a group over the summer months — but he plays little role in social cohesion: the groups are not true harems and the social order is essentially matriarchal.

The social structure of New Forest ponies is thus somewhat atypical, with the basic social unit most usually observed being a mare-foal assembly (Tubbs 1968, Tyler 1972). The fundamental elemental unit within such an assemblage may be recognised as an adult mare with her offspring of the current and previous years. Larger groups may be formed as associations of these basic units, but the size, cohesiveness and persistence of these social groups differ markedly. Stephanie Tyler, who published in 1972 the results of a classic study of the New Forest ponies and their social behaviour, recognised three main group types: (i) 'simple family units consisting of one adult mare alone or with one or more of her own offspring'; and (ii) groups of 'two adult mares with a varying number of offspring', or (iii) more rarely of 'three to six mares and their offspring'. The various proportions of these different types of grouping found by Tyler are shown in Table 3.1; the relative proportions of the different group types were not significantly different from those encountered by Pollock (1980) or in our own studies (Putman *et al.* 1981).

The most frequently observed social unit in New Forest ponies is Tyler's type (i): a family group consisting of a mature adult mare with one or possibly more of her past offspring and her current year's foal. Such a unit may vary in size from two, i.e. adult mare and

Table 3.1: Social Organisation of New Forest Ponies
Relative frequency of the different social groups of New Forest ponies recognised by Tyler (1972)

Winter	1 adult mare + offspring	2 adult mares + offspring	3 or more adult mares + offspring	Total no. of groups
1965/66	60.7	27.8	11.5	122
1966/67	59.0	28.2	12.9	124
1967/68	64.5	25.0	10.5	124

Due to the formation of false family groups (see p. 29), the true mother of some of the 'offspring' may be in doubt.

juvenile (plus perhaps a foal), to three or four where older offspring remain with the matriarch even as adults. This category blends almost imperceptibly into Tyler's type (ii) if older offspring, now themselves mature, remain with the group. Groups of more than three or four adult mares are, however, rare. This may be due in part to management intervention: many of the younger animals are sold each year, disrupting the social group and perhaps preventing the formation of larger groups of related adults. But this is not the complete answer. Tyler has shown that even under natural conditions most fillies do eventually leave their mothers' groups, usually at between two and four years of age. Such younger mares either join up with another group or with a single mare, or remain on their own to form the basis of a new family unit (Tyler 1972).

Not all larger groupings are of related animals; as noted earlier, association may be formed between unrelated animals belonging to the same owner. Such animals are generally kept in the same stable or paddock while off the Open Forest and relationships developed then will hold when the animals are released. Groups of two or three such individuals are common, and owners are widely aware of the extent of the bond between them and will rarely separate such groups deliberately. Such a phenomenon is also recorded by Arnold and Dudzinski (1978).

Nor are all associations as permanent as those described. Clearly, other more temporary associations are also observed. Many groups may co-occur on favoured feeding grounds, associating casually, merely because they happen to be in the same place at the same time. Further, groups of animals occupying the same home range frequently seem to use the range in the same way on a diurnal basis. This may give rise to what appear to be larger 'groups' of animals, but such aggregations are not in fact true social groups and do not persist for long. These aggregations of groups into larger units appear to be more common where larger home ranges are occupied.

Stallions, when on the Forest, may also form the focus of larger aggregations of mares, ranging in size up to 10-15 adults and juveniles. These harems, too, are simply composed of assemblages of mare groups and the group structure described above holds fast within the herd, i.e. members of a group will all form part of a single herd, and if one should leave then the others will also. Stallions' herds vary seasonally in size, being largest at the height of the breeding season, i.e. May to July, when the stallions actively herd as many mares as they are capable of holding together. Later in the summer, and particularly in winter, the stallions are less active and the groups comprising the herd appear to remain together on a more voluntary basis, individual groups coming and going from the herd as they please.

No significant differences in mean group size or in the distribution of Tyler's three group types is apparent within different geographic regions of the Forest. While size of social group does not vary over the Forest, group cohesiveness, however, does show marked variation. In a study of group size and cohesiveness in New Forest ponies all over the area, O'Bryan *et al.* (1980) showed that cohesiveness (defined in terms of permanence of association and median distance of each animal from others in its group) was markedly higher in a coarse-grained environment (areas where vegetational communities occur in large homogeneous blocks) than in finer-grained areas (where component vegetation types occur in a much smaller-scale mosaic). These areas were defined quite rigorously using Simpson's diversity index to delimit fine-grained (diversity index > 5.8) and coarse-grained ($2.8 <$ diversity index > 3.5) environments. Differences in group cohesiveness are highly significant: groups in coarse-grained environments remained in much closer association, both spatially and temporally, keeping fairly close together, and over longer periods of time than did ponies of fine-grained areas. Ponies in fine-grained areas operated far more independently, even within a group, and in many cases the true group was established only when the animals gathered together at dusk to move off into woodland areas for the night; median distance between group members was significantly higher in these areas. Further, it was noted that actual social interaction was reduced in fine-grained areas, social reinforcement in whinnying or allogrooming being far more evident in groups in coarse-grained environments. (Some possible reasons for the difference in group cohesiveness are presented by O'Bryan *et al.* 1980.)

Social and Sexual Behaviour

A hierarchical system, apparently based primarily on age and to a lesser extent on the character of the individual, operates both within social groups and between individuals of different groups (Tyler 1972, Gill 1980). Within a group, aggressive interactions are relatively uncommon and are used primarily only to reinforce the dominance hierarchy or if, for example, a third party takes too much interest in a newborn foal. Ear-threats and head-threats (a movement of the head towards the opponent, often with ears laid back) are the most common postures of aggression; if these displays prove ineffective, a mare may turn and threaten to kick her opponent with the hind legs (Gill 1980). Most disputes are resolved without need to press home an attack; but mares not infrequently bite, or kick out at each other in such encounters. Curiously, neither the degree nor the frequency of aggression shown by individual mares relates in any simple way to dominance position. Gill (1980) studied the frequency of the different aggressive encounters among 20 individual mares of one of the few permanent harem groups on the Forest, relating aggression to dominance hierarchy. There is little obvious relationship between dominance rank and frequency of aggression, although high-ranking animals are more inclined to press home a full-scale attack.

Friendly interactions within groups are much more common, with members of a given social 'unit' frequently nuzzling each other or grooming. Mutual grooming (allogrooming) is particularly common when two individuals have been feeding some distance apart for a period of time, or have been separated for some other reason, and seems to serve the social function of reinforcing the bond between the two, as well as the practical one of grooming those regions of the body which the individual cannot reach. (The importance of this last is suggested by the high incidence of allogrooming in the spring months at the time of the moult: Tyler 1972, Putman *et al.* 1981.) Few other direct interactions were observed between the members of a group, but such animals generally kept close company and would rest or lie up close together during periods of inactivity. If separated, the individuals show clear signs of distress, whinnying and searching actively for their associates.

Social facilitation was apparent in pony behaviour, although less so than in cattle, and applied both within and between groups. This partly explains the association of individuals which have overlapping home ranges, since a move by one individual will often precipitate several others to follow. Eliminatory behaviour appears to be influenced to some extent by social facilitation, although this more commonly within groups, and this is particularly true of eliminatory behaviour of stallions, who may 'cover' dung or urine of harem mares (Tyler 1972, Gill 1980).

Another important feature of group behaviour clearly influenced by facilitation is the curious phenomenon known locally as 'shading', and first described by Tyler (1972). Shading is a behaviour in which individual animals and even separate social groups come together and stand inactive in large congregations (commonly of 20-30 animals, occasionally numbering over 100) (Plate 5). It is particularly common in mid- to late summer, when shades may be formed as early as 9 a.m. and can last up to six hours. While shading, the animals stand close together, and remain largely inactive, merely whisking their tails or occasionally shifting position; no feeding occurs. When the shades break up, the animals drift away in their original social groups — but they may re-form the shade later the same day or on the following day; they re-form it on the same traditional site. The term 'shade' is in fact a misnomer, since animals do not necessarily seek shade from the sun (although many sites are in woodland, under rail bridges etc.) but may congregate on the brow of a hill to take advantage of any breeze which is blowing, or on such areas as roads or car parks which lack vegetation cover. The function of the behaviour is uncertain, but it is believed to be an adaptation to avoid the attacks of biting insects. (A similar behaviour in response to insects is described for Camargue horses by Duncan and Vigne 1979, and certainly the choice of 'shade' sites away from vegetation or in windy exposed locations would be compatible with such a function.)

A recent review of sexual behaviour in New Forest ponies has been presented by Gill (1980).

Home Range

The majority of pony groups on the Forest may be ascribed to a population whose focus of activity is a re-seeded or streamside lawn or other area of rich grazing. Since such lawns are relatively few in number, the populations are fairly discrete and few individuals or groups move regularly between populations. Each group of ponies occupies a well-defined home range — an area of the Forest to which it restricts its activity over the course of the day. Groups seldom venture outside this home range unless disturbed, although apparently exploratory forays were observed and mature mares may leave the area if no stallion is present when they come into oestrus. The home ranges of different groups *within* a population overlap extensively and are in many cases identical, and the pattern of habitat use of groups with overlapping ranges is often very similar. As noted above, this may lead a temporary aggregation of animals into large 'apparent' groups.

The size of the home range varies markedly over the Forest, and is usually determined by the proximity of four necessary components: a re-seeded area, streamside lawn or some other suitable grazing area;

a water supply; shelter, usually provided by woodland or gorse brake; and a shade. In the north of the Forest, where vegetation communities occur in relatively large, homogeneous units (coarse-grained), these components of the home range may be widely separated, and consequently large ranges often exceeding 1,000 ha may be found. Further south, a more heterogeneous vegetation cover results in smaller home ranges, down to 100 ha or less. No correlation has been observed between group size and home-range size, but the aggregation of groups into larger units appears to be more common in the north.

Pollock (1980) presents no data for range size of mares or mare groups, although he cites areas of 140 ha and 125 ha for two stallion groups. Range sizes of other free-ranging horses are reported in the literature (e.g. Klingel 1974; Feist and McCullough 1976). For two harem groups of free-ranging Exmoor ponies, Gates (1979) records range sizes of between 240 ha and 290 ha but notes that most activity was in fact restricted to a core area of only 45-60 ha.

Within this home area we can appreciate a general pattern of range use. Over the summer period, each animal generally spends most of its time feeding on the grazing area(s) included within its range. During the course of the day there is at least one move off the lawn to drink, and at this time the animals usually make some use of the lusher forage available in the wetter vegetation types. At dusk, there is a general move away from the more exposed grazing areas to vegetation types offering more cover, and the ponies usually spend the night in, or in the shelter of, areas of woodland or extensive gorse brake. Those groups of animals left on the Forest over the winter make somewhat less use of the exposed grassland communities, although these are still extensively grazed despite the fact that there can be little useful forage left. Over the winter, however, the ponies make greater use of those areas of their range offering shelter from the more severe weather, and emerge from the woodlands and gorse scrub for shorter periods during the hours of daylight. Range size is often reduced during the winter months, since water is more readily available and there is no need to 'shade'.

Such a description of range-use patterns is of course crude, over-general and over-simplistic. Detailed studies of the pattern of use of different vegetation types by the ponies were one of the main objectives of the programme of research sponsored by the Nature Conservancy Council between 1977 and 1981, and will be discussed at more length below.

Cattle

Cattle are now run on the Forest mainly for beef production, and are mostly Friesian/Hereford or more mixed crossbreeds. Individual herds of Red Devon and Highland also occur, however. Herds are composed predominantly either of one- and two-year-old heifers or mature beef cows and calves, these groups being commoned separately by their owners.

In contrast to the ponies, the basic social unit in cattle on the Open Forest is the herd, although the cohesiveness of this unit varies seasonally. The dispersion of herds over the Forest to some extent reflects the location of farmsteads, since many owners release their animals onto the common land during the daylight hours only and allow them to return to the enclosed fields at night. The majority, however, common their herds on the Open Forest throughout the summer months, generally releasing them on areas of good grazing such as re-seeded or streamside lawns. As in the case of pony 'populations' therefore, cattle herds belonging to different owners are dispersed over the Forest as relatively discrete units, each unit having a home range focused on one or perhaps more primary grazing areas.

Over the summer months, the herd fragments into a number of smaller groups which range independently but encounter one another frequently on grazing sites. These small sub-units are not cohesive and individuals regularly move from one to another on these encounters. The pattern of habitat use over this period is similar to that of the ponies, with cattle spending most of the daylight hours on the primary grazing areas, although making more frequent visits to water supplies. Less use is made of the wetland vegetation resources, and cattle appear to avoid particularly boggy areas. By night there is again a general move to those vegetation types offering cover, particularly deciduous woodland, although groups will often spend the night on open dry heathland when weather conditions are mild.

Cattle range more widely than ponies on the Open Forest over the summer-month period, often occupying home ranges with more than one primary grazing site. These areas may be used in turn over the course of several days, and daily movements of 4 km or 5 km are not uncommon. Movement is particularly common at dusk and dawn, groups often moving 2 km or more from the feeding site to the overnight area. These distances, although large in comparison to most pony-group movements, are less than have been recorded in similar studies on cattle in extensive ranges. Schmidt (1969) followed

Shorthorn cows over 9.2 km per day on open range, while Shepperd (1921) recorded daily movements of 8.9 km in a 260-ha paddock. This difference may be explained by the relative heterogeneity and availability of water in the Forest habitat. Cattle have been recorded as travelling 26 km daily in drought conditions on rangeland (Bonsma and Le Roux 1953).

Over the autumn months the majority of cattle are removed from the Forest in a series of 'drifts' or round-ups, and those few herds which are overwintered on the Open Forest are generally fed to a considerable extent with hay and straw. The practice of supplementary feeding radically alters the activity and ranging behaviour displayed by these animals. Sub-units of the herd are drawn together daily at the feeding site, and herd cohesiveness is markedly higher over the winter months. Feeding usually takes place on a fixed site one or two hours after dawn. The animals congregate at the site over this period, and stand around ruminating or inactive; little feeding takes place. With the arrival of the fodder the cattle will feed steadily until it has been consumed, this often taking as long as two hours or more, and will then lie ruminating over much of the remaining morning and early afternoon. Sporadic feeding may occur over this period, but total intake is negligible. During the mid to late afternoon the number of animals foraging or standing around increases, until there is a general departure from the feeding site around one hour before dusk. A steady movement then follows as the herd makes its way to the favoured overnight area. While the route taken may vary, the site selected to spend the night is almost always the same and this often involves considerably greater movement than occurs during the summer months. Round trips of over 11.5 km were recorded regularly. While en route, animals feed extensively, mostly on *Calluna* on dry and wet heath communities; they generally arrive at the night site as dusk falls. Feeding may continue over the first hour or two of darkness, particularly where a grazing area is available, before the animals bed down for the night. Despite the long hours of darkness, there is still very little feeding during the night. At first light the herd begins the return march, on which little foraging takes place, animals arriving at the feeding site in good time for the daily supply of fodder.

The reasons for this diurnal 'migration' are not immediately apparent. It may be assumed that the requirements of the overnight site are more rigorous during the winter months, and that relatively few suitable sites are available. Supplementary feeding generally takes place somewhere in the vicinity of the owner's farmstead in order to minimise the transport involved, and in selecting the site little consideration is given to the pattern of animal movement over the remainder of the 24 hours. Feeding sites and suitable overnight sites may therefore be widely separated, a daily 'route march' thus being enforced (Putman *et al.* 1981).

Patterns of Habitat Use

Within each animal's home range is contained a diversity of different vegetational types. The animal's use of its home-range area (pages 34-5 and 36-7) is clearly a consequence of the way in which it uses different vegetational communities which are available to it; the pattern of use of these various habitat types often shows considerable subtlety. In order to explore in more detail the way in which a cattle-beast or pony uses its vegetational environment and exploits the diversity of pattern, we must first examine more carefully the various different habitats available within the New Forest.

The Different Vegetational Communities of the Forest

In the first chapter of this book we presented a brief description of the Forest vegetation, emphasising that, despite its name, this is no forest of trees, but a hunting Forest of a wide diversity of habitats: some wooded, others open tracts of heathland or grassland. The diversity of habitat is indeed tremendous: a rigorous botanist would define any number of distinct formations, any number of separate systems with different characteristic associations of species. For our purposes, however, we must generalise to some degree (after all, reduced to its extremes no one square metre of the area is exactly the same as another), and in our studies of habitat use by the large herbivores of the Forest we have recognised 15 *major* habitat types. These may be grouped into gross categories as grasslands, heathlands and 'cover' communities (Table 3.2).

The natural *acid grasslands* of the New Forest, established for the most part on plateau gravels, comprise a total of about 3,000 ha of the Forest, mainly as small, irregularly shaped areas bordering deciduous woodland or heath and interspersed with clumps of gorse or thorn scrub. They are rather species-poor areas, generally dominated by the bristle bent (*Agrostis setacea*) and purple moor-grass (*Molinia caerulea*) and usually invaded to a greater or lesser extent by bracken (*Pteridium aquilinum*) and heather (*Calluna vulgaris*). On the more

Table 3.2: The main vegetation types recognised within the New Forest in this study, with details of soil, plant species composition, management history and extent in Forest as a whole (modified from Pratt *et al.*, 1985)

Vegetation type (code)	Approx. total area in N.F. (ha)	Area of individual sites (ha)	Soil characteristics Parent material	Soil type	pH	Most abundant plant species	Remarks (History, Management, etc.)
GRASSLAND							
a) Reseeded lawn (RL)	350	10-30	S,G	B	5.0-5.5	*Agrostis capillaris, Festuca rubra, Bellis perennis, Hypochaeris radicata, Plantago lanceolata, Trifolium repens.*	Established late 1940s on former A.G. Ploughed, fenced. Fertilised and sown with grass/clover mixtures. Fences removed 1-2 years after reseeding. (Browning 1951)
b) Commoners' improved grassland (CI)	270	3-10	S,G	B	5.2-5.7	*A. capillaris, F. rubra, Poa compressa, Hieracium pilosella,* sparse *Pteridium aquilinum.*	Established early 1970s on former A.G. Bracken swiped and limed.
c) Streamside lawn (SL)	320	< 10	A,C	B(Gl)	4.5-5.0	*A. capillaris, Agrostis canina, Carex panicea, Juncus articulatus, H. radicata, T. repens.*	Narrow lawns on flood-plains of partially canalised streams. Invasion of scrub (*Prunus spinosa*) common.

Key A = Alluvium, C = Clay, G = Gravel, H = Peat, S = Sand, B = Brown Earth, G1 = Gley, P = Podsol.

Table 3.2 continued

Vegetation type (code)	Approx. total area in N.F. (ha)	Area of individual sites (ha)	Soil characteristics Parent material	Soil type	pH	Most abundant plant species	Remarks (History, Management, etc.)
d) Roadside verge (RV)	170	2 m wide strip	G	—	6.0-7.0	*A. capillaris, F. rubra, P. compressa, Plantago coronopus, Trifolium spp., H. pilosella, Taraxacum officinale.*	Strips bordering unfenced roads; more extensive near car parks. Usually alongside heath-lands with associated band of gorse (*Ulex europaeus*). Very shallow soils.
e) Acid grassland (AG)	3,400	< 10	S,C	P,B	3.9-4.6	*A. capillaris, F. rubra, Sieglingia decumbens, Agrostis setacea, Molinia caerulea, Carex spp., Potentilla erecta.* May be associated with *Pteridium, Calluna vulgaris* or *Ulex europaeus.*	Very variable. Gorse abundant on some sites. Bracken-dominated areas characteristic of better soils between dry heath-land and deciduous woodland.
BOG	820	10-30	C,A	H	3.7-4.8	Eutrophic bogs: *Alnus* or *Salix* carr.	Variable depending on nutrient status of

Key A = Alluvium, C = Clay, G = Gravel, H = Peat, S = Sand, B = Brown Earth, G1 = Gley, P = Podsol.

						Oligotrophic bogs: *Sphagnum* spp., *Molinia, J. articulatus, Eriophorum angustifolium* etc.	catchment area. Often flooded and access difficult. Some sites modified by drainage, chiefly in 19C.
HEATHLAND a) Wet heath (WH)	2,020	5-20	C,A	P(Gl)	3.9-4.1	*Calluna, Erica tetralix, Molinia.* Scattered *Pinus sylvestris* saplings.	Often forms gradation from dry heath to bog or or streamside lawn. Increasing proportion of *Molinia* and *E. tetralix* in wetter parts.
HEATHLAND b) Dry heath (DH)	7,380	10-50	G,S	P	3.7-4.0	*Calluna, Molinia.* Locally *Erica cinerea.* May be associated with *Pteridium, Ulex* etc., especially around margins.	Managed by burning on c.12-yr rotation. Recently burnt acres have high proportion of *Molinia* (regenerating heath-RH).
WOODLAND a) Deciduous woodland (DW)	7,940	< 50	C,S	B(Gl)	4.0-4.5	*Fagus sylvatica* and *Puercus robur* with *Ilex aquifolium* understorey. *Betula pendula* and	Uneven age structure reflecting 3 main periods of regeneration associated with low

Key A = Alluvium, C = Clay, G = Gravel, H = Peat, S = Sand, B = Brown Earth, G1 = Gley, P = Podsol.

Table 3.2 continued

Vegetation type (code)	Approx. total area in N.F. (ha)	Area of individual sites (ha)	Soil characteristics Parent material	Soil type	pH	Most abundant plant species	Remarks (History, Management, etc.)
						Crataegus monogyna around borders with A.G.	grazing pressure. (1) 1650-1750, (2) 1850-1920, (3) 1935-1945 (Tubbs 1968). (See page 153)
b) Coniferous woodland (CW)	4,840	< 20	G,S	P(G1)	3.8-4.1	*P. sylvestris* with understorey of *Ilex* and *Crataegus.* Sparse ground flora of *Pteridium* and *Calluna.*	*P. sylvestris* introduced in late 18C. Largely self-sown and spreading on heathland and woodland margins. Partially controlled by cutting.
c) Gorse brake (GB)	1,040	< 50	S,G,C,	B	4.0-4.5	*Ulex* with ground flora of *A. setacea, Molinia, Calluna* and A.G. species.	Narrow belts near lawns, roads, trackways, etc.; more extensive tracts associated with some A.G.s. Indicator of past human activity (Tubbs & Jones 1964). Partially managed by cutting or burning.

Key A = Alluvium, C = Clay, G = Gravel, H = Peat, S = Sand, B = Brown Earth, G1 = Gley, P = Podsol.

fertile alluvial soils at the margins of the many small rivers which traverse the Forest, these acid grasslands are replaced by the more lush grazing of *streamside lawns* (Plate 6). The flora of these grasslands is variable, dependent on the degree of waterlogging/frequency of flooding etc. The dryer areas are dominated by grasses such as *A. tenuis, Cynosurus cristatus, Lolium perenne* and *Anthoxanthum odoratum*, whilst the damper parts have a greater abundance of *A. canina, Alopecurus geniculatus*, rushes (*Juncus bulbosus* and *F. articulatus*) and sedges (*Carex panicea* and *C. flacca*). Moisture-loving species such as creeping buttercup (*Ranunculus repens*), lesser spearwort (*R. flammula*), marsh ragwort (*Senecio aquaticus*), and marsh pennywort (*Hydrocotyle vulgaris*) are common, with plants such as water mint (*Mentha aquatica*), water-pepper (*Polygonum hydropiper*) and floating sweet-grass (*Glyceria fluitans*) abundant in drainage ditches and oxbows.

As noted earlier, the Forest also offers to the grazing herbivores a variety of artificial grasslands. A number of different management techniques have been attempted over the years, resulting in the formation of three distinct types of 'improved' grasslands: Commoners' improved grasslands, Verderers' improved areas, and re-seeded lawns (Plate 7).

About 20 areas of acid grassland, gorse and heathland totalling 350 ha were ploughed, fenced, cropped for several years and re-seeded with grass during the period 1941-59 under the New Forest Pastoral Development Scheme. This increased the agricultural productivity of the Forest during wartime and provided improved grazing for the rapidly increasing population of commonable animals when the fences were removed. The seed mixtures and fertilisers used in creating these *re-seeded lawns* varied considerably in different areas, with rye-grass (*Lolium perenne*), cock's-foot (*Dactylis glomerata*) and Timothy (*Phleum pratense*) particularly common, accompanied by heavy applications of lime, chalk and superphosphate. Other grasses frequently used included red fescue (*Festuca rubra*) and crested dog's-tail (*Cynosurus cristatus*).

Since re-seeding, the lawns have been colonised by indigenous grasses, e.g. *Agrostis tenuis, F. rubra*, and a wide variety of other grassland species, such as selfheal (*Prunella vulgaris*) and procumbent pearlwort (*Sagina procumbens*) which are abundant on the dryer soils, with buttercup (*Ranunculus repens*) etc. common in the damper areas. There is also an abundance of leguminous species, e.g. *Trifolium repens* (white clover), *T. dubium* and *T. micranthum* (yellow trefoils) and *Lotus corniculatus* (bird's-foot-trefoil). Many lawns show some signs of recolonisation by bracken (*Pteridium aquilinum*), heather (*Calluna vulgaris*) and gorse (*Ulex europaeus*), although these species are not usually vigorous.

Commoners' improved areas (a total of about 260 ha within the Forest as a whole, mainly in the south) were derived originally from native acid grassland. The grasslands were swiped to cut back the bracken, and then treated with lime, phosphate and other fertilisers as part of a limited scheme of pasture improvement carried out between 1969 and 1971. The flora of these areas reflects that of the acid grasslands from which they are derived, but with a greater species diversity as a result of improved soil conditions and reduced bracken cover. Several grasses are abundant, including *A. tenuis*, *Festuca rubra*, *Sieglingia decumbens* and *Poa annua*, and numerous herbaceous species are found, in addition to those also common in acid grassland, e.g. yarrow (*Achillea millefolium*), plantains (*Plantago lanceolata* and *P. major*) and dog-violet (*Viola canina*). Composites, such as the daisy (*Bellis perennis*), autumn hawkbit (*Leontodon autumnalis*), ragwort (*Senecio jacobaea*), cat's-ear (*Hypochoeris radicata*) and thistle, are common, and heathland species (*Calluna*, *Erica* spp., *Ulex* spp., etc.) are also found. A second series of improved grasslands (Verderers' improved areas) was established in the early 1960s, in the form of firebreaks totalling about 100 ha around ten Forestry Commission plantations. Although fenced, these strips are generally accessible to commonable animals via gates, and can be used to hold animals after rounding-up etc., unlike the main plantation firebreaks from which cattle and ponies are excluded.

Rotovating and re-seeding of these areas was less successful than the re-seeded lawn improvements, and many now support only a limited grassland cover. Drainage is often poor, with the damper areas dominated by sedges (*Carex* spp.), *Ranunculus* spp., composites (e.g. *Leontodon autumnalis*), and streamside-lawn species.

Although not deliberately 'improved', the Forest contains one other important type of artificial grassland in the many *roadside verges*, a grassland type quite widespread and extensively used by commonable animals. The verge normally occupies the 1.5-m-wide strip of disturbed ground between the road and the ditch (c. 30 cm deep) excavated to restrict the access of cars to the Open Forest. In areas of high human and animal pressure, e.g. around car parks and camp sites, the verge may, however, extend to about 50 m from the road before merging gradually with the surrounding vegetation.

The flora of the roadside verges is generally similar to that of the shorter areas of the re-seeded lawns, although more sparse and with a higher proportion of *F. rubra* and *S. decumbens* grasses. Composites and rosette plants such as yarrow (*A. millefolium*) and buck's-horn plantain (*P. coronopus*) are particularly common.

The transitional zone between the verge and the neighbouring vegetation (frequently heathland) comprises mainly composites (*H. radicata* and *Leontodon* spp.) with *Molinia* sp. and *Sieglingia* sp. grasses. The dwarf gorse (*Ulex minor*) is at its most abundant in this zone, whilst clumps of gorse and bramble are also typically found.

The other main vegetation types of the Open Forest are the various *heathlands* and the *valley bogs*. Vegetation dominated by heather (*Calluna vulgaris*) extends over about 4,500 ha of the Open Forest, mainly on the nutrient-poor podsolised sands and gravels (Plate 8). Growing in association with the heather are *Molinia* sp., dwarf gorse (*Ulex minor*), heaths (*Erica cinerea* and *E. tetralix*) and bristle bent (*Agrostis setacea*), these species being more abundant in the younger-aged communities. Mature dry heath is often invaded by bracken, particularly in the north of the Forest, whilst gorse bushes (*U. europaeus*) are common on disturbed soils, e.g. near roads, tracks, railway cuttings, etc.

The vegetation of these *dry heaths* is maintained as a 'fire-climax' by periodic (every 6-15 years) controlled burning. Alternatively, the mature heather may be removed by mechanical cutting and baling for use in road construction etc. These operations reduce the risk of accidental fires, and fulfil the statutory requirement of the Forestery Commission to prevent scrub invasion of the grazing lands, whilst supposedly increasing the amount of forage available for stock. The structure of the heathland clearly varies enormously with the time interval since it was last burnt or cut. Accordingly, we distinguished a series of different dry-heath communities within the Forest: recently *regenerated heath* and dry heath of ages 3, 7 and 12 years.

Humid and *wet heathland*, in which *E. tetralix* is abundant and often replaces *Calluna* sp. as the dominant dwarf shrub, occupies a further 2,000 ha (approx.) of the Forest. It frequently occurs as a transition zone between dry heath and bog or streamside lawn. *Molinia* is generally more abundant than in the dry-heath community, and other moisture-loving plants are also found, e.g. *Juncus* spp., *Carex* spp., and sundew (*Drosera* spp.). Such communities grade, in wetter areas, into the *valley bogs* so characteristic of the New Forest system (Plate 9). These bogs occupy about 900 ha of the

New Forest, the most extensive tracts occurring in the southern part of the area. The bog vegetation is very variable, depending on the acidity of the water flowing through it. Thus, water derived from the relatively alkaline Headon Beds and Barton Clays gives a central alder-carr vegetation (although the edges may be acidic), whilst soil water from the acid gravels and sands supports vegetation dominated by *Sphagnum* mosses and *Molinia caerulea*, with some local development of grey willow carr (*Salix cinerea*).

Cover communities are recognised within the Forest as woodlands and gorse brakes. Unenclosed woodland, of which more than 80% is deciduous, extends over about 3,350 ha of the New Forest, whilst the Statutory Inclosures, with 30-40% hardwoods, occupy a further 8,300 ha. The *deciduous woods* are dominated by mature oak (*Quercus robur*) and beech (*Fagus sylvatica*), usually with an understorey of holly (*Ilex aquifolium*), although hawthorn (*Crataegus monogyna*) occurs in a few places. The ground flora consists of glades of *A. tenuis* grass (recognised as a distinct vegetation type '*woodland glade*'), with species such as bracken, foxglove (*Digitalis purpurea*), bramble (*Rubus fruticosus*), wood spurge (*Euphorbia amygdaloides*) and wood-sorrel (*Oxalis acetosella*) locally common. The more densely shaded areas (particularly under beech), however support very little herbage. The unenclosed coniferous woods are dominated by Scots' pine (*Pinus sylvestris*) with a scant understorey of holly and hawthorn bushes etc. The ground flora is composed of bracken, bramble, honeysuckle (*Lonicera periclymenum*), ivy (*Hedera helix*) etc., with very little grass except on paths and rides. Finally, shelter is also offered to the grazing animals of the Forest in the extensive *gorse brakes* of *Ulex europaeus* which develop on acid-grassland or heathland communities or around the edges of the various improved grasslands.

Patterns of Habitat Use

Animals must seek from their environment satisfaction of a number of different requirements — food, shelter, water, access to mates — and the patterns of habitat use we observe reflect their attempts to satisfy these various needs from the limited range of communities and habitats available to them. The needs change, and the resources offered by different habitats themselves change; the observed pattern of use of habitat tracks and reflects these changes in resource quality and animal requirements. The Common animals of the New Forest have available to them a complex mosaic of vegetational communities as described above. Their use of this mosaic is best understood if we appreciate that their primary requirements from the Forest environment are food (and to a lesser extent water) and shelter. Differences in emphasis of these separate priorities at different times (a lesser need, perhaps, to worry about daytime shelter during summer than in winter) or changes in availability of the required commodity in different communities (as, for example, relative availability of forage may alter seasonally between the different habitats of the home range) will result in changes — diurnal or seasonal — in the observed use of habitat.

The resultant distribution of cattle and ponies across the various communities of the Forest are summarised in Figures 3.1 and 3.2.

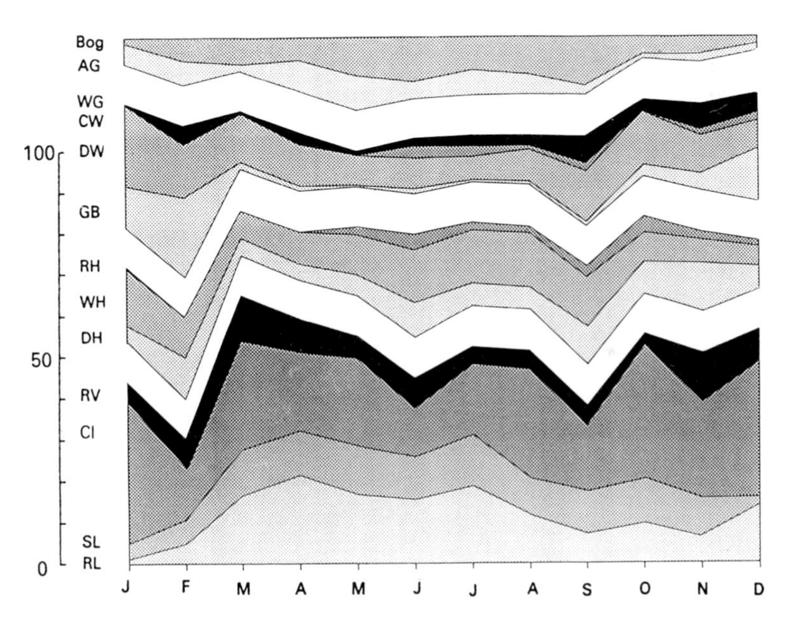

Figure 3.1: Seasonal changes in the distribution of ponies across the different vegetation types of the New Forest. The vegetation types have been grouped as follows: Grasslands: *re-seeded lawns (RL), streamside lawns (SL), Commoners' improved grassland (CI), roadside verges (RV);* Heathland: *dry heathland (DH), wet heathland (WH), regenerating heathland (RH);* Cover communities: *gorse brake (GB), deciduous woodland (DW), coniferous woodland (CW), woodland glades (WG);* Other vegetation types: *acid grassland (AG), bog. Data are expressed as percentages of total observations in each month. The scale on the left shows divisions of 10% total observations*

Source: Pratt et al. *(1985).*

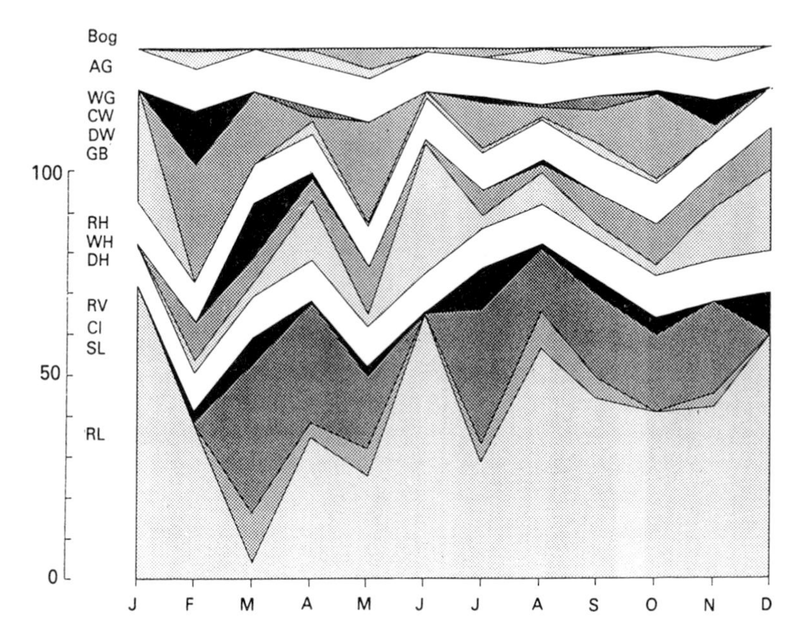

Figure 3.2: Seasonal changes in the distribution of cattle across the different vegetation types of the New Forest. See caption to Figure 3.1 for key to vegetation types

Source: Pratt et al. *(1985).*

Data presented here are drawn from detailed studies of the patterns of habitat use by cattle and ponies in the New Forest published by Putman *et al.* (1981, 1982) and Pratt *et al.* (1985). The data themselves derive in the main from two observational techniques.

1. *Static watches.* Two replicate sites were chosen for each of the vegetational communities defined above. In each month, one replicate of each community was observed over a 24-hour period (alternate replicates being observed in the following months). The observer recorded number and activity of each species present on the site at fixed intervals throughout the 24-hour period.
2. *Transect watches.* Transect routes, selected to run through as many different vegetation types as possible within a small area, were established in two main parts of the Forest. Set points, commanding a clear view of a particular vegetation type, were fixed for each route and observations were made at these predetermined points every two hours during the 24 hours of a day. Each transect route was watched

for a full 24-hour period every fortnight. (Methods throughout are described in detail elsewhere: Putman *et al.* 1981).

Results from all observational techniques confirm that, for the *ponies*, the various different grasslands of the Forest (particularly re-seeded lawns, streamside lawns and Commoners' improved grasslands) are tremendously important communities throughout the year (Table 3.3). Proportion of animals on improved grasslands and streamside lawns (as a proportion of total numbers observed on all communities) stays for the most part in excess of 40-50%. Increasing stock numbers, often coupled with the effects of drought, result in reduced availability of grass over the summer months. Declining use of the grasslands is compensated for by increased use of the bogs and both wet and regenerating heath; in autumn and winter, increased use is made of deciduous woodland and gorse brake. To a large extent, pattern of habitat use for feeding matches pattern of total occupance, suggesting that habitat use may be influenced primarily by foraging. During winter periods, however, and in hours of darkness at all times of year, extensive use is made of deciduous woodland edge and/or gorse brake for shelter. Feeding continues, but at this time choice of forage is dictated mainly by what is available within that community rather than by the converse.

Cattle were observed in all community types used by the ponies, although on all methods of observation far fewer animals were recorded (most are removed from the Forest during winter months). The pattern of habitat use revealed by transect watches show, as for the ponies, the importance of grassland communities. The various improved grasslands, together with streamsides, sustained in excess of 40% of cattle records for most of the year, with over 70% use in the summer months (July-September). Heathland communities were also used extensively throughout the year, except in midwinter months. Deciduous woodland showed peak use in spring and autumn; regenerating heath and bogs showed peak occupancy from March to May. Overall, the patterns of community use throughout the year shows relatively little variation, and displays considerably less flexibility, than that of the ponies (see Table 3.4).

Over the autumn months the majority of cattle are removed from the Forest in a series of 'drifts' or round-ups, and those few herds which are overwintered on the Open Forest are generally fed to a considerable extent with hay and straw. The practice of supplementary feeding radically alters the activity and ranging behaviour

Table 3.3: Use of habitat by New Forest ponies

Percentage of ponies seen in each vegetation type each month													
Jan	Feb	Mar	Apr	May	June	July	Aug	Sept	Oct	Nov	Dec		
2	5	17	21	14	17	19	12	8	10	11	16	RL	(Re-seeded Lawn)
5	6	12	9	19	12	13	9	9	11	13	3	SL	(Streamside Lawn)
33	15	27	22	14	11	18	27	18	34	23	27	CI	(Commoners Improved Grassland)
4	8	11	6	6	6	4	5	5	2	9	6	RV	(Roadside Verge)
3	9	1	10	3	4	6	4	3	2	3	2	AG	(Acid Grassland)
4	10	4	3	14	11	7	7	12	11	18	11	DH	(Dry Heath)
15	9	7	7	14	14	13	13	13	7	5	8	WH	(Wet Heath)
2	5	7	9	9	12	8	10	12	4	4	4	Bog	
20	12	12	10	6	6	6	9	13	13	7	6	DW	(Deciduous Woodland)
0	0	0	0	1	3	2	1	2	0	0	2	CW	(Coniferous Woodland)
0	4	0	2	0	2	3	2	6	2	4	5	WG	(Woodland Glades)
12	16	2	1	0	1	0	0	0	3	3	11	GB	(Gorse Brake)

Table 3.4: Use of habitat by New Forest cattle

Percentage of cattle seen in each vegetation type each month													
Jan	Feb	Mar	Apr	May	June	July	Aug	Sept	Oct	Nov	Dec		
51	29	19	18	24	73	31	55	43	42	30	52	RL	(Re-seeded Lawn)
24	0	8	6	16	0	5	9	5	0	2	11	SL	(Streamside Lawn)
2	0	40	17	19	0	37	16	21	16	34	0	CI	(Commoners Improved Grassland)
0	3	4	1	3	0	11	1	3	4	0	4	RV	(Roadside Verge)
9	4	0	2	2	1	0	4	0	1	3	4	AG	(Acid Grassland)
0	0	12	5	3	24	2	9	3	3	17	11	DH	(Dry Heath)
0	5	5	16	9	1	6	2	7	10	7	6	WH	(Wet Heath)
0	1	0	4	5	0	2	0	2	0	0	11	Bog	
0	39	10	10	16	0	9	2	8	24	2	0	DW	(Deciduous Woodland)
1	0	1	0	0	0	0	1	6	0	0	0	CW	(Coniferous Woodland)
0	18	0	1	0	1	1	0	0	1	5	0	WG	(Woodland Glades)
13	0	0	1	2	1	1	0	2	0	0	0	GB	(Gorse Brake)

displayed by these animals. Sub-units of the herd are drawn together at the feeding site daily, and herd cohesiveness is markedly higher over the winter months (see page 37).

Both herbivore species display a pronounced diurnal shift in habitat choice within the overall pattern of community use described, with daytime habitats selected primarily for feeding and night-time selection for communities offering some degree of cover. This shift appears to be correlated with the daily patterning of activity in the two species: foraging is undoubtedly the major daylight activity for both cattle and ponies, with the time devoted to resting (or rumination) increasing after dark (Pratt *et al.* 1985).

Diurnal movements between vegetation types by *ponies* are very much more marked over the summer months than in winter. Throughout the year, ponies show preferential night-time use of vegetation which provides cover or shelter. In winter, however, these 'shelter' communities are also used more extensively during the daylight than they are during the summer. Thus, over the winter months, the ponies spend much of the time in sheltered areas, making periodic forays onto the more open vegetation communities, and radical day/night changes in dispersion are uncommon. In summer, however, most of the daylight hours are spent in open habitats (grasslands, heathlands and bog communities), and diurnal movements into sheltered areas are very much more obvious.

Day/night differences in cattle dispersion are even more marked than those observed in ponies and persist throughout the year. In all seasons, cattle spent most of the daylight hours on the open vegetation types, and particularly the grasslands, but made a deliberate move onto other communities at dusk and back at dawn. Vegetation types most commonly used by night were deciduous woodland and dry heath.

Clearly, animals are not evenly distributed across the various communities available to them within the Forest. This, together with the obvious movement at dusk and dawn to and from communities offering shelter, makes it clear that animals positively select which habitats to occupy in their home range at a given time. Animal *preferences* for habitats may be considered in terms of a preference index calculated as the degree to which dispersion between different community types differs from a purely regular distribution (with numbers in each habitat merely reflecting relative areas of that habitat in the environment as a whole). Thus, an index

$$\frac{\text{Number of animals observed in community A}}{\text{Area of A}} \times \frac{\text{Total area surveyed}}{\text{Total number of observations in } all \text{ vegetation types}}$$

will equal 1 *if* animals are evenly spread across all vegetation types, will be > 1 if positive preference is shown for a particular community, and < 1 if the community is avoided. In Table 3.5, this index is presented as a logarithm; this adjusts the values to a more easily visualised scale so that a score of zero now represents no preference, while any positive number is positive preference and any actual avoidance will appear as a negative value in the tables. These data confirm the seasonal trends in pattern of use described above, but reveal preferences for individual community types which are not necessarily apparent from the pure analysis of animal distribution (Figures 3.1, 3.2). Among ponies there is a strong positive preference for grasslands throughout the year ($P_i = 0.410$ February to $P_i = 0.755$ March), but the index shows that streamside lawns and roadside verges are in fact preferred to Commoners' improved and re-seeded lawns despite the fact that these latter grassland types support more animals. In contrast, all heathland communities are under-exploited (greatest preference was in June and September, when P_i was −0.420). No strong preference for or against the woodland and gorse-brake communities is apparent over the summer months, but they are increasingly favoured during the winter when shelter is more important (max. 0.505 February). Conversely, the lowland bogs are avoided over the winter months but show positive preference in summer. Preference indices calculated for *cattle* confirm a more confined use of the habitat than occurs in ponies and with even greater preference being shown for the improved grasslands. In this case, however, preference for individual grassland types corresponds to the observed distribution of actual observations, with re-seeded lawns and Commoners' improved areas favoured over streamside lawns and roadside verges. Positive preference is also shown for the cover communities, primarily deciduous woodland, in several months of the year. As with ponies, the heathland communities are relatively under-exploited.

Habitat preferences and thus individual patterns of habitat use differ between individual animals. What we have described here is an overview of the pattern of use of vegetation by a whole population;

Table 3.5: Preferences shown for different vegetation types by New Forest Ponies and Cattle

PONIES

Vegetation	J	F	M	A	M	J	J	A	S	O	N	D
GRASSLANDS	0.544	0.410	0.755	0.716	0.694	0.629	0.689	0.654	0.522	0.633	0.590	0.635
HEATHLANDS	−0.538	−0.495	−0.770	−0.745	−0.585	−0.420	−0.509	−0.495	−0.420	−0.538	−0.481	−0.744
WOODLAND and GORSE	0.394	0.505	0.107	0.134	−0.108	0.127	0.041	0.053	0.290	0.199	0.228	0.336
BOG	−0.721	−0.194	−0.143	−0.187	0.025	0.053	−0.060	0.041	0.143	0.292	−0.252	−0.602
ACID GRASSLAND	0.057	0.155	−0.469	0.250	0.299	−0.027	0.146	0.004	−0.237	−0.409	−0.187	−0.469

CATTLE

Vegetation	J	F	M	A	M	J	J	A	S	O	N	D
GRASSLANDS		0.524	0.702	0.777	0.653	0.774	0.839	0.843	0.792	0.710	0.712	
HEATHLANDS		−0.721	−0.444	−0.481	−0.620	−0.292	−0.921	−0.770	−0.770	−0.699	−0.469	
WOODLAND and GORSE		0.566	0.193	−0.268	0.377	−1.000	0.111	−0.523	0.100	0.286	−0.187	
BOG		−0.921	—	−1.222	−0.276	—	−0.602	—	−0.602	—	—	
ACID GRASSLAND		−0.013	—	−0.201	−0.161	−0.602	—	−0.036	–	−0.745	−0.114	

within this there is considerable individual variation (Howard 1979, Gill 1984). In part, such differences are imposed upon the animals because of differences in the relative availability of the various vegetational communities in different geographic regions of the Forest. Thus, pony populations in areas without extensive gorse brake will make more use of woodland for winter shelter, and ponies in areas with no improved grasslands will make more use of stream-sides or natural acid grassland and heath. But this is not a complete explanation: even among individuals who share the same basic range there are subtle differences in the extent of use of different habitats; such individual differences clearly reflect true differences in habitat preferences, and may have considerable implications in relation to differences in the ability of individual animals to maintain body condition (Gill 1984).

In practice, however, such differences in habitat use are rather slight; indeed, the overall impression is of surprising uniformity in community use. Studying patterns of habitat use by Forest ponies in four different sites chosen specifically because they *did* differ markedly in vegetation and standing crop, Gill found that habitat use by the ponies in the four areas was fundamentally rather similar. Ponies in areas which showed marked differences in relative availability — and juxtaposition — of different habitats, none the less end up displaying very similar patterns of habitat use.

Comparison of Patterns of Habitat Use of Cattle and Ponies

When the patterns of use of different vegetational types by ponies and by cattle over the course of the year are compared, a number of clear differences may be noted.

We may remark once more on the relative constancy over the year of the pattern of use of the different communities by cattle, which contrasts markedly with the flexible pattern of community use displayed by the ponies. Thus, cattle spend 60-70% of their time on grassland communities and 10-20% of their time in woodland (feeding on grasslands and heathland during the day and moving into woodland for shelter at night), and this pattern changes little over the year. To some extent this constancy of habitat occupancy — together with the choice of the actual communities themselves — may be a result of the strongly herding pattern of social organisation of the cattle: restricting activity to relatively few communities and making major changes in patterns of habitat occupancy hard to bring about. (Pony social grouping, like habitat use, is far more flexible.) It may also reflect the fact that cattle on the Forest are, in general, subject to a far greater degree of management by their owners than are the ponies. Patterns of habitat use may thus be a result, in large part, of owners' management policy.

Not only does the pattern of use of communities by the cattle show relatively little seasonal variability, but the actual range of community types occupied is also more restricted than that utilised by the ponies. Certain communities which are exploited extensively by the ponies are used but little or not at all by the cattle. While the ponies, during the course of a year, show some degree of occupancy of a wide range of community types, cattle restrict most of their time to 'improved' grasslands, heathland and woodland (primarily deciduous). Acid grassland, bog and gorse brake are used hardly at all by comparison, while regenerating heath and woodland glade (at certain times of year extremely important to the ponies) are virtually excluded.

Even where certain communities are used in common, they are not necessarily of equal importance to the two species. Cattle show a far greater use of grasslands (particularly areas of re-seeded lawn) in winter than do the ponies and, while ponies make more extensive use of wet heath throughout the year than do cattle, their use of dry heath is much lower. Further, even *within* grasslands, the two species show distinct preferences. Analyses of preferences above (Table 3.5) show marked preference by ponies for streamside lawns and roadside verges among the grasslands and also woodland glades. Cattle also favour grasslands, but are found predominantly on re-seeded and Commoners' improved areas. A partial explanation for this comes from the observation that cattle appear unwilling to occupy any grassland area which is less than 10 ha in size. Strongly social animals that move around their home range as a distinct herd, it may well be that areas less than 10 ha in area are insufficient to support the full herd. Such a constraint restricts them from use of road verges and streamsides, which are characteristically of a small available area. This difference in social behaviour between the strongly herding cattle and the essentially individualistic ponies may also underlie the observed differences in the pattern of range use of the two species. Ponies act very much as individuals and drift around their home ranges as they feed. The cattle by contrast act as an integrated herd; in addition the herd does not merely move as it feeds but makes quite distinct 'route marches' between different areas, moving from one feeding area to another — or to the cover of woodlands at night. These route marches are markedly purposeful and the animals may cover considerable distances (2-4 km). Such moves are quite distinct from the random drifting of movements during grazing. The animals have an almost inflexible circuit which they tour daily, as a herd, from shelter areas of overnight rest to daytime feeding sites, and back. The daily circuit is fixed and repeatable, encompasses the same areas and same vegetation types, and is followed faithfully. (These same factors also perhaps help to explain the clear differences between the two species in terms of seasonal variation in habitat use, already described. The pattern of habitat use by cattle changes but little, entrained as it is on the daily circuit; habitat use by ponies is, as we have seen, far more flexible, and shows great seasonal change in communities selected.)

Despite these various differences in the pattern of habitat use shown by cattle and ponies, the overriding impression is actually one of striking similarity. Both species are preferential grazers, feeding primarily on the various improved grasslands of the Forest and, overnight, seeking cover in woodlands or gorse brake. Even in winter, when the ponies are making more use of other foodstuffs (Chapter 4), 40-50% of observations are still recorded from areas of improved grassland; in the summer, both species again spend the majority of their time on grasslands (particularly RL,SL, CI) and wet heath. The various differences between the species to which we have just drawn attention are actually relatively *minor* differences in an overall picture of tremendous similarity. Calculations of overlap in the patterns of habitat use shown suggest an overlap of somewhere in the region of 78% (see Chapter 4). To what extent, then, may we consider cattle and ponies in the New Forest in competition?

From our considerations to date, it is clear that the major overlap in habitat use between cattle and ponies throughout the year is in the use of the various improved grasslands of the Forest. Both species rely heavily on these grazing areas — which are in fact a relatively limited resource within the Forest as a whole. In excess of 40% of all observations for either species are from these various grassland areas — in whatever month — yet these improved grasslands account for only some 1,140 ha in total area. We have already noted some degree of separation in the type of improved grassland favoured. Perhaps for social reasons (page 56), cattle make far less use of the Commoners' and Verderers' improved areas (and the natural streamside lawns) than do ponies and concentrate primarily upon re-seeded areas. But ponies also favour re-seeded lawns, and many such lawns support large populations of both species simultaneously. Closer examination reveals, however, that despite this overlap — or perhaps because of it — ponies and cattle *are* separated even in their use of a single community.

The most detailed information we have on this spatial segregation within communities is for the Forest grasslands, and particularly for re-seeded lawns. In these areas, it is clear that cattle and ponies feed in and occupy distinct areas on the lawns; and that these areas differ botanically, both in structure and species composition (for fuller details see Chapter 8). Each lawn thus comprises a mosaic of distinct patches — which we term 'short grass' (< 20 mm) and 'long grass' (20-50 mm) — occupied by ponies and cattle respectively.

The patches derive from the selective eliminatory behaviour of, particularly, ponies. Most large herbivores avoid feeding in the close vicinity of their own fresh dung, an adaptation presumed to reduce the risk of parasitic reinfestation: if an animal returns to graze an area on which dung has been deposited only after a considerable period (during which the dung has largely decayed), the risk of increasing its permanent parasitic burden to unacceptable levels is considerably reduced. Horses are particularly 'hygienic' in this regard and recognise specific latrine areas within their range, to which they move to defaecate. Such behaviour is a well-known feature of horses in pasture fields (Archer 1973, 1978; Odberg and Francis-Smith 1976, 1977), but it was originally thought that it occurred in response to enclosure and would not be shown by free-ranging horses in areas such as the New Forest. Indeed, Tyler (1972) suggested that there *was* no such pattern of selective dunging and grazing by the Forest ponies. More recently, however, Edwards and Hollis (1982) have clearly demonstrated that a mosaic of 'latrine' and 'non-latrine' areas does develop on New Forest lawns in response to the same pattern of selective feeding and dunging in the free-ranging ponies of our studies.

When feeding on lawns, many animals move off the area into adjacent vegetation types to defaecate. This is not, however, always possible, and other latrine areas are defined within the grassland community itself. Since the animals will not feed in these latrine patches, the grass grows longer in these areas (aided by the fact that the latrines obviously become considerably better fertilised than the rest of the sward). Thus a pattern is established, of ponies feeding on 'short-grass' patches (which are cropped as close as their teeth will allow) and defecating in 'long grass'. Since the mechanics of the way in which cattle obtain their food prevent them from feeding on very short grass, cattle on these same communities are forced to restrict their feeding activity to the 'long-grass' patches on the lawn. Cattle, unlike ponies, defaecate at random, so to speak; since they spend almost all their time in 'long-grass' areas, however, their faeces remain in these patches, and so the mosaic is maintained. As a result, use of grasslands is divided, with cattle occupying and feeding within 'long-grass' areas and ponies feeding on 'short-grass' patches.

This account is of course over-simplified to a degree. Over the winter, when animal numbers on the Forest are reduced (so that faecal accumulation in any area is also reduced), and when available grazing is also at a minimum, ponies do begin to graze into the long-grass areas. By the end of the winter, the mosaic pattern is less apparent and ponies feed and defaecate freely all over the lawns; indeed, at the beginning of the growing season they feed preferentially in the more nutritious long-grass areas. The pattern begins to re-establish itself, however, from April/May, and by June/July complete separation of cattle and pony feeding areas is achieved (Edwards and Hollis 1982). Although this pattern of spatial segregation is well established only on lawn communities, we have reason to believe that it may also occur, to a lesser degree, in most other communities (Putman *et al.* 1981). Its influence is in any case most important in the grassland types for which it is described here, since it is here that the greatest overlap in feeding use of communities between cattle and ponies is experienced.

In summary, pattern of use of the various communities differs markedly between cattle and ponies, in terms of the range of communities occupied and seasonal variation in community use. Both species are preferential grazers, however, and during the summer patterns of habitat use become very similar, with animals feeding in grassland and wet heath and moving into woodland at night. Some separation in feeding habitat at least is still maintained at this time, owing to increased use of wet heath by ponies and to the establishment of 'pony-feeding' and 'cattle-feeding' sub-divisions of the grazing communities. The patterns of habitat use described here for both cattle and ponies, restricted though they may be in the case of cattle by constraints of social organisation, may clearly be explained in large part by the animals' requirements from their environment in terms of food and shelter. Seasonal differences in community use reflect seasonal changes in the need to worry about shelter, and also reflect changes in the quality and quantity of forage available in different habitats. This relationship between habitat use and forage availability and quality will be discussed in detail in the next chapter.

Chapter 4

Food and Feeding Behaviour of Domestic Stock

Without a doubt, grazing by domestic animals constitutes at present one of the dominating factors in the ecological functioning of the New Forest system. Grazing has always played an important part in the ecology of the Forest; in the past the deer populations for which the area was primarily set aside perhaps played the major role (populations were estimated at 8,000-9,000 animals during their peak). But, since the decimation of deer populations following the Deer Removal Act of 1851, numbers have never exceeded 2,000-2,500, and from that time on commonable animals have probably been the most significant herbivores.

Coincident with the passing of the Deer Removal Act was the

Table 4.1: Proportions of different grazing animals (expressed in thousands) in the period 1789-1981

Historical period	Deer	Ponies	Cattle	Total grazers
1789-1858	9	3	6	18
1858-1900	3	2.5	3	8.5
1900-1930	2	1.5	2	5.5
1930-1943	1.5	.5	1	3
1943-1965	2.0	1.5	1	6.5
1965-1981	2.5	2.75	1.75	7

Source: Modified from Countryside Commission Report (1984).

increased enclosure of land for silviculture. Although the wild deer are still free to roam over the entire Forest area, the grazing area available for common pasturage was curtailed by such enclosure, and the effective *densities* of grazing stock upon the Forest have thus been further increased.

Over recent years the unenclosed Forest — approximately 20,000 ha of open waste — has been supporting on average some 2,000 cattle and 3,000 ponies (on 1970-80 figures; numbers have now declined somewhat and 1981 figures suggested a total pasturage of only 1,100 cattle and 2,500 ponies). While the area also supports some 2,500 deer (as a maximum estimate), clearly cattle and ponies must have the most significant impact upon the environment just in terms of sheer biomass. Of these two species the ponies probably have the major influence: in terms of numbers, in view of the fact that most of the cattle are pastured on the Forest only during a small part of the year *and* because of the ponies' nutritional physiology. The only non-ruminant among the herbivores, the monogastric horse is adapted to a foraging strategy founded on a very large throughput of but poorly digested forage. This massive throughput, coupled with an ability to crop very much closer to the ground than other species, must make the ponies the species with the greatest single impact upon the Forest vegetation.

In this chapter we shall consider the foraging behaviour and diet of the two main commonable species of the New Forest, the cattle and ponies, and examine the way in which they exploit the varied vegetational resources of the Forest. Seasonal changes in foraging behaviour and dietary composition will be considered in relation to availability and quality of the herbage; we shall also look at differences in the feeding strategies of the cattle and ponies as two different large herbivores living in the same vegetational system.

Foraging Behaviour

The pattern of use of the various habitats of the New Forest by ponies and cattle for feeding is shown in Figures 4.1 and 4.2 which summarise the percentage use of the various community types by ponies and cattle in different seasons.

Results are similar to those presented in the last chapter for patterns of habitat occupancy overall (see also Pratt *et al.* 1985). Feeding use of the various habitat types by *cattle* shows a fairly constant pattern through the year, with heavy emphasis on lawns and improved grasslands and also extensive use of heathland: wet heath in summer, drier areas during the winter months. Feeding use of other communities is not extensive, although deciduous woodland is exploited during the winter, and acid grassland also used for most of the year.

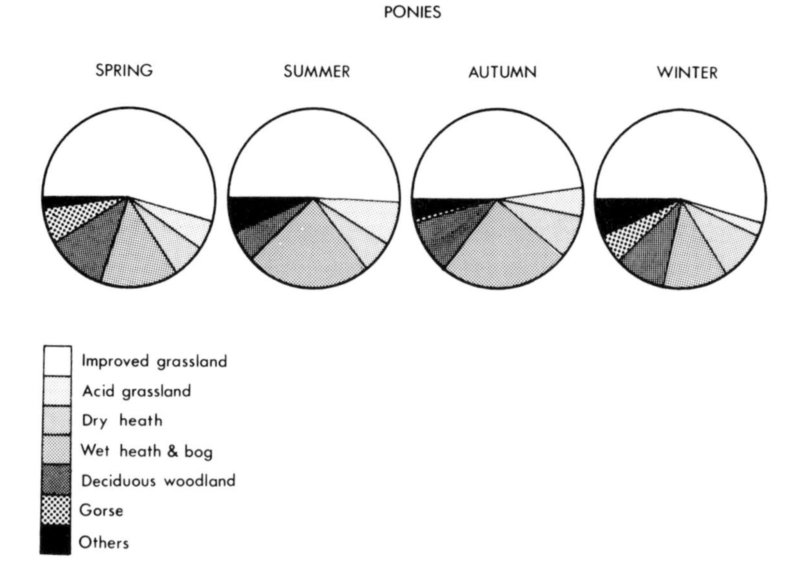

Figure 4.1: Use of different habitats for foraging by New Forest ponies

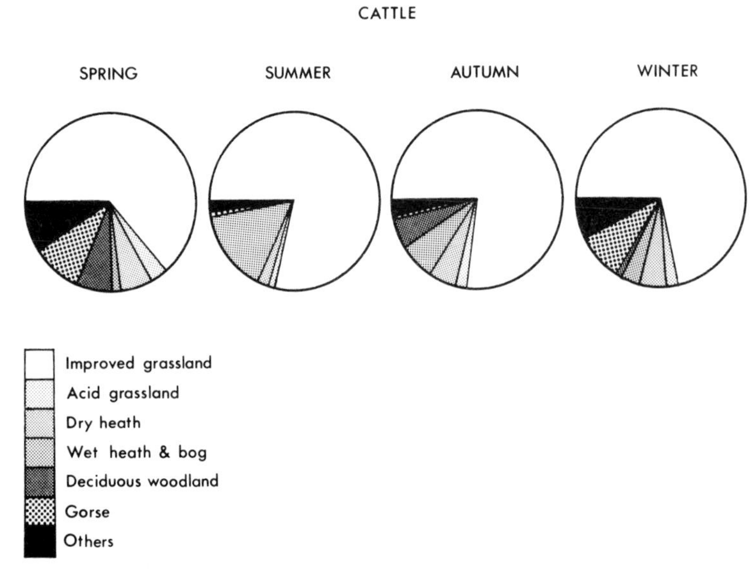

Figure 4.2: Use of different habitats for foraging by Forest cattle

Ponies exhibit a much more marked seasonality in their use of vegetation types. While improved grasslands are still emphasised throughout the year and acid grassland is also used consistently, wet heath, bogs and regenerating heathlands show a clearly seasonal pattern of use, as do deciduous woodland and gorse brake. Feeding use of bogs, while quite high throughout the summer, shows marked peaks of use from April to June and during August and September; use of wet and regenerating heath also shows a clear summer peak between May and September. Use of these latter habitats correlates closely with time of growth of the purple moor-grass (*Molinia caerulea*) in each particular community: *Molinia* is the 'main' forage species taken in bogs, wet and regenerating heaths, and, as will be seen later, is an important element of pony diet at this time, providing abundant forage from these wetter communities even in the drought months of summer when the preferred grasslands dry out and growth is halted. Feeding in gorse brake and deciduous woodland is restricted largely to the winter period, when it may become quite extensive, although woodland feeding continues throughout the year at night, when animals retire to these communities for overnight shelter.

It is quite apparent from these results that feeding by cattle and ponies is not evenly distributed across the different communities of the New Forest. Both species are preferential grazers and for most of the year concentrate the majority of their feeding activity on the various types of improved grassland and on streamside lawns (at all times more than 34% of pony feeding observations and 54% of cattle feeding is concentrated upon these areas, which together total only some 1,460 ha, less than 7% of the Open Forest area). In the spring, growing season (March, April, May) and in the summer (June, July, August), use of grasslands is still higher (ponies 61.5% spring, 47.2% summer; cattle 67.5% spring, 83.4% summer); even outside the main season of production of the Forest grasslands (May-September), feeding occupancy of these communities by both cattle and ponies still remains remarkably high, despite the fact that the standing crop of available forage declines to 25-30g/m^2 towards late summer.

Cattle, indeed, show relatively little flexibility in foraging behaviour throughout the year, and even in midwinter feed extensively on open grasslands. With the decline of available forage on the improved grasslands in autumn, however, the ponies make increasing use of the *Agrostis setacea* of acid grasslands. As winter advances further, they make more use of woodland browse and gorse brake, and this increased feeding use of woodlands and gorse continues through the winter until the first spring growth begins in the more open communities.

Both during the grass-growing season and outside it, cattle make more extensive use of wet and dry heathland than do ponies, and it is primarily an increased use of these communities during the autumn and winter that for them compensates for reduced food availability on grasslands at this time. As noted earlier, however, the majority of cattle are removed from the Forest during the winter. Those that remain are fed, and most feeding sites are concentrated on grassland or heathland areas. Thus, management too may be in part responsible for the apparent lack of flexibility in feeding style.

While this analysis offers information on the relative distribution of feeding animals across the vegetation types of the New Forest, that distribution is, once again, influenced in part by the relative availability — in terms of unit area — of the different vegetation types. If data are corrected for differences in available area of different communities by calculating preference indices of the same form as those presented for habitat use overall (page 53), we may determine more accurately preferred feeding communities. While, for cattle, relatively little selectivity emerges, beyond the gross pattern of occupancy already described with emphasis throughout the year on improved grasslands, ponies show heavy selection for particular communities at different seasons. Among the grassland communities marked preference is demonstrated for streamside lawns, and roadside verges also show consistently higher use per unit area than might be expected. The importance of *Molinia* communities (wet and regenerating heaths and bog) in summer is stressed, while feeding pressure in gorse brake during winter is also extremely high. Other points highlighted are the increased selection for coniferous woodland (probably also deciduous woodland, but available area is so high that the trend is concealed) and woodland glade during the height of summer, presumably as a response to feeding where possible in shade.

Winter use of gorse is one of the most distinctive characteristics of the New Forest ponies. As already noted in Chapter 3, gorse brake is positively selected over the winter months as one of the few communities offering good shelter from deteriorating weather conditions, but it is clear that it is also a preferred feeding community. As a legume, gorse clearly provides a highly nutritious forage at this time (page 75) despite its not inconsiderable defences (Plate 16). The ponies have developed a special technique for dealing with this rather spiny foodstuff: they eat only the tips (which are in any case the most nutritious part) and, when they have selected a shoot, roll back their sensitive upper lip and push the incisors as far forward as possible before carefully biting through the stem. The behaviour is unique to the Forest, and appears to be learnt. It should, however, be noted that not all Forest ponies feed on gorse; use of this foodstuff, and the behaviour required in order to exploit it, seem to be characteristic of particular local populations, and seem to be acquired by individual foals by copying their mothers or others in the immediate social group. There is a considerable local tradition, and *some* objective evidence (Gill 1984), that those ponies which *do* feed on gorse maintain a far better body condition over winter than those which do not have the behaviour.

The link between increased occupancy by ponies of bogs and wet and regenerating heath and feeding use of *Molinia* has been noted. Feeding use of *all* the different communities may indeed prove to be determined by forage availability. The degree of relationship between feeding use of the different grazing communities (in this case: numbers feeding per unit area) and monthly standing crop and productivity of forage has been examined (Putman *et al.* 1981 and 1985) to see if there is some correlation between forage availability and feeding use of an area.

Correlation overall, of feeding occupancy with standing crop, is not close, either for cattle or for ponies. Cattle show slight positive correlation between use of a community and standing crop during spring (March-May) and summer (August-September), but this is not statistically significant. Ponies show a consistent *negative* correlation with standing crop which is significant between May and December. Clearly, feeding occupancy of a community is not influenced by standing crop of available vegetation as we hypothesise; rather, standing crop is reduced in those communities with high feeding activity by the ponies. Relationships with productivity are even less clear, and no consistent correlation between feeding use of a community and its monthly productivity was established (Putman *et al.* 1985).

Such a result conflicts with a clear subjective impression that the timing of animal use of various forage communities is indeed closely marked to productivity, with ponies seeking at any one time the vegetation types showing maximum production. Certainly, at the beginning of the growing season in early spring, community use *is* closely related to production. As soon as the grasses and herbs begin to grow once more, the ponies at once forsake the browse communities in which they have foraged over winter and return to the more open 'grazing' communities. Clearly, they retain a marked preference for grazing rather than browsing and they move back onto these open communities at the earliest possible opportunity. At this time, the pattern of habitat use is closely related to the *onset* of production in each vegetational type. The first community types to flush are the bogs and streamside lawns, and it is noticeable that feeding occupancy of these habitats by ponies is relatively higher during early spring (February/March) than at other times of year (except during the drought months of midsummer when, once more, these wetter communities are the only ones showing significant growth). As the drier grasslands of re-seeded or Commoners' improved areas begin to flush a few weeks later, cattle and ponies begin to move onto these more preferred communities, once again matching their selection of communities to the timing of onset of new growth, until the distribution of both species shifts towards the more uniform pattern of grassland use characteristic of the main growing season.

It is here, however, that the relationship between habitat selection and productivity breaks down, for the factors determining habitat occupancy are complex — and they relate only in part to food availability. As we noted in the last chapter, habitat selection is the end result of a complex interplay of a host of different needs. Thus, in practice, the animals do not in fact select particular communities specifically for foraging. They feed within an overall pattern of habitat use which is determined by all considerations in complement: food, shelter, and whatever else. Early in the season, when palatable food is in short supply, foraging needs may indeed take priority, and a close relationship may obtain between observed patterns of habitat use and forage quality — as noted here, and also reported by Duncan (1983) for horses in the Camargue. But, as the growing season advances and food becomes more readily, and more uniformly, available, other factors — the need for water, cover or 'shade' — may predominate in habitat choice and the animals feed within a pattern of habitat use determined by these other factors.[1]

Patrick Duncan's data on habitat use and activity patterns of Camargue horses (Duncan 1983) offer interesting comparison with patterns reported here for the ponies of the New Forest. Although no woodlands are available to the Camargue horses, and the habitat is

---

[1] One especial consideration here is the purely social phenomenon of 'hefting'. As we have discovered, the ponies are preferential grazers, and indeed must focus their whole home area on a suitable grazing site. So important are these grassland areas, and so strong is the site attachment, that the ponies spend a disproportionately high amount of their time on these favoured lawns. From the time they show the first signs of growth on to well into the late autumn, the ponies heft to these areas, showing remarkable reluctance to leave the lawns even when they are virtually grazed out and other adjacent communities offer acceptable forage in much greater abundance.

overall far wetter than that of the New Forest, many of Duncan's vegetational types have approximate equivalents in the New Forest, and comparisons of patterns of habitat selection reported are quite enlightening in explaining the underlying requirements that the animals are seeking to fulfil from these two very different systems. Duncan reports an increase in use of richer grasslands for feeding from early spring (c. 22%) through to an August peak of 67% use. Feeding use declines through the autumn to lower overwinter levels. The main complement to this pattern of grassland use is reflected in use of heathland areas, which show peak use in early spring (January-March 80%) and a decrease in feeding occupancy over the summer, to 35% in August. Use of coarse grasslands is low (4-7% for most of the year), but peaks over the winter from November to March at approximately 30%.

In the New Forest, use of the coarse acid-grassland vegetation type is relatively constant throughout the year, at the same levels of 4-7% of all feeding observations, but it also shows a slight increase in autumn (August/September) and late spring (April/May). Use of the various improved grasslands and stream banks ranges from around 30% (late winter) to a midsummer peak of nearly 50% (cf. 22%-67%); in both systems ponies are preferentially grassland feeders, and move onto these vegetation types as soon as the first flush appears at the start of the growing season. Use of heathlands and bog communities in the New Forest is, however, nowhere near as extensive as in the Camargue, and shows a reverse pattern of seasonal use: contributing only approximately 20% to total vegetational use from January to March and *increasing* in use through the summer (30% use in August).

This apparent reversal is readily explained. In both the New Forest and the Camargue, grasslands are the favoured feeding areas as soon as forage is available. Over winter, when forage is exhausted, the animals favour other communities. In the Camargue, heathland vegetation types offer the best alternative feeding areas (maintaining the highest standing crop of available vegetation over the winter). In the New Forest, woodland communities are available, offering both shelter and forage; in the Forest, therefore, these are preferred and the more exposed heathlands do not suffer greatly increased use. Use of grasslands in the Camargue increases throughout the summer; in the New Forest, the real peak of use is in early summer, a slight decline in late summer being compensated for by increased use of bogs and wetter heathlands. Once again, the explanation is simple. In the wetter conditions of the Camargue, growth continues throughout the summer on the grasslands; in the New Forest, the grasslands dry out beyond midsummer, growth is suppressed, and more nutritious forage may be found in those wetter communities where growth may be maintained.

Feeding Rates

Rates of feeding in the different communities were recorded as bites per minute during follows of individual animals. Mean values for each month are presented in Tables 4.2 and 4.3. Not all communities will be represented in all months, since feeding use varies.

Bite rates differ markedly between vegetation types, and, within vegetation types, seasonal change is also observed. In a preliminary comparison of feeding rates of ponies on different vegetational communities in November and December, 1979, Putman *et al.* (1981) showed a clear, negative relationship between bite rate per minute and vegetational standing crop (Figure 4.3). A similar relationship is observed *within* communities, with bite-rate changes between months clearly correlated with changes in forage standing crop (and such relationship has previously been shown for both cattle and ponies on *agricultural* pastures by Arnold and Dudzinski (1978)). Such a result could be due to two complementary, but converse, effects. First, if standing crop is very low, bite rate must increase if the animal is to maintain a constant rate of bulk food intake. (In addition, if standing crop is low, chewing time per mouthful is decreased, permitting a higher bite rate.) Secondly, since grazing pressure on the Forest is heavy, a sustainedly high standing crop in any vegetation type can be due only to a high standing crop of non-forage species. Since the animals must therefore take time to select preferred species from among this high biomass of non-forage plants, bite rate will decrease. (In this latter case, bulk intake will also decrease. It is to be noted that, for example, dry-heath communities are rarely fed on by the ponies: time spent feeding in such communities in proportion to their percentage contribution to total area of range is extremely low.)

Such relationship between feeding rate and standing crop may also explain most seasonal changes of feeding rate *within* communities. Pony feeding rate on re-seeded lawns, as an example, increases through the winter, December-April, until the start of the growing season. As standing crop increases (productivity exceeds offtake during the summer months: Putman *et al.* 1981), bite rates decline. At the end of the summer, when productivity falls, and offtake starts to remove 'accumulated capital' in reducing standing crop (August/September),

Table 4.2: Feeding rates of New Forest ponies

	Mean bite rate, as bites per minute, in different vegetational types.											
Community	**Jan**	**Feb**	**Mar**	**Apr**	**May**	**June**	**July**	**Aug**	**Sept**	**Oct**	**Nov**	**Dec**
Improved grasslands (RL & Cl)	57	51	63	63	63	59	54	53	55	61	52	52
SL	—	51	60	57	72	52	49	43	46	61	51	51
AG	51	—	39	42	48	37	34	38	32	37	35	—
DH	—	—	13	17	—	44[a]	44[a]	23[a]	—	20	21	22
WH	—	—	41	—	39	42	36	29	29	34	—	—
Bog	—	33	22	41	40	32	37	31	35	38	31	—
DW	43	46	43	39	51	—	43	43	48	43	48	—
CW	—	—	41	—	—	45	42	—	35	—	33	—
WG	57	—	44	63	63	65	64	60	58	69	63	—
Gorse	8	8	6	—	—	—	—	-	—	—	-	5

Note: [a] Dry-heath figures for June, July and August are for feeding on *Molinia;* April, October, November, and December for feeding on *Calluna.*

Table 4.3: Feeding rates of New Forest cattle

Community	**Jan**	**Feb**	**Mar**	**Apr**	**May**	**June**	**July**	**Aug**	**Sept**	**Oct**	**Nov**	**Dec**
Improved grasslands (RL & Cl)	—	64	67	65	70	66	71	77	79	70	71	63
SL	—	57	—	—	—	—	61	58	77	56	76	—
AG	47	—	37	43	44	44	33	51	53	44	73	62
DH	40	38	—	36	30	32	39	44	56	39	41	—
WH	22	—	23	23	26	21	—	54	59	—	46	—
WG	—	—	63	—	—	—	—	—	—	—	—	—

Note: [a] Dry-heath figures for June, July and August are for feeding on *Molinia;* April, October, November, and December for feeding on *Calluna.*

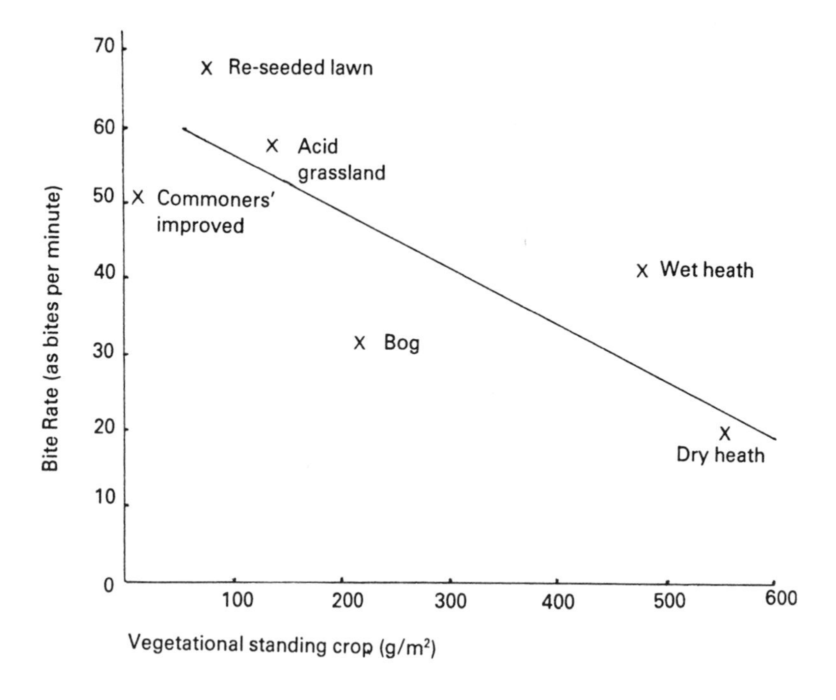

Figure 4.3: Relationship between pony feeding rate and vegetational standing crop

bite rate gradually rises once more. Similar patterns can be observed in other communities.

Diet

The above patterns of feeding behaviour may be more clearly interpreted with a more precise knowledge of actual foods taken. Dietary composition of cattle and ponies may be assessed by standard techniques in analysis of plant residues found in faecal material.

Results presented here for dietary composition of cattle and ponies are derived from microscopic analysis of faecal material. Fresh faecal material, collected monthly during 24-hour follows of individual animals, was analysed by microhistological means, and forage fragments contained were identified by reference to a prepared reference collection of known forage species. Samples of faeces were mixed in water; sub-samples were removed to a petri dish, and scanned with a binocular microscope at magnification × 3000 (methods after Stewart 1967, Stewart and Stewart 1970, and Hansen 1970). Solutions of cow dung were first centrifuged down at 600 rpm to concentrate the larger, more easily identified fragments (Putman 1984; Putman *et al.* 1985).

Tables 4.4 and 4.5 present figures for percentage species composition of identifiable fragments in cattle and pony dung. For both species, separate analyses were carried out for faecal matter collected in two different study areas within the Forest. Dietary composition did not differ significantly between the two areas in any month, and thus all data are bulked for analysis here (see also Figures 4.4 and 4.5).

Cattle

The diet of the cattle remains remarkably similar throughout the year: virtually constant in terms of species composition, varying only in minor changes in relative proportion of those components.

Cattle in both our study areas were fed hay from November to March (as is indeed usual for all cattle on the Forest). Hay and 'fresh' grass were indistinguishable in our analyses: together they made up some 70-75% of the diet throughout these winter months; the balance was made up with heather (both *Calluna* and *Erica*), which comprised a further 20-25% of the total. Other items were rarely more than 1-3% of the diet at this time. There was no significant difference in dietary composition between any of these months.

Diet in April did not differ statistically significantly from that in March, but it is clear that the animals are taking less heather and more grass at this time. Further, although in April 1980 (at the time of our study) hay was fed to a small extent in some areas, the amount available was far less than over the winter months. Thus, had we been able to distinguish between hay and grass in our analysis, differences between March and April would no doubt have been enhanced. That there is a gradual change, however, can be seen quite clearly by comparison of, for example, January and April, where a clear, statistically significant difference is observed in dietary composition, rather than as here between month and succeeding month.

Throughout the summer (May-August) the diet again remains constant, with no differences between months, however paired. Here, about 80% of the diet is composed of grass; heather still provides about 14% of the diet. Nor is any significant difference seen between this summer diet and that of September or October, although there is some

Table 4.4: Percentage composition of the diet of New Forest ponies at different months of the year

Jan	Feb	Mar	Apr	May	June	July	Aug	Sept	Oct	Nov	Dec	
49	37	43	65	90	90	92	87	83	79	69	51	grasses
0	0	0	0	0	0	0	0	0	0	0	0	herbs
0	0	0	0	0	0	0	0	0	0	0	0	conifers
20	26	25	13	0	0	0	0	0	3	11	13	holly
0	0	0	0	2	1	0	1	1	0	0	0	other broadleaves
7	7	5	3	1	1	1	1	2	3	5	10	heather
0	0	0	0	0	0	0	0	0	0	0	0	bramble/ rose
0	0	0	0	0	0	0	0	0	0	0	0	ivy
12	13	10	1	0	0	0	0	0	1	3	10	gorse
0	0	0	0	0	0	0	0	0	0	0	0	fruits
12	17	17	18	7	8	7	11	14	14	12	16	other

Table 4.5: Percentage composition of the diet of New Forest cattle at different months of the year

Jan	Feb	Mar	Apr	May	June	July	Aug	Sept	Oct	Nov	Dec	
75	67	71	80	80	83	82	70	66	70	70	66	grasses
0	0	0	0	1	0	0	0	1	1	0	1	herbs
0	0	0	0	0	0	0	0	0	0	0	0	conifers
0	0	0	0	0	0	0	0	0	0	0	0	holly
1	1	1	1	0	0	0	1	2	1	2	1	other broadleaves
21	27	24	9	14	12	13	18	23	19	21	22	heather
0	0	0	0	0	0	0	0	0	0	0	0	bramble/ rose
0	0	0	0	0	0		0	0	0	0	0	ivy
0	0	0	0	0	0	0	0	0	0	0	0	gorse
0	0	0	0	0	0	0	0	0	0	0	0	fruits
3	5	4	10	5	5	5	11	8	9	7	10	other

reduction in the percentage of grass taken in autumn, heather intake begins to increase again towards winter levels, and *Molinia* is consistently found in the diet for the first time.

Hay is again fed from November. Without this, no significant difference is recorded in the diet between autumn and winter. Indeed, without the distinction between hay feeding in the winter and grass feeding in the summer months, there is no significant change at all in cattle diet throughout the year, merely a slight change in emphasis between grass/hay and heather.

Ponies

By contrast, pony diet shows marked seasonal change, with characteristic winter and summer forages, and marked spring and autumn 'change-over periods'. Summer diet (May-August), like that of the cattle, is predominantly of grass, which comprises between 80% and 90% of the diet at this time. In contrast with cattle, the ponies feed extensively on *Molinia* at this period: *Molinia* makes up nearly 20% of the total diet, 22% of the total grass intake at this time.

During September and October, *Molinia* intake declines, to only 3% (*Molinia* is a deciduous grass and becomes sere and unpalatable at this time); total grass percentage, however remains relatively constant at about 80%, with a greatly increased intake of *Agrostis setacea* balancing the *Molinia* decline. There is also considerable use of bracken as forage over this period. Overall, the diet may be seen to differ significantly from that over the summer, although this is a progressive change.

As autumn changes into winter, the percentage of grass in the diet declines, to 50% of total intake. Correspondingly, there is a progressive increase, from October right through until February/March, of the amount of gorse and tree leaves (mostly holly) which are taken. The proportion of moss fragments in the faeces also increases over this period (4.5% to 12.7%) and heather intake, too, is increased in winter (mean 6.0%, November-February).

This is a progressive change right through the winter; again, no significant difference is recorded between successive months, but diets of September vs November, October vs December, and November vs January differ significantly. The diet of this winter period also differs significantly from that of autumn and obviously also from that of summer.

At the end of the winter, a 'spring' diet may be determined in the change-over period back to the summer mixture. This diet, with an increase in grass intake and with declining, but still significant levels of

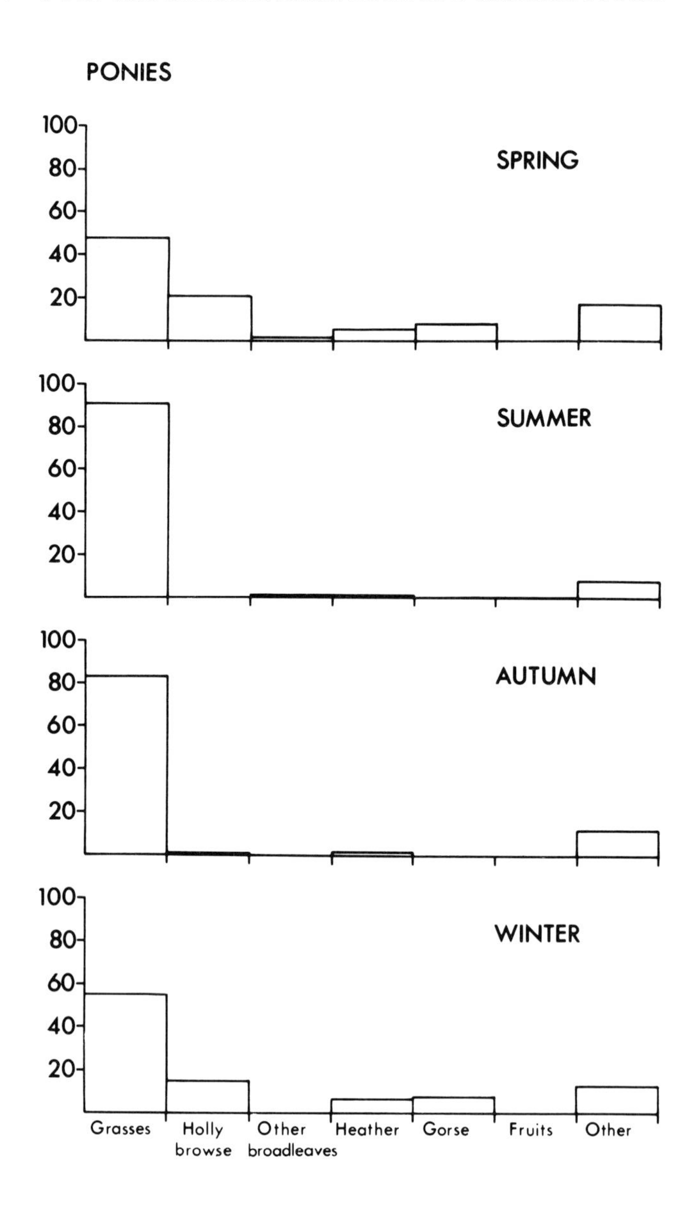

Figure 4.4: Diet of New Forest ponies. The relative proportional contribution to the diet of different foodstuffs (as percentage of total diet) is shown for different seasons

Source: Data from Putman et al. *(1985).*

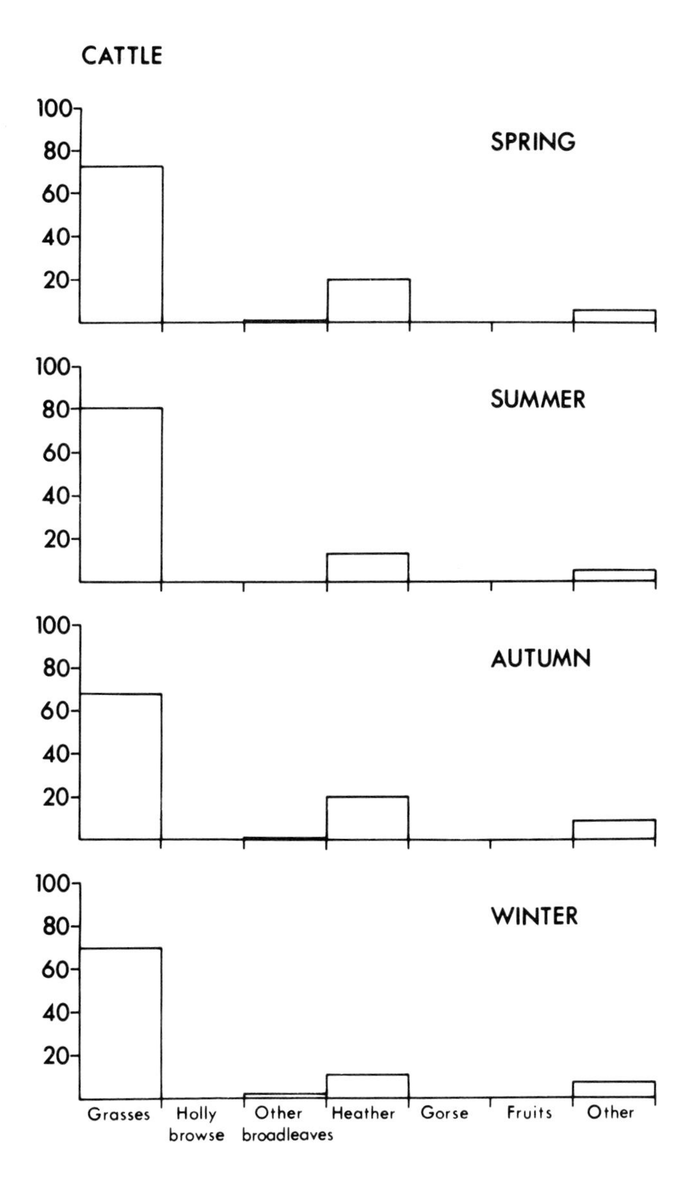

Figure 4.5: Dietary composition of New Forest cattle

gorse, moss and tree leaves, differs significantly from both true winter and summer diets.

Pony diet is clearly much more flexible than that of the cattle, adapting to changes in forage availability and quality with major changes in dietary choice. Even within the grass component of the diet there are striking seasonal changes. Use of *Agrostis tenuis* is greater in the summer than in the winter; *Molinia* is used extensively in the summer but over a very restricted period from May to August, while *Molinia* and *Agrostis setacea* appear to be complete analogues: with major use of *A. setacea* when no *Molinia* is present in the diet, and greatly reduced intake when *Molinia* is being exploited.

Overall, despite the superficial similarity between the diets of these two herbivore species — both are, after all, preferential grazers — dietary composition is shown to be significantly different in all months. As already noted, cattle are fed hay or straw throughout the winter (November-March). Although ponies do use the hay, it does not form so significant a part of their diet as for cattle. Even without this difference, however, diets of ponies and cattle still differ significantly over this period, with cattle taking more grass/hay and heather than do ponies and with ponies taking considerably more tree leaves, moss and gorse (gorse was never recorded in cattle faeces from any area). Even over the summer (May-August), when both species are eating a great deal of grass (cattle about 80%, ponies about 90%), diets of the two species still differ: the ponies are feeding extensively on *Molinia*, with only 50% of the diet made up of other grasses. Through September and October, diets remain distinct: ponies are still feeding on *Molinia* more than cattle, and the increase in tree leaves and gorse is just becoming evident.

The only other comparable information available on the diet and dietary preferences of a free-ranging population of horses is that deriving from Susan Gates's studies of Exmoor ponies (Gates 1982). The dietary patterns she describes are strikingly similar to those presented here for the New Forest animals; both species composition of the basic diet and seasonal changes in the relative emphasis of different forages closely mirror those observed in the New Forest (Figure 4.6). Like the Foresters, Exmoor ponies graze extensively throughout the year, showing strong preferences for *Agrostis* and *Festuca* swards but feeding to a significant extent upon *Molinia* during summer and autumn. When the availability of such grazing declines in the winter, the ponies compensate with increased intake of gorse and heather. Such a pattern of forage use is indeed strikingly similar to that observed for New Forest ponies, and suggests very similar strategies and responses in the two different

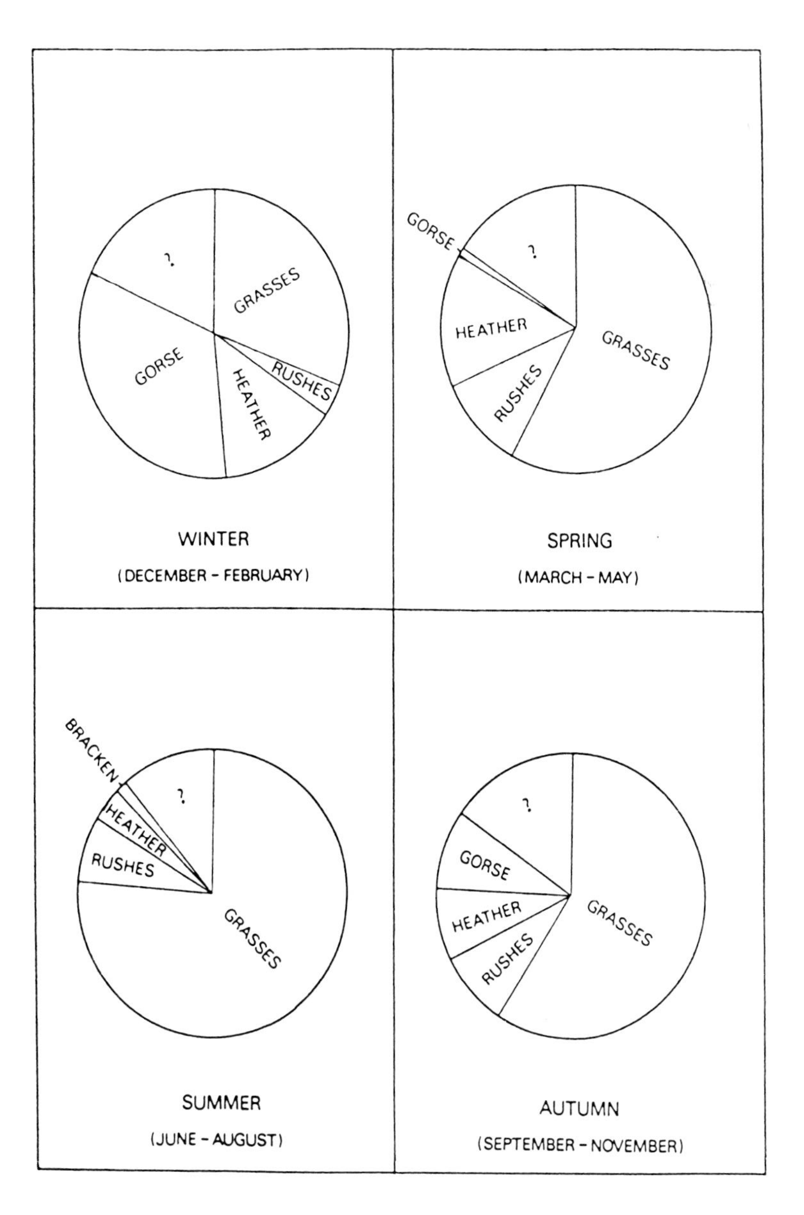

Figure 4.6: Dietary composition of a group of Exmoor ponies for comparison with New Forest animals

Source: Data from Gates (1982).

populations for maintaining themselves on the impoverished grazing of free range.

Diet Composition in Relation to Forage Availability and Forage Quality

Although an analysis of the diet of cattle and ponies ranging on the Open Forest in itself offers interesting information, it is clearly much more useful if we can attempt to *explain* dietary composition and seasonal changes in diet. For, if in so doing we can identify exactly what the animals seek from their vegetational environment at different times, we can perhaps explain the patterns of habitat use discussed earlier — and we approach a more fundamental understanding of the interrelationship between grazers and vegetation. In an attempt to identify the factors determining diet at any time, species composition of the diets of both cattle and ponies in each month of the year was related to various parameters of the availability and quality of the various forage species. Dietary composition was tested against standing crop of vegetation (per m^2), productivity, calorific value and dry-matter digestibility of the various forage species and the nutrient levels of N, P, K and Ca contained (identified as potentially the key nutrients).

Freshly plucked samples of each forage type were collected from the field each month and returned to the laboratory for chemical analyses. Where animals appeared to feed relatively unselectively, samples were taken of the complete sward; in communities where the cattle and ponies selected particular food plants, only these specific species were sampled. Methods used to determine available standing crop of a particular food type or sward, and its productivity of edible forage, are described in more detail in Chapter 7, but these were based essentially on weight yielded over a given time period from regular clipping of sample plots in each vegetation type.

Surprisingly, in neither cattle nor ponies was overall dietary composition found to correlate with any of the vegetational parameters measured. In cattle, such lack of correlation is due to the fact that, as already noted, dietary composition shows little seasonal change in any case and diet is therefore unlikely to be determined by relative quality of different forages. Lack of correlation in ponies is due to the extensive use in winter of tree leaves and gorse, and increased intake of heather: all forages being taken outside their season of maximum productivity or quality. These forages, like the grasses, show their peak productivity and quality during the summer, at which time of year, however, grass is also relatively abundant. The ponies are clearly preferentially grazers and when grazing is available feed predominantly on such forages, turning to browse only when grazing runs short. As a result, their use of browse correlates with lack of availability of grass, rather than peak availability or quality of the browse itself.

If, therefore, browse is excluded from our analysis, and composition of 'graze' species in the diet is related to availability and quality of those species, we find that the relative proportion of dietary intake by ponies of graze forages *does* correlate weakly with relative *productivity* and *digestible nitrogen content* of those forages. Rank order of species in the diet within each month correlates with rank order of productivity and digestible N.* This correlation holds only during the growing season between April and September; no correlation is observed with other vegetational parameters.

In summary, therefore, cattle diet shows no marked seasonal change and is not correlated with any vegetational measure recorded in our survey. Ponies are seen to be preferential grazers, turning to browse in winter when all available grass has been used. Within the grazing component of the diet, use of different forage species is related to their productivity, digestibility and digestible N content — except in the case of *A. setacea*, which once again is taken in any great amount only when other grasses run short.

Total Intake

Once dietary composition has been established, it is theoretically possible to determine not only species composition, but actual intake by weight. Percentage species composition of plant remains in the faeces may be multiplied by total weight of faeces produced during the day, to offer a figure for total weight of faecal fragments of each forage species. If this is further corrected for digestibility of each forage species, we arrive at a figure for total daily intake, by weight, of each of the various dietary components.

Such a calculation makes a number of gross assumptions and compounds a number of sources of error.

(i) Figures for total faecal output per day are themselves subject to a

* These two vegetational parameters are not necessarily independent of each other.

number of errors. Such figures are derived by multiplication of mean faecal weight per defaecation (in itself probably quite accurate) by number of defaecations per 24 hours. It is, however, possible to collect data on faecal events (weights, or times and number of defaecations) only during daylight. Our figures for number of defaecations per 24 hours are thus derived as mean number of defaecations observed per hour x 24, and clearly make the assumption that defaecation is regular throughout the 24-hour period.

(ii) In calculation of total daily faecal output, by weight, of plant remains by combination of faecal weight with percentage composition — by number — of plant fragments, we make the implicit assumption that all fragments are of equal weight.

Such are these assumptions that it would be misleading to present here results for total intake of individual forages. The possible errors discussed here will, however, *not* markedly influence summary values for total nutrient intake. Absolute values presented for daily nutrient intake are associated of course with considerable standard error, but the general trends are quite clear (Table 4.6).

Total dry-matter intake by *cattle* is lowest during late winter, rising through spring to a peak in the summer of 9.6 kg/day. Levels decline once more through late summer and autumn, to return to winter levels of 3.5-4.0 kg/day. Intake reflects, in part, relative availability of forage. Since the cattle do not change their diet significantly through the year, they are unable to compensate for decreased availability of a particular forage by switching to other foodstuff. Thus, in winter, when availability of their chosen forage is reduced, dietary intake must fall, while in summer, when forage is relatively more plentiful, dry-matter intake may be higher. Winter decline in dry-matter intake is, however, probably also due to physiological inappetence, a phenomenon characteristic of all free-ranging ruminants over the winter months whereby voluntary dry-matter intake is reduced considerably below levels of intake during the rest of the year (e.g. McEwen *et al.* 1957; Short *et al.* 1969; Kay 1979; Kay and Suttie 1980).

Associated with this pattern of dry-matter intake, actual *digestible* dry matter ingested also rises in summer and falls in winter. The relatively high digestibility of the grazing in summer, and low digestibility of both grass and heather during the winter, exaggerate the trend, however, and digestible dry-matter intake at its peak in midsummer (4.1 kg/day) is a factor of some five times that calculated for late winter.

Table 4.6: Dietary intake summary: daily intake of nutrients

	PONIES				CATTLE			
Month	Total dry-matter intake (g)	Digestible dry matter (g)	Digestible protein (g)	Energy (KJ)	Total dry-matter intake (g)	Digestible dry matter (g)	Digestible protein (g)	Energy (KJ)
Jan	5,079	1,991	236	91,176	*4,016	965	110	71,480
Feb	7,031	3,090	339	125,931	*3,840	1,934	84	70,148
March	4,478	1,712	191	81,800	*3,560	840	77	65,746
April	5,157	1,623	213	88,821	*6,183	1,591	224	111,750
May	6,104	2,612	449	105,690	9,657	4,010	536	179,930
June	5,993	2,483	387	106,205	9,613	4,121	507	178,447
July	5,819	2,372	327	104,376	7,191	3,148	357	133,795
Aug	5,836	2,158	336	105,820	7,745	2,953	350	147,060
Sept	6,978	2,312	349	127,720	7,795	2,608	301	150,122
Oct	7,237	2,419	313	136,111	4,846	1,600	165	92,660
Nov	7,858	2,837	350	139,225	*5,848	1,902	212	109,552
Dec	5,611	2,022	237	99,112	*5,029	1,476	165	91,264

*See text, page 82.

It should be noted, however, that these figures have been calculated using digestibility values for grass, during winter, for the entire grass-like forage (grass and hay) intake. The feeding of hay and straw is, however, extensive over the winter period, as we have already mentioned, and, if we recalculated our figures using the digestibility values for hay over this period for the grass/hay component, values for total dry-matter intake and digestible dry-matter intake rise to new totals as

	Nov	Dec	Jan	Feb	March	April
Dry-matter intake (g/day)	6,271	7,181	6,731	5,790	6,059	10,573
Digestible dry matter (g/day)	3,490	2,872	2,701	2,362	2,376	3,943

with both dry-matter intake and digestible dry matter still somewhat below summer levels, but digestible dry-matter levels now maintained at approximately 2.4 kg/day. Such assumption will of course offer an outside maximum, since much of the supplementary forage offered, as straw, has a far lower nutrient value; the true values probably lie somewhere between these two sets of figures.[3]

Total dry-matter intake and digestible dry-matter intake by *ponies* remain remarkably consistent through the year. By changing their diet, as already discussed, the animals are able to maintain a dry-matter intake of between about 5.0 kg and 7.5 kg per day throughout the year. Only in March and April (when the animals start to forsake tree leaves and gorse for, relatively poor, grazing) does daily dry-matter intake fall

[3]. Correction of total intake figures for cattle by using values for hay rather than grass during the winter months also changes figures for digestible protein and total energy intake, which become

	Nov	Dec	Jan	Feb	March	April
Digestible protein (g/day)	277	250	226	197	202	378
Energy (kJ/day)	136,130	121,436	107,345	99,646	98,010	160,585

below 5.0 kg. Digestible dry-matter intake also remains relatively high throughout the year (2.0-3.0 kg/day; 1.6-1.7 kg/day in March and April), while the remarkably constant ratio which is maintained between digestible dry matter and total intake highlights the ability of the animals to adapt their diet at any particular time of year towards an appropriate range of digestible forages.

Although these calculations for total forage intake are clearly somewhat speculative, and compound a number of initial assumptions, the possible errors should not affect results too markedly if these are expressed in terms of total dry-matter intake or total intake of particular nutrients. Further, changes in relative values between months will reflect real difference in forage intake.

It is clear from Table 4.6 that, while there is considerable seasonal variation in total dry-matter intake by ponies, intake of digestible dry matter is maintained at a relatively constant level throughout the year (although when digestibilities of all forages fall to their lowest levels at the end of winter — March/April — total intake of digestible dry matter does start to decline slightly). Cattle show far wider fluctuations in total intake of both digestible dry matter and protein. With a much less flexible feeding strategy/much more constant dietary composition, they are less able to compensate for changing quality/availability of their staple foodstuffs. When availability, digestibility and nutrient status of grasses fall over autumn and winter, cattle total intake also declines dramatically until, through January-March, they must be feeding at well below maintenance.

Feeding Behaviour of Cattle and Ponies: different strategies of exploitation

One of the most striking features of all these data is this marked difference in the flexibility of patterns of habitat use and diet, over the course of the year, between cattle and ponies. While both species are preferential grazers, feeding on the improved grasslands of the Forest so long as forage is available, cattle show little flexibility of habitat use — or of diet — and maintain essentially the same pattern of feeding behaviour throughout the year. By contrast, ponies show pronounced adaptation to changing circumstances. Although they, too, linger on the favoured grasslands for some time after production is exhausted, they eventually drift towards exploitation of other forages during autumn and winter, when food becomes scarce. Increased intake of gorse and other browse (notably holly) from November to January is a striking feature of pony diet (Table 4.4).

It is difficult to account fully for this pronounced difference between the two species. In part, the greater apparent versatility of ponies may simply reflect greater physiological specialisation for grass feeding in cattle. Although both species are *preferential* grazers, the direct-throughput system of the monogastric horse does enable it, if *necessary*, to cope with alternative foodstuffs even if of low digestibility; by contrast, the whole design of the cattle rumeno-reticulum is geared for bulk fermentation of grasses (Hofmann 1982) and perhaps does not permit it to exploit efficiently a wider range of forage. The difference in foraging pattern between the two species is, however, also related to differing social structures. The basic social unit of ponies within the New Forest is the individual — a single mare, or mare plus current foal (Chapter 3). These individuals drift independently around their home range, moving freely from one vegetation type to another. The cattle by contrast are strongly herding, and groups of 30-40 animals move as a unit around a defined range. The pattern of range use is almost stylised, with each herd following a fixed circuit each day, moving purposefully from one vegetation type to another, from night-time shelter in the deciduous woodlands to daytime grazing areas. Further, social cohesion is such that they are unwilling to occupy vegetational communities whose extent is less than about 10 ha, since on smaller areas it may be impracticable to accommodate the entire herd; as a result, many of the vegetation types of the Forest which occur only in small patches are socially unavailable to them: the herd is restricted to woodlands, grasslands and heathlands.

During the summer, growing season, the pattern of habitat use for feeding and the diet are remarkably similar for the two herbivore species. Both concentrate their feeding upon the various improved grasslands of the Forest, with peripheral use of wet and dry heathland, acid grassland (predominantly ponies), bogs and deciduous woodland. In both species, dietary composition at this time is over 80% grass. Are they in competition at this time?

Closer investigation of the data reveals that, although feeding use of grasslands by both species is extremely high at this time, the two herbivores concentrate on *different* grassland types. Such separation is in fact apparent throughout the year and not just in the season of highest apparent overlap. Thus, by way of example, cattle tend to concentrate most of their feeding activity throughout the year on re-seeded lawns; Commoners' improved grasslands are exploited during March, April and May (the peak growing season), while streamside lawns and road-side verges are used at far lower intensity. Ponies feed more evenly over all grassland types, but none the less show a preferred use of Commoners' improved grassland for much of the year (particularly from October to January/February), more consistent use of streamside lawns and relatively high use of roadside verges. Two other grassland types surveyed in this way (acid grasslands and the *Agrostis tenuis* glades of woodland clearings) are used virtually exclusively by ponies through-out the year. Figures for animal density (feeding animals per unit area) show even stronger preferences by cattle for re-seeded lawns, and by ponies for streamsides, road verges and Commoners' improved grass-land.

These results indicate some separation in feeding use of the different grassland communities by the two herbivores, with some spatial separa-tion (cattle do not use woodland glades) and some difference in the importance of shared communities to the two species.

Further *dietary* separation is also evident from Tables 4.4 and 4.5. Throughout the summer months, ponies make extensive use of the deciduous grass *Molinia caerulea*, gleaned from acid grasslands, bogs and heathlands around the edges of improved grasslands. Between May and August, *Molinia* may comprise 20% of the total diet, 22% of grass intake of ponies. This high *Molinia* intake coincides with a time when the rest of the diet most closely approaches that of cattle, and perhaps preserves a separation between the two species at the most critical time. Cattle were observed feeding on *Molinia*.

The degree of overlap between the two species in terms both of habitat use and of diet may be more formally examined in calculation of actual 'niche overlap'. The niche-overlap index of Pianka (1973) examines objectively the extent to which a number of species overlap in their use of a particular ecological resource (food, habitat, time, etc.). It calculates for each species what proportion of its needs in a given resource is met by different parts of that resource and then examines the extent to which this overlaps the pattern of use of those same resources by the other species (see e.g. Pianka 1973, Putman and Wratten 1984). The index may assume values from 0 (total ecological separation) to 1 (total overlap). Calculated for our New Forest data in terms of diet and habitat use, these indices allow a more objective assessment of overlap. Surprisingly, they suggest considerable dietary overlap throughout the year. Despite the extensive use made by ponies of *Molinia* during the summer and of holly and gorse during the winter months, and despite the extensive use of heather by cattle, calculated indices of overlap

Table 4.7: Niche overlap between cattle and ponies in the New Forest in use of habitat and in diet

	Spring	Summer	Autumn	Winter	Whole year
Overlap in habitat use	0.89	0.80	0.65	0.59	0.78
Overlap in dietary composition	0.87	0.95	0.96	0.93	0.96
Combined overlap	0.77	0.76	0.62	0.55	0.75

Table 4.8: Niche overlap between cattle and ponies in the New Forest

	Diet	Habitat use	Time	Combined overlap
Spring	0.87	0.89	0.82	0.63
Summer	0.95	0.80	0.92	0.70
Autumn	0.96	0.65	0.93	0.58
Winter	0.93	0.59	0.77	0.42
All months combined	0.96	0.78	0.93	0.70

Overlap as: $\dfrac{\sum p_{ia} p_{ja}}{[(\sum p_{ia}^2)(\sum p_{ja}^2)]^{\frac{1}{2}}}$ (Pianka, 1973)

never drop below 0.8 in any month. Overlap in habitat use, however, is high only during the first flush of growth of early summer, being 0.6 or below for most of the rest of the year. The actual ecological separation achieved in practice by both these separate effects acting in combination may be derived as the product of the two separate indices (Table 4.7). On theoretical grounds it has been calculated that, where overlap is restricted to a level less than 0.54, the animals are unlikely to be experiencing severe competition (MacArthur and Levins 1967). From Table 4.7 we may deduce that the animals risk competition only through the growing season between March and July. Further, actual overlap is probably considerably lower than is apparent here, for a number of other factors may contribute to effective separation in practice. The time of day when the animals exploit the different habitats and food resources may be crucial in avoiding competition, and, although we may also take this into account (Table 4.8; Putman 1985b), we are still not allowing fully for observed spatial or 'geographical' separation. While cattle and ponies may both make extensive use of improved grasslands, they do not necessarily use the same areas, or the same areas within them. As we noted in the last chapter, cattle rarely make use of grasslands whose total area is less than about 10 ha. In addition, where both species *do* occur together on the same favoured grasslands they occupy quite distinct and separate patches within them (page 58-9). In practice, therefore, quite a high degree of separation may be achieved between the two species.

Chapter 5

Ecology and Behaviour of the Forest Deer

Fallow Deer

Fallow deer (*Dama dama*) have long been the dominant deer species in the New Forest; indeed, William I's declaration of the area as a Royal Forest was chiefly to conserve hunting interests for this species. It is difficult to assess what numbers may have occurred on the Forest at that time; the earliest reliable census is that of 1670, when the Regarders gave a return of 7,593 fallow deer and 357 red deer within the Forest boundaries. A government report of 1789 gives an average number of fallow deer present each year as 5,900, and numbers seem to have been maintained at roughly this level until the 1850s. In 1851, the New Forest Deer Removal Act provided for the 'removal' of deer from the Forest within three years of the enactment. Total extermination of such a large population of animals, scattered over so large an area, proved of course totally impracticable, but numbers were certainly dramatically reduced and population estimates in 1900 (Lascelles 1915) gave a figure of 200 head. From then on the population has gradually expanded, and is now maintained by the Forestry Commission at a level which has been estimated at about 2,000 animals (Strange 1976). The deer are free to wander over the entire Forest area, within and outside Inclosures, though most of the population are concentrated within enclosed woodlands.

Fallow are among the most widespread and best-known of all the British deer species perhaps because they are the species most commonly maintained in park herds: their delicate build, spotted coat, and the broad palmated antlers of the bucks make them an attractive decoration for many a private estate (Plate 2). This same interest by man has indeed had a dramatic influence on their world distribution. Although widespread throughout Europe some hundred thousand years ago, the species probably became extinct in the last glaciation except in a few small refuges in southern Europe. From these relict populations, reintroduction to the rest of Europe and to Britain must have been at least assisted by man; it is thought that fallow were brought to the British Isles by the Romans or the Normans (Chapman and Chapman 1976).

There should be little need to go into too much detail at this point on the general biology of fallow deer. The species has been extensively studied and excellent reviews of the general biology have been published by Cadman (1966) and, more recently, Chapman and Chapman (1976). In the present context, therefore, a brief review only will be offered merely to act as a general outline to emphasise the major points.

In the wild state, fallow deer are characteristic of mature woodlands; although the deer will colonise coniferous plantations (provided these contain some open areas), they prefer deciduous woodland with established understorey. These woodlands need not necessarily be very large, because they are used primarily only for shelter: fallow are preferential grazers and, while in larger woodlands they may feed on grassy rides or on ground vegetation between the trees (Plate 11), they will as frequently leave the trees when feeding to invade agricultural or other open land outside. As a result, although the best known populations in England are associated with larger forests such as Epping Forest, the Forest of Dean, Cannock Chase and the New Forest, fallow deer are equally at home in smaller woodlands or scattered copses in agricultural areas.

Traditionally regarded as a herding species, fallow have rather a complex social system and social organisation is closely linked to the annual cycle (Figure 5.1). In larger woodlands males and females remain separate for much of the year, with adult males observed together in bachelor groups and females and young (including males to 18 months of age) forming separate herds, often in distinct geographical areas. Males come into the female areas early in autumn to breed; mature bucks compete to establish display grounds in traditional areas within woodland, and then call to attract females. From this time on, adult groups of mixed sex may be observed through to early winter. Rutting groups then break up and the animals drift away to re-establish single-sex herds (Figure 5.1).

This picture is, however, somewhat oversimplified. Most of the detailed research undertaken on fallow deer has been done in large blocks of woodland such as the New Forest (Cadman 1966; Jackson 1974, 1977) and Epping (Chapman and Chapman 1976), and results may be influenced by this unintended sample bias. In common with that

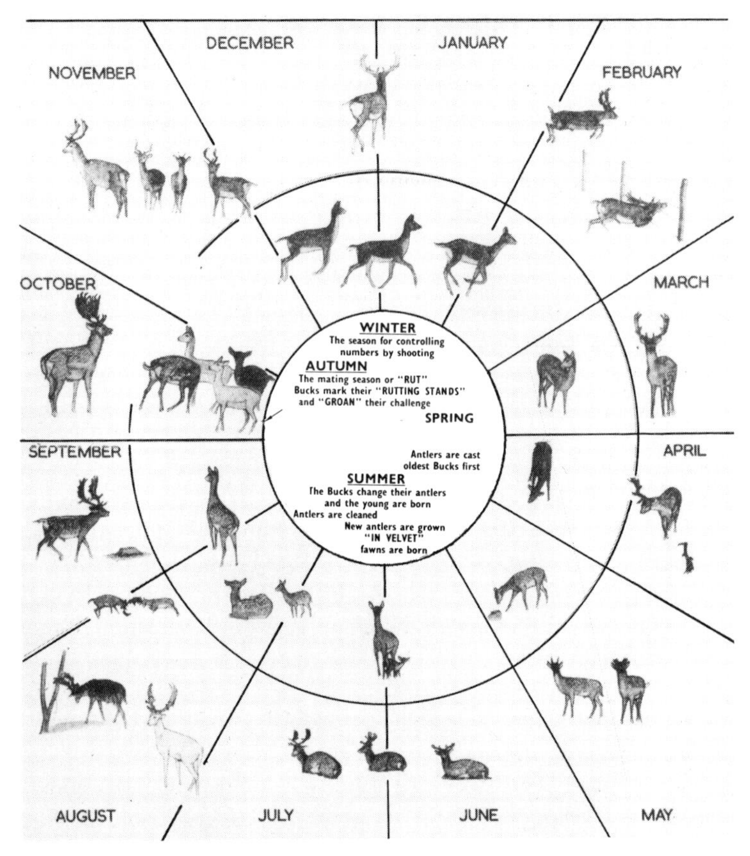

Figure 5.1: The annual cycle of life for fallow does (inner ring of diagram) and fallow bucks (outer ring)

Source: Modified from Cadman (1966). Crown copyright, reproduced by permission of the Forestry Commission.

of many ungulates, fallow deer social organisation seems to be extremely flexible and strongly influenced by environment. Group size is clearly correlated with habitat (page 100) and it appears that relations between the sexes may also differ in different environments. Thus, while in the large areas of relatively continuous woodland in which they have normally been studied, the sexes do indeed remain separate throughout the year, except for a brief and well-defined rut, elsewhere different patterns may be observed. In agricultural areas for example where small woodlands are scattered amongst arable land, group sizes tend to be larger and males remain longer in doe areas after the rut is over, with the doe groups right up to the end of April or May. These 'harem' groups do eventually break up as individual does move away when the fawning period approaches; the bucks return to bachelor groups after antlers have been shed and then remain in these male-only herds until the following rut. In other areas again, however, — usually in largely *open* habitats — mixed herds containing adults of both sexes may persist throughout the year.

Various evolutionary theories have been established to explain the tendency of all deer species to adopt single-sexed herds for part or much of the year (e.g. Clutton-Brock and Albon 1978; Staines *et al.* 1981; van Wieren 1979). Yet these cannot account for this complete continuum of strategies among fallow deer populations. We have already suggested that group size in New Forest ponies might be affected by vegetational disposition, and by size-scale of the environment (O'Bryan *et al.* 1980). I consider it likely that similar factors are influencing herd structure in fallow deer. Thus, I believe that 'single-sexing' may occur in large-scale environments — large in terms of overall available area as well as vegetational mosaic — and may be related also to the more closed nature of woodland habitat. Persistent harem groups and totally mixed populations are more characteristic of 'island' woodlands and enclosed areas such as deer parks, and may be linked both to a small-scale vegetational mosaic and to the more 'restricted' nature of the environment — which offers little opportunity to 'get away' and establish separate sex groups Further, it seems possible that the degree of sex separation may depend as much on total population density as on such vegetational cues. (A small population subdivided into component sexes might produce herds too small to be separately viable.) These explanations are, however, mere speculation, and within the large-scale environment provided by the New Forest (whether it is typical or not!) males and females do operate as independent sexes outside the brief rutting period.

Breeding is highly seasonal. Although does may be receptive at any time between September and March, and mature bucks may have active sperm in the epididymis from July or August, most conceptions are synchronised in the brief, hectic period of the rut. The timing of the rut varies from area to area; in the New Forest, it usually starts towards the last week of September and is at its peak through October; few bucks call beyond the first two weeks of November. Rutting stands are very traditional and the same buck may hold the same display ground for a number of years. Does, too, are faithful to a particular rutting stand, and will return to the area each year, frequently accompanied by their daughters. (Fallow does do not disperse widely from their natal range, and females often establish ranges which overlap extensively with that of their mother.) As a result, a buck may commonly cover his own daughters — even his grand-daughters: the implications of such regular inbreeding are considered elsewhere by Smith (1979). Does give birth to single fawns in late June or July, eight months or so after conception. For a number of weeks before and immediately after the birth, they become very independent and secretive in habit. Groups re-form only when the fawns are already a few weeks old. At this same time of year bucks, too, are somewhat solitary. Antlers are shed in April/May, and from that time the males keep themselves rather to themselves until the new antlers are quite well grown (Figure 5.1). Bachelor groups then re-form until mature animals leave the buck areas to establish rutting stands in the autumn.

Rutting stands are not always held by a single male: although mature bucks defend their stands aggressively against most potential rivals, they quite frequently tolerate the continued presence of one other, particular, male on their 'patch' (often, but not necessarily a *younger* buck). Such 'satellite' males are by no means uncommon: male deer frequently form close bonds with one other individual in their bachelor group — what one might almost describe as friendships — and where rutting stands are shared it is always between males which already have this close social bond. Most rutting stands, whether held and defended by a single male, or a pair of 'friends', are usually widely separated from each other. Occasionally, however, a whole system of stands may be established very close to each other, with as many as eight or nine adjacent rutting stands in an area of only a few hundred square metres. Such 'multiple stands' have been reported in two main areas within the New Forest, but they are not unique to this area, for they are also recorded by Chapman (1984) in Essex. The reasons for this occasional 'aggregation' of rutting stands are unclear: perhaps the does in these areas are at particularly high densities, or suitable habitat for display grounds is scarce in that particular area. Whatever the cause, competition for females in these areas must be extremely severe, and these multiple stands present a system of communal display almost analagous to the leks of birds such as ruff, blackgrouse and capercaillie.

The above review presents, with a rather broad brush stroke, a simple picture of the biology of fallow. Within this general framework, let us now turn to examine in more detail the biology of fallow deer within the New Forest itself.

Plate 1 A group of Forest ponies (*R.J. Putman*)

Plate 2 New Forest fallow bucks (*R.J. Putman*)

Plate 3 Sika hinds in summer coat (*R.J. Putman*)

Plate 4 Grazing roebuck (*R.J. Putman*)

Plate 5 New Forest ponies 'shading' (*Stephanie Tyler*)

Plate 6 Acid grassland merges into the richer grazing of streamside lawns on the alluvial soils of river floodplains (*R.J. Putman*)

Plate 7 Improved grassland (*R.J. Putman*)

Plate 8 The *Calluna*-dominated dry-heath community (*R.J. Putman*)

Plate 9 Wet heath and bogs occur wherever drainage is impeded (*R.J. Putman*)

Plate 10 Deciduous woodlands of the Open Forest have little ground vegetation or shrub
(*R.J. Putman*)

Plate 11 A group of fallow deer in the New Forest clearing (*R.J. Putman*)

Plate 12 Catching fallow deer in the New Forest to fit radio-transmitter collars (*Timothy Johnson*)

Plate 13 Temporary exclosures for measuring vegetational productivity in place on a streamside lawn (*R.J. Putman*)

Plate 14 Forest grasslands are cropped extremely close by the grazing ponies (*R.J. Putman*)

Plate 15 Browsing results in a lack of shrub layer within woodlands and a distinct 'browseline' on the woodland trees (*R.J. Putman*)

Plate 16 Gorse bushes are 'hedged' by continuous browsing pressure (*R.J. Putman*)

Habitat Use

The pattern of use by fallow deer of the various habitats offered by the New Forest has been studied by Jackson (1974) and by Parfitt (1985). Both concentrated their attention on populations based within Inclosures, rather than on the Open Forest; the deer thus had available to them a rather restricted array of habitats. Of those communities recognised in our studies of cattle and ponies (Chapter 3), Parfitt and Jackson consider that the deer had ready access only to woodland habitats (deciduous woodland, mature softwood stands, plantations, prethicket and thicket, rides and glades). Both authors describe a similar pattern of use of these available habitats (e.g. Table 5.1). In each case, the observed number of animals in each habitat differs significantly from the expected distribution if use of the range of vegetation types was uniform. Deciduous woodland was actively selected in early spring (February-April) and autumn (August-October). Woodland use remained high throughout the winter in good mast years as animals remained to feed on the abundant beech mast and acorns. Where mast was less abundant, use of the woodland blocks themselves declined over the winter and the deer made increasing use of more open habitats, grazing along rides and in clearings; rides and clearings were again used heavily in midsummer (June/July). As with cattle and ponies (Chapter 3), changes in habitat use reflect changes in the need for shelter and in the availability of favoured foodstuffs.

As noted, both Jackson and Parfitt consider the use of habitat primarily of fallow deer within Forest Inclosures. It is, however, clear that many fallow deer *do* forage out onto the Open Forest. Jackson

Table 5.1: Habitat use by New Forest fallow deer (Parfitt 1985)

	Percentage of observations in any one month recorded in each available habitat											
	J	F	M	A	M	J	J	A	S	O	N	D
RL/CI SL	not available in environment											
AG/WH	9	4	2	0	2	9	2	2	1	2	9	9
DH	0	0	0	0	0	0	0	0	0	0	0	0
Bog	0	0	0	0	0	0	0	0	0	0	0	0
DW	13	49	58	43	16	20	6	46	46	40	20	26
CW	35	38	8	32	38	33	23	28	28	26	19	23
WG and Rides	31	25	19	20	30	30	59	26	26	28	42	38

presents an analysis of habitat use by these deer as in Table 5.2. Deer near the Forest edge or whose range abuts agricultural land within the Forest may also regularly graze out on agricultural pasture, increasing their effective use of grasslands; the degree to which use of agricultural land changes the pattern of habitat use overall is, however, uncertain.

Home Range

Individual animals restrict their activities to particular home ranges (although these overlap considerably and the animals are in no way territorial). In a superb study of the social behaviour of New Forest fallow deer, Norman Rand (unpublished) has collected thousands of sightings of over 600 individually recognised animals. Rand has developed a method of identifying fallow deer from the pattern of white spots on the coats of individual animals. Although the pattern of spots on the whole coat can be used for identification, Rand has shown that animals can be distinguished just by the pattern of spots on the thigh. These spots are always larger and clearer than those elsewhere and usually occur in easily recognisable lines or groups (see Plates 2, 11 and 12). Spot patterns are, however, distinct between right and left thigh on any one individual; to avoid confusion, Rand has developed a photofile of right and left sides of all animals recognised. The pattern of spots is constant for one individual, and consistent from year to year. The spots become less clear in the winter coat, but Rand has shown definitively that each individual retains the same pattern through moults from year to year.

Since 1977, Rand has maintained a research programme within an area of some 11 km^2 in the New Forest. Using his individual-recognition

Table 5.2: Habitat use by Open-Forest fallow deer (Jackson 1974)

	Percentage observations in any one month recorded in each available habitat					
	Jan/Feb	Mar/Apr	May/June	July/Aug	Sept/Oct	Nov/Dec
RL/CI	0	58	68	16	56	0
SL	0	11	5	16	18	0
AG/WH	0	3	0	0	0	0
DH	20	18	0	9	0	0
Bog	0	0	0	0	0	0
DW	80	0	27	50	27	100
CW	0	0	0	0	0	0
WG and Rides	habitat not available					

system, he has steadily accumulated information about the deer normally resident within that area. He has always intended to base his final conclusions on an analysis of the behaviour, associations etc., of known individuals throughout their whole lives. Inevitably, this is a long-term project and he is not prepared to publish prematurely. In 1981, however, he calculated some statistics for range size. For completeness he has agreed to allow inclusion of these previously unpublished statistics in this book, but he wishes me to emphasise his own reservations about their value. They resulted from a relatively brief, preliminary examination of only a small part of the data collected in the years 1977-81. He has found fallow deer ranging behaviour to be highly complex, even intricate. As well as individual variations in range size, there are considerable variations between the years; included in the latter are long-term changes apparently of a sequential nature (there may be a cyclical element), known to be affecting considerably more than the immediate research area. When a full analysis of all the information now available has been completed, Rand believes that it will be necessary radically to revise these statistics.

With these reservations in mind, then, we may none the less offer some provisional estimates for range sizes of fallow deer within the New Forest, based on Rand's preliminary analysis of the 1977-81 data for some 60-70 does and 150 bucks (Figure 5.2). Fallow does in his study area were clearly identifiable only between May and October each year, but over this period full range size varied between individuals from 48 ha to 89 ha (with a mean of 69.2 ha). If a core range is defined on the basis of inclusion of 80% of all sightings, range is defined as 17-46 ha, with a mean of 36.9 ha.

Range sizes of bucks can be determined over a wider range of seasons. Over the same summer period (May-September, excluding records for October when the animals move to traditional rutting stands to breed) full range size had a mean of 107.3 ha with mean core range of 34.9 ha. During winter (November-February) range sizes increased: full ranges varied from 55 ha to 260 ha (mean 152.9), while individual core ranges were between 11 ha and 125 ha (mean 63.3). As foraging conditions improve again in the spring, ranges are reduced (mean 137.7 ha, in March/April (full range); 46.8 ha, 80% range).

Rand's estimates of range size can be criticised on methodological grounds: that the accuracy of identification of individuals based on their coat patterns cannot be guaranteed, and that observations may be biased towards those parts within a range where the animal is more easily visible to an observer. Having examined his data myself in considerable detail, I am left in no doubt of the authenticity of his identification; Rand's own reservations about the validity of range estimates quoted here are based on the fact that they are calculated from a relatively small sample of animals and over a four-year period only (while he himself is already aware of considerable individual variation and year-to-year change in range size). The inaccuracy of estimating home range on visual observations is a more difficult criticism to counter, but from 1979 to 1982 an independent study of the range size of New Forest fallow does was undertaken by Andrew Parfitt (Parfitt 1985) using radiotelemetry. A number of does were captured and fitted with miniature radio-transmitters on plastic collars (Plate 12); after release, the animals could be tracked by following the radio signal even when they were not visible to the observer. Parfitt's study was undertaken in a slightly different area of the Forest from that of Rand, and of course his results are based on a smaller sample of individuals and a shorter time period: 35 does were collared between 1979 and 1981. Results from this work have still to be fully analysed but will make interesting comparison with Rand's preliminary figures (Parfitt 1985).

Social Organisation

Fallow deer are traditionally regarded as one of the more social of the deer species, and certainly they may frequently be encountered in large aggregations. Such sociality, however, is in practice somewhat illusory, and derives at least in part from animals which occupy overlapping home ranges coming together in favoured areas to feed.

As noted on page 89, the separate sexes in New Forest fallow deer typically operate independently of each other for the greater part of the year, coming together only really for the brief period of the rut in late September or October. For the rest of the year we may distinguish between 'bachelor' individuals or groups, and groups of 'small-deer' (does, fawns and followers), which may even segregate in terms of geographical area: within any forest large enough to permit such separation one may recognise distinct buck and doe areas. Herds are formed by individuals from overlapping home ranges coming together, and clear seasonal changes in size and composition of such herds may be seen to be closely related to the annual cycle (Figure 5.1). Figures 5.3 and 5.4 show the frequency of occurrence of groups of particular size among male deer and small-deer in John Jackson's study areas, and seasonal changes in the frequency with which a particular group size might be encountered (Jackson 1974). Almost identical

Figure 5.2: Home-range size in fallow deer. Range boundaries may be plotted as a series of 'contour maps', enclosing increasing numbers of animal observations. In the diagram below, the outer line in each case represents the range boundary enclosing all *observations of a particular animal, the inner boundary takes in only the innermost 80% of sightings*

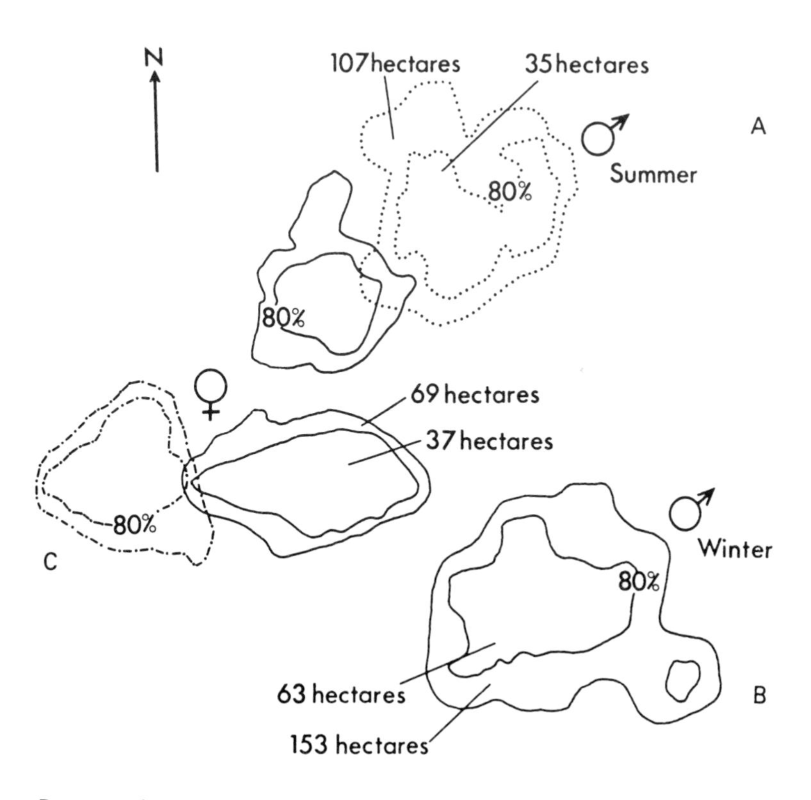

Data superimposed here are mean range areas, from Rand (1981), of
A: ♂ deer; May-September
B: ♂ deer; November-February
C: ♀ deer; May-October

patterns are reported by Parfitt for the deer in his two main study areas; for small-deer herds, Parfitt also plots the change in *mean* group size (Figure 5.5). For both sexes, groups of between one and five animals are most common throughout the year. From November to January, almost all males are encountered singly or in pairs; in February and March, groups of three to five are almost equally commonly observed, but from then on these larger groupings become progressively less and less frequent throughout the summer and autumn until, by November, 100% of males are again encountered only in ones or twos. Females show more variation in group size throughout, and are almost equally likely to be encountered in groups of one to two or three to five throughout in all seasons (Figure 5.4). Larger social groups are, however, more common in April and May; during May the groups break up, females becoming more solitary as they prepare for the birth of their fawns in mid June. By August numbers in herds increase again as does and fawns join with other family groups, and rise once more during September and October as females collect at the rutting stands. Parfitt's data for *mean* group size of such small deer highlight these same trends.

Despite the fact that it may associate into groups of up to 100 or more individuals, the fallow deer is apparently not the truly social species that has so often been claimed. Surprisingly, the basic unit is very much the

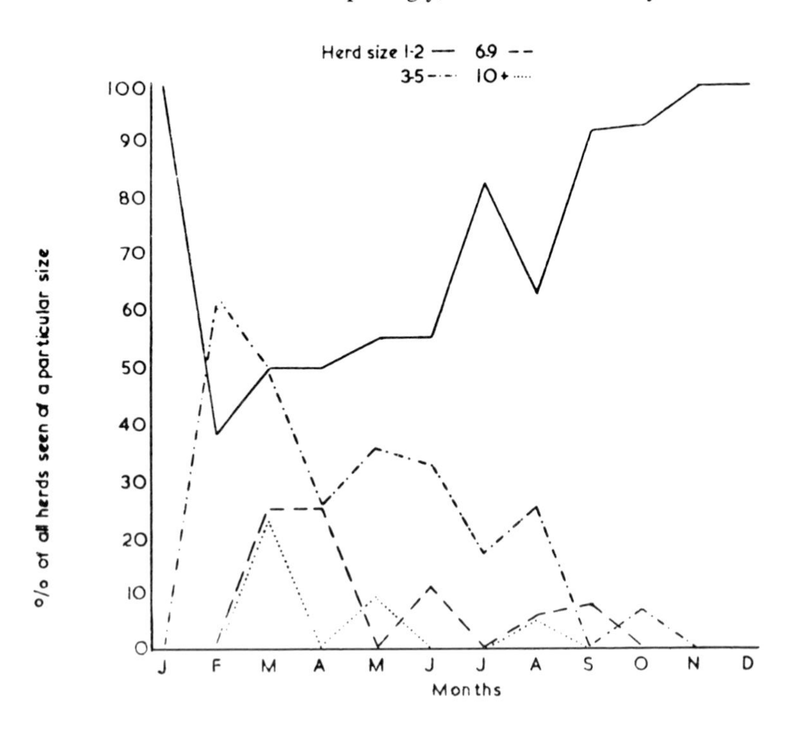

Figure 5.3: Seasonal fluctuations in the size of male deer herds of New Forest fallow deer

Source: From Jackson (1974), with permission.

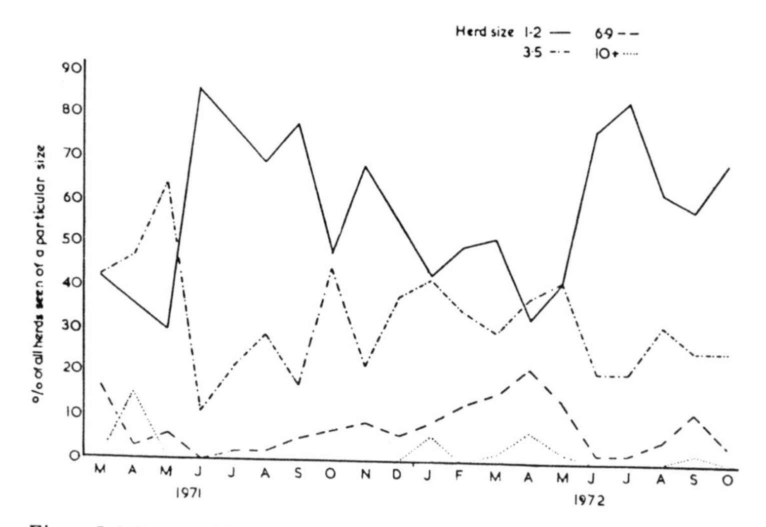

Figure 5.4: Seasonal fluctuations in the size of small-deer herds of New Forest fallow deer

Source: From Jackson (1974), with permission.

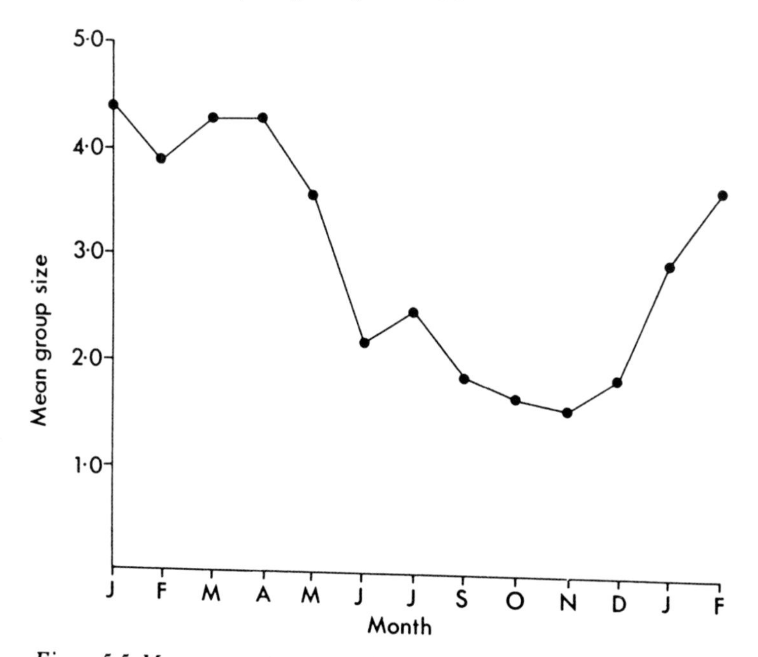

Figure 5.5: Mean group sizes of small-deer-herds observed by Parfitt (1985)

individual; Jackson and Parfitt both make it clear that groups larger than five are infrequent, and it is further clear that even these groups are formed only temporarily and are extremely fluid in composition.

The idea that fallow deer were essentially individualistic was first tentatively suggested by Putman (1981), an idea which was freely admitted to be based on few data and founded largely on hunch. At the time I noted that it was important to distinguish between strict, fundamental social groupings — what might properly be referred to as the social unit — and casual associations of more than one of these units to form larger groups. Thus, while herds of fallow deer of up to 70 or 100 animals may occasionally be observed on favoured feeding grounds in the New Forest, closer observation reveals that this is in fact a coincidental aggregation of a number of separate smaller social units, occurring together merely because they happen to be using the same area simultaneously. If such a 'herd' is observed for a long period, it may be seen that it is not of constant constitution: small groups may be seen to join the aggregation, others to leave. In effect, we must recognise that there are two levels of social organisation; that the feeding groups are no more than casual associations of a variety of sub-units is clear from the fact that, as the group disperses from the feeding grounds, it fragments once more into groups of up to seven or eight individuals, moving off in separate directions. Further, while the smaller social groups are distinct male or female parties, such feeding aggregations may gather both male and female units into a mixed-sex 'herd'.

These smaller groups, of between three and seven or (exceptionally) up to 14 animals, might then be considered a more fundamental social unit. They are indeed the same sort of size of group in which fallow deer are more regularly encountered in other habitats (Figures 5.4 and 5.5). It would appear, however, that these associations are little more stable than the larger aggregations, that even these groups are not persistent, nor constant in composition from day to day. In effect, these too represent only temporary groupings of *individuals*, adopting for the time being a group size adapted to the habitat in which they occur at that time: just as the larger herds are temporary associations of these groups in more open habitats. It may be shown that group size in fallow deer *is* influenced by habitat (Putman 1980), and it may well be that *all* these groupings are formed towards establishment of the optimum group size for exploitation of a particular environment (Putman 1981).[1]

---

[1] The influence of habitat type upon group size has now been studied in more detail by Parfitt, who showed this effect to be relatively minor; in his analyses, no clear differ-

Sika Deer

Sika deer (*Cervus nippon*) are native to eastern Asia, and are found in the wild state in China, the southeastern corner of Russia, Taiwan (Formosa) and various Japanese islands. Populations tend to be of very local geographic distribution — many are island forms, characteristic only of their own tiny island — and this isolated development of different populations has led to the evolution of a number of distinct races or subspecies each characteristic of a small local area. Thus Formosan sika, Manchusan sika and Japanese sika may all be recognised as distinct forms; other races have been described from the mainland, and are still being reported (Zhuopu *et al.* 1978). All races interbreed freely, however, and also interbreed with the more westerly red deer (*Cervus elaphus*). Such ready hybridisation has led to the suggestion that all the various races of sika deer, together with the European red deer and the North American wapiti (*Cervus canadensis*), form a continuous cline or ring species (Harrington 1981); it also suggests that many, if not all, of the mainland subspecies of sika may be entirely of hybrid origin (Lowe and Gardiner 1975).

Sika were introduced to Britain in the 1860s, and to the New Forest in the early 1900s. They have found both climate and vegetation much to their taste and have proved tremendously successful, spreading rapidly through Scotland, Eire and southern England — wherever they have been introduced. So rapidly have they spread that they have reached pest proportions through much of their range, causing significant economic damage to commercial forestry interests in Scotland and Ireland. Their geographical range now extensively overlaps that of the native red deer: the readiness with which the two species hybridise threatens the integrity of the red deer as a distinct species. For both these reasons sika deer in Britain are receiving considerable interest, and much recent research effort has been devoted to this species, for surprisingly little is known about their ecology and behaviour even in their native Asia.

Sika deer are generally regarded as characteristic of somewhat acid soils. 'Typical' sika habitat in Britain could be considered a mix of heathland and coniferous forest (two vegetation types often associated

---

ences in group sizes of small-deer could be detected between different habitats, and the dominating influence upon changing group size remained the seasonal effects of the annual cycle (page 96).

together anyway in that many commercial plantations are established on areas of native heath). The animal is, however, clearly an opportunist and can readily adapt to a wide range of conditions, and diets. Social organisation, grouping, and reproductive behaviour have in the past been little studied, but merely presumed analagous to systems established for the congeneric red deer. Most of what we *do* now know about sika deer ecology has come from two major studies in the South of England: in the Isle of Purbeck and in the New Forest itself (Horwood and Masters 1970; Mann 1983).

Habitat Use

In the 'typical' British habitat of acid coniferous woodland, sika show a very predictable pattern of habitat use. Relatively little forage of high quality is available to the deer within the plantations: when feeding, the deer tend to leave the woodland cover for the open vegetation beyond the forest — in heathland or on agricultural fields. Use of such open habitats is primarily at night (the deer are very sensitive to human disturbance); by day the animals retire to the woodlands and lie up in the dense cover. This same shyness of any disturbance causes the deer to seek out the densest thickets for shelter.

Coniferous forests are not endless, monotonous seas of evergreens that one sometimes imagines. In commercial forests such as Wareham and Purbeck, the varied species composition of different blocks provides diversity of habitat, while the different growth stages of different plots within any forest block results in a mosaic of different structural types. Although all is coniferous forest, one can recognise distinct sub-habitats such as young plantations, prethicket, thicket and polestage forest blocks, as well as mature 'high forest'. The names are self-descriptive, i.e. young *plantations* contain small trees less than 1.5m in height, planted in open ground with extensive and varied ground cover containing several species of grass and many forbs. As the trees grow, they spread and begin to shade out the ground below: a distinct *prethicket* stage may be recognised of trees about 4.5 m in height; the lower branches meet, but do not interlock. These areas provide good shelter, and a certain amount of food: the ground is still grass-covered, and heather, gorse and bramble have often become established. As the canopy closes, and lower branches of adjacent trees interlock, the forest forms a dense impenetrable *thicket.* Ground cover is almost non-existent owing to poor light penetration. At this stage the commercial forest is thinned. Lower branches are brashed, and many trees are also removed, to open up the area and provide room for the final growth of the remaining trees. Such management changes the character of these *polestage* areas: light again penetrates the plantation, to permit some ground cover of grass and bracken; in the thinned areas, an understorey of birch and holly may develop. The conifers continue to grow. Eventually the canopy closes again high above the forest floor, as the trees reach full maturity; this stage is generally referred to as *high forest.*

Despite this variety of habitat types, Horwood and Masters (1970) and Mann (1983) found that sika in Purbeck restricted their activity primarily to one woodland type: the impenetrable thicket-stage plantations. Pattern of use of the range was simple, with the animals lying up in dense thicket throughout the day and moving out onto the heaths or onto agricultural land to feed at night. This regular daily migration was most marked and was similar throughout the year: indeed, the overall pattern of use of the available habitats changed little between seasons (Table 5.3).

In contrast to the coniferous plantations of sika habitat in Purbeck, and through most of their range in Scotland and Ireland, the New Forest offers the deer a much more varied environment — and one of predominantly deciduous woodlands. Sika in the New Forest are (curiously) restricted to a small area in the south of the Forest near Brockenhurst (Figure 5.6), an area seemingly bounded to the north by the Southampton to Bournemouth railway line, to the east and west by the waters of Beaulieu River and the Lymington River. The core of this area is wooded: mostly mature oak-beech woodland with a secondary canopy of holly, yew, hawthorn and blackthorn. There are extensive patches of birch scrub. Within the area, various blocks have been cleared and planted with conifers; all the various structural stages

Table 5.3: Use of available habitat by sika deer in Purbeck Forest (Mann 1983). Numbers are those seen in each habitat as percentage of total sika observed each month during 1981

Jan	Feb	Mar/ Apr	May	June	July	Aug	Sept	
4	4	2	0	0	0	0	7	oakwoods
46	23	40	58	53	59	41	36	thicket
3	4	3	2	6	10	10	7	saltmarsh
10	34	13	7	9	3	8	10	heathland
33	32	41	32	27	24	37	38	fields
4	3	2	2	5	4	4	2	woodland rides

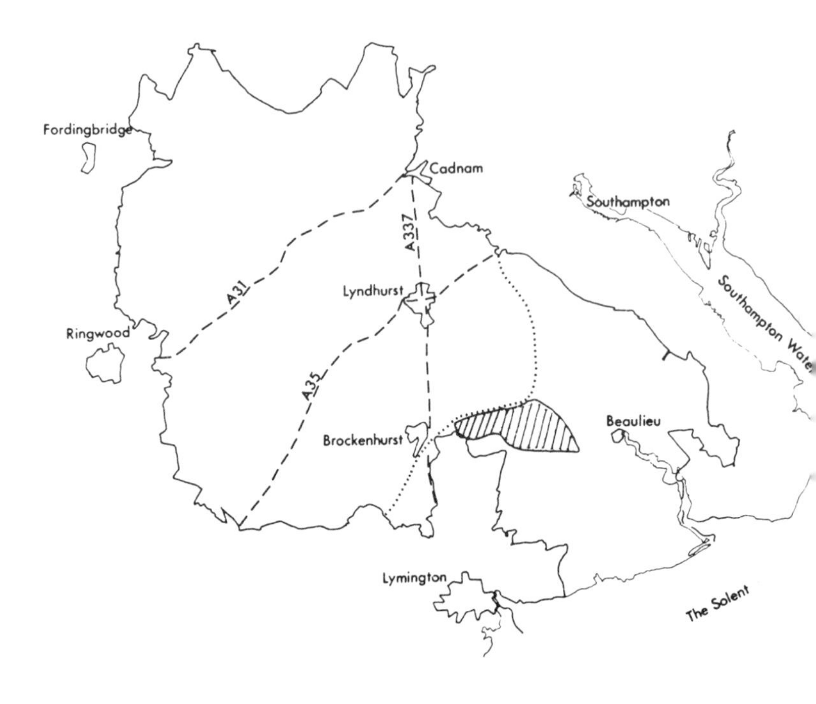

Figure 5.6: The geographical range of sika deer within the New Forest

described above for Purbeck and Wareham Forests are represented. There are many clearings and rides, and the woodlands are surrounded by extensive tracts of heath and a small amount of agricultural land.

With this more varied habitat on offer, sika deer within the New Forest show a more variable pattern of habitat use and show more pronounced seasonal variation in communities exploited. Patterns of habitat occupancy were derived from direct observation and from studies of the accumulation of dung on cleared plots established in each main vegetation type (Mann 1983). Direct observations were carried out using fixed transects, walked or driven on a regular schedule, as in studies of ponies and cattle (Chapter 3). A seasonally changing pattern of habitat occupancy was seen, with most of the animals occupying the oak-beech woodland during the entire winter period and, in spring and summer, continuing to exploit these mixed woodlands after dark but during daylight hours making extensive use of a variety of other habitats, in particular prethicket areas (Table 5.4).

Table 5.4: Habitat use by New Forest sika (Mann 1983). Numbers are those seen in each habitat as percentage of total numbers of sika observed each month during 1981-2

	Jan	Feb	Mar	Apr	May	June	July	Aug	Sept	Oct	Nov	Dec
RV					not available in environment							
SL												
AG/WH	0	0	0	0	0	0	0	0	0	0	0	0
DH	0	0	0	0	0	0	0	0	1	1	0	0
Bog	0	0	0	0	0	0	0	0	0	0	0	0
DW	50	52	50	55	48	31	27	32	35	55	74	62
CW mature	0	0	0	0	0	0	0	0	0	0	0	0
CW thicket and pole	20	17	18	16	16	23	20	20	12	18	10	11
CW clearfell and plantation	6	4	11	16	18	17	23	21	26	7	3	6
WG + rides	24	27	26	13	23	29	30	27	26	18	13	21
Gorse	0	0	0	0	0	0	0	0	0	0	0	0

Animals in Wareham and Purbeck Forests are strongly nocturnal in habit: a pattern of activity no doubt imposed upon them in large part by the fact that they must leave the woodlands to forage in the open, and are less subject to disturbance if using these exposed habitats at night. Sika deer within the New Forest are less subject to disturbance. They may find adequate feed within the woodlands themselves through much of the year, and thus may be found active throughout the 24-hour period in most seasons.

During the summer, the majority of the sika deer in the New Forest feed during the day in small groups in prethicket areas (although all habitats are used to a certain extent), exploiting the extensive forage supply and benefiting from the security of the relatively dense habitat. Diet at this time of year shows a high intake of grass and leaves (see Chapter 6), and these items are readily available in all habitats. At night, still in small groups, most of the animals lie up in the Forest oakwoods or in the shelter of the extensive polestage areas. A few deer may be found ruminating in the open.

In the autumn, when acorns and leaves fall, the characteristics of a large part of the New Forest habitat change rapidly. The prethicket still provides a good food supply, but is now inferior in quality to that offered by the oakwoods, which also provide a considerable amount of shelter. Unlike the rest of the year, when it could be suggested that the need to reduce intraspecific competition precludes the formation of large feeding aggregations in this habitat, the relative abundance of acorns and leaves now permits larger feeding groups to collect. This period, however, also coincides with the rutting season and it has been suggested that the males may also provide a focus for the accumulation of groups of small-deer at this time. In late autumn and in winter, when the acorn crop has been exhausted and most of the fallen leaves are decaying, food supplies within the Forest are more limited and the deer start feeding more extensively on coniferous browse and *Calluna* (Chapter 6). These are available in a variety of habitats; the animals still spend most of their time within the oakwoods, but may forage out in prethicket and heathland areas.

Social Organisation

In an analysis of over 3 observations, Mann (1983) found a clear annual cycle of group size in New Forest sika (Figure 5.7), similar to that discussed earlier for fallow. Once again the basic social 'unit' is the individual (male, female, or female with calf), but in sika this is far more obvious; indeed, sika are one of the less social of the deer species and

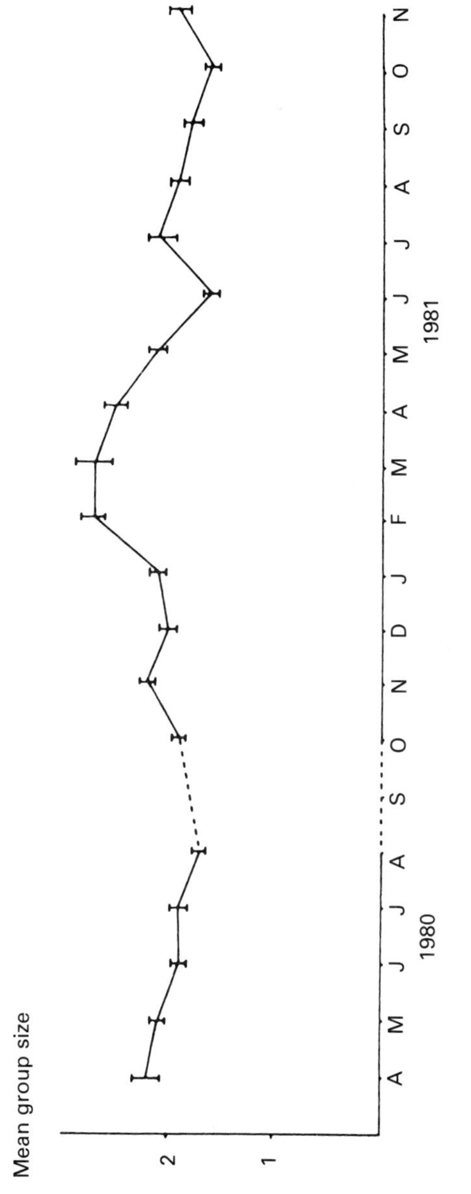

Figure 5.7: Mean group size of New Forest sika

Source: Mann (1983).

through much of the year the individuals remain completely solitary. From the end of winter through until September, animals are generally encountered alone, or as hind and calf. The rut in September causes an increase in group size, and increases the number of groups encountered of mixed sex; rich autumn food supplies allow these larger aggregations to persist through until March or April. Even at this time, however, sika never establish groups anything like as extensive as the feeding herds of 100 or more strong which may be formed by fallow. Indeed, they are rarely observed in groups of more than five or six; the largest group ever recorded in Europe consisted only of 12 individuals.

These larger feeding groups are temporary aggregations only and show little constancy of composition (though individual hinds associating with the same stag through the rut may be seen together on a more regular basis). Horwood and Masters (1970), working in Purbeck Forest, succeeded in individually marking a number of sika hinds with colour-coded collars. Successive observations showed a fluidity of group composition much as reported on page 100 for fallow. Although each hind had a relatively small home range and might be seen feeding in the same fields or same group of fields on a number of successive occasions, her companions differed. Horwood and Masters quote the example of a hind with a blue and yellow collar seen on 23 August 1969 in the company of four other hinds and five calves, one of the hinds having a red and yellow collar; two weeks later, she was seen in a different field 600 m away with three unmarked hinds and four calves. The pattern was repeated many times with different marked deer (Horwood and Masters 1970). Such observations confirm the conclusion that, at least in Purbeck, the individuals are independent members of a very loosely knit community which, since ranges show extensive overlap, may be seen feeding together in the same areas at night, but lie up independently during the day. Sika in the New Forest give the impression of similar independence.

Despite this apparent independence, all the animals within a particular population are clearly aware of each other; flexible group composition means that each individual will regularly come into contact with all other individuals sharing its range. The individuals in a particular area thus form a loose 'clan' or 'superherd' (*sensu* Putman 1981), and it seems probable that this group has a real social meaning and is more than just a passive 'envelope' of independent individuals defined together merely because of geographical coincidence. Putman (1981) noted:

Since the animals of a given area of forest will tend to use the same feeding grounds and their total range is relatively small, it is of course inevitable that one is likely to see a relatively small set of animals together fairly frequently and that they will not interact with a set of social units whose ranges are polarised on a totally different set of feeding grounds. The "clan" suggested may thus be no more than a statistical artefact. Nonetheless, it is very tempting to postulate within any one area a form of "superherd" or "clan" — an extended social network of independent but related social groups which may merge and separate and interact, but which never interact with any group outside the "clan": a system analogous to that of lion "prides". Such a supposition sweeps away other apparent inconsistencies between "herding" and "non-herding" species of deer — and may also help to explain the operation of the social units of the "herding" deer in their apparently random association into different feeding groups.

At least among sika, it would appear that each geographical population, however loosely knit, does have some social ties. Mann (1983) showed for New Forest sika evidence highly suggestive of a dominance hierarchy among hinds of overlapping home range. Direct interactions between individuals were observed too infrequently (and most hinds are indistinguishable, anyway) to allow the construction of a dominance hierarchy. That one exists, however, was suggested by the fact that some feeding hinds may be displaced by others and that in a feeding aggregation the deer seem to be relatively evenly spaced. Mann interprets such observations as suggesting that a dominance hierarchy has been established earlier and is maintained by signs that are, on the whole, too subtle for the observer to detect.

This is in spite of the fact that a high proportion of the hind's life is spent in habitats where small groups are maintained and where the opportunities to establish such a hierarchy must be limited. Mann (1983) suggests that a temporary hierarchy may be established in these feeding aggregations of sika deer, or, that if the same animals feed together on subsequent occasions as members of the same superherd or clan, then the hierarchy may be gradually established over a longer period and merely consolidated on these occasions.

Results of Horwood and Masters's work in Purbeck, and Mann's later studies in Purbeck and in the New Forest, accord well with other studies of social structure of sika in the wild. Group sizes are universally reported to be small; in Japan, the mean group size recorded is less than three, and most groups contain a hind and calf or a single male (Miura 1974). Larger groups, containing between five and 11 individuals, are sometimes formed during the winter months. In Russia, Prisyazhnyuk and Prisyazhnyuk (1974) report that the most frequently encountered groups number from two to ten animals, and in Poland, where sika deer were also introduced at the turn of the century, Dzieciolowski (1979) reports groups most commonly being between two and six animals, with the largest recorded numbering 12 individuals. All studies report largest aggregations in autumn and winter, with animals relatively solitary for most of the year; herds of mixed sex are formed during the rut and persist over winter, contributing to the increase in group size.

Part of the explanation for the observed increase in group size over autumn and winter in the New Forest is a change in food availability and habitat use. The rich mast fall provides an abundant, but highly localised food supply. It is a well-established phenomenon that not all areas of woodland, not all individual trees in any one area, crop well in any one year. While there may be generally 'good' mast years and years of poor fruit fall, in general different areas of a woodland are not at all well synchronised and mast fall is extremely patchy. As a result, this rich food source is very local in distribution; to exploit it, the animals must aggregate in relatively small areas. At the same time, in areas which have fruited, the mast is generally abundant and thus, with relatively little competition for food, larger groups can be tolerated.

Although this is a special case, habitat occupied does have an effect on group size and composition throughout the year. At any one time of year, group sizes on open habitats are larger than those of closed vegetation such as oakwood or prethicket. The difference is not particularly large (as might be expected when group sizes range only from one to, exceptionally, six individuals), but is consistent with results reported for other ungulate species (e.g. Estes 1967, Jarman 1974; and see above, page 100) where an inverse relationship may be demonstrated between group size and cover density. Relationship of group size to habitat type is precise: in habitats which are represented in both Purbeck Forest and the New Forest (oakwoods, rides, fields and prethicket), group size was the same in the two areas (Mann 1983).

As is the case with most deer species, sika hinds and stags lead very separate lives for most of the year. Although mixed-sex groups of sika may be encountered in almost any month, these are primarily chance feeding aggregations (cf. mixed-sex herds of fallow deer, p. 100) and they are most frequent around the time of the rut. Once the rut is over, mixed groups of stags and hinds may still be seen feeding together quite commonly for a month or two, but gradually the two sexes separate into their own distinct groups and remove to their own distinct ranges.

During the rut, males return once more into the main population to breed. Horwood and Masters (1970) report sika rut in Purbeck as taking place between early September and the end of October; in the New Forest the rut appears to start somewhat later, not reaching its peak until mid October. The first indication of sexual activity is the increase in calling by the males. This extraordinary noise — a shrill whistle repeated two or three time — may be heard occasionally throughout the year, but becomes more frequent during late September and October.

Published information on the rutting behaviour of sika deer (Horwood and Masters 1970, for Purbeck) suggests that the stags mark out and defend territories in the woods: territories which the hinds have to cross in their nightly migration to the fields to feed. These territories are marked by thrashed *Calluna* bushes and frayed perimeter trees, and Horwood and Masters report that the stags remain on their territories throughout the main part of the rutting season and do not resume feeding in the fields until the end of October.

At this time of year the master stag tolerates the presence of younger, attendant, stags, who may accompany him much in the same way as junior stags accompany the harem master in the red deer (Clutton-Brock and Albon 1978). Fighting between stags, Horwood and Masters report, is commonplace, both between the mature male and one of his attendants and between neighbouring territory holders.

From Mann's observations in the New Forest, however, it is not clear whether sika stags in this area mark out and defend a territory in this way or whether they collect and defend a harem as do red deer (Lincoln *et al.* 1970; Lincoln and Guinness 1973). Certainly, stags are found associating continuously with groups of females. When disturbed, the male is the last to depart and on a few occasions his actions could be described as 'rounding up' the hinds. On the other hand, males are frequently encountered alone: so either the area contains many unsuccessful males and a few dominant individuals who collect the harems, or the stags 'float' and cover receptive females as they find them.

The few animals which could be individually identified seem to spend the rutting season in a relatively small area of the woods, and return to the same area year after year, but these areas do not appear to be established as exclusive territories, as suggested by Horwood and Masters for the Isle of Purbeck. Where ranges could be plotted at this time, a considerable degree of overlap is observed (Mann 1983) and there is no evidence of boundary marking by thrashing *Calluna* bushes or fraying of perimeter trees. Some tree damage is observed (most commonly scoring of the bole or trunk with the antlers) and individual trees may be marked in several successive years; the damage is, however, unrelated to location (Carter 1981 found that in some cases over 90% of the trees in one compartment were damaged in this way) and seems unconnected to territory marking.

Overlaps in males' ranges and lack of boundary marking in the New Forest, coupled with relatively low calling rates and few observed fights, led Mann to conclude that in the New Forest sika stags do not defend a territory. Yet nor do they appear strongly to select hinds and hold them in a fixed harem. Mann concludes that the stags seem to patrol areas of superior food quality, and cover hinds in oestrus when they find them: perhaps a more appropriate strategy within the New Forest, where animal movements are more 'internal'. Territoriality at Purbeck may be linked to the regular daily migration of hinds from cover out into open fields at night to feed: territories are certainly concentrated at the woodland edge. In the New Forest, where such regular daily movements do not occur, a different strategy appears more favoured.

Red Deer and Roe

Red deer and roe are now uncommon in the New Forest, and little work has been undertaken as yet on their general ecology.

Red deer have always maintained themselves as essentially local populations — and, curiously, populations have been continuously 'subsidised' by introductions. Both James I and Charles II introduced fresh blood from France, Charles II importing no fewer than 375 deer which were released near Brockenhurst; further introductions continued throughout the nineteenth century and even into the early twentieth century.

Census records are patchy: during Henry VII's reign there were several records of red deer being killed in the New Forest, and in 1670 numbers were estimated at 357 head: 103 male and 254 female deer. By the late eighteenth century, however, the Forest's red deer population was probably extinct; certainly, returns of 1828-30 on deer in Royal Forests omit any mention of red deer within the New Forest. References to sightings in the nineteenth century probably relate to park escapes (the population of 15-20 head reported in 1892 was possibly an isolated party which had wandered across from the Wiltshire border), and in the last 200 years numbers have probably never exceeded 80-100 head.

Present-day populations are focused in two main areas of the Forest: in the northeast around Burley, and in the southeast around Brockenhurst. Southerly populations are said to be descended from three animals released by Lord Montague in 1908 into Hartford Wood. The more northern population may possibly date from much the same period. At the beginning of the present century it was not unusual to see herds of red deer in Milkham, Slufters and Holly Hatch Inclosures near Burley — in much the area still occupied to the present day. This population, too, has, however, suffered introductions, when in 1962 a number of animals escaped from Sir Dudley Forwood's estate in Burley.

Roe may be presumed to have been native to the New Forest area, but during the Middle Ages the species became virtually extinct throughout England, with the exception of the Border Counties and a few isolated sites in the South. As with red deer, the modern populations of roe in southern England have resulted from the reintroduction of animals into several places during the nineteenth and twentieth centuries (Prior 1973). Roe recolonised the New Forest from 1870 onwards, spreading across from Dorset (Jackson 1980). Census figures suggested a population of perhaps 400-500 animals in the early 1970s, but since that time populations have steadily declined. Numbers are now extremely low (estimated by the Forestry Commission as 264 in 1984), and breeding performance of those animals which do remain is noticeably poorer than average for the species: New Forest roe does never conceive before their second year and usually have only a single offspring (the national average is 1.8 kids per female: Rowe, pers comm.). The decline is unexplained, although it is generally believed that it is due to changes in habitat over the past 20 years.

Despite the fact that the roe is an opportunistic species — and may be found in a wide variety of habitats from dense woodland to open agricultural land with little or no cover (e.g. Zejda 1978, Turner 1979, Kaluzinski 1982) — it is primarily a pioneer species, associated particularly with early successional communities. Such communities are characterised by rapid growth and high production of the 'concentrated' foods that the roe favour; hence roe are themselves at their most productive in young woodlands or other disturbed habitat which offer a rich ground vegetation. Young deciduous woodlands where light can still reach the ground level offer an abundance of grasses and forbs, shrubby vegetation such as roses and brambles, and young understorey saplings of hawthorn, hazel or willow. Farmland offers rich pasture, the nutritious tips of growing cereals or an abundance of dicotyledonous weed species among the crops. Among coniferous plantations, too, younger plantings will support more productive roe populations. The wide spacing of the young trees allows development of grass and forbs in the gaps, once again with other ground-layer shrubs such as rose and bramble; the trees, too, are at a height where roe can, if they wish, browse on the growing buds (although roe do not particularly favour coniferous browse: Chapter 6). As the plantation matures, the canopy closes and cuts out the light reaching the forest floor: the ground flora is eliminated and the trees themselves are too tall to offer food. Roe *are* still found in such plantations, but use them almost exclusively for cover, foraging out from such shelter into younger woodland blocks or open vegetation at dawn and dusk. As a result, such plantations can support good populations of roe only if they are associated with open ground nearby, or other less advanced blocks of woodland. In most commercial softwood forests, trees are planted in blocks and on a rotation: so that at any one time a single forest will contain blocks of trees of all ages, from areas newly planted to stands of mature timber. The New Forest has, however, had a rather curious management history. While considerable planting was once carried out over extensive areas, there has been relatively little planting in more recent years. The earlier stock is now pre-thicket or thicket (page 102) offering little fodder, yet alternative foraging areas are rather sparse.

Roe deer are now extremely scarce in the New Forest, and indeed occur in any density only in two small areas (where open grazing is still available: around Ashurst and Godshill), and it seems highly likely that changes in habitat may be largely responsible for this decline. The change in vegetational structure, however, is not restricted just to the effects of ageing on the Forest woodlands. We have already noted that a primary influence on Forest vegetation is that of the grazers themselves, and it is possible that the decline in roe populations is also in part due to competition, direct or indirect, with other herbivores. Roe populations, as might be expected, have always been associated with the Forest woodlands. Most of these are enclosed and, at least until the 1950s, relatively free from grazing by commonable stock. More recently, however — and at about the same time as roe populations began to decline — the exclusion of, particularly, ponies from Statutory Inclosures has been less rigorously enforced and ponies are now quite commonly encountered in most Inclosures. The resultant effect on the vegetation is striking. Rides and ride edges (which used to have to be cut two or three times a year for rabbit control: Courtier, pers. comm.) are now close-cropped all year. Available browse is removed to a height of up to 2 m — and the height to which a determined Forest pony can stretch is greater than that of a roe deer. Perhap this recent influx of Common animals into Inclosures, and *their* effect on the availability of suitable forage, has also contributed to the decline of roe. Direct dietary overlap between the two species, and between roe and the other main herbivores, is considered in more detail in Chapter 6.

Chapter 6

Food and Feeding Behaviour of the Forest Deer

Much detailed research has been undertaken on the feeding behaviour and diet of deer in the New Forest (e.g. Jackson 1974, 1977, 1980; Mann 1983; Parfitt 1985) and consequently we know a great deal about the food habits of the major species: fallow, roe and sika. Patterns of habitat use in foraging are virtually identical to those described for overall patterns of habitat occupancy: foraging largely dictates habitat choice, and changes in food availability are primarily responsible for seasonal change in habitat use. As a result, we may concentrate in this chapter on the actual diet and dietary intake of the three species: comparing dietary composition in the New Forest with that recorded for the deer elsewhere in their range, and considering the degree of dietary overlap experienced by the three species where they co-occur in the New Forest — a potential source of competition in seasons of food scarcity.

Fallow Deer

The diet of fallow deer within the New Forest has been studied in detail by Jackson (1974, 1977) and more recently by Parfitt (1985). Jackson's results are based on identification of macroscopic plant remains in rumina of culled animals; Parfitt's reconstruction of dietary composition derives in the main from microscopic analyses of faecal material. This is not the place to embark on a discussion of the merits and demerits of ruminal or faecal analyses as methods of diet determination and the interested reader is referred to, for example, Staines (1976) and Putman (1984).

John Jackson's analysis of the diet of fallow deer was impressively thorough: his results came from painstaking examination of rumen samples from 325 deer over the period November 1970 to March 1973.

His results (Table 6.1) show that through most of the year the deer are primarily grazers. Throughout the growing season — from March to September — grasses form the principal food (comprising in the region of 60% of total food intake), with herbs and broadleaf browse also making a significant contribution. Acorns and mast are a characteristic food throughout autumn and early winter, although their importance in the diet varies from year to year with variations in the mast crop. Other major foods through autumn and winter, and on which the deer rely more heavily when the year's mast is exhausted, are bramble, holly, ivy, heather, and browse from felled conifers. Even at this time, however, grass still makes up more than 20% of the diet. It is evident from this that the deer are preferential grazers throughout the year, and take increasing amounts of browse over autumn and winter merely to compensate for lack of graze materials outside the growing season.

Jackson's data present an extremely thorough analysis of diet of New Forest fallow. The amounts and proportions of different food items ingested have not, however, been related to the *availability* of those foodstuffs in the environment: we do not know how selective the deer may be in their feeding. Further, since we have no knowledge of how the abundance or *quality* of the various forages may change at different times of year, we cannot fully appreciate either why the animals choose the forages they do or what causes them to change their diet in different seasons. Parfitt's later analyses of the diet of New Forest fallow were, however, undertaken within the context of detailed knowledge of changing availability and quality of forage over the course of the year.

Despite the fact that Parfitt's analyses were based upon different methods (the examination of faecal rather than ruminal materials) and represented the diet over a totally different time period, some ten years after Jackson's study, basic results were strikingly similar to those of the earlier work, suggesting an identical pattern of forage use. With some confidence that the results obtained thus present a realistic picture of fallow diet within the Forest, Parfitt then went on to test the diet selected against a number of measures of availability and nutrient quality of the different forages taken (available standing crop, energy value, and nitrogen, calcium, potassium, phosphorus and magnesium content). The one consistent relationship that emerged was with digestible nitrogen: over the winter months the deer appeared to select those forages highest in available protein. Interestingly enough, this correlation between nitrogen content of forages and their importance in the diet seems to hold only between November and March. Over the sum-

Table 6.1: Percentage contribution of the main forage components to diet of New Forest fallow deer (Jackson 1977)

Jan	Feb	Mar	Apr	May-July	Aug	Sept	Oct	Nov	Dec	
21	25	59	67	63	57	58	33	25	21	grasses
1	1	1	6	6	12	7	2	2	1	herbs
14	14	7	1	0	0	0	0	8	17	conifers
12	17	9	7	4	3	1	0	2	7	holly
1	1	0	5	14	11	6	4	12	4	other broadleaves
16	24	16	3	4	3	2	1	8	16	heather
17	7	2	3	3	12	7	10	10	8	bramble/rose
8	4	1	1	4	1	2	0	4	4	ivy
0	0	0	0	0	0	0	0	0	0	gorse
2	0	0	0	0	0	14	41	22	15	fruits
8	7	5	7	2	1	3	9	7	7	other

mer growing season — when perhaps *all* types of forage offer adequate protein? — the relationship breaks down; it is only during the winter months, when we may assume that there may be greater variability in forage quality, that there appears this positive selection for foodstuffs richest in digestible nitrogen.

Parfitt's data for diet composition are based on faecal analysis. If we know the total faecal output of each fallow deer, we can estimate total daily intake of forage by the deer in much the same way as we calculated total dietary intake for cattle and ponies in Chapter 4. Adult fallow deer on open range are observed to defaecate between ten (Bailey 1979) and 12 (Putman, unpublished) times during a 24-hour period. From theoretical considerations (density of fresh pellet groups in relation to known densities of deer: Bailey and Putman 1981), a suggested mean figure of 11.33 defaecations per day is derived for use here. Mean dry weight of 20 pellet groups from adult deer (males and females) in the New Forest each month has more recently been determined (Putman 1983). If these figures are multiplied by 11.33 (number of defaecations per day), total faecal output per animal per day in each month may be estimated. Such material has, however, suffered loss from digestion and the total defaecated weight does not in itself equate to ingested weight.

Parfitt's estimate of diet is, however, based on proportion of forage fragments *in faecal material.* If these figures are thus multiplied by the assumed daily faecal output, we may derive a figure for total *faecal* output, by weight, of each forage species as on page 79). If these figures are themselves further corrected by forage-specific digestibilities (multiplied in each case, by 100/100-DMD% — see page 79, and for fuller justification see Putman 1983), we may arrive at an estimate for total *ingested* weight of each forage material (Table 6.2). Summed over a month, such results allow derivation of total dry-matter intake. Results are interestingly of the same order of magnitude as voluntary food intake recorded in feeding trials on captive fallow deer (Putman 1980), thereby offering some confidence in our reconstruction.

Table 6.2 presents daily dry-matter intake of adult fallow deer, broken down by specific forage categories. Chemical composition of each forage species (digestibility, energy content, N, P, K, Ca) in each season is in this case known (Parfitt 1985; and see above, page 78). In any one month, therefore, the estimated ingested weight of each forage species may be multiplied by the laboratory-determined value for content of each nutrient (derived for that particular forage, in that particular area in that month). Summed within any month, over all forage species, these figures offer an estimate of total dietary intake of each of the

Table 6.2: Total ingested weight of different forage classes by New Forest fallow deer from ruminal analayses of Jackson (1977) (Values in g, dry weight): see text for derivation

	Dietary category													
Month	Broadleaved trees	Mast	Conifers	*Ilex*	*Rubus* and *Rosa*	*Hedera*	*Ulex*	Heathers	Other shrubs	Grasses	Ferns	Mosses	Forbs	Others	Total (dry-matter intake)
January	3.7	7.3	51.2	43.9	62.2	29.3	0	58.6	7.3	76.8	7.3	3.7	3.7	10.9	365.9
February	3.7	0	51.2	62.2	25.7	14.6	0	87.8	7.3	91.4	3.7	7.3	3.7	7.3	365.9
March	0	0	21.5	27.7	6.1	3.1	0	49.2	12.3	181.3	0	3.1	3.1	0	307.4
April	12.5	0	2.5	17.4	7.5	2.5	0	7.5	10.0	166.8	2.5	2.5	14.9	2.5	249.0
May-July	26.7	0	0	8.2	6.2	8.2	0	8.2	4.1	129.7	0	2.0	12.4	0	205.7
August	28.3	0	0	7.7	30.9	2.6	0	7.7	2.6	146.8	0	0	30.9	0	257.5
September	14.4	33.6	0	2.4	16.8	4.8	0	4.8	0	139.2	0	0	16.8	7.2	240.0
October	8.9	91.3	0	0	22.3	0	2.2	2.2	0	73.5	0	0	4.5	17.9	222.8
November	26.2	48.0	17.4	4.4	21.8	8.7	0	17.4	10.9	54.5	2.2	2.2	4.4	0	218.1
December	8.5	32.1	36.3	14.9	17.1	8.5	0	34.2	4.4	44.8	2.1	0	2.1	8.5	213.1

nutrients studied. Results are presented in Table 6.3. Comparisons of these data with results obtained for nutrient intake of captive deer on diets of different qualities (Putman 1980, 1983) suggest that, for most of the year, New Forest fallow deer attain levels of food intake equivalent to those sufficient to keep captive deer in extremely good condition. *Winter* intake of digestible nitrogen (perhaps the most crucial of all nutrients), however, fell to rather low levels (1.4-1.6 g), barely sufficient to maintain condition. Comparisons of actual and required intake of all nutrients considered, in relation to animal body condition, suggest a lower desirable daily intake for fallow deer of each nutrient as shown in Figure 6.1. (The rationale behind this conclusion is developed in full in Putman 1983.) It appears that New Forest fallow deer attain at least minimum intake of all nutrients in all seasons: a result which emphasises the success of their flexible feeding strategy and seasonally changing dietary composition, in enabling them to maintain adequate

	Digestible energy	N	Digestible P	K	Ca
Per animal	1,600	1,500	300	2,300	1,200
Per kg body weight$^{(0.75)}$	70.7	65	15	100	50

Figures in mg/day, except energy (kJ/day)

Figure 6.1: Suggested minimum required intake by fallow deer of key nutrients

Table 6.3: Estimated daily nutrient intake of New Forest fallow deer (Dietary profile from Jackson 1977) Figures are presented as intake of each nutrient (in grams) and energy intake (KJ)

	N	digN	P	K	Ca	E	digE
January	6.46	2.64	0.61	3.19	2.78	7,270	2,994
February	5.99	2.38	0.67	2.90	2.73	7,290	2,880
March	5.46	1.77	0.58	2.45	1.54	5,772	1,804
April	5.11	1.66	0.57	2.44	1.30	4,550	1,406
May-July	6.06	2.02	0.35	3.42	1.28	3,680	1,672
August	5.66	2.16	0.40	3.96	2.11	4,806	1,909
September	4.44	1.78	0.38	3.21	1.88	4,520	1,926
October	3.66	1.61	0.32	2.36	0.99	4,378	2,135
November	3.63	1.43	0.27	1.91	1.49	4,380	1,826
December	3.29	1.31	0.29	1.70	1.52	4,326	1,706

nutrition throughout the year despite the changes in quantity and quality of available forage.

Sika Deer

Mann (1983) used both ruminal and faecal analyses in complement, in derivation of a picture of diet for sika deer in southern England. The complete annual diet of both Wareham Forest and New Forest sika was determined through identification of plant cuticular remains in fresh faecal pellets; information on winter diet was supplemented by analysis of rumen samples from culled animals.

At Wareham, the diet was shown to be relatively constant throughout the year, with a high intake of grass (30-40%) and *Calluna* (40-50%) in all seasons; a variety of other dietary components contribute to the remaining part of the diet, but no single item comprised more than about 8% at any time (Figure 6.2). Grasses consumed were principally *Molinia caerulea* (50% of all grass), *Agrostis setacea* and *A. tenuis* (Mann 1983).

This dietary profile — a diet composed primarily of grasses and heather and seasonally unchanging — is not peculiar to Wareham, but is in fact rather generally characteristic of British sika. While investigating the diet of New Forest and Wareham animals, Mann *also* undertook an analysis of diet of Scottish sika deer from a variety of locations for comparison. Rumen samples were collected over two culling seasons (1979-80 and 1980-81) from five commercial coniferous forests in Scotland, where sika deer are becoming increasingly abundant. Results did not differ significantly between forests — nor between years — and results have been pooled to present a profile of general winter diet for Scottish sika deer (Mann 1983). The diet is extremely similar to that described for Wareham sika except that the Scottish animals take less *Calluna*, with grasses comprising 70% or more of their diet.

In general, then, sika deer in Britain may be considered primarily grazers, a result compatible with the limited data available for their diet in their native range. Prisyazhynuk and Prisyazhynuk (1974), writing of sika deer on Askold Island, USSR, report that the bulk of the food is of grasses; Furubayashi and Maruyama (1977) found that sika in the Tanzanawa Mountains in Japan consume 106 plant species but feed primarily as grazers. Takatsuki (1980) used faecal analysis and direct examination of feeding site to investigate diet of sika on Kinkazan

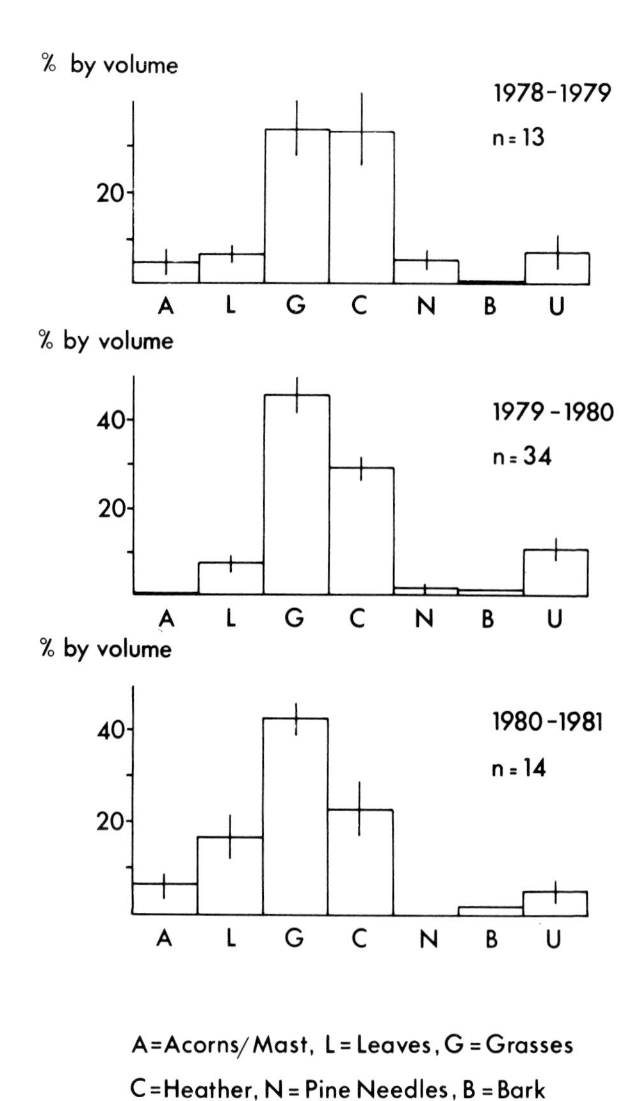

Figure 6.2: Winter diet of sika deer in Wareham Forest, Dorset

Source: Mann (1983), with permission.

Island, off Japan, and again concluded that the main dietary component was always grasses (specifically *Zaysia japonica, Miscanthus sinensis,* and *Pleioblastus chino*, a dwarf bamboo which by virtue of being an evergreen is taken all year round).

By comparison, diet of the New Forest animals appears somewhat unusual, for New Forest sika take considerable quantities of deciduous and coniferous browse, particularly in the winter. This curiosity was initially pointed out by Horwood and Masters (1970). Although their data were based on examination of 200 rumina from the Isle of Purbeck but only six New Forest rumina, Mann's later analysis (1983) confirms that New Forest animals rely heavily on browse over winter, when it may comprise up to 23% of total dietary intake. In addition, the animals show striking seasonality in diet, feeding very opportunistically on foods as they become available (Table 6.4).

In spring and summer the New Forest sika do feed extensively on grasses (about 30% during spring, up to 40% in summer) and *Calluna* (30% in spring, 35% in summer), as do Purbeck and Scottish populations. But the diet is far more varied, and includes significant amounts of other forages, forbs, deciduous-tree leaves, gorse and conifer needles. In autumn only 25% of the diet is formed of heather and grass, the remainder being composed of pine needles (from coniferous browse), gorse, holly and acorns. Winter sees an increase in intake of needles and, as might be expected, a decline in intake of deciduous browse and forbs; at this time, less than 20% of the diet is made up of grasses.

Such a heavy emphasis on browse throughout the year is unusual, and cannot easily be explained in terms of changing forage availability or quality. Analyses were undertaken to see if there is any correlation between the proportion of any item in the diet in a particular season and the area of habitat likely to contain that feedstuff (Mann 1983), and in most seasons there is good correlation. It is clear, however, that the deer are still preferentially grazers — and in the summer, when grass is more abundant, correlation fails because few coniferous needles are taken despite their abundance. Further, dietary intake does *not* correlate with nutrient status in the same way that was demonstrated for fallow deer or for New Forest ponies (page 79). Mann concludes that in the New Forest sika deer may be forced by competition with other herbivores to take large quantities of browse material. Although roe and fallow deer are uncommon in that part of the Forest where the sika occur, horses are abundant. As we have noted, New Forest ponies are officially excluded from the Forestry Inclosures, but in practice they are free-ranging and consume so much of the ground vegetation that the practice of cutting

Table 6.4: Percentage contribution of different foodstuffs to diet of New Forest sika deer (Mann 1983)

Jan	Feb	Mar	Apr	May	June	July	Aug	Sept	Oct	Nov	Dec	
25	25	22	39	38	40	39	50	44	31	28	27	grasses
0	0	0	0	0	0	0	0	0	0	0	0	herbs
20	19	23	13	2	0	0	1	0	6	8	16	conifers
1	3	1	1	1	1	1	1	2	2	1	1	holly
11	11	10	13	14	14	16	10	19	25	14	14	other broadleaves
24	23	30	23	35	37	35	29	27	23	24	25	heather
0	0	0	0	0	0	0	0	0	0	0	0	bramble/rose
0	0	0	0	0	0	0	0	0	0	0	0	ivy
7	14	8	7	6	5	6	6	7	4	7	5	gorse
6	4	2	0	0	0	0	0	0	6	14	9	fruits
6	1	4	4	4	3	3	3	1	3	4	3	other

the grass on the rides to reduce the fire risk (common until a few years ago) is no longer necessary. In the summer, forage productivity is sufficient to support offtake; in the winter, however, food resources become more limited. The ponies, by virtue of their opposed incisor teeth, can graze closer to the ground than the deer and, as their intestine relies upon a rapid throughput of large volumes of material, it might be expected that large amounts of grass will be rapidly consumed. It seems possible that the deer cannot compete effectively at this season and this may be why they feed on coniferous browse, a resource not exploited by sika in other areas (Mann 1983).

Roe Deer

Although, owing to their low density within the Forest, little work has been done on the behaviour and general ecology of roe deer in the New Forest, their diet has been well researched. While carrying out his classic studies of fallow deer (Jackson 1974), when he was collecting regular samples from the Forestry Commission cull for analysis of dietary composition, John Jackson took the opportunity to collect in addition samples from any roe deer which were shot, killed in road-traffic accidents or otherwise died within the Forest: 105 rumen samples were collected between November 1970 and March 1973. Jackson's results (Table 6.5) showed that young browse material (bramble, rose, and new growth of deciduous woodland trees and dwarf shrubs) formed the bulk of the diet through the year. Bramble and rose together formed between 25% and 45% of the diet throughout, and comprised the largest food fraction in all months except January and April. From January to March, foliage from felled conifers or from young plantations, *Calluna* and ivy were major foods, with lesser amounts of grasses, herbs and fungi. Over the summer, growing season, herbs and grasses became more important, and new growth of deciduous trees and shrubs was also favoured. (Although browse from these plants was seen in the ingesta all through the year, amounts present from October till March were minimal; from April to September, however, such browse formed between 10% and 30% of the diet.) During the autumn, acorns were a characteristic food, when available; fruit and nuts formed up to 15% of the diet from September through till midwinter.

Browse species used through the summer included birch, beech, oak, hawthorn, willow and buckthorn, as well as holly and shoots of bilberry. Twigs and foliage of coniferous species (primarily Scots pine and Nor-

Table 6.5: Percentage contribution of different foodstuffs to the diet of New Forest roe deer (data from Jackson 1980)

Jan	Feb	Mar	Apr	May	June	July	Aug	Sept	Oct	Nov	Dec	
4	5	5	10	7	8	8	8	8	9	10	4	grasses
5	2	2	30	16	16	16	17	17	4	4	6	herbs
33	22	22	5	1	8	0	0	0	12	12	13	conifers
0	2	0	0	0	14	1	0	0	0	0	0	holly
2	4	4	14	30	14	15	15	14	5	5	2	other broadleaves
6	14	14	14	5	4	7	7	5	5	4	7	heather
31	26	26	20	35	32	40	40	38	38	37	46	bramble/rose
12	22	22	7	6	4	3	3	3	11	11	2	ivy
0	0	0	0	0	0	0	0	0	0	0	0	gorse
1	1	1	0	0	0	0	0	8	8	17	7	fruits
6	2	3	0	0	0	10	10	7	8	0	13	other

way spruce) were taken throughout the year, but were most important from October to March. Grasses taken included *Agrostis tenuis*, *A. canina*, *Deschampsia flexuosa*, *Poa annua*, *Holcus lanatus* and *Sieglingia decumbens*. Roe deer seemed to avoid *Agrostis setacea*, *Molinia caerulea*, *Deschampsia cespitosa* and *Brachypodium sylvaticum* (making interesting comparison with New Forest ponies, page 73, and sika deer, page 123).

Roe have long been known as highly selective feeders who pluck small and highly nutritious morsels from a wide variety of plant species. Although it is often assumed that such a feeding style necessarily equates with browsing habit, such an assumption is unfounded: close examination of all published data shows that roe have always been known to take considerable quantities of grasses and forbs when these are at their most nutritious. Their diet in the New Forest, as described by Jackson, is very similar in general 'shape' to the diet found by other workers in Great Britain and elsewhere in Europe. Because food availability and precise species composition varies so much between different localities, detailed comparisons of actual species eaten or precise composition of diets from different areas are of limited value. Perhaps the most useful comparisons which may be made here, just in illustration, are between the diet of New Forest roe deer as described by Jackson (1980) and that of roe deer in other primarily deciduous woodland areas elsewhere in the South of England (Hosey 1974, Johnson 1984). Hosey's data for Dorset show that, as in the New Forest, bramble was the main food in all seasons; broadleaved deciduous browse was important in the summer, and again use of herbs and grasses peaked in the spring and summer growing seasons. In Hosey's study, herbs formed one-fifth of total intake in March and about 15% in November and December; consumption of grass peaked in May and November. Intake of coniferous browse was restricted primarily to the winter, as in the New Forest, but reached its maximum value in November at only 9% of total intake. Perhaps the most notable difference in diet between New Forest roe and those elsewhere in Britain is that ivy figured substantially in the winter diet only in the New Forest (Jackson 1980).

Dietary Overlap

Diets have been described in detail for the three major deer species of the New Forest; and it is clear that they are markedly different. Fallow deer are primarily grazing animals: from March to September grasses and forbs form the principal food; during autumn and winter this diet is supplemented with acorns and beech mast when available, and increased intake of browse (coniferous browse from felled conifers, bramble, holly, *Calluna*), but even over the winter grasses are still the major food. Roe have been shown to have a mixed diet, selecting in any season small morsels of highly nutritious foodstuffs; such a feeding style causes them to rely in the New Forest primarily, though not exclusively, on browse materials, particularly bramble, rose and other 'shrubby' species. Sika seem to fall somewhere between the two. Although their diet elsewhere suggests that they are preferential grazers, in the New Forest they take a considerable amount of browse and throughout the year the diet is about 30% grass, 40% heather, gorse and coniferous browse, although detailed proportions vary.

Such differences in feeding strategy between these three deer species are reflected in, and themselves reflect, actual anatomical differences in gut structure. In an extremely exciting investigation of the physiology of digestion of various ruminants, Hofmann and Stewart (1972) discovered a very clear relationship between feeding style and gut structure. Length of the gut, relative proportional size of different regions of the gut (in particular relative size of the four different chambers of the ruminant 'stomach'), even the fine structure of the lining of the rumen itself are all carefully adapted to the diet. Thus certain species have rather small rumina, and short hind guts, and the ruminal lining is resistant to high protein concentration: such animals were characteristically found to feed on small morsels of highly nutritious foodstuffs. By contrast, animals which feed unselectively, ingesting large quantities of relatively indigestible foodstuffs, require a very large rumen: reliant for the bulk of their nutrient supply on the breakdown of the structural carbohydrates of plant cell walls by symbiotic bacteria within the rumen, they must provide a huge fermentation chamber for this process. Absorptive linings to the gut walls are adapted to low nutrient levels and are rapidly damaged by 'concentrated' foodstuffs.

Recognition of these structural and anatomical differences, and of their precise adaptation to foraging strategy, led Hofmann and co-workers to draw up a comprehensive classification of large ungulates — based on their ruminal structure — as 'bulk-feeders', 'concentrate-selectors' and 'intermediate feeders' (Hofmann and Stewart 1972). On such a scheme, roe are expected to behave as true concentrate-selectors, fallow as bulk-feeders, and sika as just more 'intermediate' than the pure grazers (Hofmann *et al.* 1976; Hofmann 1982): predictions amply borne out by the ecological analyses of feeding style presented here.

Despite these overall differences in diet and feeding 'strategy' between the three deer species, however, many types of forage are taken by all three species, and there is in fact considerable dietary overlap at some/many times of year. The different deer diets *also* overlap with the foods taken by cattle and ponies in the same area (Chapter 4). How serious is this overlap — and is there evidence for competition between the five herbivore species?

We may examine the degree of overlap — and potential for competition — using the same analysis of 'niche overlap' developed in comparison of diet and habitat use between cattle and ponies on page 85. Here, we calculate overlap purely in terms of diet: for completeness, cattle and ponies are included in the analysis as well as roe, fallow and sika (Figure 6.3).

For most of the year little dietary overlap is observed between roe and the other two deer species. Overlap between roe and sika is consistently low, but in winter, when food is restricted both in quantity and variety, overlap between roe deer and fallow increases significantly. Jackson (1980) also noted that diets of roe and fallow deer within the Forest showed greatest overlap in winter and early spring. Concentrating on this period of the year on the assumption that, if there *is* any competition for food between the two species, it is likely to be at its most intense at this time, when food is shortest, he none the less concluded that widespread competition is unlikely to occur. Staple winter foods common to both species were coniferous browse, dwarf shrubs, fruits, bramble, rose and ivy, but there are clear distinctions in the relative importance that each food has in the total intake. By contrast, diets of fallow and sika deer show significant overlap throughout the year: both species are intermediate feeders on Hofmann's classification, and clearly both select the same types of food. Actual competition within the New Forest is unlikely, however: in those areas where sika deer are abundant, fallow are heavily culled specifically to reduce the potential for competition; the policy is successful and fallow numbers in these areas are extremely low.

Dietary comparison between the deer and the Common animals of the Forest reveals once again little overlap between the concentrate-selecting roe and the essentially bulk-feeding cattle and ponies. Overlap of cattle and ponies with the intermediate or bulk-feeding sika and fallow is higher, but is at its most intense from March to July — over the main part of the growing season when food is relatively more abundant. We should remember, too, that there is considerable habitat separation between the species. Although Common animals *do* gain access to Inclo-

		Cattle	Ponies	Fallow	Sika	Roe
	Cattle	*				
SPRING	Ponies	0.87	*			
(February-	Fallow	0.96	0.92	*		
April)	Sika	0.77	0.66	0.81	*	
	Roe	0.20	0.14	0.39	0.53	*
	Cattle	*				
SUMMER	Ponies	0.95	*			
(May-	Fallow	0.94	0.94	*		
July)	Sika	0.80	0.72	0.78	*	
	Roe	0.14	0.14	0.35	0.32	*
	Cattle	*				
AUTUMN	Ponies	0.96	*			
(August-	Fallow	0.86	0.88	*		
October)	Sika	0.90	0.80	0.79	*	
	Roe	0.20	0.17	0.45	0.31	*
	Cattle	*				
WINTER	Ponies	0.93	*			
(November-	Fallow	0.65	0.63	*		
January)	Sika	0.77	0.69	0.87	*	
	Roe	0.16	0.14	0.68	0.37	*
	Cattle	*				
ALL MONTHS	Ponies	0.96	*			
COMBINED	Fallow	0.92	0.91	*		
	Sika	0.84	0.75	0.84	*	
	Roe	0.18	0.15	0.43	0.37	*

$$\text{Overlap } O_{ij} = \frac{\Sigma p_{ia}\, p_{ja}}{[(\Sigma p_{ia}^2)(\Sigma p_{ja}^2)]^{1/2}}$$ (Pianka 1973)

Figure 6.3: Niche overlap among New Forest herbivores for food use

sures, they occur there only in low numbers; really high densities are found only on the Open Forest. By contrast, the deer restrict much/most of their activity to the Inclosures and make relatively less use of the commonable ground.

Further, these analyses of overlap should be interpreted with caution. High overlap does not necessarily imply competition; indeed, almost by definition, if high overlap *is* observed the animals must be exploiting superabundant resources. By converse, low overlap should not *necessarily* be seen as evidence of lack of competition; it could equally be that competition for shared resources has resulted in an ecological separation, that the competing species have been forced to adopt different diets to *avoid* competitive conflict. The fact that sika and fallow populations from different parts of the Forest show high dietary overlap reveals a high *potential* for competition. But, if fallow deer were allowed to establish in the sika areas, it is probable that analyses of the diet of that particular fallow population would reveal a great reduction in overlap.

In fact, the diets of horses, cattle, roe and fallow deer in the New Forest are much as would be expected elsewhere, suggesting that few 'adjustments' have had to be made to permit them to coexist within the Forest — and thus implying little direct competition. In sharp contrast, the diet of New Forest sika differs markedly from that which they appear to eat when allowed to 'do what they want' in isolation. Sika in Dorset and in five major forests in Scotland all have much the same diet (Figure 6.2). Only in the New Forest does the diet seem to change, with increased intake of browse and lower reliance on grasses. Nor is the direction of the change what one might expect in terms of the difference in habitat: the sika deer of Dorset and Scotland are animals of coniferous plantation and heathland; the New Forest offers a wider diversity of vegetation types with, in principle, *better* opportunities for grazing. Such an unexpected shift in diet may well then be the result of competition, and we suggested above (page 125) that in the New Forest there may indeed be real competition for forage between sika and the Forest ponies.

Chapter 7

The Pressure of Grazing and its Impact Upon the Vegetation

Earlier sections in this book have described various facets of the use of the available vegetation by the different Forest herbivores, in discussing patterns of habitat use, activity, feeding behaviour and diet. Each such usage exerts a certain pressure upon the vegetation. In this section I wish to try to pull together all the various uses of the Forest made by the different animals, in order to assess total impact on the vegetation.

We have already discussed the changing pressures of grazing on the Forest since its 'creation' in the eleventh century, and it is clear that fluctuations in numbers — and dominant species — of grazers over the years have had a marked effect on the Forest vegetation (woodland regeneration for example has been able to occur only sporadically — at times when grazing pressure was unusually low — a fact reflected now in a curious age structure of the mature trees of the Forest woodlands, page 153-4).

In the early history of the Forest, fallow deer were undoubtedly the dominant herbivores. Records are difficult to interpret, but through the seventeenth, eighteenth and nineteenth centuries numbers seem to have been steady at between 6,000 and 7,500 animals (page 88). Following the Deer Removal Act of 1851, populations were decimated: by 1900, censuses suggested a total count of only 200. Although the population has recovered from this time to its current maintained level of around 2,000 animals, the dominating grazing impact upon vegetation has clearly been that of commonable stock (Chapter 4). Densities of these Common stock have also fluctuated: Figure 7.1 summarises the changes over the past 30 years.

Estimates of numbers of grazing animals on the Forest as a whole — or even estimates of biomass — while they may highlight changes in grazing pressure over the years, are not a particularly sensitive reflection of the actual pressure sustained by the vegetation itself. The effect

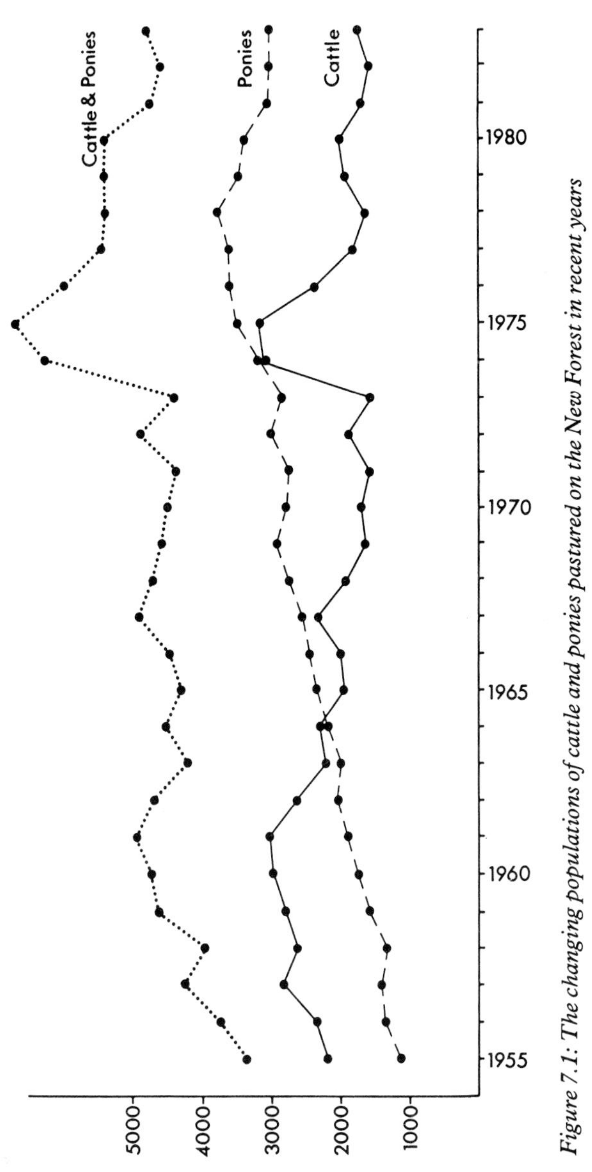

Figure 7.1: The changing populations of cattle and ponies pastured on the New Forest in recent years

Source: Modified from Tubbs (1982).

of absolute numbers is channelled through differential patterns of habitat use, differential use of different community types for forage, in determining actual grazing pressure upon the vegetation. It is clear from Chapters 3 and 5 that patterns of habitat use both by cattle and ponies and by the various deer species within the Forest are extremely complex, and that animal pressure is thus by no means distributed evenly over the environment. Certain vegetation types are but rarely used; others are particularly favoured for shelter, and thus subject to trampling; and others again are preferred feeding sites, subject to pressure by grazing and trampling. Thus, certain communities receive little herbivore pressure, whereas favoured sites may show extremely heavy use indeed. Yet the effects of these herbivores upon the vegetation are clearly going to be influenced not only by animal density on that particular vegetation type, but also seasonality of that pressure (a given grazing pressure at a particular time of year may have a very much greater or lesser impact than would that same pressure at some other stage of the vegetation's annual cycle). A more objective measure of animal pressure would take account of the temporal and spatial patterning of community use by the Forest herbivores to assess more precisely herbivore pressure on a particular area. Our studies of habitat usage by the different Forest animals enable us to arrive at such an assessment of grazing pressure, to translate a blanket figure of 'number of herbivores over the total Forest area' into a more meaningful figure of *intensity of use of particular vegetation types at particular times of year,* at least for the present day. We may calculate current herbivore pressure on an area quite precisely: in terms of animal-minutes, or animal-grazing-minutes, per unit area per unit time. These figures thus offer an objective measure of actual intensity of use of different vegetation types.

The importance of such a measure, in allowing us to quantify *effective* pressure of grazing animals, may clearly be seen if we remind ourselves of the pattern of pony use of different vegetation types within the Forest (Figure 3.2). This illustrates very clearly by way of example that intensities of use of different vegetation types are by no means equivalent. Heathlands, acid grasslands and bogs sustain the lowest pressure year round. Woodlands (particularly deciduous woodlands) are used consistently throughout the year for shelter, as well as feeding. But clearly the most heavily used communities are the various improved grasslands, which sustain at times an extremely high level of use. In fact, the ponies, as preferential grazers, spend nearly half their time on these grasslands over the whole year. The proportion of pony observations (as a percentage of total numbers observed on all communities) never drops below 35% in any month. The same picture, somewhat intensified, results from analysis of pony feeding observations. Thus, nearly 50% of all pony grazing pressure for the Forest as a whole is in fact concentrated on a mere 1,460 ha of grasslands. Such patchy use of different communities is also observed among the other herbivore species and emphasises the importance of analyses of habitat-use patterns in studies of grazing pressure. Figures for 'animal-minutes per unit area per 24

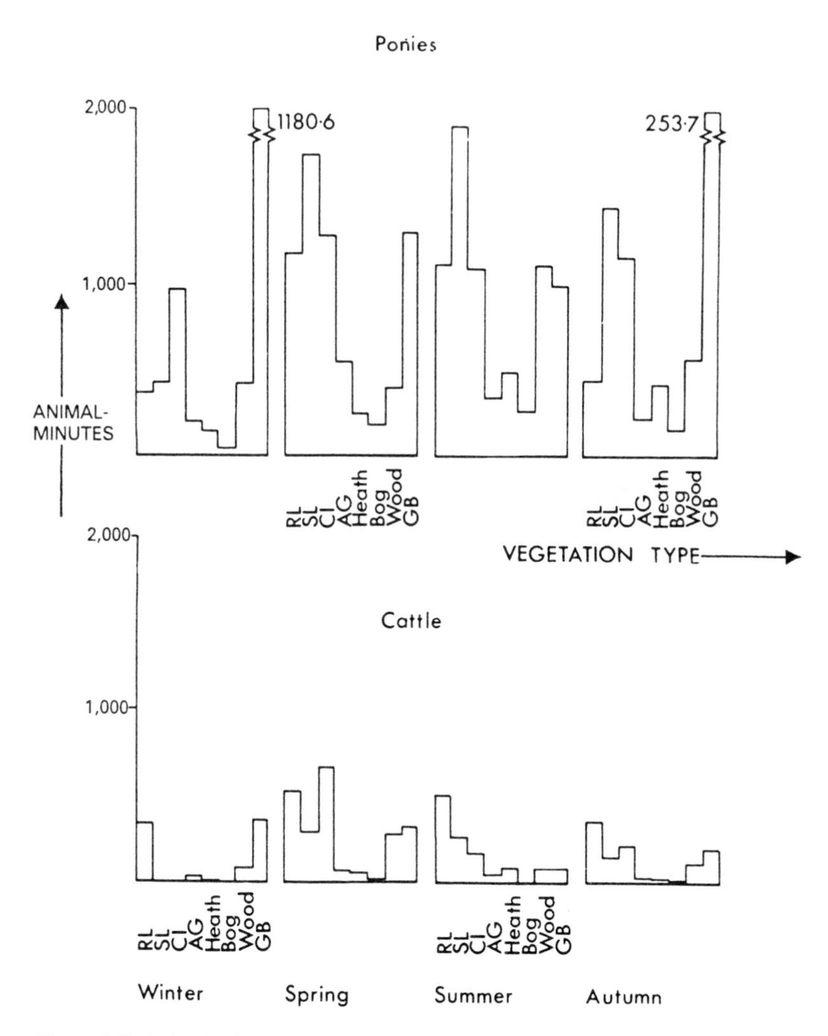

Figure 7.2: Animal-minutes sustained per hectare in an average 24-hr period

Table 7.1: Herbivore pressure: animal-minutes sustained per hectare in a 24-hour period (cattle and ponies only)

	Jan	Feb	March	April	May	June	July	Aug	Sep	Oct	Nov	Dec
Re-seeded lawn	201	888	763	1,978	2,007	1,575	1,793	1,421	717	960	634	838
Streamside lawn	362	745	2,598	1,898	2,646	1,611	2,590	1,883	1,195	1,798	1,488	141
Commoners' improved grassland	1,062	464	2,382	1,586	1,964	700	1,627	1,410	950	1,687	1,288	1,326
Acid grassland	190	412	48	675	1,074	274	480	329	338	98	356	60
Dry heath	13	56	34	58	78	130	60	64	66	48	116	40
Wet heath	81	81	177	88	164	119	197	139	130	76	73	30
Regenerating heath	33	—	680	53	266	406	283	189	224	395	314	159
Bog	18	74	124	151	345	242	258	313	276	72	151	36
Woodland glade	55	2,987	66	857	304	719	1,521	756	1,853	829	2,202	1,531
Deciduous woodland	329	594	476	350	581	230	580	264	385	425	237	149
Coniferous woodland	—	—	167	419	70	1,116	930	295	674	—	349	543
Gorse brake	11,455	13,091	1,933	2,259	702	1,870	624	883	936	2,961	4,026	11,963

hours' have been calculated for cattle and ponies on the major vegetation types of the Forest, and these highlight the tremendous pressure which some of these communities may sustain (Table 7.1, and Figure 7.2).

An alternative approach to estimating pressure of animal usage, this time restricted to consideration purely of actual *feeding* pressure, is in analysis of above-ground vegetation production and percentage animal offtake in each community. Table 7.2 presents data for this comparison of vegetational productivity and offtake for a number of the more favoured communities of the New Forest.

Measurements of forage standing crop, estimated above-ground productivity and animal offtake were made over a period of two years in most of the major vegetational communities of the Forest. In the first year of analyses (1977-8) measurements of standing-crop biomass, production and offtake were made for entire communities (the complete mixed sward of grasslands, or complete species mixture of bogs or woodland floor); such measures, while relating closely to total vegetational productivity, combine forage and non-forage species as they occur in the sward. Later analyses (1978-9) repeated these measurements, but extended them to examine in more detail the dynamics of particular individual forage species.

The productivity and offtake of available plant material within *grassland, heather,* and *bog* communities were estimated using temporary exclosures to prevent grazing (Plate 13). In this method the amount of forage material present per unit area (standing crop) is assessed by cutting and weighing samples, both when the exclosures are erected and at the end of the exclosure period. Thus:

$$\text{Production} = \underset{\text{(inside pen)}}{\text{Final standing crop}} - \underset{\text{(outside pen)}}{\text{Initial standing crop}} \quad (1)$$

$$\text{Offtake} = \underset{\text{(inside)}}{\text{Final standing crop}} - \underset{\text{(outside)}}{\text{Final standing crop}} \quad (2)$$

The length of time for which the exclosure is left is clearly critical. Vegetation inside the pen is of necessity protected from grazing; yet the vegetation it is supposed to represent normally grows under continuous grazing pressure. With the relief from grazing afforded by the exclosure (necessary for us to take any samples at all), we cannot be sure that the pattern of growth of the ungrazed vegetation within accurately reflects the growth pattern of the vegetation under its normal regime of heavy grazing. Relief from grazing must therefore be for as short a period as

Table 7.2: Total annual production[a] and offtake in the different vegetation types (figures in g/m²)

Vegetation types	Growing-season production 1977/8	Growing-season production 1978/9	Annual offtake 1977/8	Annual offtake 1978/9	Offtake as % of production 1977/8	Offtake as % of production 1978/9[b]	
Re-seeded lawns	349	226	221	296	63.3	131.0	
Streamside lawns	473	492	432	523	91.3	106.3	
Commoners' improved grasslands	321	329	313	371	97.5	112.8	
Acid grassland	190	158	132	250	69.5	158.2	Graze types
Bog	549	—	475	—	86.5	—	
Holly	—	27.3	—	1.9	—	6.8	Browse types

Notes

[a] Production is measured over growing season only.[b]

[b] 1978/9 figures for offtake/productivity are frequently > 100%. Offtake during this period of very heavy grazing did indeed exceed production, and standing crop declined during the year.

Note, however, in addition that measurement of offtake as loss of standing crop attributes all such loss to animals. In practice, loss of standing crop may also include a component from vegetation senescence.

possible, yet cannot be too brief or the growth of vegetation within the pen is too small to measure. In practice, pens on grasslands, on bogs and on woodland floors were sampled on a two-monthly rotation; heathlands were clipped every six months.

Some care is needed in interpreting the results from these grazing exclosures. The problems of altered growth patterns as soon as grazing pressure has been removed have already been discussed, but, in addition, an underlying assumption of the method is that all changes in standing crop are due either to growth of the vegetation or to grazing. During the growing season this is probably a reasonable premise, but in the winter months the standing crop may decline owing to death and decomposition of plant material; a negative production value is quite possible and merely reflects this decline. Provided the loss in standing crop owing to death and decomposition is the same inside and outside the pen, then offtake can be calculated as in equation 2. In fact these losses are probably not the same, especially where there is trampling and disturbance outside the exclosure, and estimates of offtake may thus be inaccurate over the winter period.

Different methods need to be employed in estimating standing crop, production and offtake of shrubs and trees, since heterogeneity of the 'sward' at this size-scale means that no one site can be selected as an exact 'control' for any other area of woodland, however large the sample exclosure. Accordingly, the density per km of each of the common tree species within Forest woodlands was assessed by counting the number of trees of each species along a number of survey transects. Then, several individual trees were selected of each species, and the available browse (all material within 2 m of the ground) was clipped with secateurs to determine the weight of available browse per tree and the proportion of twigs showing browsing damage. Samples were taken in March and October: October measurements (at the end of the growing season) give a first estimate of available standing crop, and a measure of growing-season offtake; measurements in March offer confirmation of standing crop, and allow estimation of winter offtake. Actual offtake, by weight, was more difficult to determine. Each sample clipped was sorted into browsed and unbrowsed twigs; and it would appear simple to determine the mean weight of an unbrowsed twig and the mean weight of a browsed twig, and then calculate weight of offtake by difference. There is, however, such variance in the weight and length of the shoots that there is in practice little difference in mean weights of browsed and unbrowsed samples. Accordingly, unbrowsed twigs were sorted into size categories (on the basis of basal diameter) in an attempt to reduce the variance within each sample. *Then*, when variation around the mean weight of each class of twig was reduced to an acceptable level, weight of browsed samples could be compared with the weight of the unbrowsed shoots of an *equivalent size.*

A summary of results for forage production and offtake of the major communities and of the main forage species is presented in Tables 7.3 and 7.4. These data may be used in examination of animal feeding patterns; we have already seen that grazing by ponies and certain of the Forest deer is closely related to the timing of maximum production of each forage type. Results may also be used to investigate the effects within a given vegetation type of different intensities of grazing (as offtake) on productivity (page 163). In the present context, the data are used to assess grazing intensity through relationship of offtake to vegetation production. This relationship of offtake to forage productivity is widely applicable, in offering an independent estimate of grazing pressure in any system.

It is immediately apparent from Table 7.2 how high a level of offtake is sustained by the various Forest grasslands. Offtake in most graze communities is extremely high, amounting to nearly 100% of above-ground production for the year as a whole. Even during the summer growing season, month by month, 80-100% of the productivity may be removed at once by the grazing animals. Offtake of heather and woodland browse is far lower in proportion to their availability. (Data for holly are presented here since it is by far the most abundant browse species in woodland and has already been shown to be highly palatable to most of the Forest's herbivores: Chapters 4 and 6.)

Offtake figures in graze communities are tremendously high. To some extent such figures may be overinflated. As we have already noted, the method by which offtake is measured — effectively by change in standing crop — does not differentiate between losses due to grazing and those due to senescence and decomposition. During the growing season it is probably fair to assume that the majority of vegetation losses *are* due directly to grazing, but in the winter, when the vegetation is no longer growing rapidly and may be suffering considerable die-back, such an assumption may be unjustified. However, even if we recalculate the values of Table 7.2 for the main grassland types, but using data from the main growing season only (April-September), offtake during those months alone is still seen to be a *massive* proportion of production (Table 7.5). And we *know* that offtake calculated in this way is an underestimate: animals continue to graze in these same vegetation types throughout the year, and grasses and other

Table 7.3: Above-ground production of major vegetational communities and forage species. Figures, in g/m² are for the 1978-9 season

	Vegetation										
Month	**RL**	**SL**	**Cl**	**AG**	**Bog**	**WG**	**Ilex**	**Molinia[a]**	**Juncus[a]**	**Rubus[a]**	**Ulex[a]**
January	—	13.5	9.2	6.5	13.9	—	↑				
February	0.9	30.0	16.1	11.7	17.0	—					
March	5.7	25.3	0	0	51.2	—					
April	12.2	27.5	0	0	84.7	—					
May	27.1	49.6	46.0	8.6	95.5	—	24.4	May-Sept 226	May-Sept 1,228	May-Sept 318	May-Sept 722
June	39.0	72.0	78.9	26.2	118.0	28.4					
July	59.6	129.0	81.2	45.2	111.0	47.2					
August	50.3	117.0	65.5	33.9	42.4	26.7					
September	26.5	77.3	28.1	26.9	16.7	3.1					
October	5.2	43.7	6.3	22.7	11.6	—					
November	—	33.4	—	—	5.2	—					
December	—	38.3	—	16.8	—	—	↓				

Note:[a] Results for these single-species samples are calculated per m² of plant area only. (Figures for whole communities are g /m² total ground area.)

Table 7.4: Monthly offtake of major vegetational communities and forage types 1978-9 (g/m²)

Month	Vegetation										
	RL	SL	CI	AG	Bog	WG	Ilex	Molinia[a]	Juncus[a]	Rubus[a]	Ulex[a]
January	15.3	46.2	14.0	14.3	—	9.7		0	36.3	37.1	
February	11.1	40.9	12.5	7.8	—	6.5	Oct→	0	28.6	16.1	Oct→
March	8.7	33.8	6.7	3.7	—	2.3	April	0	9.7	8.4	April
April	14.5	43.5	14.3	18.4	—	0.4	1.9	3.2	23.3	4.2	364.0
May	27.0	54.8	42.0	20.6	—	4.9		16.2	106.0	9.5	
June	38.7	64.5	61.5	13.7	—	13.2		26.8	35.5	15.6	
July	54.4	94.5	64.2	22.1	—	8.8	April→	35.1	187.0	19.7	April→
August	48.1	71.5	50.1	37.9	—	11.9	Oct	20.7	132.0	20.8	Oct
September	29.3	42.7	36.0	54.0	—	16.8	0	4.8	116.0	20.5	0
October	19.9	41.7	23.9	40.9	—	20.9		1.9	61.2	20.2	
November	20.2	24.6	13.7	16.3	—	10.9		0	29.6	27.5	
December	21.7	19.9	12.1	11.5	—	9.1		0	25.4	36.3	

Note:[a] Results for these single-species samples are calculated per m² of plant area only.

Table 7.5: Total production and offtake of Forest grasslands during the growing season of 1979 (figures in g/m²)

Vegetation types	Growing-season production	Growing-season offtake	Offtake as percentage of production
Re-seeded lawns	215	212	98.6
Streamside lawns	472	372	78.8
Commoners' improved grassland	310	268	86.5
Acid grassland	151	167	110.6

ground vegetation contribute a significant proportion of the diet of both cattle and ponies outside the immediate growing season (Chapter 4 and Figures 4.4, 4.5).

Such a calculation, however, makes another important point. We noted that the *timing* of defoliation was perhaps as important a factor to the plant as the actual amount (page 136). Yet it is clear from Table 7.5 that, in the grasslands at least, the majority of material produced is removed at the time, during the actual growing season. While offtake from heathlands and woodland browse is restricted to a relatively short period, and for the most part outside the growth period, close examination of data for grasslands reveals that offtake is more or less continuous. Not only is most of the growing season's production removed during the growing season, but the majority of it is removed *at once*— and on most communities figures for production and offtake balance even within *months* (Putman *et al.* 1981).

The Impact of Grazing upon the Vegetation

The effects of this tremendous grazing pressure are obvious. At a purely qualitative level one merely has to glance at the close-cropped turf of the Forest lawns, or to take in the absence of understorey in the Forest woodlands — and the clear browse line on the forest trees demarking the level to which the browsing ponies and deer can reach the foliage (Plate 15) — to appreciate just how significant the grazing pressure must be in the ecological balance of the Forest ecosystem. The influence on both structure and species composition is so obvious that it hardly needs quantification, but closer examination reveals even more wide-reaching effects.

The potential effects of heavy grazing upon vegetation discussed in the first chapter of this book may be summarised as:

1. Change in species composition through loss of graze-sensitive species, or encouragement of graze-resistant species, and through changes in nutrient availability.
2. Change in physical structure: grazing results in a loss of structural diversity.
3. Alteration to patterns of nutrient cycling both within communities and in transfer of nutrients from one community to another.
4. Stimulation or suppression of vegetation productivity.

Now that we have introduced the herbivores themselves, we shall examine the evidence for each of these various effects in the New Forest grazing system.

Changes in Physical Structure

Graze Communities. The patchy turf of Forest grasslands boasts no plant material higher than a few millimetres — apart from the occasional stem of ragwort! (Plate 14). Such grasslands clearly lack many of the possible structural layers of mature, ungrazed grasslands. On heathlands, too, the effects of heavy grazing are clear in their reduction of structural diversity. Heathlands in the south Hampshire area support dense stands of purple moor-grass (*Molinia caerulea*) among the *Calluna* and *Erica* heaths. Outside the Forest boundary the tall flower spikes of *Molinia* tower above the canopy of the heather plants and provide a whole additional structural element within the vegetation. Inside the Forest this whole stratum is missing. *Molinia* plants are just as abundant within the heathlands, but *Molinia* is an important component of pony diet (Chapter 4) and the plants are always heavily grazed — kept well below the heather canopy. This loss of a structural element within the heathland community is dramatically illustrated on the boundaries of the Forest where a simple fenceline divides a single heathland. On one side — inside the Forest — ponies graze undisturbed; from the area outside the Forest boundary, ponies and cattle are totally excluded. The geology, soil type and weather conditions are identical on either side of the fence; the only difference is the intensity of grazing — and the vegetation itself.

But, while grazing appears to reduce the structural diversity of vegetation on the vertical scale, it actually increases diversity of the sward itself, promoting discontinuity and patchiness within a homogeneous community. Perhaps the clearest example of this last effect is the recognition within the improved grasslands of the Forest of distinct sub-communities corresponding to those areas where animals graze and those where they defaecate (see page 58-9). In their use of such grasslands, the Forest ponies establish traditional latrine areas to which they move to defaecate. No large herbivore feeds in the immediate vicinity of fresh dung, and as a result there is a mosaic established within these grasslands of short-grass areas where the ponies graze, and relatively ungrazed, but nutrient-rich, latrine patches.

This effect might not be expected to persist in a multi-species system; but the action of the other herbivores does not counteract this mosaic pattern. The other major herbivores of these grasslands are the cattle; their mouthparts are such that they cannot feed on the close-cropped turf of pony-feeding areas, but are perforce restricted to latrine areas. (Since they have no traditional latrines of their own, but defaecate where they happen to be, the dung remains in these pony-latrine areas and is not returned to the short-turf patches.) As a result, owing to the grazing and eliminatory behaviour of the herbivores, distinct sub-communities are established and maintained within the improved grasslands which differ both in species composition *and* in overall intensity of grazing; these two factors contribute to a distinct difference in physical structure (although a subtle one: areas of pony grazing have a maximum turf length of about 5 mm; vegetation in latrine areas rarely exceeds 2 or 3 cm). Although these sub-communities are most apparent on the Forest lawns, we have data to suggest that the same latrine behaviour may lead to heterogeneity in other communities as well — certainly on acid grasslands and heathland, where the development of gorse brakes seems to correlate well with patterns of animal elimination.

These effects on vegetational structure might strictly be considered effects of eliminatory behaviour rather than direct effects of grazing. But grazing patterns, and different intensities of use of different areas within a community, can also lead to the development of structural diversity. In Chapters 3 and 5 we have shown that grazing pressure by deer and domestic stock is not evenly distributed across all vegetation types, but that some are more favoured and subject to greater intensity of use than others. This holds even *within* a single vegetation type: ponies grazing pasture on abandoned farmland in the Netherlands show clear preference for different parts of their range, despite apparent vegetational uniformity (Oosterveld 1981). Oosterveld has shown clear differences in the scale of use of a single 100-ha grassland system by free-ranging Iceland ponies (at a density of 0.3 per hectare) to the extent that 3% of the area is heavily used, 70% only moderately or highly used, and 27% is subjected to a very low level of impact or never used at all. Presumably, some areas are more sheltered than others, or are favoured for some other, perhaps social, reason; whatever the reason, distribution of animals is uneven — and the uneven grazing pressure that results leads ultimately to structural diversity within the habitat (Table 7.6).

Woodlands. In woodland communities of the New Forest, once again, some of the effects of animal usage on physical structure of the vegetation are immediately apparent. New Forest woodlands virtually lack *any* ground flora or shrub layer. The woodland floor is essentially bare, and indeed the whole structural 'layer' between ground level and 1.8 m — the extent of a pony's reach — is missing: most of the Forest woodlands display a marked browse line at this level (Plate 15). Under continuous browsing pressure, palatable shrubby species such as hawthorn, blackthorn and hazel are eliminated and fail to regenerate. Even species relatively resistant to grazing, such as holly or gorse, are heavily used: taller holly trees are thoroughly browsed up to the 1.8-m browse line and have little vegetation below this level; shrubs of both holly and gorse which fall entirely within the reach of the herbivores are severely stunted and 'hedged' by the continuous browsing (Plate 16). At the ground level, brambles, ivy and other low vegetational species are completely eliminated; the only species which gives any structure at this

Table 7.6: The effects of uneven grazing pressure by ponies in creating different structural patches within an area of disused farmland

	Development of structural diversity within the sward (%)			
Grazing intensity	Short-grazed (25 cm)	Rough stands (25 cm)	Bushes, woody plants (2.5 m)	Wood (2.5 m)
Heavy	100			
Moderate	80	20		
Light	10	50	30	10
Very light	5	10	25	50

Example: A moderate use of the area means that you get 80% short-grazed and 20% rough vegetation as a result; at moderate stocking rates, 60% of the area is moderately used.
Data from Oosterveld (1981).

level is bracken (*Pteridium*), which, although eaten by the ponies at certain times of year, is not particularly palatable.[1]

A direct analysis of the overall effects of grazing on woodland structure may be found in a unique experiment carried out within one small area of Forest woodlands. In 1963, 11.2 ha of mixed (oak/beech) woodland in Denny Lodge Inclosure were fenced to create two separate, but adjoining, compounds of approximately equal area. All grazing animals were removed from one pen; in the other, a population of fallow deer was maintained at known density (averaging one per hectare over the years). The pens were surveyed after six years and again after 15 years, to consider changes in vegetational character in the two areas (Mann 1978). Strictly speaking (since both pens were originally heavily grazed), differences between the two pens should be considered due to secondary succession within the ungrazed pen on relief from grazing, rather than viewed the other way round. Since the vegetation of the ungrazed pen is secondarily developed from an area once heavily grazed, it cannot be assumed equivalent to the primary vegetation cover of the area before grazing; it is therefore not legitimate to use a comparison of the two pens to discuss in detail the effects of grazing in the grazed pen — because the vegetational characteristics of the ungrazed plot are not a good control. None the less, results do clearly demonstrate the effects of *continuation* of grazing in preventing secondary succession, and thus offer some indication of the effect of *current* grazing on community structure in preventing this successional development elsewhere in the Forest.

The two 5.5-ha pens were compared six and fifteen years after the exclusion of all the large herbivores from one pen. In the first survey (intended as an interim measure to evaluate some of the more major differences between the areas and to offer an earlier 'time point' to assist with interpretation of a later, more complete, analysis) records were taken of tree numbers and species composition of each pen and biomass of ground vegetation. The second survey repeated these basic measurements, and in addition recorded species diversity and structural diversity of the vegetation in each pen. Although carried out on only a relatively small scale, the study was tremendously detailed and its results are worth considering here in some depth.

A 50-m grid was established within the area by recognised survey

---

[1] Such an impression, however, striking, is none the less subjective. A more formal analysis of the structure of the 'ground layer' in woodlands inside and outside the Forest boundary, based on hemispherical photographs, is presented by Hill (1985).

techniques. Within each of the 2,500 m^2 squares resulting, a 1-m^2 quadrat was located. Twenty such quadrats were identified in this way in each pen and these were used as sampling points. Vegetational parameters recorded in each quadrat were:

a) Tree numbers	(6 and 15 years)
b) Species composition	(6 and 15 years)
c) Biomass	(6 and 15 years)
d) Structural diversity	(15 years only)
e) Species diversity	(15 years only)

In the 1969 survey all the trees in each pen were plotted on a map (scale 50″ to the mile) by reference to the grid, and a list was compiled of the species and the numbers of each species occurring in each pen. In 1978, an analagous result was obtained by establishing a circular quadrat (radius 10 m) centred on each quadrat, and recording all the trees within that area. Species composition of the ground vegetation was assessed by counting all the plant species present in each of the 40 quadrats. In 1978, relative abundance of the various species within the quadrats was recorded in addition, using a 'pinframe'. (The three-dimensional pinframe is an extension of the point-quadrat method of sampling whereby the recording of the number of hits on a species by pins lowered into the vegetation allows a number of quantitative measures to be extracted from the records.)

In 1969, and again in 1978, the standing-crop biomass of each quadrat was clipped as close to the ground as possible and the vegetation was weighed fresh, as a crude estimate of structural diversity based on vegetational density. Vegetational structure may be more precisely defined, however, in terms of relative density of vegetation and the way this changes with height above the ground, and in 1978 measures of actual vegetation densities at various heights were made, once again employing a pinframe. The pinframe used for this experiment consisted of a single bar, 1 m long, which could be moved across the quadrat in stages (in this case 10 cm). Holes were drilled in the frame at 10-cm intervals and the pins were lowered in 10-cm stages. Pins terminated in a needle in order to approach as closely as possible the requirement of a 'true point', and records were made of the contacts between the pin tip and the herb species for each point on the matrix. In this instance detailed records were made of the number of hits per species per quadrat, at every 10-cm stage above the ground, giving an index of vegetation density at 10-cm intervals up to 1.5 m.

Published regressions relating the girth at breast height (or sometimes at ground level) to timber volume, leaf weight and area, and leaf and timber production (Whittaker and Woodwell 1968) were used to estimate vegetation densities above 1.5 m. Finally, measures of species diversity of the ground flora of two pens were made during the 1978 survey. (These are discussed later, page 155-6).

The gross morphological and vegetational differences between the areas are very clear. In one pen there is a distinct browse line and vision is relatively unimpeded, easy progress being possible in any direction. The regeneration in the ungrazed pen, on the other hand, has now reached such a stage that passage through it is difficult; the area can accurately be described as a thicket, and it is full of young, very vigorous saplings. There is no evidence of a browse line, and ground vegetation reaches up into the canopy of the trees, forming a dense and deep layer of plant material.

Fifteen years after fencing, there were shown to be 35 times as many trees in the ungrazed pen as in the pen still grazed by fallow deer, although the difference between the compounds is not significant until the regenerating trees are included in the analysis. This last point shows the equivalence of the two areas before enclosure, and stresses the effect of grazing in prevention of such regeneration: there are in fact *no* tree seedlings or saplings recorded in any of the samples taken from the grazed pen (Table 7.7). (Regenerating trees in the ungrazed pen — that is those included here being less than 1.5 m in height — were in fact more common in the 1969 results, e.g. *Pinus sylvestris, Pseudotsuga menziesii, Larix* sp., but figure less in the later study as they have grown out of that category. The 1978 data do include an imbalance of *Salix* saplings, with again more in the ungrazed pen.) Canopy *structure* also differed markedly between the two pens. The canopy in the ungrazed pen was not distinct and reached down into the bracken and bramble clumps; there

Table 7.7: Total number of trees per hectare in the Denny pens exclosures

	Pen	Trees/ha	Mature trees	Regenerating trees
1969	Grazed	125		
	Ungrazed	106		
1978	Grazed	188	188	0
	Ungrazed	6,893	250	6,643

was no indication of a browse line. The grazed pen had no canopy vegetation below 1.5 m and the browse line was very distinct.

Figures for standing-crop biomass of *ground vegetation* are also higher in the ungrazed than in the grazed pen, as might be expected. Results from both 1969 and 1978 agree on this point. The more detailed analysis of 1978, however, suggests that this difference is barely significant (Table 7.8). Vegetational density can also be considered in terms of volume, or 'bulk'. In comparison of grazed and ungrazed pens in terms of the bulk of ground vegetation, pinframe results were used to calculate the total number of pinhits from each set of 20 quadrats, and then to find the mean number of hits per quadrat. The results are given in Table 7.8, where a comparison of means indicates that the areas are not significantly different in this parameter.

Structure is, however, three-dimensional; differences in physical structure of the vegetation between the grazed and ungrazed pens should properly consider 3-D *distribution* of this vegetational bulk. Mann presents an index of structural diversity using pinframe data. This index considers the proportion of plant material occurring at a particular height above the ground in any quadrat, and is derived as the number of pinhits at a particular height (chosen to suit the circumstances) as a percentage of the total number of hits for the quadrat, as:

$$\frac{\text{Number of pinhits at height x}}{\text{Total hits for the quadrat}}$$

In this study, the index was calculated at 0, 10, 50, and 100 cm above ground in each quadrat. The results reveal a significant disparity between the pens for the lower levels, 0 and 10 cm, but, interestingly, *not* at the higher levels, 50 and 100 cm (Table 7.9).

Table 7.8: Mean standing crop of ground vegetation in Denny pens experiments (figures in g/m²)

	1969	**1978**
Grazed pen	138	495
Ungrazed pen	605	687

Vegetation density from bulked pinhit data (1978 data)

	Grazed pen	Ungrazed pen
Total pinhits per pen	1221	857
Mean hits per quadrat	61.05	42.85

Table 7.9: Mean vegetation-density index Denny pens experiments (1978: see text, page 152)

Height (cm)	Grazed pen	Ungrazed pen
0	16.5	2.5
10	21.9	11.6
50	7.3	10.4
100	1.7	3.0

Results from this study highlight the differences in vegetational composition and structure between woodland areas free of grazing animals and those maintaining a high density of large herbivores. Differences in structure — and actual vegetational bulk in terms of pure biomass or bulk of material — are apparent both in the ground flora and in the woody vegetation. Much of the difference between the grazed and ungrazed area results from the massive regeneration of tree species which has occurred in the area free of grazing; results emphasise the lack of such regeneration in grazed areas. This suppression of regeneration may have a direct effect on physical structure of the vegetation, but in the long term has an even more significant effect — on the population age structure of the Forest trees.

Were regeneration to be prevented for ever, the woodlands would of course ultimately disappear altogether as the existing trees aged and fell and no new trees could establish in replacement. Intensity of grazing pressure fluctuates over the years, however, and at times of relatively lower herbivore numbers some regeneration may occur — enough to guarantee continuation of woodland cover. But such regeneration is spasmodic, and can occur only over brief periods when herbivore density is low enough to permit the survival of at least some of the seedlings. This alternation of periods of heavy and lighter grazing pressure, with resultant alternation of periods when seedlings may and may not establish, has a curious effect on the age structure of the populations of trees within the Forest woodlands.

Deciduous woodlands in the Open Forest in fact consist of trees of only three major age classes: those which germinated during, respectively, 1650-1750, 1860-1910 and 1930-1945 (Peterken and Tubbs 1965). These periods correspond to times when numbers of grazing animals on the Forest were reduced. In the intervening periods, animal numbers rose so high that no further regeneration could occur.

The trees of 1650-1750 — Peterken and Tubbs' A-generation — are the last survivors of that period of history when the first Inclosures were made within the Forest for timber production (page 19). The evidence suggests that this A-generation of trees arose as the result of deliberate management when, by enclosure, establishment of trees by natural regeneration was permitted on sites at that time largely cleared of their former tree cover. (It is of considerable interest to discover that the 1650-1750 regeneration in the New Forest had its counterparts in other Royal Forests: Tubbs 1968.) Relaxation in the administration and management of the Forest, and a deterioration in control exercised over the Forest deer, spelled the end of this phase of regeneration in the middle of the eighteenth century.

Herbivore numbers remained high enough to prevent further seedling establishment until over a century later. The regeneration of 1860-1910 (and the establishment within the Open Forest of Peterken and Tubbs' B-generation) followed the killing of large numbers of Forest deer when the Crown relinquished its hunting rights on the Forest in 1851 (Chapter 2); that of 1930-1945 coincided with lowered populations of domestic stock over this period as a result of low market prices in the recession.

This picture is somewhat oversimplified, for, although regeneration ceased around 1920 at the end of the B-generation, herbivore grazing pressure was no higher at this time than in the 1880s — in the middle of the regenerative phase. What, then, finally finished the period of regeneration? Tubbs (1968) points out that an investigation of the rate of growth of B-generation hollies, by means of growth-ring counts at successive heights, showed that the period of rapid height growth ceased between 1910 and 1920. Seedlings established after 1910 were heavily suppressed, so it may be inferred that the canopy closed during those ten years. On sites where the B-generation holly had developed as a scatter of trees with an incomplete canopy, regeneration was found to have continued throughout the 1920s and 1930s. This and other evidence suggested that the B-generation largely ceased because the canopy itself prevented further regeneration (Tubbs 1968), rather than as a direct result of increased grazing pressure. But, if the closing of the B-generation canopy itself inhibited further regeneration, it remains necessary to explain why the outer margins of the woods did not continue to expand and why new areas of scrub and emergent woodland did not come into being on open, unshaded heathland sites at that time. In recent years there *has* been some encroachment of birch and Scots pine on areas of open heathland, despite the extremely high numbers of cattle and ponies on the Forest — regeneration, despite a grazing pressure which is probably higher than that experienced in the 1920s. The evidence here points to the development of heath-burning practices on an unprecedented scale in the past hundred years, which severely restricted the spread of woodland; reduction of this large-scale burning in recent years has perhaps permitted the more recent scrub encroachment observed.

Changes in Species Composition

Many of the changes in physical structure which result from the heavy pressure of herbivores on the Forest vegetation are accompanied by, or are indeed *due* to, actual changes in species composition within the plant community. Thus, the lack of the understorey shrub layer of woodlands is due to the selective eradication of species such as hawthorn, blackthorn, hazel and willow.

Even moderate levels of grazing may result in loss from a community of particularly graze-sensitive species. Under heavy grazing pressure, even moderately *resistant* species may be eradicated if they are particularly palatable and thus exposed to particularly high levels of use. As a result, any species which are particularly palatable or graze-intolerant are lost from the vegetational community. By the same token, species which are resistant to grazing (possessing physical or chemical defences) or graze-tolerant (by growth form and physiology able to withstand some degree of defoliation) may be encouraged. Grazing restricts the distribution and performance of potential competitors, allowing existing graze-tolerant or resistant species to expand, or new species to enter the community. Moderate or heavy levels of grazing can thus dramatically alter relative abundances of graze-sensitive and graze-resistant or -tolerant species within the community, altering its entire species composition.

Chris Mann's study of the Denny pens (pages 149-53) showed clearly the selective loss under heavy grazing of species such as hawthorn, blackthorn and hazel, on which we have already remarked. In the secondary succession permitted within a small enclosure after relief from grazing, all these species rapidly re-established themselves. Abundance after 15 years relief from grazing is shown in Table 7.10 by comparison with abundances in the adjacent, grazed, area. While less tolerant species such as these are quickly lost when exposed to heavy use, holly (*Ilex aquifolium*), a resistant species, shows considerable expansion under these conditions and has a far greater abundance and wider distribution within the Forest than would be expected by comparison with equivalent woodland areas elsewhere.

Similar changes in species composition can be observed in the ground flora. Plant species were recorded in each quadrat at Mann's

Table 7.10: Species composition of trees in two woodland enclosures, one grazed, one ungrazed, as mean number of trees per 10-m radius circle. Data from Mann (1978)

Species	Grazed pen	Ungrazed pen
Fagus sylvatica	2.54	22.4
Quercus sp.	1.75	12.9
Pinus sylvestris	1.0	43.25
Larix sp.	.35	6.3
Betula sp.	.25	65.25
Pseudotsuga menziesii	.15	13.6
Ulex europaeus		19.66
Ilex aquifolium		17.05
Crataegus monogyna		.96
Prunus spinosa		.6
Salix sp.		23.8

study-site on a presence or absence basis, and these results provide a list of 'species occurrences' for each compound. An index was devised to emphasise the differences beween the compounds and to 'weight' those species more prolific in one pen than the other. The index was calculated from

% quadrats occupied in
ungrazed pen − % quadrats occupied in grazed pen

and it has a range of values from +100 to −100. Those species with an index of zero occur equally in both regions; those with a negative index are more abundant in the grazed pen. More species of grasses were found in the grazed pen, a shift in community structure frequently noted in response to heavy grazing. Grasses grow from the base rather than the tip, and are thus more tolerant of grazing than many other species (page 181); small ground-covering herbs, however, were similarly distributed between the two pens (Table 7.11).

In almost all vegetation types within the New Forest it is possible to demonstrate similar effects. Re-seeded lawns, established in the 1950s with pasture seed mixes containing several species of herbage plants of which perennial rye-grass and white clover were usually the most important, show now a tremendous change in species composition. Current studies demonstrate almost total loss of many of the original, more palatable forage species, with increased domination of the areas by *Agrostis tenuis.*

Table 7.11: Index of species occurrences in grazed and ungrazed woodland pens (Denny pens experiment, see text, pages 155-6)

1969 survey	Index	1978 survey
	65	*Rubus*
Calluna	30	*Calluna*
Salix	25	*Betula*
Pinus sylvestris, Betula spp.	20	
Rubus	15	*Hypericum, Rosa, Salix*
Epilobium, Ulex, Pseudotsuga mensiesii, Ranunculus, Viola	10	*Hedera, Ilex, Potentilla, Ulex*
Crataegus, Larix sp., *Erica, Oxalis*	5	*Agrostis stolonifera Anthoxanthum, Crataegus, Epilobium, Pinus, Teucrium*
Molinia	0	*Quercus*
Euphorbia amygdaloides, Juncus conglomeratus, Pteridium, Ilex	−5	*Fagus, Lonicera, Oxalis, Rumex*
Deschampsia cespitosa, Carex sp.	−10	*Pteridium aquilinum*
Hedera helix	−15	*Digitalis, Luzula campestris, Viola*
Lonicera	−20	*Molinia*
Rosa, Luzula campestris	−35	*Deschampsia cespitosa, Juncus* spp.
	−45	*Agrostis tenuis*

About 20 areas of acid grassland, gorse and heathland (totalling some 350 ha) were ploughed and fenced during the Second World War, and cropped for cereals or potatoes. After the war these areas were re-seeded with grass and opened to stock, providing improved grazing for the increasing populations of commonable animals. The seed mixtures and fertilisers used varied considerably in different areas, with perennial rye-grass (*Lolium perenne*) cock's-foot (*Dactylis glomerata*) and Timothy (*Phleum pratense*) particularly common as grass mixtures, accompanied by heavy applications of lime, chalk and superphosphate. Red fescue (*Festuca rubra*) and crested dog's-tail (*Cynosurus cristatus*) were other commonly used grasses, and many swards contained white clover (*Trifolium repens*). In most cases establishment of the sward was successful; since the re-seeded lawns were 'thrown open' to stock, however (at various times between 1948 and 1953), their species compositions have changed considerably, with a gradual invasion of native species and almost complete elimination of all those that were sown. The present species composition of a number of such lawns is shown in

Table 7.12. In general, the species composition is now remarkably uniform, both within and between lawns, with a very characteristic and constant suite of species on all well-drained sites. As noted, *Agrostis tenuis* has established itself as the most abundant species, with a percentage cover ranging between 31% and 57% (mean 45%). Other species which were invariably present (mean percentage cover in brackets) were *Bellis perennis* (5%), *Hypochoeris radicata* (5%), *Plantago lanceolata* (4%), *Sagina procumbens* (3%), *Taraxacum officinale* (2%), *Luzula campestris* (3%), and *Trifolium repens* (3%). Thus the lawns are largely composed of low-growing species and rosette plants, well adapted to heavy grazing pressure and trampling. In addition a number of annual species are usually present, especially in the short-grass areas where they take advantage of small gaps in the sward. Among the most constant of these are *Trifolium micranthum*, *T. dubium*, *Vulpia bromoides* and *Aira praecox*.

We can actually examine the *timing* of these changes in species composition since the sites were first re-seeded. Pickering (1968) provides a detailed survey of the species composition of several lawns in 1964,

Table 7.12: Species composition of New Forest grasslands. Mean values of percentage cover are based on several samples; from six sites

	Re-seeded Lawns
Agrostis tenuis	45
Cynosurus cristatus	1
Festuca rubra	8
Lolium perenne	2
Poa compressa	2
Sieglingia decumbens	1
Bellis perennis	5
Hieracium pilosella	1
Hypochoeris radicata	6
Leontodon autumnalis	3
Lotus corniculatus	2
Luzula campestris	3
Plantago coronopus	1
Plantago lanceolata	5
Prunella vulgaris	1
Sagina procumbens	2
Taraxacum officinale	2
Trifolium repens	3
Other species	3
Bare ground	4

between four and 15 years after the various different sites had been thrown open to grazing. He showed how the number of colonising species increased with time and was highest in the oldest swards. For the first few years the sown species, particularly *Lolium perenne* and *Trifolium repens*, were fairly successful. In all cases the most abundant of the invading species was *Agrostis tenuis*, which showed a very rapid increase in abundance between ten and 15 years after seeding to achieve cover values ranging between 30% and 50%. The 15 years between Pickering's survey and our own saw the almost complete elimination of the sown species, and in most cases the present species composition has almost nothing in common with the original seed mixture (Table 7.13).

It is, however, not purely through actual grazing that large herbivores may change the species composition of their vegetational environment. Trampling has a direct impact; and both urine and dung, in causing locally marked changes in soil conditions, may also cause changes in floral composition. We have already discussed the mosaic of distinct sub-communities created on the Forest lawns in response to the eliminatory behaviour of the ponies. Patterns of defaecation by the ponies, coupled with a reluctance to graze near their own faeces, result in the establishment of distinct feeding and latrine areas on the Forest grasslands. Because of the shortness of the turf in areas grazed by ponies, the other major herbivores, cattle, are forced to graze in pony latrines. Their dung also accumulates within these latrine areas (page 58-9). Within the grasslands we thus find areas characterised by high nutrient input and reduced grazing pressure (cattle cannot crop the herbage as closely as ponies) and other areas under high grazing pressure and suffering continual loss of nutrients. Characteristic of these distinct conditions, each sub-community has its own, markedly different species composition. The most obvious differences are the presence of ragwort (*Senecio jacobaea*) and thistles (*Cirsium arvense*, *C. vulgare*) only on latrine areas; other differences are noted in Table 7.14.

Similar effects may be observed in different communities: heather plants (*Calluna*) are easily damaged by trampling and the shoots killed by contact with nitrogen-rich urine. Bristle bent grass (*Agrostis setacea*) and heaths (*Erica* spp.) are also species easily killed by urine. In contrast, species such as carnation sedge (*Carex panicea*), purple moor-grass (*Molinia caerulea*) and common bent (*Agrostis tenuis*), although they may be scorched by direct contact with urine, recover quickly and grow more strongly afterwards in the enriched soil. As a result, trampling and eliminatory patterns may also alter the species composition of, in this case, heathland.

Table 7.13: Comparison of species composition (percentage cover) of Longslade re-seeded lawn in 1958, 1963 and 1979. (Data for 1958 and 1963 from Pickering 1968; data for 1979, Putman *et al.* 1981)

	1958[a]	1963	1979
Agrostis canina	—	1.5	
Agrostis tenuis		47.8	51.0
Cynosurus cristatus		1.2	0.9
Dactylis glomerata	58	6.8	2.4
Festuca rubra			2.9
Holcus lanatus		1.1	1.2
Lolium perenne		2.0	2.5
Poa annua		0.2	0.7
Poa sp[b]		0.9	0.9
Sieglingia decumbens		0.6	0.6
Vulpia bromoides		5.3	0.1
Bellis perennis		4.7	6.3
Cerastium holosteoides		0.6	0.2
Hieracium pilosella		0.2	0.3
Hypochoeris radicata		3.2	3.8
Leontodon autumnalis		3.7	1.8
Luzula campestris		0.8	2.0
Plantago lanceolata		0.3	7.4
Plantago major		0.2	
Potentilla erecta		0.3	0.1
Prunella vulgaris		0.9	0.6
Ranunculus sp.[c]		0.5	0.4
Sagina procumbens		1.2	0.4
Senecio jacobaea		0.3	0.1
Taraxacum officinale		0.5	1.6
Trifolium dubium		0.4	0.3
Trifolium pratense	10	0.6	
Trifolium repens	32	12.8	3.2

Notes
a. Estimated from seed mixture.
b. Identified as *P. pratensis* in 1963, as *P. compressa* in 1979.
c. Identified as *R. repens* in 1963, as *R. acris* in 1979.

If such changes in species composition of vegetational communities caused by grazing, trampling, or patterns of elimination continue over a protracted period, we may observe gross shifts in community structure, even conversion of one entire community type to another. Continued use of woodlands with no respite for regeneration would result, in extremes, in the establishment of grasslands — as parkland savannas in

Table 7.14: A summary of the main differences in plant-species composition between 'short-' and 'long-grass' (latrine) areas of re-seeded lawns

			% Cover of	
Status	**Species**	**Proportion of sites**	**Latrine areas**	**Short-grass areas**
Confined to long grass	*Cirsium arvense*			
	Cirsium vulgare			
	Senecio jacobaea			
More abundant in latrine areas	*Hypochoeris radicata*	6/6	7.7	4.8
	Lolium perenne	5/6	2.2	0.3
	Trifolium repens	5/6	5.3	1.3
More abundant in short grass	*Poa compressa*	4/6	1.8	2.8
	Sagina procumbens	4/5	1.2	2.2

Africa are prevented from completing their succession to woodland by fire and the periodic ravages of elephant. On a shorter time-scale, however, we may appreciate more subtle changes in community structure. Urine scorch of heathland, discussed above, causes loss of certain base-intolerant species such as *Calluna*, *Erica* spp. and *Agrostis setacea*, while encouraging the growth of other species, e.g. *Agrostis tenuis*. Use of heathland areas bordering the Forest grasslands as latrine sites by the ponies thus encourages the outward spread of the grassland itself. Ponies feeding on the edges of grasslands regularly move off the areas into adjacent ones to defaecate. If acid-loving plants are eliminated in these communities and replaced by grasses, these areas will eventually be used for grazing; the continued grazing maintains the grassland sward and prevents return to a heathland vegetation, and the heathland is gradually converted to grassland. Similar factors influence the development of patches of gorse (*Ulex europaeus*) around the edges of the same grasslands (Putman *et al.* 1981).

The effects of large herbivores upon their vegetation, through grazing, trampling or elimination, can clearly dramatically influence the species composition of that vegetation. This has an immediate and dramatic effect on the functioning and development of these communities, for it alters not only the species structure of the community, but also its physical structure (section one here). We can begin to appreciate what a dominating role the large herbivores must play in the creation and maintenance of the many curious vegetational formations characteristic of — and peculiar to — the New Forest.

Alteration to Patterns of Nutrient Cycling and Productivity

Since the Forest supports such an enormous biomass of grazing herbivores, a large proportion of its available nutrient resources must, at any one time, be bound up in animal tissue. Nutrient cycles become tighter and the system becomes more fragile (Putman and Wratten 1984). In addition, soils and vegetation are nutrient-impoverished.

In addition to these overall effects, actual patterns of feeding and elimination by large herbivores alter the detailed pattern of nutrient cycling, and can result in gross shifts of nutrients around communities or from one community to another. A clear illustration of this may be drawn from the curious pattern of use of improved grasslands shown by ponies and cattle, mentioned earlier.

Establishment by the ponies of traditional grazing areas and different, latrine, patches within these grasslands leads to a gross shift of nutrients from grazed to ungrazed portions of the lawn. Nutrients are continually deposited in the latrines, continually removed from the grazed areas. Nor does the action of the other main herbivores (cattle) redress this balance, since they both feed and defaecate primarily within the pony latrines. As a result, there is continued transfer over the years of nutrients from areas grazed by ponies to latrine patches, and a continued impoverishment of the grazing. The most consistent differences are in potassium and phosphorus content of the soil, this being higher in latrine areas by a factor of some 1.2 (phosphorus) to 1.7 (potassium) (Putman *et al.* 1981). Organic-matter content of the soil in latrine patches is also consistently a little higher. Only where periodic flooding spreads the dung more evenly over the area (as on stream banks) is this continued gross shift of nutrients reversed. (This difference in nutrient status of the soil is a major contributing factor to differences in species composition between these different sub-communities, noted earlier.) Clearly, disruption of nutrient cycles, or continuous depletion of available nutrients in an area, will result in not only a change of plant-species composition, but a reduction of productivity. Grazing and eliminatory patterns may thus markedly affect patterns of vegetational production.

In addition, grazing itself may have more *direct* effects on vegetational productivity. Various studies in the literature (page 3) have established that grazing and browsing affect productivity of the vegetation exploited: with low levels of grazing stimulating regeneration and enhancing productivity, and higher levels of foraging suppressing growth by reducing effective photosynthetic area below an efficient threshold. Clearly, the extremely high levels of grazing in the New Forest, particularly on grasslands, must be having an affect on forage productivity directly, and not just through soil-nutrient impoverishment. Offtake in browse communities is only a relatively low proportion of production, and too low for us to detect any changes in productivity as a result. As we have seen, however, the various Forest grasslands sustain an extraordinarily high grazing pressure.

The effects of this grazing pressure have been examined in the natural situation, by comparing productivities of lawns under different grazing pressures. In addition, a number of experimental plots were set up, grazed *artificially* at different intensities. (Since different intensities of natural grazing will actually be reflected in the time interval between successive bites suffered by any one plant, clipping plots by hand at different time intervals simulates quite accurately different levels of use.) Plots were established in 'long-grass' areas (pony latrine) and on the shorter turf of pony-grazing areas, and clipped fortnightly, monthly and at two-monthly intervals. Plots in latrine areas were clipped back to 4 cm, the normal height of the surrounding vegetation subjected to continued natural grazing; those in short-turf areas were clipped to 2 cm. Preliminary results (Table 7.15) indicate that, on improved grasslands at least, the current pressure of grazing reduces vegetational productivity to a level considerably below that which it would achieve if ungrazed.

If total potential production of an ungrazed sward — one that has *never* been grazed — is assumed to be closest to that of a latrine sward (normally grazed relatively lightly, by cattle alone) exclosed for a full two-month period, then yields of other grazing regimes may be compared against this as a base-line. (In fact such a figure is probably an underestimate of total ungrazed production, so conclusions are conservative.) Heavy grazing, of 'pony-type' (i.e. cropped every two weeks to a height of 2 cm), reduces production to 0.45 of this 'maximum' yield (Table 7.15). Even if relationship to 'latrine swards' is not strictly legitimate (in that

Table 7.15: Effects of different levels of artificial grazing on grassland forage yield. (Values are total yields in g/m² obtained over 16-week period in 1979)

	Frequency of clipping		
Grazing regime	**2 weeks**	**4 weeks**	**8 weeks**
Short-grass areas (cropped close by ponies)	204	240	270
Latrine areas (grazed as by cattle)	324	372	456

animal grazing has been so heavy for so long on these Forest grasslands that the basic soil conditions of latrine and pony-grazed areas are not truly equivalent), so that analysis is restricted to comparisons 'within type', production can be seen to fall — in both types of swards — as grazing pressure increases (yield for swards clipped fortnightly being about 70% of the yield of the same swards clipped at two-month intervals). All levels of grazing *reduce* yield. There is no evidence that on these Forest grasslands light grazing stimulates productivity; alternatively, all levels of grazing intensity imposed are above this 'stimulation' threshold.

Chapter 8

The Effects of Grazing on the Forest's Other Animals

Grazing and the Forest's Smaller Herbivores

As with the larger herbivores, the smaller animals of the Forest must find in their environment food and shelter if they are to survive. Different species differ in their food habits and in the type and degree of cover they prefer, but distribution of each species is restricted to those habitats which provide their particular requirements in food source and vegetative cover.

The various common British species of small rodents, for example, Wood mouse (*Apodemus sylvaticus*), yellow-necked mouse (*A. flavicollis*), field vole (*Microtus agrestis*) and bank vole (*Clethrionomys glareolus*), have all been shown to exhibit predictable variation in relative numbers — even in whether they occur at all — in relation to the available cover in a habitat. Thus, the field vole is the only species which prefers to live in open environments, unshaded communities of uniform and short ground vegetation; bank voles clearly prefer dense cover, and tend to be characteristic of hedgerows or woodland edge, habitats where both ground flora and shrub layer are at their most dense. Neither wood mice nor yellow-necked mice appear to display any particular preference for cover (Hoffmeyer 1973, Corke 1974) and, particularly in areas where population densities are high, may be found in a variety of different habitats. Thus, the wood mouse, while essentially a woodland species, is commonly found in 'bank vole habitat', or even open fields, when competition there is low. Although intrinsic *preferences* are thus less clear-cut than those of field voles or bank voles, the mice do not survive equally well in all types of habitat, and in the face of competition from other species may be seen to be more restricted in distribution. Hoffmeyer (1973) showed that the wood mouse restricts itself far more tightly to its optimal habitat in the presence of the more aggressive yellow-neck; in the same way, while wood mice may colonise 'bank vole habitat' where voles are rare, they avoid the same habitats when bank vole populations are more abundant (Corke 1974).

Habitat preferences are not determined just by vegetative cover. While we have noted that the yellow-necked mouse shows no apparent preference for any particular degree of cover (Hoffmeyer 1973, Corke 1974), it is none the less found predominantly in woodlands (even in the absence of competitors). Further, throughout most of Europe it is most abundant in woodlands with a high proportion of beech or hazel (Schröpfer 1983). In autumn and winter, *A. flavicollis* feeds extensively on seed crops: beech mast and hazel nuts are staple foods. In feeding trials with other types of seeds such as acorns, the animals were found to be unable to maintain body condition; Schröpfer suggests that their preference for beech or hazel woodlands is thus a direct response to food needs. In the same way, *Microtus*, as a grazer for preference, feeding on shoots and roots of green vegetation, is probably restricted to open habitats not because it actively chooses to live without any cover from weather or predators, but because only in such environments can it find adequate food. *Clethrionomys* (grazer and fruit eater) chooses dense vegetation for cover, and hedgerows or woodland edge because of high food density, and the reason why the wood mice can afford to be unfussy in their choice of habitat is probably that they are catholic feeders, taking shoots, seeds, even insect food, as it is available.

Within the New Forest, the availability of cover and food is determined not just by the availability of particular vegetative associations or habitats, but is directly limited by grazing. Grazing by cattle, ponies and deer has a dramatic effect on the structure of particular communities, affecting the degree of cover offered — in canopy closure, extent of shrub layer or ground cover — as well as availability of food. How does this affect the smaller mammals, which, we have just discovered, appear to have quite clear and specific habitat requirements? What are the rodents of the Forest and how do their diversity and abundance compare with what we might expect in similar areas not subject to such heavy grazing? From 1982 to 1985, Steve Hill surveyed the rodent populations of the Forest woodlands, heathlands and open grasslands, comparing numbers and diversity with those obtained on equivalent vegetation types just outside the Forest boundary. Results were striking: total density of small mammals within the Forest was far, far lower than in adjacent ungrazed areas, though this was due not to a general lowering of population numbers across the board, but rather to a reduction in the number of species. Somewhat naively we had assumed that, if heavy grazing reduced both cover and food availability in all vegetation types, then certain species, whose requirements were highly specific, would indeed be excluded, but that those which 'hung on' would be surviving in suboptimal habitats and would thus show poor performance in comparison to populations in habitats more closely approximating to optimum conditions. What emerged from Hill's studies was that the diversity of species within the Forest was indeed much lower than in equivalent areas elsewhere: whole species were missing from the Forest woodlands and heathlands which occurred in abundance outside. Yet those species which *did* occur rather unexpectedly performed just as well inside the Forest as outside.

Heathlands and grasslands are obviously profoundly affected by grazing. The physical structure of these open vegetation types displays marked contrast with heathland or acid-grassland sites not subject to such heavy grazing (Chapter 7, Plate 8). On heathlands, the only animals caught in 1,200 trap-nights were two wood mice; adjacent heaths beyond the Forest boundary hold large permanent populations of wood mice and large populations of harvest mice (*Micromys minutus*). The story is very similar within acid-grassland communities: wood mice were more abundant, but again were found to use the area irregularly, and even then only in the summer, when bracken grows dense in these areas and offers some cover; only three field voles (*Microtus agrestis*) were caught during the complete grassland survey, and all of these were in a single area where bracken and litter cover is unusually thick all year round. The absence of harvest mice and field voles in trapping records does not by itself mean that the species are absent from the Forest: Hirons (1984) has found remains of both species in pellets cast by kestrels feeding within the Forest. Densities are, however, obviously *extremely* low; even those individuals caught by kestrels may come from the few ungrazed areas that remain within the Forest, such as railway embankments, fields and fenced roadside verges (see below, page 175).

Within the Forest *woodlands*, diversity of small mammals is again low. Most woodland sites had resident populations of wood mice, but bank voles (*Clethrionomys*) were almost totally absent from all sites surveyed and only one site held a permanent population of yellow-necked mice (*Apodemus flavicollis*). Bank voles were abundant in woodland areas outside the Forest, and the fact that they reappear within the Forest when grazing is prevented makes it clear that their absence from Forest woodlands is not just geographical accident. As part of an experiment to investigate the secondary succession within Forest woodlands after relief from grazing, two 11-acre (4.5-ha) plots of mixed deciduous woodland were fenced in 1963. All grazing herbivores were excluded from one pen; in the other, fallow deer were maintained at a constant density (Chapter 7). Hill exploited this ongoing experiment, surveying mammal populations in the two, adjacent, pens. The ungrazed pen showed a mammal community identical to those established in woodlands outside the Forest, with healthy populations of both bank voles and wood mice; 50 m away in the grazed pen, bank voles were never recorded (Table 8.1).

All these changes in species composition and species occurrence of small rodents within the various habitats of the Forest are exactly as one might predict from knowledge of the habitat preferences in terms of cover and food for each of the species (pages 165-6). None the less, the results emphasise that the effects of heavy grazing on the ecological functioning of the Forest are not confined to its direct effects upon the vegetation.

In Hill's studies, populations of *Apodemus sylvaticus* appeared little affected by the heavy grazing pressure. True, the mice were largely restricted to woodlands and made little use of more open habitats, but, within the woodlands, wood mice maintained healthy resident populations. Populations seemed to be of similar density to those outside; numbers inside and out showed the patterns of seasonal fluctuation typical of most woodland rodents. Indeed, on whatever criteria differences were sought — in age or sex structure of populations, average weights, fecundity, patterns of survivorship — no such differences emerged (Hill 1985); the wood mice of the New Forest appear unaffected by the dramatic vegetational changes inflicted by grazers. This result in itself is fascinating, considering the dramatic effects on other rodent species, and it is perhaps significant that the one species that *does* survive in the Forest and appears to perform as normal is that

Table 8.1: The number of individuals of the two species of small rodent, the wood mouse (*Apodemus sylvaticus*) and bank vole (*Clethrionomys glareolus*), caught during ten days Longworth trapping within the two deciduous woodland Denny pens, June 1982 (from Hill 1985)

Species	Grazed pen	Ungrazed pen
Wood mice	8	7
Bank voles	0	15

single species which we might have selected from previous knowledge as an opportunist: unfussy about cover and catholic in its diet.

The Forest Predators

These changes in distribution and abundance of small mammals presumably will have repercussions on the various predators of the Forest who rely for all or part of their diet on rodent prey. A very preliminary analysis has been made of the winter diet of New Forest foxes (*Vulpes vulpes*) by Senior (unpublished), who analysed the contents of a number of stomachs obtained between October and February. The foxes obviously took vertebrate prey whenever it was encountered (and 28% of guts examined contained the remains of birds or mammals), but such prey items were rare and the foxes clearly relied heavily on invertebrate material. Almost all guts contained significant quantities of beetles or earthworms, as well as leaf litter which had presumably been ingested accidentally while foraging for invertebrate prey of this kind. Perhaps more surprisingly, most of the guts examined contained quantities of plant material which would appear to have been taken deliberately: between 10% and 15% of the food contents of most stomachs looked at consisted of fruit, fungi or other edible vegetable material. While it is tempting to suggest that such a diet is unusual for an acknowledged carnivore and that the heavy dependence on invertebrate and vegetable material is a direct consequence of the 'distorted ecology' of the New Forest system, such claims would be excessive. In fact, the diet revealed by Senior's analyses is not particularly unusual; and analyses of fox diet in more 'typical' rural environments, while they do perhaps show a lesser reliance on beetles and other insect prey, none the less show equivalent quantities of fruit and other plant materials taken and also mirror the relatively low contribution made by vertebrate prey items (Table 8.2).

Badgers (*Meles meles*) are specialist earthworm feeders. In a study of badger diet at other sites within Hampshire, Packham (1983) found that 100% of all scats collected in all months contained earthworm chaetae, and that earthworms almost always constituted a high proportion of the diet of individual animals. Badgers in the Forest remained earthworm specialists despite the fact that earthworm densities within the Forest woodlands and heathlands are exceptionally low (Table 8.3, Figure 8.1). As a result, badger populations are little affected by low rodent numbers. Their ecology *is*, however, markedly affected by the

Table 8.2: Preliminary data on the overwinter diet of New Forest foxes. Equivalent data for the diet of woodland foxes in Oxfordshire (from Macdonald 1980) are shown for comparison

	Percentage frequency of occurence	
	New Forest foxes	**Oxfordshire foxes**
Lagomorphs and small rodents	28	20
Birds	—	11
Insects	56	21
Earthworms (and soil or leaf litter)	70	60
Fruits	28	20
Fungi	42	—

Table 8.3: Densities of earthworms available to badgers in different habitats inside the New Forest and elsewhere in sourthern Hampshire (from Packham 1983)

	Earthworm density: Mean numbers per m^2 (S.D. in brackets)	
Habitat	**New Forest**	**Itchen Valley**
Pasture/lawn	150 (26.6)	192.2 (89.4)
Heathland	0	—
Deciduous woodland	18.4 (15.9)	116 (27.9)
Coniferous woodland	1.2 (1.9)	26.8 (15.1)
Mixed woodland	3.2 (3.7)	64.4 (31.1)
Rides and glades	45.2 (33.9)	—

low densities of the earthworms on which they specialise, which in itself may be an additional effect of the Forest's heavy grazing pressure.

In studies of the social structures of badgers in areas of Hampshire outside the New Forest, Packham demonstrated an apparently constant relationship between the number of badgers in a territorial group ('clan'), territory size and earthworm density, such that the number of badgers shows a constant relation to total number of earthworms within the range:

$$(\text{Range area} \times \text{earthworm density}) \div \text{number of badgers} = \text{Constant}$$

A similar relationship has also been demonstrated for Scottish badgers

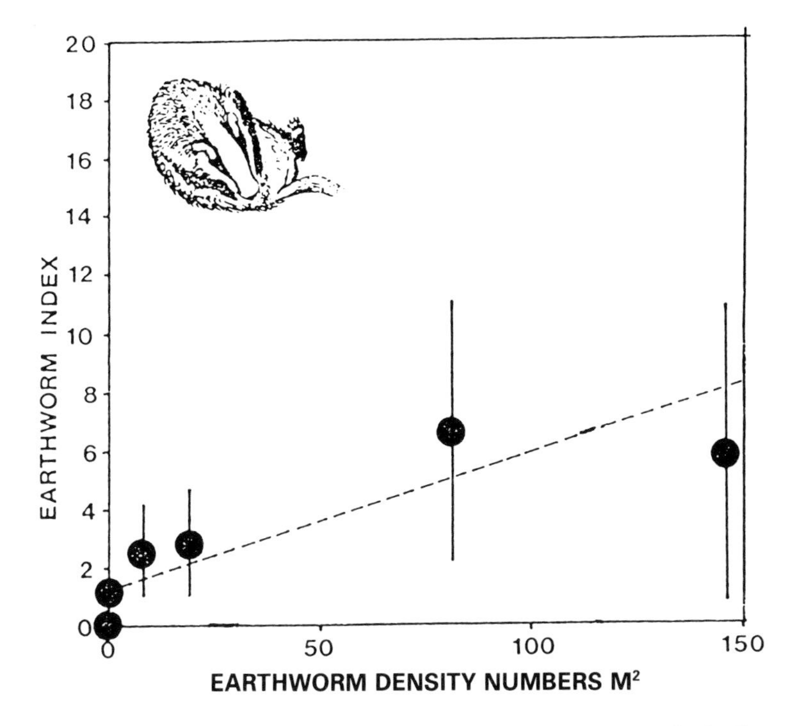

Figure 8.1: The relationship between relative earthworm abundance in the diet of New Forest badgers and density of earthworms in the habitat in which the diet was obtained (bars indicate one standard deviation around the mean)

Source: Packham (1983).

by Kruuk (1984). If Packham's 'formula' is extended to the New Forest, in calculation of territory size required to support a single badger, the low earthworm densities obtaining within the Forest result in predictions of a range so large that it would be impossible to defend. Linear dimensions increase only as the square root of area: as larger and larger range areas are predicted for clans of increasing numbers of badgers, so the length of the range boundary does not increase in the same proportion. But, even with 'theoretical clans' of eight or nine badgers, territory sizes predicted would have boundaries far too extensive to be effectively defended. Packham concluded that within the New Forest badgers should be more or less solitary, with extensive, overlapping home ranges which were not defended. Studies of the badgers themselves provided some empirical support for this: the badgers appeared to make no attempt to delimit range boundaries, and the usual marker latrines (concentrated around territorial limits in other areas) were not restricted to any clear perimeter lines; indeed, within the Forest, latrines appeared to be distributed randomly within the animal's range (Packham 1983).

In addition to its mammalian predators, the New Forest supports a remarkable diversity of birds of prey; populations of raptors may also be tremendously influenced by changes in the availability of rodent prey. Indeed, Colin Tubbs (1974, 1982) was the first to suggest that Forest predators in general might be influenced by grazing pressure and its effects on rodent numbers, drawing his conclusions from his now classic study of long-term changes in the populations of buzzards (*Buteo buteo*) within the New Forest. Tubbs noted that, in parts of England where rabbits are not readily available, buzzards appear to rely heavily on small rodents as prey; breeding success in these areas is directly related to abundance of such rodent prey. Although Tubbs had no direct data on abundance of small mammals within the New Forest, he none the less showed a clear correlation between reduced breeding success of the buzzards themselves and increasing grazing pressure from Common stock (Figure 8.2). Tubbs explains this relationship in terms of rodent numbers, suggesting that under high grazing pressure rodent populations are reduced, and buzzard breeding success declines accordingly: a not unreasonable presumption.

More recently, Hirons (1984) has examined diet and breeding success of other birds of prey of the New Forest whose diets would normally be expected to include high numbers of rodents. His studies of the diets of kestrels (*Falco tinnunculus*) and tawny owls (*Strix aluco*) produced surprising results (Hirons 1984). Small rodents contributed 66% by weight to the prey taken by tawny owls in the New Forest, comprising 42% of all prey items taken (Table 8.4). Studies of the diet of tawny owls elsewhere in Britain suggest that small rodents would normally constitute between 60% and 70% of prey items identified (Southern 1954, Hirons 1976). Owls in the New Forest thus seem to rely as heavily on small mammals here as they would elsewhere, despite reduced densities. In the tawny owl populations studied by Southern and Hirons, wood mice and bank voles each comprise about 30% of identifiable remains, with field voles contributing up to a further 13%. In the New Forest, voles are unavailable; but, within the Forest woodlands, *Apodemus* are present in normal densities. The owls respond to the reduced availability of voles by increasing their predation on wood mice and becoming wood mouse specialists. Thus, in the New Forest, as elsewhere, small

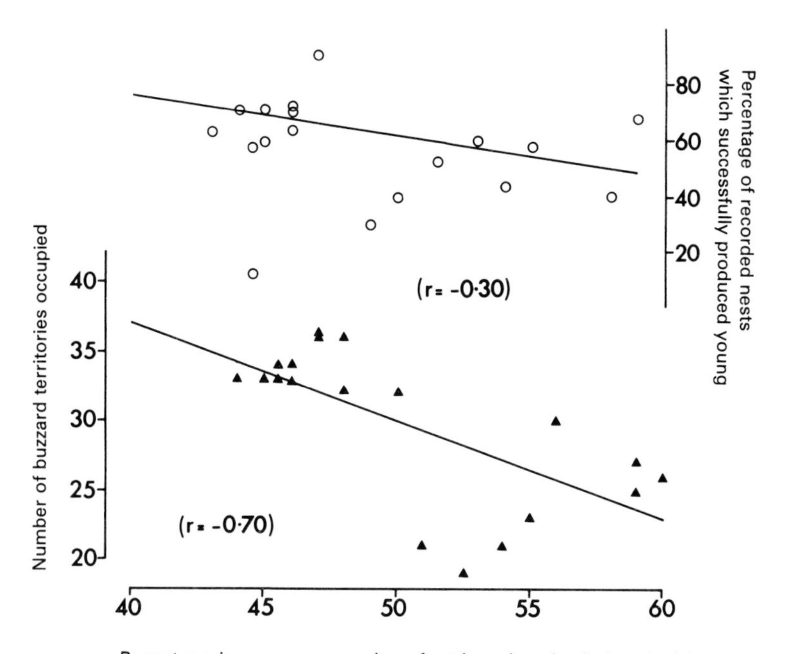

Figure 8.2: Relationship between grazing pressure on the New Forest and breeding success of buzzards. Data from Tubbs and Tubbs (1985). Note that, while success of nests, once a breeding attempt has been started, shows no significant effect from grazing pressure, the number of buzzard pairs attempting to breed declines as grazing pressure increases

rodents are the most important component of the diet of tawny owls, but this is achieved by taking very high numbers of wood mice: *Apodemus* accounted for 57% of the vertebrate prey items identified, while voles (*Clethrionomys* and *Microtus*) together constituted less than 10%.

Rodents still make up only 42% of the owls' diet; less in summer, when vertebrate prey constitutes only 33% of total intake. The balance is made up with a diversity of invertebrate food. Surprisingly, despite the importance of earthworms to tawny owls in other areas (where chaetae are found in nearly 100% of all pellets), *no* earthworm remains were identified from any pellets recovered from the New Forest (cf. badgers, page 170 above); it is clearly not these which compensate for reduced availability of vertebrate food. Dor beetles (*Geotrupes* and *Typhoeus*), however, contributed a surprisingly high proportion of the

Table 8.4: Analysis of 65 tawny owl pellets collected in the New Forest in 1982-3 (from Hirons 1984)

	May-October			November-April			Overall		
Prey	No. items	% items	wt	No. items	% items	wt	No. items	% items	wt
Wood mouse	26	35.1	45.0	21	27.3	55.4	47	31.1	49.1
Bank vole	2	2.7	3.1	6	7.8	14.1	8	5.3	7.4
Field vole	2	2.7	4.0	6	7.8	18.5	8	5.3	9.8
Grey squirrel	1	1.4	6.7	—	—	—	1	0.6	4.0
Rabbit	1	1.4	9.6	—	—	—	1	0.6	5.8
Common shrew	4	5.4	3.1	2	2.6	2.3	6	4.0	2.8
Birds	10	13.5	24.0	1	1.3	3.7	11	7.3	16.0
Slow worm	1	1.4	1.9	—	—	—	1	0.6	1.2
Coleoptera	27	36.5	26.0	41	53.2	6.6	68	45.0	3.9

prey of New Forest Tawny Owls, particularly in winter when they accounted for over 50% of items taken and formed 60% by weight of the total diet. The abundance of these beetles in the diet may reflect the large amounts of dung deposited in the New Forest by domestic herbivores and deer.

Field voles and bank voles have also been found to be an important element of the diet of kestrels (Ellis 1946; Davis 1960, 1975; Yalden and Warburton 1979). In the population studied by Yalden and Warburton, the two vole species together accounted for 73% of all vertebrate prey. In the New Forest, although bank vole remains are uncommon in kestrel pellets, the birds still feed extensively on field voles, which contributed 27% of identifiable vertebrate prey items recovered. Although the importance of *Microtus* in the diet is clearly lower in the New Forest than elsewhere in Hampshire (Hirons 1984) or in the studies mentioned above, the number of vole remains recovered from kestrel pellets is still higher than would be expected from Hill's independent surveys of rodent numbers: which suggested that field voles were completely absent from the closely grazed acid grasslands of woodland rides within the Forest. Most of the kestrel ranges from which pellets were collected in the Forest, however, contained roadside verges or railway embankments, which would provide habitat suitable to support at least small *Microtus* populations; in addition, many of the kestrels extended their range over adjoining farmland, which would provide an alternative source of voles as prey.

Overall, mammalian prey constitutes only 30.5% to the weight of vertebrate prey taken by New Forest kestrels (Table 8.5), as against the 73% estimated by Yalden and Warburton (1979). This shortfall is made good by a greater reliance on birds and by extensive predation on common lizards (*Lacerta vivipara*), which occurred in 45% of all pellets analysed (Table 8.5). Such heavy predation on lizards (which occurred in 59% of all pellets in April/May, when they spend much time basking in the open) appears to be a curiosity of the New Forest (in the study by Yalden and Warburton, only 12 lizards were recovered from the 219 pellets examined) and is probably a direct response to low densities of small mammals in open habitats within the Forest.

Tawny owl territories have been mapped in various areas within the Forest. Rather strikingly, there are no significant differences in average territory size between woodlands inside and those outside the Forest (Table 8.6), and territory sizes are similar to those found in deciduous woodland elsewhere (e.g. Southern 1969). Such a finding fits well with the previous discovery that the diet of New Forest owls is also essentially

Table 8.5: The frequency of occurence of different prey in kestrel pellets collected in the New Forest in 1982-3 (from Hirons 1984)

Period	April-May	June-July	Aug-Sept	Overall
No. of pellets examined	104	57	29	190
Prey	No. (%)	No. (%)	No. (%)	No. (%)
Wood mouse	2 (1.9)	3 (5.3)	1 (3.4)	6 (3.2)
Bank vole	2 (1.9)	3 (5.3)	—	5 (2.6)
Field vole	28 (2.0)	16 (28.1)	1 (3.4)	45 (23.7)
Harvest mouse	19 (18.3)	1 (1.8)	5 (17.2)	25 (13.2)
Pygmy shrew	7 (6.7)	6 (10.5)	—	13 (6.8)
Common shrew	22 (21.1)	8 (14.0)	3 (10.3)	33 (17.4)
Small mammals				137[a] (72.1)
Birds	27 (26.0)	8 (14.0)	2 (6.9)	37 (19.5)
Common lizard	61 (58.6)	17 (29.8)	8 (27.6)	86 (45.3)
Earthworms	1 (1.0)	—	—	1 (0.5)
Lepidoptera	—	—	24 (82.8)	24 (12.6)
Coleoptera	69 (66.3)	26 (45.6)	20 (69.0)	115 (60.5)

Note a. This figure includes unidentified mammals indicated by counts of upper incisors or fur.

Table 8.6: Mean territory size of tawny owls in different areas

Area	Territory size (ha)	No. of pairs
New Forest (Inclosure)	23.4 ± SE 1.9	11
New Forest (Open Forest)	17.5 ± 2.1	10
Outside Sides in Hampshire	17.9 ± 1.9	12
Wytham (Southern 1970)	18.2 ± 2.1	31
Forest of Ae (Hirons 1976)	46.1 ± 7.1	12

not dissimilar to that of tawny owls in general in that they are able to make up from 60% to 70% of their diet from small rodents, much as do owls elsewhere (page 173). Tawny owls are essentially woodland birds. We have already noted that New Forest woodlands support populations of wood mice of equivalent density to woodlands in areas less heavily grazed: it is thus not too surprising to find the owls behaving much as do owls elsewhere. But there is a difference. Hirons (1984) monitored breeding success of the owl populations through two years (1982 and 1983). In the early spring of 1982, rodents were scarce all over the country following a severe winter and a failed mast crop. Whether inside or outside the Forest, most owls did not attempt to breed at all, and no difference in breeding success could thus be detected between areas. In 1983, however, when breeding was attempted, the proportion of owl pairs successfully rearing young was much higher in ungrazed woodland outside the New Forest (65%) than within the Forest (25%).

Owls are woodland birds and, of all the Forest communities, woodlands showed the least difference in rodent densities from ungrazed areas outside the Forest; in more open habitats of the Forest, small rodents were almost totally absent. In 1982 the density of breeding kestrels was only one pair per 16 km^2; the density was lower still in 1983 (Hirons, pers. comm.). This compares with a figure of one pair per 4 km^2 in farmland outside the Forest.

Chapter 9

The New Forest — Present, Past and Future

The Present

Our studies over the past few years have shown how closely the various animals of the Forest community interact with one another and with their common vegetation, and how dominant are the effects of grazing at all levels of the system. Ponies, fallow, sika deer, and almost certainly roe — although as yet we have not the data to prove it — adjust their diet so that at any season they are feeding on those foodstuffs which offer maximum availability of nutrients, particularly digestible nitrogen (Chapters 4 and 6). Foraging behaviour is a major influence on patterns of habitat use: habitat selection (preference) and observed occupancy of different communities reflect the changing choice of foodstuffs and their changing availability (e.g. Chapter 3). Different vegetation types suffer markedly different pressure of use, and within communities seasonal changes in foraging are reflected in different levels of grazing pressure at different times (Chapter 7).

Animals' feeding behaviour and habitat use are affected not only by the vegetation. Through their own effects upon the vegetation they influence each other. Thus cattle and ponies, both preferential grazers, divide between them the different grassland types — and different zones within any common grassland (Chapter 3). Fallow, roe and sika deer show clear overlap in diet at certain times of year, and in seasons of food shortage potentially compete; throughout the year sika are clearly affected in terms of diet by interaction with ponies (Chapter 6).

The combined impact of all these larger herbivores upon the Forest's vegetation is colossal: the effects are dramatic (Chapter 7). Grazing acts to eliminate particularly favoured or sensitive species from a community. Species diversity falls in woodlands and open vegetation types alike: grassland swards are remarkably species-poor, dominated by a few resistant species. Physical structure is altered: in woodlands, heathlands or grasslands there is little or no vegetation between 0.1 and 1.8 m — the whole structural layer is missing. Grazing prevents regeneration: the age structure of Forest trees reflects past periods of lesser and heavier grazing (Peterken and Tubbs 1965), and current grazing pressure *still* prevents regeneration within heavily used woodlands. Finally, productivity and nutrient status of the vegetation is affected. In theory, relatively light grazing may stimulate productivity (e.g. McNaughton 1979), while heavier grazing pressure will reduce vegetative production. In the New Forest, all vegetation types surveyed are so heavily grazed that productivity is suppressed: no sites showed increased production as a result of grazing (Chapter 7). At the same time, alteration to patterns of nutrient cycling within and between communities, combined with the fact that the greater part of a rather limited nutrient pool is bound up into animal tissue, result in continued impoverishment of soils subjected to heavy grazing.

It is perhaps appropriate to examine the more detailed results of our analyses in relation to comparable figures which are available from similar studies elsewhere. Bobek *et al.* (1979) estimated that red and roe deer in deciduous forest in southern Poland removed 46 kg/ha of browse material, some 28% of annual production; herbivore consumption of holly (*Ilex aquifolium*) in the New Forest, at approximately 7% (Table 7.4), is far lower and clearly has little immediate effect upon the holly itself. Offtake of deciduous browse species today, is low, but this should not be interpreted as of low impact: the reason why offtake is so small is that no understorey browse species other than holly persist within the Forest in any great abundance, having been eliminated by browsing pressure in the past. Impact of grazing on New Forest grasslands is colossal, with between 70% and 130% of growing-season production removed by grazing herbivores almost as soon as it is produced, far in excess of that recorded in the literature for almost any other *natural* grazing system. The open grasslands of the East African plains support a diversity and abundance of grazing ungulates perhaps unrivalled anywhere in the world, yet equivalent estimates for offtake (Wiegert and Evans 1967) suggest that from 30% to 60% of primary production only is removed.

This tremendous pressure on New Forest grasslands is of course a function of the distinct habitat preferences shown by the herbivores and the extremely uneven pattern of grazing pressure that results. Thus, some 50% of all animal grazing time is concentrated on grasslands which together comprise only 7% of the available Open Forest area (page 63), while other communities such as heathlands, despite their enormous acreage within the Forest, sustain but little grazing pressure (less than 9% overall of the pressure taken per m^2 of grassland). The net effect on the grasslands is a dramatic decline in productivity. Typical daily offtake for the improved grasslands of the Forest during the main growing season (May-July) is calculated as 3 g/m^2 per day, and conservative estimates suggest that over-grazing suppresses production of these grasslands to less than half of their assumed 'ungrazed' yield.

An increase in grazing intensity on sheep pastures in Australia (Vickery 1972) resulted in an increase in production up to stocking rates of 20 sheep per hectare. Production was least at ten sheep per hectare, and declined again with stocking densities higher than 20. A similar pattern of initial increase in production followed by gradual decline is well documented by other authors. McNaughton (1979a) presents a detailed analysis of the relationship between grazing and above-ground primary production for the Serengeti. It is notable that in the wet season (=main growing season) production increases only slightly with light grazing, up to offtake levels of 5 g/m^2 per day.[1] Stimulatory effects of light grazing were, however, never demonstrated by our studies of New Forest grasslands, and all levels of grazing observed by us reduced yield.

It is clear that patterns of dunging and urination also affect vegetational dynamics. Crawley (1983) quotes Harper in noting 'that to be covered by a cow pat is disaster for most plants; if they are not killed by the darkness or by the concentrated nutrients, they may have their living space usurped by creeping plants which grow over the top of the dung' (and others whose seeds may have been deposited in the dung itself) (Harper 1977). A cow pat has a long-lasting impact upon the vegetation, both because fouled areas are avoided and thus released from grazing and because a different community of plants establishes in the nutrient-rich conditions. But more significant perhaps is the effect that herbivore *behaviour* may have on nutrient distribution over the area as a whole: by grazing in some areas and defaecating in others, the herbivore may cause major discontinuities and translocation of nutrients (pages 6-7). Within the New Forest, clear redistribution of nutrients occurs on the Forest grasslands: the grazing and eliminative patterns of ponies and cattle result in a complex mosaic of patches

[1] Although this is higher than the figure quoted for offtake from New Forest grasslands (3 g/m^2 per day), values for forage production were also *extremely* high: consistently above 20 g/m^2 per day (McNaughton 1979a).

within the community. There are marked differences in the nutrient status of pony-grazing areas and cattle-grazing areas (= pony latrines), with continued loss, particularly of potassium and phosphorus, from the main pony-grazing areas (page 162).

These differences in nutrient status *and* differences in grazing pressure and 'style' cause noticeable differences in the species composition of these areas as well (Table 7.14). Other changes in species composition can be noted in grasslands and woodlands by comparison with equivalent sites beyond the Forest boundary and subjected to lower grazing pressure, or by comparison with historical records of past species structure. The main changes are recorded in detail elsewhere (Chapter 7) and there is no need to repeat them here. It is clear, however, that the changes in species composition and diversity of the different Forest communities are due to (i) altered nutrient status, (ii) grazing elimination of preferred or particularly sensitive forage species, and (iii) relief of competitive pressure on more graze-resistant or graze-tolerant species. Within woodlands, loss of more palatable species such as willow, hawthorn, blackthorn and birch shows clear parallels with the effects of grazing by cattle and sheep in deciduous forests in Europe (e.g. Adams 1975). In the patchy flora of the woodland floor, grasses increased in both abundance and diversity where grazed (Denny pens experiment, page 156). Similar results were shown by Spence and Angus (1971) in an analysis of the effects of large herbivores on *Combretum-Terminalia* woodland in Uganda: in plots protected from grazing, palatable species of shrubs and trees, such as *Desmodium* and *Indigofera* spp., increased from an abundance of perhaps 40 plants per hectare to nearly 654 per hectare; in the same plots, abundance of grasses such as *Sporobolus pyramidalis*, *Hypparhenia filipendula* and *H. rufa* declined when grazing pressure was removed.

Heavy grazing of New Forest grasslands causes a marked decline in diversity and a clear shift in species composition. Sensitive species such as clovers (*Trifolium pratense* and *T. repens*) are eliminated from the sward; there is a dramatic increase in the dominance of resistant grass species (notably *Agrostis tenuis*) and more prostrate, rosette-shaped forbs, which by their growth form are better able to escape severe defoliation (Table 7.13). Exactly comparable changes are recorded for grasslands in the Seregenti by McNaughton (1979), who notes that the shift in species composition between grazed and ungrazed areas is the result of replacement of tall, stemmy species, abundant in areas where grazing is prevented, by prostrate, low-growing species resistant to grazing. McNaughton suggests more generally that there is an adaptive trade-off among the species between ability to sustain sufficient leaf area and photosynthetic potential under intense grazing (facilitated by a prostrate growth form) and upward growth required in competing with other plants for light and space (facilitated by investment in stem production). In the Serengeti system studied by McNaughton, species like the *Pennisetum*, with a heavy investment in stems, are very susceptible to damage by grazing, but win at light competition when grazing is reduced (page 5); by contrast, prostrate species like *Andropogon greenwayi*, quickly out-competed in the battle for light in ungrazed swards, dominate the community in areas more heavily grazed.

Within the more immediate context of the New Forest, it is particularly interesting to consider the fate of those lawns re-seeded in the 1950s and 1960s with results of studies by Jones (1933) on the long-term effects of grazing sheep upon sown pastures of different compositions at Jealott's Hill in southeast England. These experiments are peculiarly relevant because they demonstrate so clearly the effects of different grazing patterns: pastures sown with exactly the same seed mix and thereafter managed according to exactly the same cultural practice were simply subjected to different grazing regimes.

In one trial where a rye-grass/clover (*Lolium perenne/Trifolium repens*) mixture was sown, changing only the timing of grazing produced an almost pure rye-grass sward in one case (6% clover) and a predominance of clover (62%) in the other. This was brought about by grazing regimes which tipped the competitive balance in favour of the grass in the first case and of the prostrate-growing clover in the second. To produce a rye-grass sward, the animals were kept off the pasture in early spring so that the rye-grass tillers produced vigorous root and shoot systems. When the sheep were introduced to the pasture, the grass plants had substantial reserves of root and shoot and were able to recover rapidly from defoliation. Despite the attentions of the sheep, they were able to over-top and out-compete the clover plants. The clover sward was created by introducing the sheep to the grassland very early in the year (during March and April), so that the first rye-grass leaves were grazed off as they developed. The sheep were removed before they grazed down to the lower clover plants. By the peak growth period in May and June the clover plants were in fine condition, but the rye-grass had poorly developed root and shoot systems, and suffered from the combined effects of defoliation and competition from the clover (Jones 1933). Clearly, management has a profound effect on the results of grazing: the New Forest grasslands perhaps represent a third permutation of Jones's experiments, one where grazers were permitted early in the year but not 'removed before they grazed down to the lower clover plants'!

In the long term, heavy grazing at this level can affect successional processes, and ultimately result in vegetation loss and desertification. Certainly within the Forest, current and past levels of grazing have been sufficient to prevent significant regeneration of woodland; yet other communities, though somewhat 'distorted' by the continuing pressure from herbivory, seem none the less relatively stable.

The Forest — Past

All the effects of grazing upon vegetation described are markedly influenced by *intensity* of grazing. Over its 900-year history, the Forest has seen marked fluctuations in the numbers and type of grazing stock (Table 4.1); such changes in grazing pressure will have meant differences in 'immediate' ecological function of the Forest system at different times — and also leave a legacy for the future. The New Forest today is affected strongly by current grazing pressure, but is also markedly influenced by its past.

As noted in Chapter 2, deer, and particularly fallow deer, were a significant and probably fairly constant pressure on the Forest from its establishment in the eleventh century until the Deer Removal Act of 1857, perhaps having been present in numbers up to 8,000 or 9,000. From then numbers suddenly declined, and have only recently regained their current levels of about 2,500. Changes in the numbers of Common stock grazing on the Forest are harder to determine. From the eleventh to the eighteenth centuries there must have been considerable populations, but no formal records of animal numbers exist prior to 1789. Since that time there have been dramatic fluctuations in overall numbers as economic climates have changed; and, more subtly, there have also been changes in the relative balance of the two main species depastured, cattle and ponies.

Between 1789 and 1858, marking fees were paid for between 2,000 and 3,000 ponies, but for up to 6,000 cattle. In the latter part of the nineteenth century, numbers of cattle and ponies depastured were more nearly equal (between 2,000 and 3,000 head). From then onwards numbers of cattle have continued to decline; numbers of ponies have increased steadily until the late 1970s (with exceptional decreases of all stock during the war years). Thus, there has been a shift in commonable stock from cattle — bulk-feeding ruminants, which lack the ability to crop close — towards monogastric horses, which crop close, digest poorly and rely on a rapid throughput of forage. Among the cattle that remain, there has also been a shift from predominantly dairy herds towards beef and store cattle. Initially cattle turned out on the Forest grazings were predominantly dairy animals, with many Commoners turning out a few 'house' cows and heifers. More recently this practice has declined, while a few Commoners have started to run relatively larger herds of store cattle upon the Forest. Such herds are more likely to be kept upon the Open Forest throughout the year (with the aid of supplementary feed), and this may ultimately lead to a more significant effect upon the vegetation. Changes in the relative balance of cattle and ponies, or store and dairy cattle, closely reflect changes in market prices. At the same time they are influenced by a social change in commoning. Over the years we have witnessed a shift in use of Commons, from supporting a subsistence cottage economy to supporting a form of commercial livestock agriculture.

Such social and economic changes are clearly of tremendous significance in the Forest's ecology. Common stock *now* represent the most significant grazing influence upon the Forest and their changing fortunes, past and future, will clearly have a powerful effect upon the Forest.

Colin Tubbs has done more than any other to tease out the social history of commoning in the New Forest, painstakingly picking over hundreds of reports, notebooks, official and unofficial records, and piecing together the few scraps of information culled from so many sources to try and gain a picture of the changing fortunes of Common agriculture in relation to the changing ecology of the Forest itself (Tubbs 1965, 1968). His study is a masterpiece of patient historical detection, in researching the records and in interpretation; it seems hard that we may now draw on the results of all his labours and initiative so painlessly.

Although records before the initial Registers of Claims of 1635 and 1670 (page 23) offer only fragmentary evidence about commoning in the years leading up to the seventeenth century, it is clear from what evidence is available — presentments at Forest Courts in respects of abuses of Common Rights, or petitions from Commoners alleging unreasonable infringement of these rights — that the area supported a vigorous pastoral economy. Sheep, cattle and ponies were grazed upon the Forest from the thirteenth to sixteenth centuries. Some of the biggest graziers were the large landowners — the big religious houses at Beaulieu, Christchurch and Braemore — but at the same time it is clear that *all* Rights of Common were also extremely important to occupiers of smaller holdings: cottagers who could make a subsistence cottage economy considerably more comfortable by exercising their rights to fuel, animal bedding and fodder for their few livestock. With the dissolution of the monasteries in the middle of the seventeenth century, and the accompanying changes in land ownership, patterns of commoning also began to shift. In the Registers of Claims compiled in 1635 and 1670, most claims (whether entered individually, or as part of a 'block application' by a large estate on behalf of many tenants) were in respect of smaller holdings. The Register shows three main classes of people exercising Rights of Common over the Forest: Lords of the Manor with land within or adjacent to the Forest; his tenant farmers and copyholders; and a class of freeholders or yeomen with their own small farms. Most claims were in respect of cottagers with only a few acres; even the holdings of tenant and yeomen farmers were generally smaller than 50 acres (20 ha) and more usually between 15 and 30 acres (6-12 ha). Now, too, Commoners were exercising their rights not just in order to support a cottage economy, but to support agriculture; Tubbs notes that the majority of holdings were too small in themselves to be self-supporting, and we may deduce that particularly grazing rights were being used to support an economy of small livestock farmers, entirely reliant on the use of Common lands.

The picture remains essentially unchanged throughout the eighteenth and nineteenth centuries. The majority of claims submitted for Rights of Common were for freeholdings of less than 80 acres (30 ha). A number of claims were entered by large estates, but once again each of these submitted a general claim on behalf of numerous tenants who occupied holdings usually of less than 50 acres. Of claims allowed in 1858 for freeholders, 629 were for properties of less than 30 acres, and 400 of these for cottages with between 1 and 4 acres (0.4-1.6 ha). In 1944, Kenchington quoted the total number of holdings with Common Rights on the Forest as 1,995; of these, 731 were below 5 acres (2 ha) and only 396 were above 50 acres. Kenchington notes that even the tiny holdings were equipped not as smallholdings proper but as miniature farms. Two-thirds of the holdings represented the main source of livelihood for the occupiers. Clearly, the shift in use of Common Rights away from merely assisting a cottage economy towards supporting grazing agriculture on a large scale, a shift that we identified even in the seventeenth century, had continued still further. Evidence given before Select Committees in 1868 and 1875 noted that 'The Right of Common of grazing enabled a Commoner to maintain three times as many cattle as he would be able to without that right.'

Numbers of cattle and ponies depastured on the Forest today are as high as those of the peak years of the 1880s, but the actual number of Commoners exercising grazing rights is considerably smaller. In 1858 1,200 holdings were registered; today Rights of Common are exercised only by a few hundred. A survey published recently by the Countryside Commission (1984) allows more objective analysis of the social and economic context of commoning today. The picture is still one of many small farmers with small properties turning out small numbers of stock, rather than a Common dominated by a few landowners turning out huge herds. The majority of commoners (75-80%) turn out 20 head of stock overall, or less, onto the Forest grazings; 52% of all practising Commoners work properties of less than 20 acres, and only 8% own more than 50 acres (Tables 9.1, 9.2). The pattern of grazing has not changed — but the social context has. Only 10% of the holdings are now worked full-time, and, for 50% of Commoners surveyed, their 'agricultural' enterprise contributes less than 10% of total income. More fundamentally, only 44% of existing Commoners are agricultural or manual workers within the Forest; 24% are professional or clerical. Commoning, which was once an intricate part of the livelihood of the New Forest, 'has become basically a grazing option offering an interesting way of life and supplementary or seasonal pasture to a few small-holding Commoners' (Countryside Commission 1984).

In summary, it would appear that the socio-economic status of commoning has changed more markedly in the last few years than over seven centuries before that. From the thirteenth to sixteenth centuries, Common pasturage was exploited by two very distinct classes: by big landowners, for free-range pasturage of large herds of grazing stock; and, almost at the opposite extreme, by cottagers with a small plot of

Table 9.1: Sizes of herds pastured by today's Commoners (from Countryside Commission Report 1984)

	Commoners involved	
Size of herd put out	**Number**	**Percentage of all Commoners**
0-25	298	84
26-99	49	14
100+	6	2
Totals	353	100

Table 9.2: Size of holdings owned by present-day New Forest Commoners

Acreage	Percentage of Survey
< 1	8
1-5	21.3
6-10	12.0
11-15	5.3
16-20	5.3
21-30	2.7
31-50	9.3
51-100	5.3
> 100	2.7

land, as additional grazing for their few domestic animals. Through the seventeenth, eighteenth and nineteenth centuries, there was a gradual shift in the use made of the Forest Commons; Common Rights were used primarily to support small farming enterprises, allowing the owners of relatively small holdings to practise as economic livestock farmers on a scale far beyond what would be expected of their limited acreage, because of the extra grazing offered by the Open Forest. This, in part holds true today. Much of the property around the edge of the Forest, together with those pockets of agricultural land within the Forest boundary, consists of small or comparatively small holdings (mostly still less than 50 acres). Yet many of these are run as miniature farms, whose profitable management rests largely on exercise of their Common Rights. Yet things *have* changed: commoning is less central to the life of all but a few of today's Commoners, and the 'social' context of commoning is clearly far removed from that of earlier centuries.

The extent to which Common Rights other than those of pasturage have been exercised has also changed over the years, and the changes reflect in part the increasingly 'agricultural' use of Common grazing. Through the thirteenth to seventeenth centuries, every claim for recognition of Commons was for all rights — pasture, mast, estover, turbary and marl; where Commons were supporting subsistence economies, these other rights made a considerable difference in reducing the costs of running a home. But, as the relative balance of different 'types' of Commoner changed, with the majority using Common pasturage to support their holdings more and more as miniature farms, so the direct value and importance of the other Common Rights declined. Various other factors also reduced the actual economic value of such concessions. Successive Acts of Parliament restricted the quantity of fuel each commoner might take under Rights of turbary and estover; Rights to estover were bought out by the Office of Woods as deliberate policy in order to try to check unbridled exploitation of the Forest timber; and the importance of all concessions dwindled as social circumstances changed, the Forest was opened up and access improved by road and rail.

From the thirteenth to the seventeenth centuries all claims for Common Rights included claims for estover, turbary and marl; by the end of the nineteenth century, Rights of marl and estover were little exercised. In the 1858 Register of Claims 1,500 holdings claimed Right of turbary; by the turn of the century even these rights were rarely claimed. Turves are cut by fewer than a dozen holders of turbary rights in the Forest today, and Rights of estover and marl are exercised only sporadically (Table 9.3).

The Future?

Clearly, both the social context and the economics of commoning are changing, and fluctuations in the numbers of stock pastured on the Forest will continue unpredictably as markets rise and fall. But Commoners' animals will probably always remain a significant influence on the ecology of the Forest; indeed, the maintenance of some level of grazing is absolutely essential if we are to retain the peculiar vegetational structure of the Forest as we know it. In addition, commoning has more than a purely ecological relevance; it is in itself a curious social and historical phenomenon, worthy of conservation in its own right. So, what of the future? How can the Forest continue to offer all things to all men — continue as an economic silvicultural enterprise, support Common agriculture and maintain its unique ecological value? What should the balance of interest be, and how should it be maintained?

Table 9.3: Current use of Common Rights (from Countryside Commission Report 1984)

	Number
Pasture	73
Mast	23
Pasture (sheep)	5
Turbary (turf cutting)	9
Estovers (fuel wood)	8
Marl	4

It is frequently claimed that with its present densities of Common stock the Forest is currently 'heavily overgrazed'. Over-grazing is an emotive but rather nebulous concept. Do we mean that grazing pressure on the Forest is so high that animals are no longer able to maintain condition, or that from a vegetational point of view grazing is such that productivity is suppressed? Or do we mean that grazing pressure is causing a decline in the 'ecological value' of the area through reduction in diversity of species composition and structure, or that it is preventing woodland regeneration? According to which criterion we adopt, we may feel that certain areas of the Forest may or may not be considered overgrazed.

Equally importantly, different community types may differ in the degree of 'over-grazing' experienced. We have stressed already that grazing animals do not use all habitats equally and that some sustain much higher impact than others (Figure 7.2). Further, different patches of the same community type suffer different grazing pressure dependent on their juxtaposition with other communities or geographical position within the Forest. Thus, certain areas of the Forest or certain specific communities may indeed be overgrazed, while at the same time others could be claimed to be *under*grazed.

The maintenance of some level of Common grazing (with, in consequence, some degree of vegetational management aimed at supporting the grazing stock) is clearly a crucial factor, but what should that level of grazing be in order to create and maintain ecological diversity and not diminish it, and how far should vegetational management be directed at improving the grazing for Common stock?

From the data we have presented so far, it is possible to attempt to calculate a crude estimate of the Forest's capacity to support grazing livestock. Such calculation is blatantly agricultural in its approach and considers the capacity of the Forest's vegetational resources to support stock; such a stocking density may *not* be that which would be 'desirable' in terms of maintenance of ecological diversity. None the less figures for required forage offtake per day per individual cow or pony may be related, in theory, to measured forage productivity, in order to determine the potential stocking density of the Forest for Commoners' animals and other herbivores.

Of course, such a calculation is grossly oversimplified. As discussed elsewhere (de Bie 1982, van der Veen 1979), one cannot expect merely to divide total annual forage production of an area by required animal offtake and come up with a figure for environmental carrying capacity. Forage availability is not the only factor affecting animal density;

animal numbers may be limited by other factors such as availability of shelter, etc. Even within the forage component, forage availability — and thus carrying capacity — changes in different seasons. Nor is 100% of forage produced available to the animals; a proportion may be inedible or unpalatable, while even whole areas of potential forage may also be relatively under-used because they lie outside the animal's normal range. In any case, animals do not require just a 'total energy intake' irrespective of how it is derived, but may require different things from different foodstuffs, and may require a specific mixture in the diet (and are thus limited not by total forage availability, but by the availability of the most limited of these resources). Other major factors affecting carrying capacity are purely behavioural. Animal densities may be restricted because of considerations of range use and social organisation. Some of these factors may be taken into acount. Our data for the New Forest present information for seasonal changes in productivity, while the fact that 'not all vegetation types are alike' to the animals may be accounted for in that seasonal productivity is calculated separately within each vegetation type and may thus be related directly to figures for required offtake also derived for that particular vegetation type. With carrying capacity thus derived in any season in terms of that vegetation type offering minimum supply in relation to demand, full account is taken of the specific requirements of cattle and ponies from their different forage types in areas of mixed vegetation (Putman *et al.* 1981).

Calculations are presented here, by way of example, for New Forest ponies. For each two-monthly period of our vegetational measures we have a figure for the productivity of each community type, or for forage species within them. For the same two-monthly period we have a figure for amount of offtake required per animal *from that particular community type,* or *forage species* (calculated as offtake of that community type per grazing animal, or total intake from that community: data from Chapter 7). For each season, it is then possible to calculate *total usable productivity over the whole Forest* for each community type/forage species by multiplication of its productivity per unit area by known available area of that forage type in the Forest as a whole (Table 3.2). This can be divided by total offtake required per animal of that particular community/forage species in that season; the lowest figure (for productivity/requirements of animals from that particular community) for whatever community must represent the maximum numbers of animals the Forest can support at that season, even if other communities show surplus productivity (Table 9.4).

Table 9.4: Carrying-capacity calculations: total production of each vegetation type over the Forest as a whole is divided by required offtake of the same vegetation type per pony to derive an estimate of the potential number of animals supported (see text, page 190)

	VEGETATION TYPE				
Month	Re-seeded lawn	Streamside lawn	Commoners' improved grassland	Acid grassland	Woodland glades
January	—	113	139	182	—
February	? 47	424	423	392	—
March	1,228	164	—	—	—
April	1,310	493	—	—	—
May	594	218	376	290	—
June	832	384	622	1,812	116
July	1,510	556	770	1,503	344
August	892	450	244	510	133
September	459	198	280	430	8
October	152	172	66	434	—
November	—	186	—	—	—
December	—	203	—	—	—

(While this calculation does take into account seasonal changes in forage productivity and offtake and also allows for differing importance of different forage types, we are to use in our calculations total vegetational area of a particular community calculated for the Forest as a whole. We are not able to take into account the fact that not all areas of the same vegetation type are used equally intensively, that some are perhaps used hardly at all owing to the fact that they do not fall within the home range of many animals. Nor can we take into account the effects of social interaction; or the fact that not all *apparently* available forage may be acceptable or palatable. Our calculations thus assume that all forage produced within the Open Forest is (a) available to the animals and (b) equally available to them. With these reservations, then, we may still view the calculation as some form of approximation.)

Before we even examine the results of such calculations, certain facts are immediately apparent from raw figures for vegetational production and offtake. Available heathland, both wet and dry heath, far exceeds the extent of animal use of heather or other heathland plants. Even on wet heath, which becomes an important community in midsummer, when grass is in short supply, production consistently exceeds offtake. Similarly acid grasslands, woodland glades and woody browse are all in superabundance within the Forest. By contrast, there is severe competition for the favoured communities and forage types. Demand for brambles and gorse exceeds supply in certain areas in appropriate seasons, while offtake from all 'sweet' grassland communities (RL, SL, C1) consistently equals or exceeds production, even to the extent that in these communities productivity is depressed. These, then, are the vegetation types limiting the densities of animals upon the Forest.

Examination of Table 9.4 would suggest that, in fact, acid-grassland communities and woodland glades may also be potentially limiting after all; we must, however, remember that carrying capacity was estimated by relation of offtake to *production*, rather than production + standing crop (because measured standing crop in any one month is so close a function of *current* grazing pressure). We know from other elements of the study that in both woodland glades and acid grassland there is a high standing crop the year round: carrying capacity is therefore probably considerably higher than might be suggested by considerations of production alone.

Where true availability *is* more closely related to forage production (RL, CI, SL) it is clear that resources are indeed limited, as we have concluded from other evidence. No one of these communities, however, is likely to be limiting in its own right: the animals are well able to use them as alternatives. Thus, *if* we assume from these, and other, considerations that grasslands are likely to be limiting to stocking density, and *if* we assume that our calculations here are reasonable, we might derive a final, seasonal figure for carrying capacity as the sum of the separate capacities of RL, SL, CI (Table 9.5).

Such a calculation is fraught with assumption, and compounds many sources of error. Moreover, any conclusions drawn from it *assume* that the animals will continue to display in the future, and under altered conditions, the same pattern of community use as they do at present, for the calculations are based on current patterns of community use. The calculation, while of interest, is thus *highly* speculative and its conclusions should be interpreted with caution!

Belief that the Forest is currently overstocked (in relation to Commoners' agricultural interests, and thus in the sense of its capacity to support stock) has repeatedly led the Commoners to press for management aimed at increasing the available grazing on the Forest. The grazing improvements traditionally sought include clearance of scrub and gorse from grasslands, and draining of bogs and streamside lawns. Such measures clearly run counter to conservation interests within the Forest, threatening structural diversity, threatening the integrity of whole communities, and if carried through would perhaps shift the delicate balance of conflicting interests in the Forest too heavily towards commoning.

In the past, there was a real need for Common Rights — and a proper use of those rights — in sustaining the livelihood of the cottagers and smallholders of the Forest. It was appropriate at that time that the

Table 9.5: Final estimates of carrying capacity for ponies

Month	Stocking capacity
January	—
February	c 900
March	1,800
April	2,200
May	? 1,200
June	1,840
July	2,840
August	1,600
September	940
October	400
November December	uncertain: 200-300

interests of commoning should be upheld and that those interests should be taken into account in structuring management policy. Now perhaps, with the new social role of commoning, there is not the same justification if commoning conflicts with other interests. But, in any case, the measures suggested could in practice prove counter-productive even to the Commoners themselves. We have noted that both cattle and ponies are both socially and ecologically so 'hefted' to the Forest grasslands that they are reluctant to leave them even when other communities offer more food. This remains true in autumn and winter, when animals remain on the exposed Forest lawns even in appalling weather. Gorse brakes and scrub offer some degree of shelter in such areas — and may be crucial in preventing excessive heat loss in the worst of winter weather. The effects of draining bogs and streamside lawns are equally dubious. The beneficial effects of periodic flooding, which flushes the ground with nutrients, make streamside lawns among the most productive communities in the Forest; they also begin to offer fresh grazing slightly earlier in the season than the 'dry' grasslands. The precise changes in vegetation following drainage of valley bogs are more difficult to predict because of wide variation in wetland conditions. Nutrient-poor bogs dominated by purple moor-grass (*Molinia caerulea*) and cottongrass (*Eriophorum* spp.) growing on oligotrophic peat are likely to develop an extremely short and scanty turf composed largely of *Carex* species if drained and heavily grazed. It is doubtful whether such vegetation represents an improvement in food resources, since the total yield of such areas is probably lower than in the bogs, where *Molinia* grows luxuriantly. Further, the effect of both drainage proposals is to increase the availability of grassland. By reducing the diversity of community type, asynchrony of production is also diminished. All 'drier' grassland types show peak production from May to July: at a time when in fact there is alternative forage in abundance on the wetter heathlands. Drainage of bogs and streamsides merely increases vegetational production in the only season of relative abundance: useful supplies of winter forage provided by *Juncus* spp. and the early bite provided by streamside turf or the *Juncus* and *Molinia* of bog communities will be reduced.

Such comments are somewhat conjectural. We can be more objective about the likely effects of other potential forms of direct vegetational management. For the most part, we have considered the entire herbivore-vegetation complex in these pages as if it were a completely natural system. Yet many, if not most, of the vegetational communities are already man-managed, or at least susceptible to such management. Effects of the various management practices adopted in the past may be evaluated — in terms of their effects on the vegetation itself *and* on the various grazing herbivores — and used to predict the likely response to future management.

Improved grasslands of various types prove among the most attractive grazing areas for both cattle and ponies, and indeed for the fallow deer of the Open Forest. As a result, they are somewhat over-utilised: almost all production is used immediately, and both productivity and nutrient quality of the areas have been significantly reduced by the continuous heavy grazing pressure. The animals are offered an increasingly impoverished food supply; and, in response to both heavy grazing and poor soil nutrients, botanical diversity of the areas is minimal (Table 7.13). The original seed mixtures used for re-seeding contained several species of herbage plants of which perennial rye-grass and wild white clover were usually the most important (page 43). Under conditions of continuous heavy grazing, these valuable herbage species were quickly eliminated and replaced by various less desirable native species, of which *Agrostis tenuis* is predominant. (It is worth noting, however, that the most successful of the sown species was red fescue cultivar: S 59, and where this was sown it retains a cover of around 22%.) Many of the problems outlined above are a direct result of hard, continuous grazing. If access to re-seeded areas were controlled, it would be possible to manage these pastures to achieve very substantially increased yields. It would also be possible to alter the balance of plant species, favouring more productive and nutritious species such as perennial rye-grass and white clover (see also page 198).

The improvement of several acid-grassland sites in the early 1970s through swiping of bracken and liming has probably been the most cost-effective of all grassland management. The removal of tall bracken has opened these areas to grazing animals during the summer months, and many of them are now heavily used by livestock. The applications of lime raised the soil pH significantly (from about 4.5 to about 5.7), and was probably the cause of a marked reduction in the highly unpalatable bristle bent grass *Agrostis setacea.* The higher pH, coupled with closer grazing, may also explain the increased plant-species diversity of these sites compared with unaltered acid grassland.

The yield of herbage on the improved acid grassland is as high as, if not a little higher than, that of most re-seeded lawns. There is, however, a possibility that with increased use they will experience the same depletion of nutrients that has occurred on re-seeded areas as a result of the grazing and dunging patterns of ponies. It will be interesting to see whether they maintain their present levels of productivity over the next few decades.

The other notable vegetation types markedly affected by management are the dry heathlands of the Open Forest, which are burnt or cut on a regular rotation (page 45). Our studies demonstrate that only a small proportion of heather is taken by large herbivores, and that ponies in particular take very little. Most heather browsing occurs in the immediate vicinity of lawns and roadside verges. Burning or cutting of heather does, however, significantly promote the growth of *Molinia* for a few years, and this provides a grazing resource which is particularly valuable at the beginning of summer before the grasslands are fully productive.

Most of these options for management of the vegetation are aimed at maintaining densities of grazing herbivores at current high levels. Many of the measures would conflict directly with conservation or other interests; more fundamentally, it is arguable that current stocking levels of herbivores are already too high. It is apparent that, at least in some habitats and in some areas, grazing really is excessive — by *whatever* criteria.

Control of animal densities is, however, difficult. The Forestry Commission attempts to maintain numbers of the various deer species at appropriate levels, by careful census and suitable culling; the Verderers, too, have the right to restrict the numbers of Common stock depastured. Clearly, the most obvious management possibility is to restrict *overall* the numbers of animals pastured on the Forest. Restrictions could be absolute (set by considerations of carrying capacity at that time when it is at its minimum), or more flexible (different densities at different seasons, or in different geographical areas of the Forest so far as any of these are relatively defined entities). Such restrictions, however, would be difficult to impose and still more difficult to administer. In addition, we can offer here no clear idea of what may be the carrying capacity of the Forest.

The alternative strategy is to alter in some way the *distribution*, both geographically and between different community types, of existing numbers of animals on the Forest. The two most striking features about the Forest grazing system are, as already noted, that

(i) Whole areas of the Forest are relatively less used than others. Even within one community type which may be quite regularly exploited by the Common animals, one 'patch' may be far more extensively used than another.

(ii) Many particular communities of the Forest suffer considerable pressure — to the extent perhaps of over-grazing — while adjacent communities, containing perfectly palatable forage species on which the animals are quite happy to feed, are left virtually ungrazed.

The reasons for these two phenomena are, we believe, purely behavioural; in addition, we think that it may prove possible to manipulate the animals, using these natural behavioural responses, in order to achieve a more even distribution.

Clearly, the problem of differential density of animals over the Forest as a whole could be solved purely mechanically, in containing animals within certain geographical areas by fencelines and road grids. This, however, would be politically contentious, and expensive, and would spoil the open nature of the Forest. In addition, it might 'penalise' animals restricted to currently less favoured areas. We believe that the relative over-use and under-use of different parts of the Forest is a result of home-range patterns, and that an alternative solution may be adopted in exploitation of this knowledge. As already noted in this volume (Chapter 3), both cattle and ponies are preferential grazers, and base their home range upon a focal area of grassland. With a restricted number of such grassland areas available, animals congregate on those areas containing such a lawn, with resultant uneven distribution over the Forest as a whole. Cattle often make protracted route marches to and from this focal lawn and thus may range quite widely from it. By contrast, ponies feed and move less widely and tend to remain on those communities in the immediate vicinity of the focal lawn. The distribution of ponies is markedly aggregated, and ranges cluster around these few favoured lawns. Large areas of perfectly palatable forage remain unexploited, because they are too far from a suitable grassland, and no range includes them.

As noted, this is more of a problem with regard to ponies than to cattle. But we suggest that a more *even* level of grazing within given communities over the Forest as a whole could be achieved by the artificial production of small areas of new grassland in appropriate regions of the Forest — to open up areas of acid grassland and heathland at present under-exploited. These grasslands need *not* be particularly large (approx. 10 ha) and would thus pose no threat to conservation of the communities into which they are introduced; but, by evening up grazing pressure on the Forest as a whole, they could in fact benefit conservation by relieving grazing pressure on currently hard-pressed areas.

This preference for lawns also lies behind the second problem, that of the differential grazing pressure between *different* communities within a home range; and lawns are very decidedly a two-edged weapon. As soon as grass production starts on the lawns, the animals forsake other feeding communities and move onto these grasslands. Instantaneous offtake virtually equals production, so that in practice there is almost at once very little grazing to be had. Once on the lawns the animals are loath to leave, even though abundant, palatable and nutritious forage is available in adjacent communities. Both cattle and ponies may move off the lawns to forage on wet heath, bogs or even dry heath around the lawn edge, but they will not willingly penetrate deep into these communities and soon return to the focal lawns. Huge congregations of animals thus accumulate on lawns where there is little to eat, while in the surrounding communities forage resources are under-exploited. So strong is this 'hefting' that it persists even long after the growing season is over: ponies and cattle are still on the lawns in large numbers in late autumn and early winter, and only gradually drift away. Here, again, the lawn hefting results in under-exploitation of alternative forages but over-exploitation of the grasslands themselves. As a result of all this, the lawns become severely overgrazed — to a degree where productivity is suppressed through pure lack of photosynthetic tissue and through continued nutrient impoverishment. The animals, too, are suffering: the concentrations of animals on the food-sparse lawns result in a forage intake by each pony or cow far below that which it could achieve on other communities. The problem is further exacerbated in the winter months, when lawn-hefted animals (or animals attracted to any exposed community for hay feeding) stand about in open habitats, with nothing at all to eat, while fully exposed to wind and cold.

Winter loss of condition could be decreased, forage intake the year round increased, and differential grazing pressure of the various forage communities *and* the impoverishment of the re-seeded and other improved lawns could, we believe, be simply resolved, by allowing only seasonal access to these grasslands. We suggest that, were these grasslands fenced and common animals allowed onto them only in late summer (from perhaps August till October/November), a number of benefits would result:

1. By resting the lawns for eight months of the year, or by allowing only intermittent grazing on them during this time, they would be permitted time to recover and the continued impoverishment would be arrested.
2. By preventing over-grazing during the growing season, greater

total forage productivity could be achieved for the time when the lawns *were* opened for grazing.

3. Shutting off the lawns during the summer would in no way deprive the animals of forage, since the grassland growing season coincides with the peak growing season of many other communities currently under-exploited. Indeed, by getting the animals off the lawns and into these other communities, forage intake would actually be increased.
4. More even grazing of all communities would result.

In addition, we believe that a further benefit might be obtained: better overwintering performance of stock left on the Forest.

More subtle management of this kind might help to restore the uneasy balance between grazers and the vegetation, might relax over-intense pressure on threatened vegetation types and allow diversity to increase, while at the same time not altering in any fundamental way the character of the New Forest system. Grazing pressure will continue to fluctuate for its own independent reasons; already increases in marking fees and a slackening market for ponies has resulted in a decline in animal numbers from their late 1970 levels (in 1984, marking fees were paid for only 3,000 ponies and 1,760 cattle). Such factors are outside our control: we can hope to manage only *within* the framework of a given animal density to 'spread' its effects in the most appropriate way. There will always be conflict between the interests of economic forestry, conservation, recreation and commoning. Perhaps the Forest is fortunate in that each interest has its own statutory body to champion its cause: Forestry Commission, Nature Conservancy Council, Countryside Commission and Verderers Court. The New Forest has survived the vagaries and vicissitudes of 900 years of changing use: it would be interesting to follow its fortunes over the next 900 years.

References

Adams, S.N. (1975) Sheep and cattle grazing in forests: a review. *Journal of applied Ecology 12*: 143-52.
Andrzejewska, L. (1967) Estimation of the effects of feeding of the sucking insect *Cicadella viridis* L. (Homoptera: Auchenorrhyncha) on plants. In K. Petrusewicz (ed.),*Secondary Productivity in Terrestrial Ecosystems*, vol. II: 791-805.
Archer, M. (1973) Variations in potash levels in pastures grazed by horses; a preliminary communication. *Equine Veterinary Journal 5*: 45-6.
—— (1978) Studies on producing and maintaining balanced pastures for horses. *Equine Veterinary Journal 10*: 54-9.
Arnold, G.W., and Dudzinski, M.L. (1978) *Ethology of free-ranging domestic animals. Developments in Animal and Veterinary Sciences, 2*, Elsevier.
Bailey, R.E. (1979) Use of faecal techniques in analysis of population density and diet in fallow deer populations. BSc Honours thesis (Environmental Sciences), University of Southampton.
—— and Putman, R.J. (1981). Estimation of fallow deer (*Dama dama*) populations from faecal accumulation. *Journal of applied Ecology 18*: 697-702.
Berger, J. (1977) Organisational systems and dominance in feral horses in the Grand Canyon. *Behavioural Ecology and Sociobiology 2*: 131-46.
de Bie, S. (1982) Carrying-capacity. Paper to meeting '*Begrazing door Vertebraten*', Rijksuniversiteit Groningen — Landbouwhogeschool, Wageningen. 1982.
Bobek, B., Perzanowski, K. Siwanowicz, J., and Zielinski, J. (1979) Deer pressure on forage in a deciduous forest. *Oikos 32*: 373-9.
Bonsma, J.C., and Le Roux, J.D. (1953) Influence of environment on the grazing habits of cattle. *Farming in South Africa 28*: 43-6.
Browning, D.R. (1951) The New Forest pastoral development scheme. *Agriculture 58*: 226-33.
Buechner, H.K., and Dawkins, H.C. (1961). Vegetation change induced by elephants and fire in Murchison Falls National Park, Uganda. *Ecology 42*: 752-66.
Bullen, F.T. (1970) A review of the assessment of crop losses caused by locusts and grasshoppers. *Proceedings of the International Study Conference on Current and Future Problems of Acridology*: 163-71. London.
Cadman, W.A. (1966) *The Fallow Deer*. Forestry Commission Leaflet No. 52. HMSO: London.
Carter, N.A. (1981) A study of bole damage by sika in England, with particular reference to bole-scoring. BSc thesis, University of Bradford.
Chadwick, M.J. (1960) *Nardus stricta* L. *Journal of Ecology 48*: 255-67.
Chapman, D.I., and Chapman, N. (1976) *The Fallow Deer*. Terence Dalton.

Chapman, N. (1984) *Fallow deer.* Mammal Society/Anthony Nelson.
Clements, R.O. (1978) The benefits and some long-term consequences of controlling invertebrates in a perennial rye grass sward. *Scientific Proceedings of the Royal Dublin Society, Series A, 6*: 335-41.
Clutton-Brock, T.H., and Albon, S.D. (1978) The roaring of red deer and the evolution of honest advertisement. *Behaviour 69*: 145-70.
—— (1979) Sexual differences in ecology between male and female red deer. Unpublished paper to the Ungulate Research Group Winter meeting, Edinburgh, December 1979.
Corke, D. (1974) The Comparative Ecology of the two British species of the genus *Apodemus* (Rodentia: Muridae). PhD thesis, University of London.
Countryside Commission (1984) *The New Forest Commoners.* Countryside Commission: Cheltenham.
Coupland, R.T. (1979) *Grassland Ecosystems of the World: Analysis of Grasslands and their Uses.* Cambridge University Press: Cambridge.
Crawley, M.J. (1983) *Herbivory: The Dynamics of Animal-Plant Interactions.* Blackwell Scientific Publications: Oxford.
Davidson, J.L., and Milthorpe, F.L. (1966) Leaf growth in *Dactylis glomerata* following defoliation. *Annals of Botany* (London) *30*: 173-84.
Davis, T.A.W. (1960) Kestrel pellets at a winter roost. *British Birds 53*: 281-4.
—— (1975) Food of the Kestrel in winter and early spring. *Bird Study 22*: 85-91.
Detling, J.K., Dyer, M.I., Proctor-Gregg, C., and Winn, D.T. (1980) Plant-herbivore interactions: Examination of potential effects of bison saliva on regrowth of *Bouteloua gracilis* (H.B.K.) Lag. *Oecologia 45*: 26-31.
Dinesman, L.G. (1967) Influence of vertebrates on primary production of terrestrial communities. In K. Petrusewicz (ed.) *Secondary Productivity in Terrestrial Ecosystems*, vol. 1: 261-6.
Duncan, P. (1983) Determinants of the use of habitat by horses in a Mediterranean wetland. *Journal of Animal Ecology 52*: 93-109.
—— and Vigne, N. (1979) The effect of group size in horses on the rate of attacks by blood-sucking flies. *Animal Behaviour 27*: 623-5.
Dzieciolowski, R. (1979) Structure and spatial organisation of deer populations. *Acta Thieriologica 24*: 3-21.
Edwards, P.J., and Hollis, S. (1982) The distribution of excreta on New Forest grasslands used by cattle, ponies and deer. *Journal of applied Ecology 19*: 953-64.
Ellis, J.E.S. (1946) Notes on the food of the Kestrel. *British Birds 39*: 113-15.
Ellison, L. (1960) Influence of grazing on plant succession of rangelands. *Botanical Review 26*: 1-78.
Estes, R.D. (1967) The comparative behaviour of Grant's and Thompson's gazelles. *Journal of Mammalogy 48*: 189-209.
Feist, J.D., and McCullough, D.R. (1976) Behaviour patterns and communication in feral horses. *Zeitschrift für Tierpsychologie 41*: 337-71.
Furubayashi, K., and Maruyama, N. (1977) Food habits of sika deer in Fudakake, Tanzawa Mountains. *Journal of the Mammal Society of Japan 7*: 55-62.
Gates, S. (1979) A study of the home ranges of free-ranging Exmoor ponies. Mammal Review 9: 3-18.
—— (1982) The Exmoor Pony — a wild animal? *Nature in Devon 2*: 7-30.
Gessaman, J.A., and MacMahon, J.A. (1984) Mammals in ecosystems: their effects on the composition and production of vegetation. *Acta Zoologica Fennica*, in press.
Gill, E.L. (1980) Some aspects of the social and reproductive behaviour of a group of New Forest ponies. BSc Honours thesis (Environmental Sciences), University of Southampton.
—— (1984) Seasonal changes in body condition of New Forest ponies. Internal report, University of Southampton.
Grant, S.A., and Hunter, R.F. (1966) The effects of frequency and season of clipping

on the morphology, productivity and chemical composition of *Calluna vulgaris*. *New Phytologist 65*: 125-33.
Gray, A.J., and Scott, R. (1977) *Puccinellia maritima* (Huds.) Par 1. *Journal of Ecology 65*: 699-716.
Hansen, R. (1970) Foods of free-roaming horses in Southern New Mexico. *Journal of Range Management 29*: 347.
Harper, J.L. (1977) *Population Biology of Plants.* Academic Press: London.
Harrington, R. (1981) Immuno-electrophoresis and the genetics of red/sika hybrids. Unpublished paper given to Ungulate Research Group Winter meeting, Southampton, December 1981.
Hill, S.D. (1985) Influences of large herbivores on small rodents in the New Forest, Hampshire. PhD thesis, University of Southampton.
Hirons, G.J.M. (1976) A population study of the Tawny Owl (*Strix aluco*) and its main prey species in woodland. DPhil thesis, University of Oxford.
—— (1984) The diet of Tawny Owls (*Strix aluco*) and Kestrels (*Falco tinnunculus*) in the New Forest, Hampshire. *Proceedings of the Hampshire Field Club and Archaeological Society 40*: 21-6.
Hoffmeyer, I. (1973) Interaction and habitat selection in the mice *Apodemus flavicollis* and *Apodemus sylvaticus*. *Oikos 24*: 108-16.
Hofmann, R.R. (1982) Morphological classification of sika deer within the comparative system of ruminant feeding types. *Deer 5*: 252-3.
—— and Stewart, D.R.M. (1972) Grazer or browser? A classification based on the stomach structure and feeding habits of East African ruminants. *Mammalia 36*: 226-40.
—— Geiger, G., and Konig, R. (1976) Vergleichend-anatomische untersuchungen an der vormagenschleimhaut von Rehwild (*Capreolus capreolus*) und Rotwild (*Cervus elaphus*). *Zeitschrift Säugetierkunde 41*: 167-93.
Hope Simpson, J.F. (1940) Studies of the vegetation of the English Chalk. VI. Late stages in succession leading to chalk grassland. *Journal of Ecology 28*: 386-402.
Horwood, M.T. (1973) The world of the Wareham sika. *Deer 2*: 978-84.
—— and Masters, E.H. (1970) *Sika Deer.* British Deer Society: Reading.
Hosey, G.R. (1974) The food and feeding ecology of roe deer. PhD thesis, University of Manchester.
Howard, P.C. (1979) Variability of feeding behaviour in New Forest ponies. BSc Honours thesis (Biology), University of Southampton.
Jackson, J.E. (1974) The feeding ecology of Fallow deer in the New Forest, Hampshire. PhD thesis, University of Southampton.
—— (1977) The annual diet of the Fallow deer (*Dama dama*) in the New Forest, Hampshire, as determined by rumen content analysis. *Journal of Zoology* (London) *181*: 465-73.
—— (1980) The annual diet of the roe deer (*Capreolus capreolus*) in the New Forest, Hampshire, as determined by rumen content analysis. *Journal of Zoology* (London) *192*: 71-83.
Jameson, D.A. (1963) Responses of individual plants to harvesting. *Botanical Review 29*: 532-94.
Jarman, P.J. (1974) The social organisation of antelope in relation to their ecology. *Behaviour 48*: 215-66.
Johnson, T.H. (1984) Habitat and social organisation of roe deer (*Capreolus capreolus*). PhD thesis, University of Southampton.
Jones, M.G. (1933) Grassland management and its influence on the sward. *Journal of the Royal Agricultural Society 94*: 21-41.
Kaluzinski, J. (1982) Composition of the food of roe deer living in fields and the effects of their feeding on plant production. *Acta Theriologica 27*: 457-70.
Kay, R.N.B. (1979) Seasonal changes of appetite in deer and sheep. Agricultural Research Council. Research Review 1979.

—— and Suttie, J.M. (1980) Seasonal cycles of voluntary food intake in red deer. Unpublished paper presented to the Ungulate Research Group winter meeting, Cambridge, December 1980.

Kenchington, F.E. (1944) *The Commoners' New Forest.* Hutchinson: London.

Klingel, H. (1974) A comparison of the social behaviour of the Equidae. In V. Geist and F. Walther (eds.), *The Behaviour of Ungulates and its relation to Management.*

Krefting, L.W., Stenlund, M.H., and Seemel, R.K. (1966) Effect of simulated and natural browsing on mountain maple. *Journal of Wildlife Management 30*: 481-8.

Kruuk, H. (1984) Group sizes and home ranges of carnivores. In R.H. Smith and R.M. Sibly (eds.), *Behavioural Ecology: the ecological consequences of adaptive behaviour.* British Ecological Society Symposium 25.

Lascelles, G.W. (1915) *Thirty five years in the New Forest.* London.

Laws, R.M. (1976) Elephants as agents of habitat and landscape change in East Africa. *Oikos 21*: 1-15.

Lincoln, G.A., and Guinness, F.E. (1973). The sexual significance of the rut in red deer. *Journal of Reproduction and Fertility,* Supplement *19*: 475-89.

—— Youngson, R.W., and Short, R.V. (1970) The social and sexual behaviour of the red deer stag. *Journal of Reproduction and Fertility,* Supplement *11*: 71-103.

Lowe, V.P.W., and Gardiner, A.S. (1975) Hybridisation between red deer and sika deer, with reference to stocks in North-west England. *Journal of Zoology* (London) *177*: 553-66.

MacArthur, R.H., and Levins, R. (1967). The limiting similarity, convergence and divergence of coexisting species. *American Naturalist 101*: 377-85.

Mann, J.C.E. (1978) An investigation into the vegetational differences between an ungrazed and a grazed portion of the New Forest, with special emphasis on the species composition, species diversity and productivity of the areas. BSc Honours thesis (Biology), University of Southampton.

—— (1983) The Social Organisation and Ecology of the Japanese Sika deer (*Cervus nippon*) in Southern England. PhD thesis, University of Southampton.

McEwen, L.C., French, C.E., Magruder, N.D., Swift, R.W., and Ingram, R.H. (1957) Nutrient requirements of the white-tailed deer. *Transactions of the 22nd North American Wildlife Conference*: 119-32.

McNaughton, S.J. (1979a) Grazing as an optimisation process: grass-ungulate relationships in the Serengeti. *American Naturalist 113*: 691-703.

—— (1979b) Grassland-herbivore dynamics. In A.R.E. Sinclair and M. Norton-Griffiths (eds.), *Serengeti: dynamics of an ecosystem.* University of Chicago Press, 46-81.

Milthorpe, F.L., and Davidson, J.L. (1965) Physiological aspects of regrowth in grasses. In F.L. Milthorpe and J.D. Ivins (eds.), *The Growth of Cereals and Grasses.* Butterworth: London, 241-55.

Miura, S. (1974) On the seasonal movements of sika deer populations in Mt. Hinokiboramu. *Journal of the Mammal Society of Japan 6*: 51-66.

Nicholson, I.A., Paterson, I.S., and Currie, A. (1970) A study of vegetational dynamics: selection by sheep and cattle in *Nardus* pasture. In A. Watson (ed.), *Animal Populations in Relation to their Food Resources.* BES Symposium *10*: 129-43.

O'Bryan, M.K., Palmes, P., and Putman, R.J. (1980) Factors affecting group size and cohesiveness in populations of New Forest ponies. Unpublished MS, Southampton University Library.

Odberg, F.O., and Francis-Smith, K. (1976) A study on eliminative and grazing behaviour: the use of the field by captive horses. *Equine veterinary Journal 8*: 147-9.

—— (1977) Studies on the formation of ungrazed eliminative areas in fields used by horses. *Applied Animal Ethology 3*: 27-34.

Oosterveld, P. (1981) Begrazing als natuurtechnische maatregel. Unpubl. ms.

Rijkinstituut voor Natuurbeheer, The Netherlands.
Packham, C.G. (1983) The influence of food supply on the ecology of the badger. BSc Honours thesis (Biology), University of Southampton.
Parfitt, A. (1985) 'Social Organisation and Ecology of fallow deer'(*Dama dama*) in the New Forest, Hampshire. PhD thesis, University of Southampton, in preparation.
Parsons, K.A., and De La Cruz, A.A. (1980) Energy flow and grazing behaviour of conocephaline grasshoppers in a *Juncus roemerianus* marsh. *Ecology 61*: 1045-50.
Peterken, G.F., and Tubbs, C.R. (1965) Woodland regeneration in the New Forest, Hampshire since 1650. *Journal of applied Ecology 2*: 159-70.
Pianka, E.R. (1973) The structure of lizard communities. *Annual Review of Ecology and Systematics 4*: 53-74.
Pickering, D.W. (1968) Heathland reclamation in the New Forest: the ecological consequences. Unpublished MSc thesis, University College, London.
Pollock, J.I. (1980) *Behavioural Ecology and Body Condition Changes in New Forest Ponies.* Royal Society for the Prevention of Cruelty to Animals. Scientific Publications No. 6.
Pratt, R.M., Putman, R.J., Ekins, J.R., and Edwards, P.J. (1985) Use of habitat by free-ranging cattle and ponies in the New Forest, Southern England. In press.
Prior, R. (1973) Roe deer: management and stalking. Game Conservancy Booklet No. 17, The Game Conservancy: Fordingbridge.
Prisyazhnyuk, V.E., and Prisyazhnyuk, N.P. (1974) [Sika deer on Askold Island.] Bulletin Moskow o-va ispyt. Priv. otd. Biology 79: 16-27 (in Russian).
Putman, R.J. (1980) Consumption, protein and energy intake of fallow deer fawns on diets of differing nutritional quality. *Acta Theriologica 25*: 403-13.
—— (1981) Social systems in deer: a speculative review. *Deer 5*: 186-8.
—— (1983) Nutrition of Wild and Farmed Fallow deer. Paper to 1st International Conference on the Biology of Deer Production, Dunedin, New Zealand.
—— (1984) Facts from Faeces. *Mammal Review 14*: 79-97.
—— (1985a) Effects of grazing mammals on ecosystem structure and function: a review. In press.
—— (1985b) Competition and Coexistence in a multispecies grazing system, *Acta Theriologica*, in press
—— and Wratten, S.D. (1984) *Principles of Ecology.* Croom Helm: Beckenham.
—— Edwards, P.J., Ekins, J.R., and Pratt, R.M. (1981) Food and feeding behaviour of cattle and ponies in the New Forest: a study of the inter-relationship between the large herbivores of the Forest and their vegetational environment. Report HF 3/03/127 to Nature Conservancy Council: Huntingdon.
—— Pratt, R.M., Ekins, J.R., and Edwards, P.J. (1982) Habitat use and grazing by free-ranging cattle and ponies, and impact upon vegetation in the New Forest Hampshire. Proceedings of IIIrd International Theriological Congress, Helsinki. *Acta Zoologica Fennica, 172,* 183-6
—— (1985) Food and feeding behaviour of cattle and ponies in the New Forest, Hampshire. (in prep.)
Rand, N. (1981) Fallow deer coat patterns: a system of identification and some results from its use in a long-term study of behaviour. Unpublished paper given to Ungulate Research Group Winter meeting, Southampton, December 1981.
Rawes, M. (1981) Further results of excluding sheep from high level grasslands in the north Pennines. *Journal of Ecology 69*: 651-69.
Reardon, P.O., Leinweber, C.L., and Merrill, L.B. (1972) The effect of bovine saliva on grasses. *Journal of Animal Science 34*: 897-8.
—— (1974) Response of sideoats grama to animal saliva and theamine. *Journal of Range Management 27*: 400-1.
Russell, V. (1976) *New Forest Ponies.* David and Charles: Newton Abbott.
Schmidt, P.J. (1969) Observations on the behaviour of cattle in a hot dry region of the Northern Territory of Australia, with particular reference to walking, watering

and grazing. MSc thesis, The University of New England, Armidale, New South Wales.

Schröpfer, R. (1983) The effect of habitat selection on distribution and spreading in the yellow-necked mouse (*Apodemus flavicollis*). Proceedings of IIIrd International Theriological Congress, Helsinki. *Acta Zoologica Fennica.*

Schultz, A.M. (1964) The nutrient-recovery hypothesis for arctic microtine cycles. In D.J. Crisp (ed.), *Grazing in Terrestrial and Marine Environments.* Blackwell: Oxford, 57-68.

—— (1969) A study of an ecosystem: the arctic tundra. In G.M. van Dyne (ed.), *The Ecosystem Concept in Natural Resource Management.* Academic Press: New York, 77-93.

Schuster, M.F., Boling, J.C., and Morony, J.J. (1971) Biological control of rhodesgrass scale by airplane release of an introduced parasite of limited dispersal ability. In C.B. Huffaker (ed.), *Biological Control.* Plenum Press: New York, 227-50.

Shepperd, J.H. (1921) The trail of the short grass steer. North Dakota Agricultural College Bulletin *154*: 8.

Short, H.L., Newson, J.D., McCoy, G.L., and Fowler, J.F. (1969) Effects of nutrition and climate on southern deer. *Transactions of the 34th North American Wildlife Conference*: 137-45.

Simenstad, C.A., Estes, J.A., and Kenyon, K.W. (1978) Aleuts, sea-otters, and alternative stable-state communities. *Science 200*: 403-10.

Sinclair, A.R.E., and Norton-Griffiths, M. (1979) *Serengeti: Dynamics of an Ecosystem.* University of Chicago Press: Chicago.

Smith, R.H. (1979) On selection for inbreeding in polygynous animals. *Heredity 43*: 205-11.

Southern, H.N. (1954) Tawny owls and their prey. *Ibis 98*: 384-410.

—— (1969) Prey taken by Tawny owls during the breeding season. *Ibis 111*: 293-9.

—— (1970) The natural control of a population of Tawny Owls (*Strix aluco*). *Journal of Zoology* (London) *162*: 197-285.

Spedding, C.R.W. (1971) *Grassland Ecology.* Clarendon Press: Oxford.

Spence, D.H.N., and Angus, A. (1971) African grassland management: burning and grazing in Murchison Falls National Park, Uganda. In E. Duffey and K. Watt (eds.), *The Scientific Management of Animal and Plant Communities for Conservation.* Blackwell: Oxford, 319-31.

Staines, B.W. (1976) Experiments with rumen-canulated red deer to evaluate rumen analyses. *Journal of Wildlife Management 40*: 371-3.

—— Crisp, J.M., and Parish, T. (1981) Differences in the quality of food eaten by red deer stags and hinds in winter. *Journal of applied Ecology 19*: 65-79.

Stebbins, G.L. (1981) Coevolution of grasses and herbivores. *Annals of the Missouri Botanical Garden 68*: 75-86.

Stewart, D.R.M. (1967) Analysis of plant epidermis in faeces: a technique for studying the food preferences of grazing herbivores. *Journal of applied Ecology 4*: 83-111.

Stewart, D.R.M., and Stewart, J. (1970) Food preference data by faecal analysis for African plains ungulates. *Zoological Africana 5*: 115.

Strange, M.L. (1976) A computer study of the fallow deer of the New Forest. BSc Honours thesis (Environmental Sciences), University of Southampton.

Takatsuki, S. (1980) Food habits of sika deer on Kinkazan Island, Japan. *Scientific Reports of Tohaku University, Series 4* (Biology), *38*: 7-32.

Tansley, A.R. (1922) Studies of the vegetation of the English Chalk. II. Early stages of redevelopment of woody vegetation in chalk grassland. *Journal of Ecology 10*: 168-77.

—— and Adamson, R.W. (1925) Studies of the vegetation of the English Chalk. III. The chalk grasslands of the Hampshire-Sussex border. *Journal of Ecology 13*: 177-223.

Tubbs, C.R. (1965) The development of the small-holding and cottage stock-keeping economy of the New Forest. *Agricultural Historical Review 13.*

—— (1968) *The New Forest: an ecological history.* David and Charles: Newton Abbott.

—— (1974) *The Buzzard.* David and Charles: Newton Abbot.

—— (1982) The New Forest: conflict and symbiosis. *New Scientist* 1st July 1982: 10-13.

—— and Jones, E.L. (1964) The distribution of gorse (*Ulex europaeus*) on the New Forest in relation to former land use. *Proceedings of the Hampshire Field Club 23*: 1-10.

—— and Tubbs, J.M. (1985) Buzzards, *Buteo buteo*, and land use in the New Forest, Hampshire, England. *Biological Conservation 31*: 41-65.

Turner, D.C. (1979) An analysis of time-budgetting by roe deer (*Capreolus capreolus*) in an agricultural area. *Behaviour 71*: 246-90.

Tyler, S. (1972) The behaviour and social organisation of the New Forest ponies. *Animal Behaviour Monographs 5*: 87-194.

van der Veen, H.E. (1979) Food selection and habitat use in the red deer (*Cervus elaphus* L.). PhD thesis, Rijksuniversiteit te Groningen.

van Wieren, S.E. (1979) Sexual segregation in deer species: a possible theory. Unpublished paper to the Ungulate Research Group Winter meeting Edinburgh, December 1979.

Vickery, P.J. (1972) Grazing and net primary production of a temperate grassland. *Journal of applied Ecology* 9: 307-14.

Vittoria, A., and Rendina, N. (1960) Fattori condizionanti la funzionalita tiaminica in piante superiori e cenni sugli effetti dell bocca del ruminanti sull erbe pascolative. *Acta Medica Veterinaria (Naples)*, 6: 379-405.

Wells, S. and von Goldschmidt-Rothschild, B. (1979) Social behaviour and relationships in a herd of Camargue horses. *Zeitschrift für Tierpsychologie 49*: 363-80.

Welsh, D. (1975) Population, behavioural and grazing ecology of the horses of Sable Island, Nova Scotia. PhD thesis, Dalhousie University.

Whittaker, R.H., and Woodwell, G.M. (1968) Dimension and production relations of trees and shrubs in the Brookhaven Forest, New York. *Journal of Ecology 56*: 1-25.

Wiegert, R.G., and Evans, F.C. (1967) Investigations of secondary productivity in grasslands. In K. Petrusewicz (ed.), *Secondary Productivity in Terrestrial Ecosystems*, vol. II: 499-518.

Wolff, J.O. (1978) Burning and browsing effects on willow growth in interior Alaska. *Journal of Wildlife Management 42*: 135-40.

Yalden, D.W., and Warburton, A. (1979) A diet of the Kestrel in the Lake District. *Bird Study 26*: 163-70.

Zejda, J. (1978) Field grouping of roe deer (*Capreolus capreolus*) in a lowland region. *Folia Zoologica 27*: 111-22.

Zhuopu, G., Enyu, C., and Youzhin, W. (1978) A new subspecies of sika deer from Szechuan (*Cervus nippon sichuanensis*). *Acta Zoologica Sinica 24*: 187-92.

Index

HITLER'S FOREIGN EXECUTIONERS

One basic principle must be the basic rule for the SS man: we must be honest, decent, loyal, and comradely to members of our own blood and to nobody else ... In twenty to thirty years we must really be able to provide the whole of Europe with its ruling class.

Heinrich Himmler, 4 October 1943

The Romanians act against the Jews without any idea of a plan. No one would object to the numerous executions if the technical aspect of their preparation, as well as the manner in which they are carried out, were not wanting ... The Einsatzkommando has urged the Romanian police to proceed with more order.

Report by Einsatzgruppe D, 31 July 1941

A Jew in a greasy caftan walks up to beg some bread, a couple of comrades get a hold of him and drag him behind a building and a moment later he comes to an end. There isn't any room for Jews in the new Europe, they've brought too much misery to the European people.

Danish SS volunteer

HITLER'S FOREIGN EXECUTIONERS

EUROPE'S DIRTY SECRET

CHRISTOPHER HALE

The views or opinions expressed in this book and the context in which the images are used, do not necessarily reflect the views or policy of, nor imply approval or endorsement by, the United States Holocaust Memorial Museum.

First published 2011

The History Press
The Mill, Brimscombe Port
Stroud, Gloucestershire, GL5 2QG
www.thehistorypress.co.uk

British Library Cataloguing in Publication Data.
A catalogue record for this book is available from the British Library.

ISBN 978 0 7524 5974 5

Typesetting and origination by The History Press
Printed in the EU for The History Press.

Contents

Note on Language

I have taken a pragmatic approach to German terms. Most specialised organisational terms are given in German to begin with and thereafter in English (Special Task Force for Einsatzgruppe, for example, although there is no 'Special' in the term), unless the original has become broadly accepted – 'Der Führer' for example. I have referred to SS ranks in German to distinguish them from army ranks. 'Die Wehrmacht' in English language books has come to signify the German ground forces – strictly speaking Das Heer. After 1935, the term embraced all the armed forces in the Third Reich, including the army, navy and air force. For this reason I refer to the 'German army' rather than 'the Wehrmacht'. I have treated place names on a case-by-case basis.

Preface

Riga, 2010

Imagine Whitehall on a dank, autumn morning. A far-right British political party leader steps towards the Cenotaph, jaw set, dark suited, clutching a wreath.[1] Behind him stands the party elite sporting banners displaying back and white photographs of hard-faced men in grey military uniforms. A dense police cordon holds back jeering anti-fascists who have gathered in Parliament Square. He and his followers have come here to commemorate a handful of forgotten anti-communist martyrs who joined the German armed forces and fought against Stalin during the Second World War. After solemnly placing the wreath at the foot of Edwin Lutyen's chaste memorial to the dead of the Great War, the party leader makes a short, angry speech denouncing the post-war British government for punishing these brave, far-sighted warriors as traitors. History has proved them right! As he finishes, an egg splatters on his immaculate black coat. Then the party men march up Whitehall to Trafalgar Square, closely pursued by protestors. Scuffles erupt, banners are trampled underfoot. Tourists and passers-by scratch their heads, puzzled. Who on earth were these 'heroic' anti-communists?

In August 1942, an odious public school dropout called John Amery and his companion Jeannine Barde arrived in Berlin masquerading as 'Mr. and Mrs. Browne'. Amery was very well connected: his father was Secretary of State for India and his brother, Julian, an illustrious war hero.[2] John Amery's hosts, the 'England Committee' at the German Foreign Office, hoped that his defection would provide them with a propaganda coup. They were grossly mistaken. Rebecca West, who witnessed Amery's post-war trial for High Treason at the Old Bailey, concluded that he 'was not insane … but his character was like the kind of automobile that will not hold the road'.[3] Although his odious personal ideology perfectly fitted the

German world view, John Amery was no great catch. His grandmother had been a Hungarian Jew who had found refuge in Britain, but the Amery clan were all die-hard conservatives. Given this bigoted cradling, it is not surprising that John became a fervent anti-communist whose views, at least to begin with, mimicked those of his father and brother. But unlike them, he became an outspoken and virulent anti-Semite who was in thrall to the vicious 'Jew hatred' of French ultranationalist culture. John Amery spurned his well-to-do family and became a dedicated bohemian. He contracted syphilis at the age of 14 and by the time he arrived in Berlin was an alcoholic bigamist, burdened by massive debts. But Hitler and the German Foreign Office had a crass understanding of English social mores and it took them a long time to understand that John Amery had little to offer the Reich.

For a year, Amery and Jeannine boozed and rowed in the capital of the Reich, sending their bills to Hitler's personal office. Amery made a few radio broadcasts and narrowly escaped a manslaughter charge when Jeannine choked to death on her own vomit. Then in January 1943, the French fascist leader Jacques Doriot, who had formed the Légion des Volontaires Français contre le Bolchévisme (LVF), persuaded the German military authorities to give Amery access to British prisoners of war. Most gave Amery short shrift and he succeeded only in recruiting a tiny band of about thirty-eight turncoats, the majority of whom were former members of the British Union of Fascists (BUF). By the end of 1942, Amery's bizarre antics had exhausted German tolerance and he was effectively sidelined by the England Committee. The baton passed to one Thomas Cooper, a former resident of Chiswick – who was already serving as SS Corporal Thomas Haller Cooper. Cooper had spent time as an SS camp guard in Sachsenhausen and fought on the Eastern Front. In early 1943, he was transferred to a British POW camp at Genshagen where he busily promoted the German cause. In September 1943 Gottlob Berger, the SS head of recruitment, formerly took over Cooper's band of converts as the *Britisches Freikorps* (British Free Corps or BFC). At the end of April 1944, SS officer Hans Werner Roepke formally inspected Cooper's dozen or so men and issued them with SS identification papers and side arms. The SS provided uniforms sporting heraldic leopards and a Union Jack shield.

The contribution of the BFC to the Reich's 'crusade against Bolshevism' was not even trivial. SS General Felix Steiner reported that 'they were suffering from an inner conflict ... they were depressed'. Steiner refused to use them in combat – and last saw the sorry band shambling westwards along an autobahn. In May 1945 the relics of the BFC surrendered to American troops near Schwerin.[4] John Amery, who had inspired German recruitment of British prisoners, fled to Italy, to confer with Mussolini who was by then holed up in the fascist Republic of Salo. When

Amery arrived in nearby Como, he was captured by Italian partisans and handed over to British Military Intelligence. In November, Amery was repatriated, tried at the Old Bailey and, on 19 December 1945, hanged in Wandsworth prison. His distinguished father Leo Amery claimed that his son had been 'inspired by a desire to save the British Empire'. Thomas Cooper, who had served in an SS Death's Head unit, had his death sentence commuted to life imprisonment.

Suppose that seventy years after John Amery's fatal encounter with hangman Albert Pierrepoint, the leader of a far-right British political party proposed commemorating Amery and his ludicrous handful of followers as prescient Cold War warriors, who understood long before most British citizens that 'Uncle Joe' Stalin was a tyrant infinitely more terrible than Adolf Hitler. Did not the crimes of the Soviet Union far outstrip those of Nazi Germany? Amery, this political party claims, was no treasonous villain, but a hero whose execution in a British prison was a travesty of justice. This counterfactual scenario is by no means unimaginable.

In 2008 many of the far-right parties of Europe backed the Prague Declaration on Conscience and Communism. This was hatched up by Baltic scholars and politicians. Its authors demand that the European Union 'equally evaluate totalitarian regimes'. In other words, the crimes of the Soviet regime and the Nazi Holocaust should have equivalent moral status. This is often summed up by the slogan 'red = brown'. The Prague Declaration proposes replacing Holocaust Memorial Day on 27 January with a 'Day of Remembrance' to be held every 23 August, the day on which the German Foreign Minister Joachim von Ribbentrop and his Soviet counterpart Vyacheslav Molotov signed the Nazi-Soviet Non-Aggression Pact in 1939. This 'equal evaluation' may appear seductive. After all, how often does one hear that 'Stalin was just as bad or worse than Hitler'. But the apparently reasonable claim that 'there are substantial similarities between Nazism and Communism in terms of their horrific and appalling character and their crimes against humanity' is not what it seems.[5] The authors of the Prague Declaration grossly distort the historical record and seek ultimately to tear down the unique moral status of the Holocaust. The concept of 'double genocide' lumps together heinous Soviet practices such as summary execution, deportation, imprisonment and loss of employment with the deliberate and planned attempt to liquidate an entire human group. Soviet crimes should indeed be properly memorialised, but they are not equivalent either in intent or result to the 'Final Solution'.

The consequences of rendering the crimes of the Soviet Union equivalent to the German Holocaust are already becoming clear in many Eastern European nations. In the Baltic States, Hungary and Ukraine it is now commonplace to hear politicians imply that wartime collaboration with the Third Reich should no longer be regarded as a moral catastrophe – a stain on the nation. Instead collaboration is increasingly reinterpreted as a pragmatic means to oppose the destructive power of the Soviet Union. This inevitably means that the tens of thousands of men who volunteered to serve the German occupiers as policemen and soldiers can be reinvested as heroic nationalists – no longer vilified as collaborators in genocide. Compelling evidence that this historical lie has begun to take root in Europe can be observed every 16 March in the capital city of Latvia.

In spring 2010, I travelled to Riga to observe the annual 'Legion Day' – a parade by Latvian Second World War veterans. Nothing remarkable about that you might suppose. But you would be wrong; the veterans' parade I witnessed commemorates the 'Latvian Legion' recruited by Heinrich Himmler's private army, the Waffen-SS, in 1943. Surviving members of this SS Legion mourn their fallen comrades in Riga's cathedral, the Dom, then march to the 'Freedom Monument' that stands in central Riga close to the old town.

In 2009, the Latvian SS Legion was splashed across the front pages of British newspapers when David Miliband, then British Foreign Secretary, denounced the Conservative Party for forging links with far-right European parties – including the Latvian For Fatherland and Freedom Party that, Miliband alleged, supported the Nazi Waffen-SS. Miliband's speech provoked an international storm – from both the Conservative Party and the Latvian government. Timothy Garton Ash, the doyen of historians of Eastern Europe, weighed in: 'How would you describe a British politician who prefers getting acquainted with the finer points of the history of the Waffen-SS in Latvia to maximising British influence with Barack Obama? An idiot? A madman? A nincompoop?'[6]

William Hague, now Foreign Secretary, refused to back down. The 'Latvian Legion' had nothing to do with the Holocaust, he claimed. The old Legionaries had never been Nazis. Hague went on: 'David Miliband's smears are disgraceful and represent a failure of his duty to promote Britain's interests as Foreign Secretary. He has failed to check his facts. He has just insulted the Latvian Government, most of whose member parties have attended the commemoration of Latvia's war dead.' Hague neglected to mention that the 'Latvian Legion' refers to two Waffen-SS divisions: the 15th Waffen-Grenadier-Division of the SS (1st Latvian) and the 19th Waffen-Grenadier-Division of the SS (2nd Latvian). These war dead sacrificed their lives for Hitler's Reich – and its 'war of annihilation'. Now their surviving comrades will commemorate the memory of the legion as national heroes.

I arrive at Riga airport early on Monday morning. It is bitterly cold and wet; the sky a leaden canopy. Snow is forecast for the following day, 16 March, when the SS commemoration takes place. When I cross the grand Vanšu tilts, or 'Shroud Bridge', an hour or so later, faltering sunshine glitters on the broad expanse of the Daugava River. At first sight, Riga resembles any prosperous modern European city. Its wide boulevards are lined with imposing villas, built by a German elite two centuries ago, and swarm with gleaming Mercedes and BMWs. The skyline of the old city is pierced by spindly brick spires – also built by industrious Lutheran Germans. It is hard to escape the shadow of the Teutonic Knights who conquered the Baltic region in the fourteenth century and whose descendants dominated Riga until the end of the First World War. In one Lutheran church, I notice a wall plaque dedicated to a composer and concert meister, Johans Gotfrids Mitels (1728–88), who is also buried as Johann Gottfried Müthel. But Riga is not a fustian museum city. Although the global recession hit Latvia hard, pushing up unemployment to 23 per cent, many young Latvians conduct themselves like students all over Europe, crowding into busy new internet cafes, American-style coffee bars and McDonald's. A rather beautiful tree-lined canal flows through the centre of Riga, crossed by the Freedom Boulevard. At the intersection stands the granite-clad Freedom Monument, built in 1935 to honour the soldiers killed fighting for Latvian independence in 1919. It is a potent symbol of nationhood which has withstood three foreign occupations. Next day, on 16 March, the Latvian SS legionaries would march here from the Dom cathedral and lay wreaths to their fallen comrades.

In 1939, under the secret terms of the Nazi-Soviet Non-Aggression Pact, Soviet forces had occupied the Baltic States, instigating a reign of terror and deporting tens of thousands of Latvians. In June 1941 Hitler launched Operation Barbarossa, the invasion of the Soviet Union, and by early July had driven Stalin's armies out of the Baltic region. To begin with, many Latvians welcomed German troops as liberators – a pattern repeated elsewhere in the east. But the new masters of Latvia swiftly threw together an occupation regime whose savagery eclipsed the brutality of the Soviets. German administrators amalgamated the three Baltic States into a single entity – the Ostland – effectively abolishing them as sovereign nations. On the heels of the German armies came the Einsatzgruppe – the Special Task Force death squads that unleashed the systematic mass killing of Jewish civilians in a bloody swathe across the Baltic, Belorussia and Ukraine. As these death squads moved north towards Leningrad, the German SD (Sicherheitsdienst), an agency of Heinrich Himmler's SS, began recruiting fanatical young Latvians as auxiliary policemen and used them to murder Latvian Jews. These so-called Schuma battalions proved horribly effective. By October, at least 35,000 Jews had been murdered. In the summer of 1942, SS Chief Heinrich Himmler authorised recruitment of 'non-German' Waffen-SS soldiers in neighbouring Estonia – and extended the net to Latvia at the beginning of 1943. According to the Latvian government, more than 100,000 Latvians ended up serving in the German SS.[7] On 16 March 1944, as the Soviet Army drove Hitler's armies towards the Baltic, the two Latvian SS divisions fought 'shoulder to shoulder' against the Russians on the banks of the River Velikaya. It is these allegedly heroic events that are commemorated on Legion Day. A few brigades of the Latvian SS that survived these terrible battles ended up defending Berlin, Hitler's last 'Fortress City'. After the destruction of the Reich, the Russians rapidly consolidated their occupation of the three Baltic States and turned them into Soviet socialist republics. As Riga's Occupation Museum insists, this was the second Soviet occupation – and this time the Russians held the Baltic in an iron grip for nearly half a century. Few Latvians who endured these grim years imagined that the vast Soviet Empire would collapse with such humiliating speed – and that Latvia would once again become an independent nation and part of the European Union.

Freedom is a heady drug. But it can also be a sour blessing. In the aftermath of independence, Latvia, Estonia and Lithuania successfully applied to join the European Union. As a condition of membership, the new Baltic governments came under intense and unwelcome pressure to 'document and clarify' crimes against humanity committed on their territory during the Second World War. This necessarily implied exposing the role of collaborators – whose participation in mass murder was already well documented by scholars. The Latvian government has always insisted that historians in the west are excessively preoccupied with the Holocaust and overlook Soviet crimes. They insist that their nations had suffered equally under Soviet and German occupation. The near destruction of Latvian Jews should never be accorded an elevated moral status overshadowing the fate of other Latvians. Thus German and Soviet crimes became morally equal – and it is this historical relativism that encourages some Latvians to sanction the commemorative rituals of the 'Latvian Legion'. These veterans did not fight for Hitler – 'they defended Latvia against the Soviet army.'[8]

Shortly after I arrive in Riga, I meet Michael Freydman in the 'Peitava-Shul', the single Riga synagogue to survive the German occupation, which has recently been restored. Squeezing into a tiny, hemmed-in lot on Peitavas Street, the synagogue is exquisite. As I look for the entrance an edgy police officer watches me warily. Inside, Mr Freydman points out the Hebrew dedication from the Psalms, above the Ark: 'Blessed is Jehovah who hath not given us/A prey to their teeth.' Mr Freydman has no time for the moral sophistry that not just forgives but honours men who swore oaths of loyalty to Adolf Hitler as Waffen-SS recruits. He points out that in Latvian schools, students rarely hear the word Holocaust – instead they are taught about 'the three occupations'. This 'occupation obsession' has now become the mantra of amnesia. But the few survivors of the Latvian Holocaust cannot forget that many thousands of their fellow citizens proved all too eager to volunteer as executioners for the Reich. In 1935, some 94,000 Jews lived in Latvia – about 4 per cent of the population. After 1941, the German occupiers and their Latvian collaborators murdered at least 70,000 Latvian Jews in camps, ghettoes and in the countryside; 90 per cent of Latvia's Jews died as 'prey to their teeth'. The Legionaries made a choice – and it was the wrong one.

On the afternoon before Legion Day, I catch a train to a tiny station just outside Riga, called Rumbula. Between the railway line and the main road to Riga, there is a silent and enclosed glade of trees. Twisting paths link low concrete rimmed mounds. These are mass graves. Here, at the end of November 1941, SS general and police chief Friedrich Jeckeln and his Latvian collaborators, led by the notorious Victors Arājs, slaughtered more than 27,800 Jews in two days. Himmler thoroughly admired Jeckeln as a highly proficient mass murderer. He knew he would 'get the job done' quickly and efficiently. Jeckeln had invented a 'system' that he referred to, with grotesque cruelty, as 'sardine packing', which he had honed and refined at killing sites in Ukraine. 'Sardine packing' allowed the SS men and their collaborators to 'process' many thousands of victims every hour, ransacking their possessions then dispatching them at the edge of a pit. At Rumbula, Jeckeln applied his highly regarded 'system' with industrial efficiency – and without mercy. After each day's 'work', the SS men recycled their plunder. Clothes, jewellery, money even children's toys ended up enriching the lives of supposedly needy German families.

I am the only visitor to the Rumbula memorial site that morning. A few hundred yards away, gleaming Mercedes race along the road to Riga or pull off into a glitzy new shopping mall. Mountains of litter have washed up along the edge of the memorial site. The only sounds are the wind in the trees and the distant rumble of traffic. Latvian historians like to emphasise the macabre fact that Himmler authorised recruitment for the Waffen-SS in Latvia after the majority of Latvian Jews had been murdered. It follows, they claim, that the 'Latvian Legion' had 'nothing to do with the Holocaust'. This callous argument wass put to me on a number of occasions during my visit – most forcefully by Ojārs Kalniņš, the eloquent Director of the Latvian Institute. The claim is a puzzling one. Many of the Latvian police auxiliaries who voluntarily took part in Friedrich Jeckeln's 'special action' at Rumbula, as well as hundreds of other mass shootings of Jewish civilians, later enlisted in the 'Latvian Legion'.

Tuesday, 16 March 2010. For Latvians, this has been the worst winter for thirty years and overnight temperatures have plummeted. Heavy snow falls and long lines of traffic crawl blindly across the Daugava bridges, generating a sickly yellow haze.

A giant Baltic ferry squats in the iced-up river. Snow ploughs rumble through Riga's old town towards the Dom, where the legion will begin its march to the Freedom Monument. Ice sheaths a red granite memorial to the Latvian 'Red Rifleman', recruited by the Russians at the end of the First World War to fight the German Imperial Army – a reminder that many Latvians backed the Soviets and fought against the Latvian SS divisions.[9] Soon after dawn, police vehicles park close to the Dom, engines running to warm the police reserves still sheltering inside. From misted windows, they gaze into the swirling snow, gloomy and bored. Their comrades on duty outside in the blizzard are dressed in beetle-like black armour and helmets.

The thick snow shrouds the towering spire of the Dom, a monument to German Lutheranism. Outside the main entrance, journalists and film crews outnumber police. Cameras flash as elderly men, accompanied by wives, most clutching bunches of Easter flowers, hasten inside. The veterans are like phantoms who return here every 16 March, bringing a chill and unwelcome reminder of the past. On the corner of Doma Laukums square, a knot of old men huddle together, shivering and selling copies of a pamphlet about the 'Latvian Legion'. A few old men stop to tell their stories, evidently knowing the routine: 'Forget the SS: we fought for Latvia, for freedom.' When I arrived at Riga airport I was astonished to see that newspaper stalls sell weighty memoirs written by 'Latvian Legion' officers. Since 1991, the organisations that support the veterans have hammered out a shared historical narrative that explains and justifies joining the war on the side of Hitler's Reich. Although I have contacted *Daugavas vanagi*, the veterans association, to request an interview, it becomes increasingly clear that these old men are conveyors of the party line, not historical testimony.

The journalists and photographers shivering outside the Dom expect trouble. Inside, camera crews and photographers already gather beneath the tall, plain nave. The Legionaries and their wives fill the front rows beneath the pulpit. A few sat with tears streaming down their cheeks; others glare angrily at the flashing cameras. Outside the Dom's main entrance, snow is still falling thickly. A sinister honour guard begins to muster. Shaven-headed young men from the nationalist Klubs 415 stand in line beneath a canopy of billowing red and white Latvian flags. Standing to one side are young thugs who had travelled up from Lithuania to support the old Legionaries. They sport white arm bands – modelled on those worn by wartime Lithuanian death squads.

Soon the old Legionaries stumble from the Dom to join these guardians of Baltic national pride, who close ranks around them. The snowstorm at last began to falter. A stern-faced young man with a shaven head takes up a position at the head of the legionary column. Right behind him stands the national leader of the ultranationalist Visu Latvijai (All for Latvia), Raivis Dzintars, and his wife, both clad in Latvian folk costume. The couple add a curious, even kitsch dash of colour – like morris dancers leading a march by a far-right British political party. But there is nothing pretty about Dzintars' political views: Visu Latvijai aggressively promotes the cause of a mono-ethnic Latvia. Latvia for Latvians! It was this brand of aggressive chauvinism that led many nationalist Latvians to throw in their lot with the German occupiers in 1941.

By now, police battalions are lined up along the route of the march – they stretch like glistening black insects all the way from the Dom to the Freedom Monument, a mile or so away, where the march will end. It is here that Latvians and others opposed to the march have been corralled. At the Dom, the old Legionaries finally set off, led by the peasant couple and the skinhead, his face set hard. Banners ripple in the cold wind. The elderly Legion veterans march briskly through Riga's old town and then cross the bridge that leads to the Freedom Monument. As the column approaches, ethnic Russian communists shout obscenities: 'Fuck off! Fuck off!' As the legion veterans, now shielded by an impenetrable cordon of armed police, begin laying wreaths, high voltage arguments spark up among the crowds.

A short distance behind the police lines stands a smaller, silent group of older men and women – Latvian Holocaust survivors. Standing with them today is Ephraim Zuroff, the Director of the Simon Weisenthal Centre, who has fiercely denounced Legion Day, and Josef Koren, a former beekeeper and now leader of the LAK, Latvia's Anti-Fascist Committee. When a Legion supporter screams at Koren that 'A soldier is a soldier and all are equal!' he turns away. Another mantra of Legion defenders is that the volunteers were conscripts – compelled to join. But as Koren points out to journalists, 'At least 25% of the "Latvian Legion" were volunteers, recruited from the Latvian police who were involved in the murder of Jews and other Latvians – and the SS Legion should not be permitted a celebration of itself in the centre of our city'.

Midday. Sunlight glitters on the Pilsetas canal. The old Legionaries and their honour guard begin to disperse. Soon they have vanished – the mute ghosts of history.

Now there is a carnival atmosphere. At the foot of the Freedom Monument, groups of young Latvians take pictures of each other beside the mass of wreaths and flowers. A young man tells a BBC reporter that for him the old Legionaries are heroes. They defended Latvia. Many thousands of Latvian SS men gave their lives for the freedom of Latvia. These young Latvians look prosperous and happy. They do not shave their heads or sport provocative armbands. But their enthusiasm for the legion is troubling – and unexpected. It would seem that the old Legionaries have become a symbol not of collusion with a murderous foreign occupier but of Latvian national freedom.

It is an outcome that SS Chief Himmler, who was profoundly hostile to the national aspirations of Latvians, could never have foreseen.

I remember the words of the great Latvian poet Ojārs Vācietis:

So all forests are not like this …
I stand and shriek in Rumbula –
A green crater in the midst of grainfields

Introduction

> With Germans it is thus, if they get hold of your finger, then the whole of you is lost, because soon enough one is forced to do things that one would never do if one could get out of it.
>
> Viktors Arājs, commander Latvian Arājs Commando[1]

> I really have the intention to gather Germanic blood from all over the world, to plunder and steal it where I can.
>
> Heinrich Himmler[2]

In the summer of 1944, a racial anthropologist serving with the SS, Oberscharführer Dr Bruno Beger, received an unusual assignment. He was ordered to travel to Bosnia–Herzegovina, then part of the puppet state of Croatia, to prepare a study of 'races at war'. He would focus on Bosnian Muslims serving in a Waffen-SS division called the 'Handschar', meaning scimitar. Its official designation was the 13th Mountain Division of the SS (1st Croatian).[3] More than 10,000 Bosnian Muslims had been recruited in the spring of 1943 with the connivance of the Grand Mufti of Jerusalem, Haj Amin el-Husseini – the Arab nationalist leader then resident in Berlin. The SS issued the Muslim recruits with standard uniforms but permitted them to wear fezzes bearing the death's head and eagle of the SS. Himmler and the mufti recruited and trained divisional imams who preached the doctrine of 'Jew hatred' to the recruits. The following year, as the military situation in the Balkans deteriorated, Beger was transferred to another Muslim SS division based in northern Italy – the Osttürkischer-Verbänd, recruited in the Caucasus. The SS 'Handschar' carved a bloody trail of murder and destruction across the Balkans in the final years of the Second World War. The German invasion of Yugoslavia that began in April 1941 had unleashed both massive repression and overlapping civil wars that continue to bedevil this fractured region. The atrocities committed by German sponsored militias like the Croatian Ustasha and Bosnian 'Handschar' have never been forgotten or forgiven.

But why did the elite Waffen-SS recruit Bosnian Muslims, an inferior south Slavic people according to Nazi doctrine, to join what Hitler called a 'war of annihilation'? Why, for that matter, did they recruit Latvians, Ukrainians, Kossovar Albanians, Estonians and a multitude of other non-Germans? To be sure, the recruitment of foreign soldiers, pejoratively labelled mercenaries, has, of course, been a convention of most wars throughout recorded history. The armies mustered by the Persian ruler Xerxes in the fifth century BC, for example, sucked in fighting men from all over the ancient world, including Jews, Arabs, Indians, Babylonians, Assyrians and Phoenicians.[4] In modern European history, Napoleon's *Grande Armée* boasted divisions and brigades of German, Austrian Dutch, Italian, Croatian, Portuguese and Swiss troops recruited from all over the French Empire and its vassal or allied states.[5]

The ethnic diversity of the armed forces of the Third Reich far exceeded Napoleon's *Grande Armée*. But that is not the principal reason why the recruitment of non-German troops by the Third Reich is surprising and paradoxical. The dogma and practice of the Third Reich was racism so radical that it culminated in mass murder on an unprecedented scale. Hitler characterised National Socialism as 'a Völkisch and political philosophy which grew out of considerations of an *exclusively racist nature*.'[6] The war launched by Hitler and his generals in September 1939 was intended to begin the task of 'annihilating' the racial enemies of the Reich, usually characterised as 'Jewish-Bolsheviks', and enslaving Slavic 'sub-humans'. The outcome would, in theory, be the founding of a new German empire and the complete 'Germanisation' of vast tracts of Eurasia. In other words, the German imperial project was by definition a racial undertaking. The ambition of SS Chief Heinrich Himmler was to forge the Waffen-SS as the elite shock troops of this racial imperialism: the apostles of Germanisation. Why then did he recruit apparently non-Aryan Latvians, Ukrainians and Bosnians?

The majority of historians have explained SS recruitment strategy as an expediency that fatally compromised the elite status of the militarised SS. The most recent history of the SS by Adrian Weale asserts: 'In 1940, [the Waffen-SS] had legitimately been able to claim that it was an elite ... by June 1944 ... in no military sense could [the bulk of the organisation's combat units] ever be described as a corps d'elite.'[7] This is the latest reformulation of a view that has been repeated ad nauseam by most historians of the SS. In short, they argue, Himmler simply needed bodies in SS uniforms to hurl at the advancing Soviet armies. It was a numbers game – a necessary evil.

In this book I propose a different explanation. The recruitment of non-Germans not only complied with Nazi-sponsored race theory as it evolved during the course of the war, but was a vital component in a master plan hatched up by secretive SS 'think tanks'. Himmler was despised by many of the Nazi elite as an obsequious and petty-minded bureaucrat – a judgement echoed by many modern historians. This was a sham. Himmler's imagination was secretive, lethal and boundless. His covert master plan was to build a German empire dominated not by Hitler and the Nazi Party (NSDAP), but the SS. The construction of this SS 'Europa' required the complete physical liquidation of every racial enemy of the Reich. At the same time, Himmler and his cadre of SS experts proposed a root and branch re-engineering of European ethnicity. To enact this monstrous scheme, Himmler transformed the SS into a formidable militarised apparatus dedicated to blood sacrifice. SS police battalions and Waffen-SS divisions would become the armed agents of a perverted revolution whose outcome would be a racial utopia. Naturally, Himmler did not discuss these ideas openly, but he provided some tantalising clues about the SS plan in the course of a conversation with Avind Berrgrav, the Archbishop of Norway. SS recruitment, he makes clear, was not a matter of numbers – he wanted the best of the best, the pinnacle of the 'Germanic' peoples:

> 'Take the regiment Nordland [SS division] as an example,' Himmler says, 'Do you believe that we need these men as soldiers? We can do without them! But we mustn't block these men from freely pursuing their desires. I can assure you that they will return as free and committed supporters of our system.'[8]

This was not Hitler's plan. While Himmler dreamt of a future SS 'Europa', Hitler clung to the petty minded ideas of the barrack-room bigot. He admired, grudgingly, the British Raj and its subjugation of dark-skinned masses. He despised the Indian nationalist Subhas Chandra Bose, who fled to Berlin to seek German assistance against the British, and dismissed his Indian Legion, recruited by the German army, as 'a joke'. Whilst Himmler regarded Bose and his Indian recruits as members of an 'Aryan brotherhood'. He sponsored a German 'scientific expedition' to Tibet to look for racial connections between European peoples and Tibetan aristocrats. SS 'Europa' was just the beginning. Writing in 1943, Himmler looked forward a few decades to when 'a politically German – a Germanic World Empire will be formed'. To begin with, Himmler's master plan embraced only the Nordic peoples of Western and Central Europe. Just as Hitler did, he viewed the east as the murky domain of Slavic hordes whose degenerate blood was a mortal threat to European survival. The experience of war changed his mind – and led to a radical rethinking of long-term SS strategy.

At the end of June 1941, Hitler's armies swept into the Soviet Union. Millions of Soviet soldiers fell into German hands, and were incarcerated in vast open camps built hurriedly in occupied Poland. These camps became instruments of mass murder. More than 2 million Soviet prisoners would perish from disease, deliberate starvation or at the hands of execution squads, many because they 'looked Jewish'.[9] But for German anthropologist Wolfgang Abel, who was attached to an SS agency called the Race and Settlement Office (RuSHA), these hellish camps provided a pseudoscientific treasure trove. Inside this German camp, Abel and his team could examine and measure hundreds of living 'specimens' culled from every corner of the Soviet Empire. They soon made some startling discoveries. Abel's meticulous anthropometric examinations revealed that Germanic blood lines had penetrated far into the east through the Baltic, Ukraine and beyond. In the 'General Plan East', hatched up after the German invasion of the Soviet Union, SS scholars had proposed the complete Germanisation of conquered eastern territories. In crude terms, they envisioned liquidating native peoples and importing German settlers. The findings of the 'Abel Mission' significantly complicated matters. The simplistic distinction between Germanic and Slavic peoples began to look a lot more intricate.[10] These anthropological findings implied that some 'Eastern' peoples might possess sufficient 'Germanic' blood to qualify as future citizens of the Reich. Later, Himmler would reconsider the racial status of Balkan peoples like the Bosnian Muslims, the Bosniaks.

But how could these 'Germanic bloodstreams' (a phrase used in an SS instructional pamphlet) be exploited? Could this 'lost blood' somehow be returned to the Reich, where it belonged? In the perverse logic of German racial ideology, this Germanic blood was merely a latent quality. It was a potent substance, to be sure, but did not necessarily guarantee that its bearers would loyally serve the future Reich. Himmler had a radical solution. He would 'harvest' this lost Germanic blood through martial service and blood sacrifice. Himmler revered the pseudoscientific ideas of anthropologists like Hans F.K. Günther, who interpreted race in strictly biological terms. But he also admired ideas promoted by Günther's rival, the psychologist Ludwig Ferdinand Clauss, and his followers. In books like *Rasse und Seele*, published in 1926, Clauss had developed a somewhat heretical theory that different races possessed different 'souls'. Germans, for example, manifested the attributes of a noble Nordic soul; Jews were cursed by their materialistic 'Semitic' souls. The details of this gaseous speculation need not detain us here. But the idea of a 'racial soul' detached from merely physical attributes implied that race was to some degree malleable. For Himmler, racial identity was also a matter of will, capable in special circumstances of reshaping biological inheritance. According to this cowardly soldier manqué, the supreme manifestation of will was the warrior's acceptance of the need to kill and be killed. Himmler called the Waffen-SS the 'assault force for the new Europe'. He believed that military service, sacrifice and, above all, the zealous destruction of the racial enemies of the Reich, could provide the means to remould the racial 'souls' of non-German recruits – opening the door to membership of the greater Germanic community.

This master plan did not only apply to the Waffen-SS – the armed SS. The rapid expansion of Himmler's empire and its security division, the SD (later renamed the RSHA), had begun in the mid-1930s with the takeover of the German police services. For Himmler, there was no fundamental difference between a Reich policeman and a Waffen-SS soldier. Whether a recruit donned the green uniform of the German police or the asphalt grey of the Waffen-SS, he was a warrior dedicated to upholding the security of the Reich; his 'combat spirit' (*Kampfgeist*) would be dedicated to the 'ruthless annihilation of the enemy'.[11] Likewise, after 1939, the first wave of foreign SS recruitment drew in non-Germans as police auxiliaries – Schutzmannschaften (known as Schuma). Commanded by German SD officers, these men unleashed a campaign of mass murder directed at their fellow citizens. From the summer of 1942, Himmler began authorising the recruitment of non-German Waffen-SS units. Many former Schuma men transferred to the new divisions. Himmler's master plan had astounding consequences. In the summer of 1942, Himmler authorised the formation of an Estonian SS division – then began recruiting Latvians the following year. In 1943, at least 15,000 Bosnian Muslims were admitted to the Waffen-SS. Just over a year later, by the summer of 1944, over 50 per cent of Himmler's Waffen-SS soldiers had not been born in Germany; every SS division had foreign recruits and nineteen were dominated by non-German recruits.[12] At the end of the war, the SS absorbed over a million Soviet Osttruppen (eastern troops), many of them Muslims. Indians, Arabs, Albanians, Croats, Ossetians, Tadjiks, Uzbeks, Bosnians, Ukrainians, Azerbaijanis and even Mongolian Buddhists eventually joined Himmler's foreign legions.

At the end of April 1945, as Hitler ended his life in the Führerbunker beneath the Berlin Reich Chancellery, a few hundred yards away French, Belgian and Latvian SS men fought alongside German Volksstum and Hitler Youth brigades, vainly struggling to hold back the irresistible deluge of Stalin's armies. At the same time, at least 10,000 Ukrainian SS men fled west hoping to surrender to the

Allies and evade arrest by vengeful Soviet NKVD battalions. All over Europe, the foreign legions of the Reich had to confront the brutal reality of defeat. They cast a long shadow.

SS foreign recruitment appears to challenge Daniel Goldhagen's hypothesis that German 'exterminatory anti-Semitism' provided the motor of the Holocaust, the systematic mass murder of the Jews of Europe. Goldhagen's celebrated *Hitler's Willing Executioners* was published in 1996. Like Christopher Browning's *Ordinary Men*, published four years earlier, Goldhagen focused on the German police battalions that had carried out mass shootings of Jews in occupied Eastern Europe. Goldhagen argued that the men who served in these battalions had been typical Germans, saturated in anti-Semitic hatred which made them 'willing executioners'. He implied that any German provided with the same opportunity to kill would have done the same as the policemen he studied.

According to Goldhagen, the Holocaust was thus a German crime: 'the outgrowth … of Hitler's ideal to eliminate all Jewish power.'[13] Hitler proclaimed that he wished to kill all Jews – and set about achieving this goal with the enthusiastic connivance of German citizens. Goldhagen claimed that the majority of Germans in the 1930s and 1940s sympathised with Hitler's plan; the Holocaust was, in this sense, a 'national project'. Goldhagen characterised so-called 'good Germans' as 'lonely, sober figures in an orgiastic carnival'. He concluded that this 'set of beliefs', shared by the majority of Germans, was 'as profound a hatred as one people has likely ever harboured for another'.[14]

No other book about the Third Reich has provoked such fierce debate – and, when it was translated and published in the newly unified Germany, so much soul searching. When Goldhagen embarked on a tour of Germany, an army of journalists and photographers pursued him wherever he travelled. It was said he 'looked like Tom Hanks' and became a trophy guest on the most prestigious television talk shows. A new generation of Germans seemed to want to wallow in the guilt of their grandparents. But after the grand tour and media commotion came sober analysis. Goldhagen the historian was soon discovered to have feet of academic clay. He was accused of misinterpreting research carried out by other historians, notably Browning, and ignoring any data that did not fit his theory. Historian Eberhard Jäckel called this son of Holocaust survivors a 'Harvard punk' and denounced *Hitler's Willing Executioners* as 'simply bad'. But after more than a decade of impassioned debate, Goldhagen's 'big bang' idea still stubbornly refuses to go down quietly. *Hitler's Willing Executioners* forced historians to think seriously about the perpetrators of genocide as well as the terrible fate of its many millions of victims.

The question I want to ask in this book is quite simple. Does Goldhagen's theory of 'German exterminatory anti-Semitism' account for the mass killing of Jews and other enemies of the Reich in Croatia, Romania, the Baltic, Belorussia and Ukraine and many other regions of Eastern Europe after 1939 at the hands of local militias? How does it explain for the eagerness with which hundreds of thousands of young non-German men rushed to join the armed forces of the Reich, above all the Waffen-SS after June 1941? Were these foreign collaborators not also willing executioners? Was the Holocaust not a German crime at all but a European phenomenon? In the case of Eastern Europe, the first major pogrom of the war took place in Romania in the city of Iaşi. As German armies swept into the Baltic nations, Belorussia and Ukraine, followed by the SD Einsatzgruppen murder squads, Lithuanians, Latvians and Ukrainians seized the chance to murder their Jewish neighbours in an orgy of seemingly spontaneous mass killings. Eastern Europe was consumed by a spasm of violence that consumed the lives of more than 5 million Jews, while in France, Belgium, Scandinavia and the Netherlands, collaborating militias betrayed, arrested and deported their Jewish fellow citizens to German camps. Many Holocaust perpetrators were not German. Surely, then, we must conclude that these non-German men and women too were Hitler's willing executioners?

It would appear that Goldhagen simply got it wrong. You did not have to be German to become what French historians call a *génocidaire*. Many of these foreign collaborators had been reared in national cultures equally infused with anti-Jewish loathing as Germany. Now, the motivations of many tens of thousands of auxiliary policemen and Waffen-SS soldiers are necessarily diverse and hard to define. For every fanatic there is an opportunist or thrill seeker. Apologists for the Waffen-SS foreign volunteers argue that they were soldiers 'like any other'. Military historians tend not to be interested in ideology – and in the case of the Second World War appear loath to discuss the Holocaust. But combat in the armies of the Third Reich, whether the regular army, the Wehrmacht or the SS police battalions and Waffen-SS, meant signing up to fight in a war that was not at all 'like any other', before or since. General Erich Hoepner summed up German military ethics as follows: '[the war] is the age old struggle of the Germanic people … the repulse of Jewish-Bolshevism … and must consequently be carried out with unprecedented severity … mercilessly and totally to annihilate the enemy … no sparing of the upholders of the current Russian-Bolshevik system.'[15] The enemy was defined not as a body of hostile armed men but as 'upholders of a system'. According to the perverse ideology of the Reich, any Jew somehow 'upheld' the Bolshevik 'system' simply by being Jewish. These 'ethics' necessarily sanctioned 'collective forcible measures' – meaning, in practice, the mass murder of non-combatants whose continued existence threatened the well-being of the Reich. According to the ethics of annihilation, the killing of unarmed civilians, men, women and children was no longer to be considered 'collateral damage' but an integral part of military strategy. The foreign volunteers who joined the various agencies of the Reich clearly understood the ethics of the German war.

As Hitler's armies swept into the Soviet Union, Reinhard Heydrich, head of the SS security service, began deploying 'native' police auxiliaries to carry out the 'self-cleansing' of their homelands. By this he meant mass murder of Jews and communist officials. German SD men and their native collaborators tore through the ancient Jewish communities of the east with unrelenting savagery. As Heydrich and his subordinates understood well, Eastern European nationalists regarded their Jewish neighbours as agents of Bolshevism. This irrational merging of the Jew and the Bolshevik, which was shared by the Germans and their collaborators, was a death sentence for millions. By the end of 1941, German police and Schuma battalions had shot at least a million Jews in Eastern Europe and the occupied regions of the Soviet Union. In the course of the following year, another 700,000 perished by shooting or in the so-called Reinhardt extermination camps. Millions died in lonely, unmarked forests and meadows in the east as well as in Auschwitz.[16] And their killers were not only German SS men and solders, but Latvians, Ukrainians and other Slavic servants of the Reich. As killing centres like Treblinka, Sobibór and Auschwitz-Birkenau (which was also a labour camp) took over the business of genocide, the native Schuma battalions ran out of work. It was during this transitional period that Himmler authorised, for the first time, the formation of eastern Waffen-SS legions or divisions. When they became soldiers rather than policemen, these men did not stop murdering Jews.

War is, by definition, a bloody business – so men in uniform tend to be excused a few 'excesses'. As Ian Kershaw puts it, historians have a 'tendency to separate the military history of the [1939–45] war from the structural analysis of the Nazi state'.[17] A new cadre of historians, led by Omer Bartov, have begun to dismantle artificial firewalls that have been built between politics, ideology and mass murder. The war in the east, Bartov argues, 'called for complete spiritual commitment, absolute obedience, unremitting destruction of the enemy'.[18] 'Unremitting destruction' succinctly defines the war Germany fought between 1939 and 1945 – and fighting it irrevocably and profoundly corroded the moral decency of its practitioners whether they were German, French, Latvian or Bosniak.

This book is not a general history of the SS or the Waffen-SS, nor does it set out to provide an exhaustive 'catalogue' of every non-German police battalion or combat division. Instead, it analyses in some detail specific case histories that illuminate the recruitment of non-German collaborators as agents of genocide. Part One begins with the German invasion of Poland and the simultaneous development of both a new doctrine of warfare and an 'armed SS' charged by Hitler with maintaining security in conquered territory. In Nazi doctrine, security depended on the liquidation of the racial enemies of the Reich. From the very beginning of the war, Himmler used SS police battalions and armed SS units as the vanguard agents of systematic mass murder. During the short Polish campaign, the Germans made only limited use of non-German forces – mainly ethnic Germans and Ukrainians. After the invasion of the Balkans in spring 1941, German-backed native militias like the Ustasha in Croatia and the Iron Guard in Romania embarked on lethal campaigns directed at Jewish citizens. For the Germans, these pogroms provided crucial lessons about the deployment of non-German executioners, strongly implying that a murderous solution to the 'Jewish problem' had been hatched up, at least partially, before the summer of 1941.

The Balkan pogroms provided a rehearsal for genocide – and encouraged the SS to cultivate ultranationalist factions in the Baltic and Ukraine. We discover in Part Two that this meant that just days after the German invasion of the Soviet Union, on 22 June 1941, Himmler's Special Action squads began recruiting suitable Lithuanians, Latvians and Ukrainians to assist in the arduous tasks of mass murder. At the same time, German military intelligence under Wilhelm Canaris formed two Ukrainian combat battalions known as the 'Roland' and 'Nachtigall', which also took part in mass killings of Ukrainian Jews. Although the Germans quickly disbanded the two battalions, they demonstrated that combat battalions could also be deployed as mass murderers of unarmed civilians classified as enemies of the Reich.

Eastern Europe was a geographical locus of the worst genocide in history. This is where the SD murder squads were deployed; this is where the Germans built their camps. As a consequence, Eastern European collaborators took a direct role in mass murder, under the auspices of SS commanders. However, the western SS volunteers from France, Belgium and the Netherlands, for example, espoused the same ideological commitment to the destruction of 'Jewish-Bolshevism'. Recruitment of these 'Germanics' had begun in 1940, but gathered pace after the invasion of the Soviet Union. I examine in some detail the case of the notorious Belgian collaborator Léon Degrelle to expose the complex motivations of these 'crusaders against Bolshevism'. In the summer of 1942, Himmler began authorising recruitment of non-German Waffen-SS divisions in the east, starting in Estonia. This new phase of recruitment accelerated after the destruction of the German 6th Army at Stalingrad – but also reflected a step change in SS thinking about race. By then Himmler had begun to view recruitment as a means to facilitate 'Germanisation'. As Soviet partisans, pejoratively referred to as 'Banditen', began to successfully challenge German security in the occupied east, Himmler mainly used these eastern legions as anti-partisan units. Since the Germans referred to Jews as 'bandits', it also meant that the foreign SS divisions continued murdering Jews who had, through whatever good fortune, survived the SD murder squads and Operation Reinhardt.

Part Three opens in the summer of 1944, when Himmler's SS was a militarised state within a state that had been bloated by its recruitment of non-Germans. Despite calamitous military reversals on every front, Himmler continued to think in terms of a Greater Germanic Empire – defended by a pan-Germanic army, toughened by combat and zealous mass murder. Himmler had begun to think 'beyond Hitler'. The image of Himmler, memorably set out not long after the war ended by Hugh Trevor-Roper in *The Last Days of Hitler* as the Führer's most loyal paladin and, in his own mind at least, heir apparent, has rarely been questioned. In the final part of this book, I suggest a more complex if not completely contradictory interpretation. For Himmler, loyalty was a brand – a means to ascend in the treacherous world of Hitler's court and to fix the corporate identity of the SS. Affirming loyalty may well have been a psychological necessity for this enigmatic bureaucrat, but Himmler knew that any overt challenge to Hitler would have led to catastrophe. In a succession of barely perceivable steps, Himmler's ambition began to outstrip Hitler's. His covert master plan was grounded in an elastic pseudoscientific logic that however lunatic it now appears, inspired a future vision that left the Nazi Party and its leader far behind. 'Germanisation' implied both a massive destruction of life alongside the co-option of suitable non-Germans as the dog soldiers of conquest and occupation. For Hitler, war was a means to extract living space in Eastern Europe and impose German hegemony. For Himmler, it was merely the prelude to the ethnic transformation of Eurasia as a Nordic empire.

In March 1945, as the 'Thousand-Year Reich' collapsed under Allied hammer blows, these two very different visions finally collided. Hitler excommunicated Himmler and sentenced him to death. The final break was provoked by news of Himmler's futile contacts with the Allies. But the fuse had been lit years before, and then burned silently out of sight until the downfall of the Reich. The 'loyal Heinrich' was no more. But Himmler had little time to enjoy a world without Hitler. Spurned by the provisional new government of Admiral Karl Dönitz, he wandered aimlessly through northern Germany. At the end of May 1945, Himmler, disguised as an officer in the Geheime Feldpolizei (Secret Military Police), was captured by a British army unit. His last reported words, before biting on a cyanide pill concealed in his mouth, were 'I am Heinrich Himmler'.

Part One:
September 1939–June 1941

1

The Polish Crucible

Genghis Khan hunted millions of women and children to their deaths, consciously and with a joyous heart. History sees him only as the great founder of a state.

Hitler, August 1939

On 22 August 1939 Adolf Hitler summoned the German army high command to his southern headquarters in the Bavarian Alps, the Berghof, near Berchtesgaden. The generals and their adjutants tramped past the massed cactus plants in the entrance and assembled in the Great Hall, dominated by a giant globe and vast picture window that looked out towards Austria, now absorbed by the Reich. In his study here, Hitler spent many hours sipping tea and gazing at the rocky flanks of Untersberg Mountain where according to legend the red-bearded German emperor, Frederick Barbarossa, lies entombed, awaiting a wake-up call to rescue Germany in its hour of need. Hitler would attach Barbarossa's name to the invasion of Russia in June 1941. He should perhaps have recalled that the German emperor had not perished in battle with the infidel, but had drowned while bathing in an Armenian river.

The German top brass had come to hammer out the objectives of Fall Weiß (Case White), the plan for the invasion of Poland.[1] Against the dazzling background of the Bavarian Alps, Hitler unveiled a dizzying vision of conquest. He informed his generals that German relations with Poland had reached a political nadir. Polish provocation was 'unbearable' – the only solution was the literal destruction of the Polish nation. This meant that the success of Case White depended on waging a new kind of warfare. Germany, Hitler insisted, would not only be asserting its alleged historic rights to the Polish lands – 'an extension of our living space in the East'. The task of the German armed forces would be to eliminate a 'mortal enemy' of the German Reich: the Polish elite. Hitler clarified what he meant by this: Poland's 'vital forces' (*lebendige Kräfte*) must be liquidated: 'It is not a question of reaching a specific line or new frontier, but rather the annihilation of the enemy, which must be pursued in ever new ways.'[2] Hitler's language left no room for ambiguity: 'Proceed brutally. 80 million people [i.e. Germans] must get what is rightfully theirs.' At a later meeting he hammered home 'there must be no Polish leaders, where Polish leaders exist they must be killed, however harsh that sounds'.[3]

According to his diary account of the earlier meeting, German Army General Franz Halder eagerly concurred: 'Poland must not only be struck down, but liquidated as quickly as possible.' The Prussian elite relished this new opportunity to smash the hated Poles who all too often had risen from the ashes of defeat. Now they would be finished off once and for all. Hitler and his generals conceived the Polish campaign as a 'war of liquidation'. Poland would not simply be conquered but destroyed. 'Have no pity!' Hitler insisted. Wehrmacht generals like Halder often used words like 'liquidation' and evidently had few misgivings about the 'physical annihilation of the Polish population'.

Prussian military doctrine had long demanded 'absolute destruction' of the enemy's fighting forces ('bleeding the French white' in 1871), as well as the punitive treatment of enemy culture and civilians. But Hitler's new war strategy insisted on unprecedented 'harshness'. The problem for his generals was not a moral but a practical one. In purely military terms, liquidation of a nation's 'vital forces' was time consuming and necessarily meant diverting troops from 'Zones of Operations and Rear Areas'. SS Chief Heinrich Himmler and his oleaginous deputy SD head Reinhard Heydrich realised that Hitler's 'war of annihilation' offered astonishing opportunities. The SS would assume responsibility for liquidation, security and 'mopping-up' operations, meaning mass executions – onerous tasks best handled by specialised militias that the SS could readily supply. In return, Himmler would demand an ever expanding share of the political and material rewards of occupation.

The Polish campaign of 1939 would provide Himmler with a breakthrough opportunity to transform the SS into the vanguard force of this new kind of war. The destruction of Poland would begin laying the foundations of an embryonic plan to remould the ethnic map of Europe. Although the Germans would deploy few non-German troops in Poland, the war applied SS doctrine for the first time to actual military practice. To understand Himmler's vision of modern racial war, we need to look at the way the destruction of Poland forged the radical ideology of Hitler's 'political soldiers'.

SS Chief Heinrich Himmler was notoriously inscrutable. The dutiful son of a reactionary Bavarian schoolmaster, he had missed out on martial glory in the First World War and been educated as an agronomist. He seemed to enemies and friends alike as Sphinx-like but unexceptional, with the manners of a fussy schoolmaster, a plodding pedant obsessed by homeopathic remedies and oddball pseudoscientific fantasies. But this cold-hearted crank transformed Hitler's bodyguard, the Schutzstaffel, into a 'state within a state' that directly managed the plunder of occupied Europe and the slaughter of millions. Psychological analysis of the 'architect of genocide' has generally spawned the most banal speculation; there can be no doubt that loyalty and devotion were at the heart of Himmler's self-image and his relationship to Hitler. Hitler's craving for dog-like devotion from acolytes like Rudolf Hess is well attested. From the very beginning of his political ascent, he adroitly manipulated rival courtiers who felt obliged to continually reaffirm their devotion. Thanks to his father's assiduous cultivation of the Bavarian royal family, Himmler had developed refined skills as a disciple. He understood from very early on that the frequent affirmation of loyalty was the road to power in Hitler's competitive and treacherous court. For Himmler, such devotion was both a psychological need and a vital, thoroughly honed political skill. Hitler rewarded him with a much repeated soubriquet 'the loyal Heinrich' – which implies that he stood out from even his most sycophantic peers. And Himmler insisted that loyalty became the hallmark of SS ideology.

Himmler was a highly competent organiser and manager. Like Stalin, he made himself master of the card index file. No detail was too trifling. Himmler knew everything about everybody who mattered. He liked to deliver pompous homilies on the black art of political manipulation and fervently believed that the acquisition of power was a conspiratorial skill practised by 'wire pullers'. As 'loyal Heinrich', the manipulative Himmler put these insights to good use. The Baltic German Felix Kersten, who became Himmler's masseur and confidante, was surely right when he called his master a 'crass rationalist coldly taking human instincts into account and using them to his own ends'.[4] Although Himmler presented himself as 'loyal Heinrich', and evidently derived satisfaction from seeming dutiful, loyalty was a means to an end – one that would serve him very well in the slippery world of Hitler's court.

Unlike Hitler and many of the Nazi elite, Himmler had never experienced active service on the front line. This humiliating failure seems to have provoked in him a perverse need to embrace violence as an abstract human quality – one that profoundly shaped his world view. The Germanic or Nordic race, he believed from very early on, possessed a natural right to domination, but this racial privilege was resented and threatened by Jews and 'Asiatic' peoples. This antagonism could only be resolved through bloodshed. In January 1929 Hitler appointed Himmler Reichsführer-SS in charge of his personal bodyguards, the Schutzstaffel. This insignificant 'Gruppe' could muster just 280 men when Himmler received his appointment, but he seems to have grasped its potential very quickly. The rapid expansion of the SS is well documented. By the time Hitler seized power in 1933, membership had expanded to more than 50,000. Even more significant than these numbers was Himmler's understanding of brand and corporate identity. Drawing on very diverse models such as the Knights Templar, the Order of Jesuits as well as Italian Black Shirts, Himmler fashioned a distinctive paramilitary elite, replete with oaths and slogans, that was avowedly aristocratic. The SS that emerged after 1933 would spawn numerous agencies, militias and pseudo-academies like the Ahnenerbe, all dedicated to a radical refashioning of German imperialism. Himmler forged a political apparatus designed to enforce security on the Home Front and on the frontiers of an expanding imperial domain.

Hitler never sanctioned such profligate ambition. He could not afford to allow a single individual or agency to acquire hegemonic power. The Nazi state has often been viewed as an embattled arena in which highly aggressive power-brokers continuously jostled for favour and power. Hitler frequently handed the same apparently sovereign power to more than one of his paladins. After 1941, for example, the Reich Commissar of Ukraine, the notoriously brutal Erich Koch, waged war on his nominal superior, the 'Reich Minister for the Occupied Eastern Territories' Alfred Rosenberg. For Hitler, this wasteful duplication of powers was strategic. It allowed him to dominate squabbling competitors who would win or lose according to laws that mimicked the natural 'survival of the fittest'. Himmler understood this very well. It was essential that he disguise his master plan for the SS so that he retained his claim to be 'loyal Heinrich', not a rival. Hitler deftly exploited Himmler's anxieties concerning the intentions of his deputy Reinhard Heydrich. But Himmler rarely rose to the bait and took full advantage of the arcane mechanisms of the 'Chaos State' to pursue his own ends. His first big opportunity came in the summer of 1934.

In the period immediately after the Nazi seizure of power in 1933, Hitler was preoccupied with the thorny matter of the storm troopers (SA) and their ambitious leader Captain Ernst Röhm. A thuggish homosexual, Röhm insisted that his brown-shirted hordes deserved the lion's share of victory spoils now that Hitler had become Chancellor thanks to their hard work and fearless struggle. Now, the SA leaders insisted, a 'Second Revolution' was needed to finish the job and properly 'brown' Hitler's 'New Order'. Röhm's petulant ambition directly threatened the German army, the Reichswehr. He insisted that the SA should be acknowledged as Germany's principal armed force. By mid-1934, an indecisive Hitler, possibly unwilling to betray old comrades, had been persuaded to turn against Röhm – and to liquidate the anachronous SA leadership. Himmler had once been Röhm's deputy – but now he took a leading part in the assault on the SA leadership, the 'Night of the Long Knives'.[5] This notorious purge of troublesome former comrades marked a step change in the political fortunes of Himmler, the SS and Heydrich's SD. Himmler had both proven himself loyal and demonstrated that the new state depended on his growing security apparatus. The purge liberated the SS and SD from SA control – and simultaneously raised the public standing of the SS. It was after the violent summer of 1934 that the German middle and upper classes began to perceive the SS as a way of reinforcing their status in the New Order. Bright young men flocked to join, bringing with them the aggressive racial ideologies of the German universities. The SS now became an academy of the most reactionary kind as well as a security state within a state.

Himmler and Heydrich both understood that they had to move carefully to tighten their grip on power. They must appear to be the servants of the New Order – not its aspiring masters. The SS brand was 'loyalty'. It is surprising to discover that Hitler was unsettled by the ferocious bloodletting of the 'Night of the Long Knives', and the growing power of the SS disquieted both Hermann Göring and a relic conservative faction led by the Interior Minister, Wilhelm Frick. In the course of the next two years, Himmler cunningly discredited his opponents and cemented his own power. He did this mainly by exploiting the prejudices shared by the Nazi leadership and many ordinary Germans. He assumed that Germans were a superior people, with a natural right to hegemony in Europe and the east. Conquest and settlement of the east was of course a widespread obsession among both conservative Germans and radical nationalists like Hitler. But Himmler had acquired an emotional 'Eastern obsession' in his adolescence and it was he rather than Hitler who made this ultra-imperial aspiration such a pervasive ingredient in National Socialist thinking. Himmler's foreign policy – meaning German acquisition of eastern territories – was itself profoundly connected to his domestic thinking. German ethnic rights to natural hegemony were threatened and undermined by the enemy within: the Jews. In SS ideology, Jews, a people without a nation, naturally took on the role of 'international conspirators' with connections and kin in both Moscow and the capitalist economies. German destiny was, as ever, vulnerable to the mythic 'stab in the back'. Radical imperialism thus depended on scapegoating – and in the National Socialist mind, Slavic peoples, black people, Freemasons and Gypsies (Roma) might all take supporting roles to the Jewish leads. Himmler assiduously cultivated this dual mythology of blood-sanctioned imperialism and its shadow world of internal enemies. Himmler's allegedly eccentric fascination with German mythology was not in any sense whimsical; it was a means to reinforce the status of the SS as the standard-bearer and aggressive protector of Germanic values. It was in a sense a 'sales campaign'.[6]

After the breakthrough of 1934, Himmler played these two chords with monotonous persistence. By representing Germany as an embattled state, he drove home again and again the message that the New Order depended on its security apparatus, the SS. His efforts paid off in May 1936, when Hitler appointed him chief of the German police, thus binding together all the German police agencies under a single banner. Himmler had specified his own job description to Hitler, insisting in a private letter that he was to be 'Chief' not 'Commander' which implied a more circumscribed role. His appointment as Reichsführer-SS and chief of the German police on 17 June signalled Himmler's defeat of his main rivals, above all Interior Minister Frick.[7] Heydrich was a cunning negotiator. It was he not his boss who secured the final wording of Hitler's decree which referred to the 'unified concentration of police responsibilities in the Reich', and the responsibilities of the new chief of police as 'the direction and executive authority for all police matters within the competence of the Reich and Prussian ministries of the interior'.[8]

The consequences of Himmler's triumph were both organisational and ideological. He welded together all uniformed police into the Order Police (*Ordungspolizei*) and handed command to SS General Kurt Daluege. He appointed Heydrich chief of the Security Police (Sicherheitspolizei, or Sipo) which took over all detective police, both political and criminal. This administrative reorganisation was an astonishing feat for it yoked together the SS and German police, creating at a stroke the foundations of an SS/police state. Although the Sipo and the SD remained administratively separate, they shared a single head, namely Heydrich – and two years later would be amalgamated under his command as the Reich Security Main Office (RSHA).

To fully appreciate the ideology of this administrative legerdemain we need to understand the pernicious theoretical underpinning of Himmler's ambition. In his world view, Germany's imperial ambition depended on combating internal enemies; forces that threatened Germany's natural rights of conquest. These foes of the Reich were by definition criminals – and criminality was itself the mark of 'alien' ancestry. Accordingly, the defence of the Reich depended on liquidating any criminal element – conceived in racial and genetic terms. It was this overlap between the figure of the criminal and the racial outsider that reinforced the exclusion of German Jews and justified a radical solution to the 'Jewish problem'. Criminal behaviour was 'Jewish'; all Jews were potentially enemies of the state. This pseudo-logic implied in turn that Himmler's policemen were also soldiers – warriors tasked with defending the Reich as its borders expanded. By fusing national security with imperial ambition, Himmler prepared the way for mass ethnic slaughter. The roots of the German genocide can be traced back his appointment as RFSS and police chief.[9]

We can now return to September 1939; to the moment when Himmler's SS would be 'blooded' in the first act of Hitler's 'war of annihilation'.

For Himmler's SS, planning for Case White, the invasion of Poland, began as early as May 1939. All the major offices of the Reich participated and protracted negotiations concerning the deployment of SS paramilitary police and the embryonic Waffen-SS, the SS-VT regiments (*Verfügungstruppe*), were convened between representatives of the Gestapo, the OKH (the Army High Command), and the office of military intelligence, the Abwehr. True to form, Himmler's number two, Heydrich, secured a leading part in these preparations, and reported directly to Hitler. At SD headquarters, Prinz Albrecht Strasse 8 in central Berlin, he set up a new office to direct Operation Tannenberg: the 'Zentrastelle II P', the P referring, of course, to Poland. He appointed Franz Six, considered to be an expert on 'Jewish matters', to head the new department. SD bureaucrats under SS-Oberführer Heinz Jost began compiling target lists (*Sonderfahndungsliste*) that named some 61,000 Polish Christians and Jews, broadly categorised as 'anti-German elements' – meaning those 'elements hostile to the Reich and to Germany in enemy territory behind the troops engaged in combat'.[10] These diligently compiled file cards would provide the blueprint for mass murder.[11]

Heydrich later confided to Daluege that Hitler had given him an 'extraordinarily radical order' for the 'liquidation of the various circles of the Polish leadership', meaning clergy, nobility, Jews and the mentally ill.[12] Hitler's criminal order prefigured the notorious Commissar Order (*Kommissarbefehl*) and the *Kriegsgerichtsbarkeitserlaß* issued before Operation Barbarossa two years later, which sanctioned illegal summary mass executions. In 1939, however, nothing was put in writing – and Hitler demanded an operational smokescreen that referred to 'elements hostile to the Reich'. This obfuscation filtered down through the ranks. The main instrument of mass murder – though not as we shall see the only one – would be the Einsatzgruppen. These 'Special Task Forces' or 'death squads' had already been deployed during the Anschluss with Austria, and later in the Sudetenland and Czechoslovakia. At the beginning of July, Heydrich appointed SS-Brigadeführer Werner Best, a 36-year-old lawyer, to begin the selection of appropriate staff that would soon be sent into action in Poland. They would be recruited from every branch of Himmler's police forces and become the main agents of the Nazi genocide – and later, the first recruiters of non-German auxiliaries. Most of these dedicated killers were German law graduates in their early 30s.[13]

In mid-August, Heydrich met his Task Force commanders and informed them that Hitler had personally tasked him with combating Polish 'resistance': 'everything was allowed, including shootings and arrests,' he revealed.[14] The target lists already compiled by Jost made perfectly clear what 'resistance' meant: the word was merely window dressing for the decapitation of Polish civil society. But Heydrich refused to specify how these 'radical' instructions should be carried out; that was down to individual commanders in the field. Much would depend on the intuition and initiative of young German men. These Special Task Forces would be backed by ethnic German 'Self-Defence Corps' (*Volksdeutcher Selbstschutz*), recruited from the German minority in Poland which was saturated with fanatical National Socialists, eager to take revenge on their Polish fellow citizens.[15]

German troops began to move east towards the Polish border as early as June. Against a background of frantic diplomatic manoeuvring, Hitler ordered Heydrich to provide a suitable *casus belli* to launch his war. He claimed that ethnic Germans in Poland had been persecuted with 'bloody terror' – now he needed some evidence. Heydrich hatched up Operation Himmler. At the end of August a cadre of SS and SD men secretly assembled at the police school in Bernau where they were issued with Polish army uniforms and papers. Since the Poles had refused to provoke a war, the SS would do it for them. Heydrich assigned these 'provocation' teams a series of targets on the Polish border, including a radio station at Gleiwitz. Here they waited for Heydrich's coded signal 'grandmother has died'. Led by SS-Sturmbannführer Alfred Naujocks (the author of a post-war autobiography, *The Man who Started the War*), the first unit of provocateurs attacked the radio station, inadvertently killing a German policeman, and broadcast in German accented Polish that 'The hour of freedom has struck!' to the accompaniment of pistol shots. Another sham Polish team attacked the German customs post at Hochlinden, where they deposited six corpses dressed in Polish uniforms, referred to as *Konserve* or 'preserved meat'. These human props had been provided by Heydrich's rival SS General Theodor Eicke, the man who had shot Ernst Röhm in 1934 and become head of the Concentration

Camps Inspectorate and commander of the SS Death's Head division. Eicke had selected and poisoned the unfortunate *Konserve* for Operation Himmler at Sachsenhausen camp near Berlin. Military intelligence, the Abwehr, supplied their uniforms. Yet another SD team struck a German forestry station at Pitschen, daubing its walls with ox blood. For the benefit of the press, astonished that Poland had attacked Germany, Heydrich had ordered a model of the border which featured flashing red lights where Polish attacks had taken place. The message conveyed by these macabre theatrics was obvious: Polish forces had violated the borders of the Reich. Germany was under attack!

These staged provocations resembled a grotesque comic opera. This truly was gangster diplomacy. In the early hours of 1 September, at 4.45 a.m., Hitler broadcast to German troops massed on the Polish border: force would be met by force. An hour later, the German training ship *Schleswig-Holstein*, anchored in Danzig harbour, turned its guns on the Polish garrison and opened fire. A total of 1,500 German aircraft roared into the air and crossed swiftly into Polish airspace. Five German armies, made up of sixty divisions, comprising more than 1.5 million men swept across Polish borders, led by five panzer tank divisions. As German forces pounded the Polish armies from air, land and sea, Hitler was driven to the Kroll Opera House, which had temporarily replaced the Reichstag. Wearing his Iron Cross and dressed in a field grey uniform, Hitler slandered the Poles as warmongers and reassured the governments of France and Great Britain that he merely wished to settle the status of the Pomeranian Corridor and Danzig. It was sheer mendacity. The Germans intended to obliterate the Polish nation.

Case White delivered a powerful straight punch combined with swift, ruthless encirclement. From the north, the 4th Army drove through the Polish Corridor between Pomerania and East Prussia towards Warsaw. From East Prussia, the 3rd Army pushed south towards the Bug River, cutting behind helplessly confused Polish divisions. From Silesia, German armies struck north-east.[16] These hammer blows took full advantage of new borders created after the destruction of Czechoslovakia. In return for a promise of 300 square miles of Poland, puppet dictator Joseph Tiso granted the German 8th, 10th and 14th Armies permission to cross the Slovakian border with Poland, alongside German-trained Slovakian troops, to slice into Polish forces from the south.[17] With relentless momentum, the German forces penetrated deep inside Poland. In less than twenty-four hours, the Luftwaffe Stuka bombers had eliminated 75 per cent of the dilapidated Polish air force.

In August, Hitler's Foreign Minister Joachim von Ribbentrop had signed a non-aggression pact with his Soviet opposite number in Moscow. Its secret protocols guaranteed the division of Poland and Eastern Europe between the two dictatorships. On 11 September, Stalin had withdrawn his ambassador from Warsaw – but as Hitler's armies crushed the Poles, the Soviets prevaricated, hoping the Germans would perform much of the hard labour of conquest. On 17 September, when German victory and thus the destruction of Poland as a nation was certain, the Soviets finally struck from the east, finally snuffing out any chance that the Poles could continue to resist.

Barely noticed by the outside world, on 15 August a few hundred Ukrainians had arrived in Slovakia, a client state of the Reich, to begin training as a Bergbauenhilfe (BBH). Although Hitler was hostile to Ukrainian political demands, Abwehr head Admiral Wilhelm Canaris had been cultivating the Ukrainian nationalist faction (the OUN), and its anti-Semitic leader Andriy Melnyk since the mid-1930s. Although the Nazi-Soviet Pact, signed in August, complicated German relations with the Ukrainians, Canaris pushed ahead with a special training programme, appointing Colonel R. Sushko, a prominent OUN man, to lead the Bergbauenhilfe into action against the Poles. But when the Soviets began their occupation of eastern Poland, Canaris was forced to abandon his plans. Hitler's Bolshevik allies in Moscow naturally opposed the arming of any anti-Soviet nationalists. The BBH was reclassified as a police unit and took 'self-defence' actions against Polish troops as they fled towards the Romanian border. In other words, they murdered them. These Ukrainian recruits were the first of Hitler's foreign executioners.[18]

The Polish government, vainly hoping for French and British support, had delayed mobilisation – but in any case, their armed forces, despite putting up tremendous resistance, proved pathetically inadequate in the face of the German blitzkrieg. The astonishingly swift and co-ordinated air and ground attack had shredded Polish communications. Lines of command disintegrated. In just twelve days, German forces overran the western half of Poland, and the Polish government fled Warsaw, as the German armies threw a ring of steel and fire around the city defences.

Hitler followed the Polish campaign with rapt attention. On 3 September, his special headquarters train began steaming east from Berlin's Stettiner Bahnof. He frequently called for halts so that he and his doting entourage could tour the rapidly advancing front line in motor vehicles. In Danzig, jubilant crowds of ethnic Germans greeted Hitler and his exultant entourage. Then his train steamed on towards the beleaguered Polish capital which was ringed by 175,000 German troops. On 25 September, waves of Luftwaffe bombers and transport planes rained down fire and destruction backed by massive barrages launched from rail-mounted artillery. Exhilarated by this fiery Armageddon, Hitler insisted that the Polish government must surrender unconditionally. On 27 September, Polish forces defending the city capitulated. Hitler's blitzkrieg had killed 70,000 Polish troops and wounded

130,000. Nearly half a million had been taken prisoner. Tens of thousands of others had fled into Romania and Hungary. Poland had ceased to exist; its territory was occupied by totalitarian forces who would install two destructive but distinct reigns of terror: one animated by race, the other by class.

On 5 October, Hitler boarded a Junkers Ju52 to fly over Warsaw's empty and smouldering streets and gloat over the smoking ruins of the hated Polish capital. Five years later, a multinational SS army would finish off the job. Hitler's war against Poland and its peoples did not end with the destruction of the Polish armed forces. As Wehrmacht divisions smashed the Polish armies, an undeclared shadow war had begun. This shadow war would be waged by Himmler's paramilitary police, Heydrich's Special Task Forces and the armed SS-VT. Himmler's spectacular success in Poland meant that the SS would eventually secure the right to manage the occupation of conquered territory – and to set in motion monstrous plans for the Germanisation of the east. These plans would soon draw in non-German collaborators who would become Hitler's foreign executioners.

When it came to waging war on the enemies of the Reich, Himmler exploited a strategy that already had a long tradition in German military practice, but would now become the defining principle of SS warfare. In German, *Bandenbekämpfung* literally means the 'combating of bandits'.[19] Although the term predated the Hitler period, Bandenbekämpfung provided a strategic rationale for the systemic slaughter of any group of people deemed to be *banditen* (members of criminal gangs). As we will see, this might include unarmed civilians and Jews and genuine partisan fighters, and their alleged supporters. The term may first have been used during the Thirty Years War but it officially became part of strategic doctrine after the Franco-Prussian War of 1871, when German auxiliary troops, later called *Etappen* (from the French word *étape*, meaning stages), fought French resistance fighters known as *francs-tireurs*. But Bandenbekämpfung could embrace a multitude of sins, for it was also used to justify armed responses to acts of civil disobedience, as opposed to attacks by francs-tireurs. In the period after 1871, Bandenbekämpfung would be used to justify attacking rebellious African tribes people in the German colony of Namibia during the Herero Wars and later German communists on the streets of Berlin. It was this slippery classification of the enemy as bandits that appealed so powerfully to Himmler. The Bandenbekämpfung concept permitted the targeting of a broad cast of ethnic and ideological enemies – from armed partisans to unarmed civilians.

In September 1939 the German army was equipped to launch a sledgehammer blow against the Poles. But the Wehrmacht planners had little time to build up Etappen units in significant numbers. This neglect was Himmler's opening – and he seized it ruthlessly. On 3 September, Hitler formally appointed his SS chief to take charge of 'law and order matters' behind the front line, the so-called 'Army Rear Area'. We can be certain that Himmler was expecting to receive such an order, for on the very same day, SD Chief Heydrich issued a policy document, 'Basic principles for Maintaining Internal Security during the War', which listed potential targets to be eliminated 'through ruthless action'.[20] Himmler issued secret orders to Special Task Force commanders, sanctioning execution of insurgents 'on the spot', and the taking of civilian hostages. This order signalled that SS security forces would fight according to the doctrines of Bandenbekämpfung, which would have a profound and deadly impact on both Wehrmacht and SS tactics.[21] Anti-bandit 'actions' legitimated the murder of targeted non-combatants by both Wehrmacht soldiers and SS police. While it is, of course, true that Polish franc-tireurs harassed German forces throughout the Polish campaign, they were not the main targets of Himmler's Bandenbekämpfung. Instead, the SS exploited internal security needs to liquidate Polish leadership cadres like the intelligentsia, aristocracy and clergy, as well as communist officials and Polish Jews.

In the first weeks of the war, some of the worst atrocities took place in the town of Bydgoszcz (German Bromberg) in the Polish Corridor.[22] This region of Poland was ethnically very mixed, and in Bydgoszcz a local ethnic German militia clashed with retreating Polish troops. Attacks on ethnic Germans invariably provided an excuse for indiscriminate reprisals – backed by anti-Polish campaigns in the German press that referred to 'Bromberg Bloody Sunday' and grossly inflated ethnic German casualties. Once Brigadier General Eccard Freiherr von Gablenz had formally occupied the city on 5 September, SS police arrived and began rounding up thousands of Poles, mainly teachers, civil servants, lawyers and other members of the city's professional elite. Hundreds were executed in artillery barracks and in the old market square. When SS officer Lothar Beutel reported to Berlin that more attacks on ethnic Germans had taken place, an enraged Hitler demanded full-scale reprisals. Between 9 and 10 September, Einsatzgruppe IV and SS police (6th Motorised Police Battalion) carried out sweeps, aided by ethnic German informers, in the Schwedenhöhe district where Polish units had made their last stand. The commander of 'Aktion Schwedenhöhe', Helmut Bischoff, demanded that his police show that 'they were men'; they must be 'tough and harsh'. Most complied. Even unarmed Poles who 'looked suspicious' were shot dead. By the end of Aktion Schwedenhöhe, SS police and German soldiers killed at least 1,000 Poles, 'priests, teachers, civil servants, rail operators, postal officers, and small business owners', as well fifty students attending the Copernicus Gymnasium. Himmler and the SS consistently referred to victims as *banditen* – opportunist killers who, as 'criminals', deserved no mercy. The victims of Aktion Schwedenhöhe were nothing of the sort. The figure of the bandit would provide the mendacious rationale for the genocidal murder of targeted ethnic elites.

In 1939 the main target was the Polish elite, but the SS police battalions rarely hesitated to humiliate and attack Polish Jews. In many towns, SS commanders set up sentry posts outside synagogues to terrorise Jewish neighbourhoods. The SS men humiliated and dishonoured Jews by cutting their hair and shaving their beards; they forced them to clean streets and sidewalks with toothbrushes. These SS 'ordinary men' relished such tasks; they gloated about meting out rough justice to 'Jewish vermin'. These humiliations proved, naturally, to be a prelude to murder. In Bydgoszcz, for example, the SS had liquidated the entire Jewish population of the city by November. Walther von Keudell, a former district president of Königsberg, commended the SS police for the 'energetic use of their weapons', their 'courage and common sense'.

The bloody climax of the SS police campaign in Poland engulfed the town of Ostrów Mazowiecka on 11 November. Two days earlier, precisely one year after Kristallnacht, a fire broke out in the centre of town – and 'Jewish arsonists' were blamed. As punishment, the local Nazi leader (Kreisleiter) ordered a group of Jews to operate a water pump and enlisted German soldiers to beat the Jews as they worked. The following day Police Battalion 11 gathered all the Jews of the town together and officers convened a kangaroo 'police court'.[23] In the meantime, SS police reinforcements arrived from Warsaw. The court pronounced the Jews guilty of arson – and on the morning of 11 November, PB 11 escorted *all* the Jews of Ostrów Mazowiecka to an execution site in a nearby wood; ditches had been dug the day before. As they herded the men, women and children in groups of ten to the edges of the ditch, officers from the Warsaw police battalions ordered their men to open fire. To begin with, a few hesitated. But a kind of terrible momentum quickly built up, and SS men began firing spontaneously; no further orders needed to be given. A few SS men baulked when they saw Jewish children being led to the execution pit. But one of the officers shouted that Jews had tried to assassinate Hitler a few days earlier in Munich; after this, shooting resumed. In Ostrów Mazowiecka, SS policemen slaughtered 156 Jewish men and 208 women and children.

In the wake of Daluege's Order Police Battalions (Orpo) came Heydrich's Special Task Forces – the elite killers of Himmler's security militias. To lead these Einsatzgruppen and their sub-units, the Einsatzkommandos, Heydrich and his recruitment chief Werner Best had turned to a cadre of elite SS officers. Best especially favoured an older generation, born in Silesia, who had 'won their bones' fighting with German Freikorps against the Poles after Germany's defeat in 1918 and had ever since cultivated violent anti-Polish sentiments. During the Weimar period, many of the younger SD recruits had absorbed radical nationalist and anti-Semitic doctrines at German universities. These German students, organised in reactionary fraternities, the *Burschenschaften*, had become Hitler's most fanatical backers and, after 1933, were generously rewarded. Membership of Himmler's SS provided a fast track for academic careerists. In universities, the new powerbrokers expelled Jewish professors and impatient young *Doktoren* gratefully occupied their vacated positions. Outside the universities, SS agencies like the Race and Settlement Office, the RuSHA, and SS-Ahnenerbe (ancestral heritage, a think tank that investigated German prehistory and related topics) took on many of Germany's best and brightest. The Nazi seizure of power was a young man's revolution and the sclerotic German armed forces had already been thoroughly radicalised by this 'NSDAP generation'.

This meant that the men Heydrich recruited to lead the Einsatzgruppen were not thuggish brutes by any means. But the kind of education they had received in the Weimar period appears to have reinforced bigotry rather than encouraged genuine critical thinking. One SD recruit, Friedrich Polte, who attended a number of universities, wrote an autobiographical sketch when he gave up his doctorate and joined up. He described his academic studies as a 'revolutionary mission' that would expose the factual evidence of 'international conspiracies'.[24] In 1939, of the twenty-five Einsatzgruppen and Einsatzkommando leaders, fifteen had acquired the prestigious *Doktortitel.*[25] For example, Dr Alfred Hasselberg, Dr Ludwing Hahn, Dr Karl Brunner and Dr Bruno Müller had all studied law, and like many German lawyers had rushed to join the NSDAP bandwagon in 1933, eager to become the judicial vanguard of the New Order. For this highly politicised elite, membership of the SS or SD was highly seductive – and useful. Lawyers and other professionals soon dominated the higher ranks of the German police. Himmler's *doktoren*, as meticulous as they were dedicated, would play a deadly role in the Nazi genocide. Their fanatical commitment to mass murder, in the words of historian Joshua Rubenstein, 'staggers the imagination'.[26]

In Berlin, the Sonderrefferat Tannenberg managed every aspect of Special Task Force operations in Poland. In SD offices, 'desk killers' liaised with Task Force commanders in the field – men such as SS-Brigadeführer Bruno Streckenbach who would lead the largest group Task Force 1, which had been mobilised in Vienna. Heydrich issued all his commanders with the wanted persons lists (*Sonderfahndungslisten*) filed in custom-designed ledgers that each Task Force commander took with him into the field. These 'hit lists' named Polish political leaders, nobility, Catholic clergymen and prominent Jews. As well as Streckenbach's Special Task Force 1, Heydrich and Best assembled six other operational groups, split into smaller Einsatzkommandos and numbering between 2,700 and 3,000 men.

Himmler took a special interest in the activities of one particular Special Task Force, *Einsatzgruppe zur besonderen Verwendung* (Einsatzgruppe z.b.V.). The commander of this 'Special Purpose Operational Group' was SS-Obergruppenführer Udo von Woyrsch (b. 1895), who had served on Himmler's personal staff since 1935 and knew the SS chief well enough to address him with the familiar *Du*. Himmler. He had a high regard for aggressive radicals like von Woyrsch and Erich von dem Bach-Zelewski who had roots in Germany's troubled borderlands, like Silesia and Pomerania. For them, *Heimat* was not a bucolic world of rolling hills and farms, but a human wall thrown up to defend Germany from the Slavic east. The chaos that followed in the wake of Germany's unexpected defeat in November 1918 sharpened this instinctive contempt for treacherous eastern peoples – and their Jewish allies. Von Woyrsch proved himself a tough fighter for the Nazi cause – and when he joined the SS rose quickly through the ranks. Now his reward would be to lead Himmler's campaign of terror in Poland.

Himmler used this Special Task Force (Einsatzgruppe z.b.V.) as a kind of shock troop and once the Polish campaign was under way, he followed its progress closely and ordered von Woyrsch to send situation reports listing 'special incidents and measures' every three hours. From these it is evident that the task of the Special Task Force was to target not only Polish bandits, but Polish Jews.

On 3 September, von Woyrsch travelled to the Silesian city of Gliwice (Gleiwitz) some 60 miles south-east of his old powerbase in Breslau. At the police praesidium, he picked up orders from Himmler appointing him Sonderbefehlshaber der Polizei (Special Police Commander). His task, the orders continued, would be the 'ruthless suppression' of a local Polish uprising '*mit allen zur Verfügung stehenden Mitteln*' ('with every means available'). But von Woyrsch soon discovered that the uprising had already been neutralised. His deputy Emil Otto Rasch reported to Berlin that, as a result of executions carried out by another Special Task Force, the 'insurgency movement no longer existed'. But that was no reason for the Einsatzgruppe men to withdraw – instead they turned their attention to other kinds of 'hostile element'.

For the next three days, von Woyrsch scoured the region and soon enough tracked down his quarry. On 6 September, the Special Task Force crossed into eastern Upper Silesia and began attacking Jewish settlements in Katowice, Bedzin and Sosnowiec. They used flamethrowers to burn down synagogues and in Bedzin murdered more than 100 civilians, including Jewish children. The orgy of violence continued over five days and as the Special Task Force made its way towards Kraków, the men took every opportunity to 'terrorise' Jews. On 11 September, von Woyrsch met Bruno Streckenbach, commander of EG 1, and SD Chief Heydrich, who was 'touring' southern Poland. There is no detailed record of what the three men discussed, but according to Streckenbach's post-war testimony, Heydrich outlined a plan to expel Polish Jews eastward across the San River – the demarcation line established by the secret protocols of the Nazi-Soviet Pact – 'using the harshest measures'. Shortly afterwards, Himmler, following a meeting with Hitler, issued orders that Jews must be pushed into the Soviet sphere, rendering German-occupied Poland *Judenfrei*.

For Himmler, the Polish campaign offered a unique opportunity to experiment, to try out ways and means of securing and pacifying an occupied territory, and setting in motion its eventual Germanisation. Although he was forced to contend with a barrage of criticism from the Wehrmacht top brass and new rivals like Hans Frank, front-line experience had bolstered the SS and forged even closer bonds between its different offices. Himmler's fiefdom had been transformed into an unique paramilitary elite that shared a code of brutally simplistic values: blind loyalty and 'hardness'. These values saturated the SS police and its armed wing, the Waffen-SS.

In the summer of 1939, no one had heard of the Waffen-SS.[27] As soon as he had been appointed Reichsführer-SS in 1929, Himmler had explored ways and means of arming his elite corps. Before 1934, in the period when the SS was a junior partner to the heavily armed SA, this was a mere pipe dream. But in 1934, when the SA leadership was liquidated, Himmler earned the gratitude not just of Hitler but the German army, which had feared the SA and its ambitious leader Ernst Röhm. In the aftermath of the 'Röhm Purge', the SS was well rewarded. The SS was detached from the SA and permitted to form 'armed standing Verfügungstruppe of the strength of 3 SS regiments and one intelligence department ... subordinated to the Reichsführer of the SS'.[28] It was that final clause that should have sent shivers down the collective spines of the German high command – but it took some time for the military establishment to see the SS as a threat. Himmler had helped crush the upstart storm troopers and, in any case, Hitler could not afford to be seen encouraging SS military ambitions; so the arming of the SS necessarily proceeded covertly in fits and starts.

The slow, uneven emergence of the 'armed SS' should not obscure its vital role in Himmler's expanding empire. The SS-VT regiments developed alongside the SS Totenkopfverbände, or Death's Head units, recruited by SS-Gruppenführer

Theodor Eicke, who was Inspector of the Concentration Camps and Commander of SS Guard Formations. Military historians sometimes defend the Waffen-SS as an 'army like any other' – in other words, SS soldiers, including non-German recruits, should be viewed as combatants not agents of genocide. This defence does not stand up to scrutiny. Waffen-SS men were by definition *politische Soldaten* (ideological warriors). Take the case of Eicke himself. On 15 March 1937 this brutal and devoted SS man retreated to his office inside the Dachau concentration camp near Munich to update his curriculum vitae. This remarkable document begins 'Elementary and secondary school not completed'. After the war:

> financial resources ran out ... fought the November republic ... reactionary agitation ... unemployed ... security officer with IG Farben ... On March 21st, 1933, the Day of Potsdam, I was once again arrested ... Gauleiter Bürckel described me as a 'dangerous mental case' ... at the end of June, 1933, the Reichsführer-SS freed me and assigned me as commander of the Dachau concentration camp.[29]

Eicke was a born fighter and astute empire builder. He built up his Sturmbanne (guard units) into three Totenkopfstandarten (Death's Head regiments), headquartered at concentration camps: the 'Oberbayern' at Dachau, the 'Brandenburg' at Sachsenhausen/Orienenburg and the 'Thuringia' at Buchenwald. Eicke did not admire Himmler, and regarded his Death's Head regiments as a private army. His rigorous training programme, conducted inside the camp system, instilled in his men a uniquely savage fighting ethos – which would define the values of the Waffen-SS. And in September 1939, Eicke's Death's Head units marched out of their concentration camp training grounds to fight Hitler's 'war of annihilation'. By 7 September, a week into the Polish campaign, Eicke's SS Death's Head units had swelled to 24,000 men. Hitler ordered Eicke to deploy his men in the army rear areas, with full authority to conduct 'police and security measures'.[30] Eicke's mission deliberately blurred any distinction between combat and security – and these SS regiments operated as murder squads like the Special Task Forces. The SS 'Oberbayern' and 'Thuringen' followed the German 10th Army into the region between Upper Silesia and the Vistula River south of Warsaw; the 'Brandenburg' followed the 8th Army into west central Poland. As a Higher SS and Police Leader (HSSPF), Eicke had sweeping powers to 'pacify' areas already conquered by the Wehrmacht in the three central Polish provinces of Poznan, Łódź and Warsaw. Eicke did not trouble to visit the front line. Instead, he managed his murderous campaign from Himmler's special train, *Heinrich*, or his dedicated motor cavalcade the *Wagenkolonne-RFSS*.

The trail of blood left by the 'Brandenburg' is documented both by the unit's reports, compiled by Eicke's devoted Standartenführer Paul Nostitz.[31] As Himmler's warriors set about pacifying their allotted territory, (Nostitz reported to Eicke) they zealously shot 'suspicious elements, plunderers, insurgents, Jews and Poles' 'while trying to escape'. On 22 September, the 'Brandenburg' arrived in the city of Wloclawek, which lies on the Vistula north-west of Warsaw. Here they embarked on a vicious spree of killing and destruction that Nostitz logged as a *Judenaktion*: the SS men plundered Jewish shops, dynamited and burned synagogues, and carried out mass executions. As this *Judenaktion* continued, Eicke (on board Hitler's train) sent new orders to Nostitz to carry out what he called an 'intelligentsia action' (meaning, of course, murdering 'listed' Polish civilians) in nearby Bydgoszcz, already the site of Einsatzgruppe mass killings. On 24 September, two 'Brandenburg' storm units entered Bydgoszcz equipped with 'death lists' that named some 800 Polish civilians. They shot all of them.

Eicke's killers hunted down other 'lives not worthy of life'. At the end of October, the 12th SS Totenkopfstandarte marched into Owinska, where there was a large psychiatric hospital. The Death's Head men rampaged through the wards, dragging screaming patients into trucks. They drove them to specially excavated pits, where SS-VT squads waited with loaded rifles.

It was not only the SS Death's Head regiments that took part in such 'special operations'. The SS 'Leibstandarte Adolf Hitler' (the Führer's personal SS bodyguard) was, like Himmler's other SS-VT regiments, assigned to army divisions and corps as they advanced towards Warsaw. Commanded by one of Hitler's favourite generals, SS-Obergruppenführer 'Sepp' Dietrich, 'Leibstandarte' men crossed the Polish border just before dawn on 2 September. Dietrich had been ordered to protect the right flank of the 17th Infantry Division – but at 5 a.m. SS men on motorcycles roared into the small town of Bolesławiec. According to Karol Musialeck 'they drove around the market place three of four times and went back the same way'.[32] Less than an hour later, SS 'Leibstandarte' units returned. They began dragging Jews and Poles from their homes, and herding them into the marketplace. As this was going on, other SS men randomly began shooting Jews, often in the back at point blank range. Back in Bolesławiec market, the Germans separated villagers into two groups – Poles and Jews – and began marching them eastwards in the direction of the Soviet demarcation line. The Germans provided the Poles with basic foodstuffs, Musialeck recalled, but the Jews they starved, beat and robbed. In the meantime, the SS 'Leibstandarte' men set fire to the village.

As the SS men followed the 17th Infantry eastwards, they took every opportunity to harass and murder Jews in every village they passed through. On 3 September,

German soldiers and SS 'Leibstandarte' men arrived in Złoczew. They began burning buildings and shooting anyone still on the streets. A German soldier (not SS) smashed the skull of a baby. A teenage girl was shot and disembowelled. This opportunist barbarism soon became standard practice. The 'Leibstandarte' men, as they followed in the wake of the 17th Infantry in the direction of Łódź, shot civilians and burnt their homes, their synagogues and churches. In fact, the SS 'Leibstandarte', which was supposed to be a fast-moving motorised unit, was so preoccupied with its 'security tasks' that the SS men began to lag far behind the German infantry: vandalism and murder was time consuming. Regular army officers sent reports to headquarters criticising the SS men's sluggish progress, the 'wild firing' and 'reflexive tendency' to set villages alight. Some German army soldiers and officers also took part in the shooting of unarmed civilians and Jews. The difference was that Himmler and Hitler expected the SS men to treat Jews and Polish civilians without mercy and they did not disappoint.[33]

Throughout September and October 1939, these *Säuberungsaktionen* (cleaning -up operations) took place in scores of towns and villages behind the swiftly advancing German front line. In many such operations, local Volksdeutsche (ethnic Germans) spontaneously participated. As this carnage engulfed the Polish countryside, the special headquarters trains of the Nazi elite, including Himmler's opulent *Heinrich*, clattered towards Warsaw, loud with the sound of busy typewriters and euphoric congratulation.

The SS paramilitary police forces and the new, armed VT regiments had been blooded by the Polish campaign. A few German army commanders may have grumbled about 'excesses', but Himmler smeared complainers with the most damning word in the Nazi lexicon: disloyalty. In any case, many Wehrmacht soldiers did not hesitate to join in with cowardly attacks and murders if they had the opportunity. In Hitler's armies, hatred of Poles and Jews was pervasive. Army denunciations reflected anxiety about the rising power of the SS rather than moral outrage. Hitler had few difficulties sabotaging isolated efforts to penalise SS men accused of 'excess'. On 17 October 1939 a 'Decree relating to the Special Jurisdiction in Penal Matters for members of the SS and for Members of Police groups on Special Tasks' abrogated the power of Wehrmacht military courts to court-martial SS personnel. But still Himmler had to tread carefully. He could not afford to be openly confrontational. Even after the lightning triumph in Poland, Hitler had nothing to gain from undermining his delicate transactions with his Wehrmacht generals – even though he was commander-in-chief of the army. So when Field Marshall Walther von Brauchitsch insisted on a meeting to discuss SS tactics, Himmler proved to be more conciliatory; he assured von Brauchitsch that he wanted 'good relations' with the Wehrmacht and promised that 'special operations' would be carried out in 'a more considerate way' in future. Himmler's act of kow-towing evidently worked, for soon afterwards von Brauchitsch officially dismissed the reports of SS atrocities as mere 'rumours'. The majority of the German army top brass let the SS get on with its appointed tasks of 'maintaining security' and dealing with 'hostile elements'. This moral abdication had fateful consequences. In the mind of German commanders and front-line soldiers, it normalised the mass murder of unarmed civilians deemed to be hostile in some way to the Reich. In occupied Serbia, for instance, it was the Wehrmacht not the SS that took the lead role in the mass murder of Serbian Jews in the summer of 1941.[34]

As the victorious Wehrmacht withdrew its armies from Poland, the SS muscled in to undertake what Hitler called a 'new ordering of ethnographic relations'.[35] The Nazi-Soviet Pact had divided the Polish lands between Germany and the Soviet Union. But to begin with, Hitler dithered about what to do with his portion – until Stalin forced his hand. Although the Italian dictator Benito Mussolini tried to persuade Hitler to create a relic Polish state to placate the French and British, Stalin insisted on the annihilation of the Polish state. As enticement, the Russians offered to cede the Lublin district in return for German recognition of Soviet interests in Lithuania. The offer intrigued Hitler and Himmler. Once the SD Special Task Forces had completed the liquidation of the Polish intelligentsia, the problem of what to do about the *Ostjuden* (eastern Jews) became a more pressing concern. Himmler concluded that the Lublin region offered a solution, albeit temporary, as a 'reservation' or dumping ground for 'the whole of Jewry as well as other unreliable elements'.

These decisions foreshadowed the catastrophe that would soon engulf the former Polish territories and the coveted east. By the end of the year, the SS had set in place the most important instruments of occupation strategy. With the connivance of the Wehrmacht, Hitler had redefined warfare not merely as blitzkrieg but as the means of achieving racial dominance: the Polish nation had been destroyed and its elites liquidated. The 'Jewish Problem' was now a matter of open discussion – and many thousands of Jews had been forced into the Soviet domain. Himmler had also begun the process of Germanisation by resettling ethnic Germans 'imported' from the Soviet Union and elsewhere. The first efforts had been made to exploit the chauvinist emotions of non-German nationalists – in this case, the Ukrainian OUN, whose militia had participated in the campaign. The Nazi elite had begun to think in practical terms about the vexed questions of empire, race and nation

– and Himmler's new appointment as Reichs Commissar for the Consolidation of German Nationhood (Reichskommissar für die Festigung deutschen Volkstums, RKFDV), made in October, meant that the SS would now control the process of deportation and resettlement in occupied territories, beginning with Poland, 'to purge and secure the new German territories'. Hitler expressed nothing but contempt for the conquered Poles – as he explained to Nazi 'party philosopher', Alfred Rosenberg, he had 'learnt a lot' in Poland. The Jews were 'the most appalling people one can imagine'. The Poles, he went on, exhibited 'a thin Germanic layer underneath frightful material'.[36] Himmler saw matters differently. Occupation was an opportunity: 'It is therefore absolute national political necessity to screen the incorporated territories … for such persons of Teutonic blood in order to make this lost German blood available again to our own people.'[37] For Himmler, the successful conclusion of the Polish campaign offered an opportunity to consolidate his ideological vision.

At the end of October, Himmler published an 'SS Order', which set out the fundamental principles of the SS, and its strategy for the future.[38] He begins by citing a favourite maxim: 'Every war is a bloodletting of the best blood.' Throughout Himmler's lectures and speeches, 'blood' is repeated like a Wagnerian leitmotif. Racial strength, Himmler asserts, depends on the shedding of blood – and its replacement by fecund SS men. 'He can die at peace who knows that … all he and his ancestors demanded and fought for is continued in his children.' Himmler further developed his blood obsession in a second speech given to the new Gauleiters, who now ruled the former Polish lands. He began with a typical assertion: 'I believe that our blood, Nordic blood, is the best blood on this earth … Over all others, we are superior.' He points out that over many centuries, bearers of Nordic blood had become the rulers, experts, members of cultural elites when settled among lesser races. Inevitably, they had mixed with their inferior hosts and polluted the Nordic bloodline. This was dangerous, for Nordic blood conferred tremendous power even when it was diluted. He noted that in the recent war, the gallant defender of Warsaw had been General Juliusz Rommél – evidently from Teutonic stock. If Germanic or Nordic blood was so threatening in the wrong veins, as it was, what was to be done? One solution was simply to liquidate the elites and subtract their contaminated bloodline from the national stock. But mass murder was just one possible solution. 'While we are strong,' Himmler proclaimed '*we must do our utmost to recall all our blood, and we must take care that none of our blood is ever lost again.* [my italics]'[39] For Hitler, racial admixture or miscegenation was an irreversible catastrophe. Himmler took a strikingly different view: lost Germanic blood might somehow be recovered.

Himmler went on to explain what he meant by 'recalling our blood'. He assumed that, with the exception of Jews, race was not fixed – it was to some degree fuid. The execution of unarmed civilians is a cowardly act. But according to Himmler ruthlessness or 'hardness' was character forming: 'An execution must always be the hardest task for our men … but they must do it with "a stiff upper lip"', he once said. Since good character was an expression of racial inheritance, it followed that the cultivation of 'hardness' through voluntary participation in violent actions offered individuals with some measure of Germanic blood the chance to ascend the rungs of the racial ladder. Soldiers, of course, not only kill – they get killed. For Himmler, sacrifice was another means by which an ethic group could elevate its racial status. Recruits who laid down their lives as SS warriors guaranteed the racial values of their comrades. Borrowing from a garbled version of Lamarckian inheritance, Himmler asserted that these racial characteristics acquired through violent action and sacrifice would be inherited by future generations that would be progressively 'Germanised'. Himmler was not troubled by the abundant contradictions of this twisted, semi-mystical rationale for mass slaughter. Instead he looked forward to building a 'Germanic blood wall' to guard 'Germanic, blond provinces'.

This was the first preliminary sketch of an evolving master plan – and its depraved sophistication fundamentally contradicted Hitler's petty-minded bigotry. The problem, naturally, was implementation. How was the German or Nordic blood to 'be recalled' in practice? By the beginning of 1940, Himmler had at least the rough outline of a solution. His police battalions and armed SS units would offer 'Germanic' recruits the chance to 'top up' their racial qualifications through blood sacrifice. Himmler's next task would be to refashion the new Waffen-SS as a receptacle of reclaimed Germanic blood, 'wherever it might be found'.

In the aftermath of the Polish campaign, Himmler energetically impressed on Hitler the heroic part played by the SS 'Leibstandarte'. Realising that the Wehrmacht remained squeamish about fully embracing a 'war of annihilation', Hitler agreed to expand the 'armed SS' from one to three combat divisions: the Totenkopfdivision, the SS-VT and the SS Polizei Division. By now, Hitler and his generals had begun planning Fall Gelb (Case Yellow), the invasion of Western Europe and Himmler faced an unexpected dilemma. In the short term, he had no idea how to acquire the manpower to fill these new divisions. Army recruitment had drained the well close to the bottom, and the Wehrmacht high command, thoroughly rattled by SS aggression, would do whatever it could to cut off supplies of men and materials to the SS. The new 'armed SS' made only an insignificant contribution to the attack on Western Europe.

Nevertheless, on 19 July 1940 Hitler stood once again in the Kroll Opera House to announce the successful completion of the latest blitzkrieg. 'The German armoured corps,' he proclaimed, 'has inscribed for itself a place in the history of the world. The men of the Waffen-SS have a share in this honour.' He then acknowledged a beaming Himmler: 'Party comrade Himmler, who organised the entire security system of our Reich as well as the units of the Waffen-SS.'[40] By that summer, Waffen-SS combat units could muster some 100,000 men. The manpower problem had been solved, for the time being at least, by one of Himmler's most forceful henchmen. While Himmler waffled about Germanic empires and Nordic bloodlines, Gottlob Berger, a bluntly spoken wedge of man with a talent for making enemies, got on with the job. That summer, as SS administrators tightened their grip on Hitler's European empire, Berger embarked on a new campaign to streamline his command organisation and channel fresh recruits into SS divisions. Although the term 'Waffen-SS' had been officially in use since March, Himmler bound his private army even closer to the 'general SS' by setting up the Kommando der Waffen-SS inside the SS Main Office. Like any Reich agency, the SS was a battleground of aggressive egos and empire builders. Ambitious military types despised the 'schoolmasterly' Himmler and made the dangerous mistake of thinking he could be bullied. Ambitious generals like Theodor Eicke and 'Sepp' Dietrich had, with Hitler's tacit approval, treated their SS divisions, the 'Leibstandarte' and Totenkopfdivisions, as personal fiefdoms. Himmler and Berger would use the new Kommando to tame these malcontents and promote their own tame placemen. Himmler was determined that the armed SS – the Waffen-SS and its kin the police – would take a vanguard role in the renewed assault on the east. He knew that this could not be put off for much longer. His destiny was the conquest of the east.

As Himmler streamlined the SS, Hitler was becoming preoccupied with Britain's refusal to 'knuckle under' (accept defeat) and with the 'Russian problem'. The two issues were closely connected. Hitler assumed that the British were convinced that his fickle ally Stalin would eventually join the war – against Germany. This may have been a rationalisation on Hitler's part designed to appease his nervous generals, since his principal war aim was ultimately the destruction of the 'Jewish-Bolshevik' enemy in Moscow and the German resettlement of the east – an ambition fully endorsed by the impatient Himmler. According to Goebbels, Hitler saw the coming war with Russia in completely Manichean terms: Bolshevism was 'enemy number one'.[41]

On 31 July, Hitler called his senior military advisors to the Berghof, and informed them that he had made a 'final decision' to 'finish off Russia' in the spring of 1941.[42]

Hitler rationalised this by arguing that 'England is counting on Russia … if Russia is beaten, there is no more hope for England'. As his plans matured, Hitler would abandon this kind of rationalisation, even when he discussed strategy with the Wehrmacht generals. SS Chief Himmler understood completely the implications of renewing the National Socialist 'war of annihilation' that Hitler characterised as 'deliberately racial'. As his eastern plans took shape in the winter of 1940/41, Hitler openly combined strategy with ideology. This meant that the Wehrmacht must 'fight an ideological war alongside the SS'. Instead of merely taking on 'special tasks', SS values would shape German invasion strategy. This meant that the Waffen-SS would need to acquire a lot more clout – and that meant recruiting.

For Himmler, the planned attack on the Soviet Union, initially codenamed 'Otto', presented both another opportunity to reinforce the status of the SS and a fearsome challenge. In 1940, Berger still depended on a pool of native German citizens for recruitment. The Germany army, represented by the OKW, still controlled the flow of military-age manpower to both the Wehrmacht and the Waffen-SS. The army did everything in its power to starve the Waffen-SS. On 7 September, in a speech to the officers of the SS 'Leibstandarte', Himmler announced that he had a solution to the manpower problem. 'We must,' he declared, 'attract all the Nordic blood in the world to us, depriving our enemies of it.'[43] His proposal was expedient – but fitted perfectly with his developing pseudo-biological philosophy.

And the SS recruiters would begin by tapping the German *diaspora*.[44]

It was estimated that some 13 million Volksdeutsche lived outside the Reich, mainly in Hungary, Romania and Russia – a number comparable, as Valdis Lumans points out, to the population of medium-sized state.[45] The lost Germans had fascinated Himmler for some time and he knew that the key to exploiting this tantalising human reservoir was the Volksdeutsche Mittelstelle (Ethnic German Liaison Office, VoMI), a Nazi Party organisation founded in 1935 to look after the interests of ethnic Germans living outside the Reich and, of course, promote National Socialist ideology. By controlling VoMI, Himmler could influence the Volksdeutsche leadership. In 1937, Himmler engineered the appointment of SS-Obergruppenführer, Werner Lorenz, as VoMI chief. As an NSDAP agency, VoMI officially came under the aegis of deputy party leader Rudolf Hess and the NSDAP treasurer. But in 1938, Hitler granted VoMI state authority as well – meaning that it was no longer simply a party organisation but a kind of hybrid. In theory, Hitler's decision should have made the Foreign Minister Joachim von Ribbentrop an equal partner with

Hess, but Himmler swiftly exploited VoMI's 'mixed' status and ordered Lorenz to saturate its staff with SS placemen. By 1940, the SS dominated VoMI, and through its hundreds of offices wove a web of connections to German minorities scatted across the Balkans, Poland, the Baltic States, France, Belgium, Denmark, Austria, the former Czech republic and the Netherlands.

Himmler's plan to exploit German Volksdeutsche communities as a recruitment reservoir was not as straightforward as it might appear. German anthropologists like Hans F.K. Günther, the so-called *Rassenpapst* (Race Pope) argued that ethnic Germans had become excessively contaminated by intermixing with their Slavic neighbours. Günther's many books were widely read – and although Nazi propaganda often celebrated the typical Volksdeutscher as a heroic Aryan paragon, many Germans regarded them as second- or third-rate, 'not quite' Germans. The Austrian-born Hitler, arguably a Volksdeutscher himself, thoroughly despised most ethnic Germans as 'degenerates'. There were in any case, according to specialists, different Volksdeutsche 'species'. Some ethnic Germans inhabited territories that had been separated from the Reich as recently as 1919. Others, like the Sudeten and Carpathian Germans of Czechoslovakia, had once been subjects of the German-speaking Austro-Hungarian Empire. Other Volksdeutsche, the relics of medieval Germanic empires, had much older roots. Germans first immigrated to Hungary in the tenth century; 200 years later, the Teutonic Order and Hanseatic merchants colonised the Baltic region. German migrants tended to form business or cultural elites: the oldest of all German universities was founded in Prague in 1348. The so-called 'Volga Germans' had originally come to Russia in the mid-eighteenth century at the invitation of the German-born Catherine the Great. After the Russian Revolution, Lenin declared the Volga region the 'Volga German Autonomous Soviet Socialist Republic', with its capital at Engels – commemorating another notable German exile.[46]

Whatever their history and origins, these Volksdeutsche communities aggressively celebrated their German roots and identity. Ethnic Germans rarely displaced indigenous, usually Slavic peoples; they tended to become state officials, academics and landowners. Like the British ruling class in India, ethnic Germans proscribed fraternisation and sheltered inside what they called *Sprachinseln* (language islands). This isolationism encouraged disdain for both Slavs and Jews. After the First World War, with the collapse of the Austro-Hungarian Empire, this long-cultivated aloofness underwent a poisonous and reactionary transformation, and many ethnic Germans would prove themselves the most fervent Nazi ideologues. In 1939, when Hitler charged Himmler with a massive resettlement programme, he would turn to these islands of German speakers to fill the new *Gaue* carved form the vanquished Polish lands.

Naturally, human nature being what it is, many Volksdeutsche could not resist the temptations of their Slavic or Jewish neighbours, however zealously ethnic German communities protected their ethnic boundaries. Gene flow is unstoppable. This meant that in the new Reich, Aryan status could not be conferred automatically on every Volksdeutsche. So when the German resettlement programme got underway, every ethnic German applicant had to be rigorously screened by the Oberste Prüfungshof (Highest Court of Examination) and then classified in the *Volksliste*: an exacting hierarchy that descended from Category 1 ('pure and politically clean specimens') to Category 4 ('renegades' with 'alien blood'). German 'proofing' officials were shocked by the 'racial quality' of the Volksdeutsche they examined. They complained frequently that many ethnic Germans behaved just like Poles and Ukrainians; they lacked the proper German values. Worse, they confessed to sleeping with Polish and Ukrainian women. As Doris L. Bergen succinctly puts it, the 'Volksdeutsche notion was always tenuous'.[47]

Although Himmler's fastidious race experts might question the right of some ethnic Germans to join the Nordic club, the average Volksdeutscher, for his or her part, shared the xenophobic prejudices of the 'Master Race'. During the Weimar period, scores of German support organisations had sprung up to promote the interests of the Volksdeutsche, especially those regarded as 'victims of Versailles'. After 1933, these contacts deepened. Hitler's frequently renewed promises that he would 'roll back Versailles' ignited the aspirations of a new ethnic German generation. Across the German *diaspora*, Nazi agitation cells proliferated, especially in southern Russia and eastern Poland where ethnic German communities had long been riddled with the most virulent anti-Semitism. According to Valdis Lumans: 'National Socialism was even more attractive to the average Volksdeutscher than to his Reich counterpart.' In his book about the German occupation of Greece, Mark Mazower confirms that ethnic Germans eagerly rallied to the cause of ethnic destruction. Following the German occupation of Greece in 1941, SS-Standartenführer Dr Walther Blume recruited middle-aged Volksdeutsche as concentration camp guards – and they soon became feared for their extreme cruelty. These men had been recruited in Hungary and Romania and had few illusions about their less than exalted place in Hitler's New Order; they understood well enough that Germans from the 'old Reich' despised them and their kind. This sense of exclusion fuelled their merciless treatment of camp inmates, especially Greek Jews. Camp commandant Sturmbannführer Paul Radomski, an ethnic German recruit, was described by his superiors as 'energetic and made of iron'. He was in fact a murderous brute.[48]

Himmler eyed these millions of German exiles greedily. Since race rather than citizenship qualified someone to join the Waffen-SS, Himmler pressured the VoMI to begin recruiting ethnic Germans. In Germany, VoMI officials arranged physical training programmes and athletic visits for young ethnic Germans from abroad – and, once they were on Reich soil, pressurised them to volunteer for service in the Waffen-SS. Guided by the SS, VoMI became a recruitment agency. Friedrich Umbrich (b. 1925) recalled his first encounter with emissaries of Himmler's SS in a memoir called *Balkan Nightmare*. Umbrich was an ethnic German, born in Transylvania he grew up in the little village of Belleschdorf – today Idiciu in modern Romania.[49] Saxons had lived in this lush, green valley between the Carpathians and Transylvanian Alps for six centuries. In the aftermath of the First World War, after the signing of the Treaty of Trianon in 1920, the Transylvanian Saxons woke up to find they had become Romanians. This sparked feelings of profound anxiety and resentment. Umbrich admits that after 1933, the Romanian-Saxon community was caught up in the 'hysteria sweeping Europe'; in other words, National Socialism. VoMI officers soon arrived in Transylvania, bringing rousing songs and propaganda films extolling the virtues of the new Germany with its *Tüchtigkeit und Einigkeit* (efficiency and unity). Saxons had a long tradition of making do and compromising with their neighbours. All the Umbrichs spoke fluent Hungarian. They tolerated the few Jews who lived in the village – including a childless couple who would vanish 'unexpectedly' in 1940. But the men from VoMI promised a bright new future as part of a greater Germany.

Sometime after September 1940, Umbrich tells us, 'three tall, handsome SS men' appeared in Belleschdorf. Their bearing was proud and erect, their uniforms were crisply pressed, their long leather boots polished and shining. The SS men politely requested to speak to a village leader and were directed to the Umbrich household, where they spoke with Friedrich's father. They brought a message: '*Der SS-Röntgenzug ist unterwegs!*' ('The SS x-ray train is coming!') The SS had come to show off German medical prowess and, as Friedrich later understood, to check their physical suitability to serve in the SS. The 'day of the x-rays' was a festive occasion – and the SS men praised the Saxons' excellent German and gobbled down their food and slurped their best schnapps. Sixteen-year-old Friedrich was impressed by the intimidating German machines and the strapping SS men who set them up in the village church; it was the beginning of a spectacular feat of seduction. Two years later, Friedrich was fighting Serbian partisans in the Balkans.

Many of SS recruitment chief Gottlob Berger's kin were ethnic Germans scattered all over Europe. In Romania, where he launched his recruitment drive, he had close kin among the Transylvanian Saxons. Andreas Schmidt, the head of the German minority, was Berger's son-in-law. Schmidt was a radical Nazi and a Volksgruppenführer with close ties to the NSDAP in Berlin. At the end of the

1930s, Schmidt had brought together the ethnic German group of Romania, the GEGR and the local NSDAP. He was a brutish fanatic and wholeheartedly devoted to the Nazi cause. He enthusiastically embraced his father-in-law's campaign to recruit for the Waffen-SS in Romania. But Romanian leader Ion Antonescu insisted that his government would regard service in a foreign army as desertion. So Schmidt and his father-in-law smuggled more than a thousand ethnic German Waffen-SS recruits, disguised as labourers, across the border for training in Prague.[50]

It was a logical next step to look beyond the ethnic German world to the Nordic nations like Denmark and Norway that had been overwhelmed by the German army in 1940. It was becoming increasingly evident that the expansion of the Waffen-SS would depend on Berger's foreign recruitment drive. It was a strategy that had been forced on the SS by the Wehrmacht but Himmler embraced it with a passion. In occupied Europe, the 'Almighty' Berger would need to negotiate some thorny obstacles. To entice foreign recruits into the Waffen-SS, he would have to overcome natural scruples about serving an occupying power. The Hague Convention, still accepted by Germany, made conscription illegal in any occupied nation; any SS recruits thus had to be 'volunteers'.[51] Even some of the European pro-German radical nationalist movements like the Dutch National Socialists (NSB) were bitterly divided between those who longed to serve Hitler's Reich and others who, rightly, feared that their own national cultures would be extinguished. Berger nevertheless set to work and set up SS recruiting offices in Denmark, Norway and the Netherlands. Recruiting criteria were identical for both foreign and German applicants: Dutch and Flemish men who joined the SS 'Westland' and 'Nordland' regiments, for example, had to be over 17 and under 40, possess 'Aryan racial characteristics', be in good health and meet the minimum SS height requirement of 165cm. But Berger's first efforts yielded very modest results. By the summer of 1941, the new SS Standarte 'Nordland' and 'Westland' had, between them, attracted only a few hundred Danish, Norwegian, Dutch and Belgian volunteers.[52] But Himmler was not discouraged. He was increasingly obsessed with building a pan-European army – and by the end of that fateful year, the tally of 'Germanic' volunteers would look very different.

The German invasion of the Soviet Union on 22 June 1941 would draw in tens of thousands of foreign collaborators who, as SS police or Waffen-SS volunteers, would play a vile part in the destruction of European Jewry. These 'willing executioners' shared a common ideological language with the Reich: a political Esperanto founded on a lethal hatred of a completely mythic entity, the 'Jewish-Bolshevik'. Hitler's declaration of a 'Crusade against Bolshevism', which concomitantly implied a war on Jewry, welded together a broad alliance of radical nationalists who pledged allegiance to the Reich.[53] They would take on what Himmler called the 'hardest task' of mass execution. In return for this service, he would promise some of them a place at the Nordic table of honour. This shadow war fought by Hitler's foreign executioners would ravage the *shtetls*, fields and forests of the east – a region Hitler claimed would become Germany's 'garden of Eden'.[54] But the tragedy of this 'Holocaust by bullets' had already been rehearsed before 22 June 1941 – in the 'Forgotten Holocaust' that overwhelmed Romania and the Balkans.

2

Balkan Rehearsal

> We are not going to wait for any declarations of loyalty by the new government but to carry out all preparations for the destruction of the Yugoslav armed forces and of Yugoslavia itself as a national unit ... It is especially important, from the political point of view, that the blow against Yugoslavia should be carried out with the utmost violence.
>
> Hitler, War Directive 25[1]

The shadow cast by Hitler's foreign executioners is a long one. On 4 March 1999 an elderly man called Dinko Šakić stood in a Zagreb courtroom accused of war crimes. He had been tracked down in Argentina by Ephraim Zuroff, Director of the Simon Wiesenthal Centre in Jerusalem. A senior lieutenant in the wartime Ustasha militia, Šakić was accused of murdering and torturing prisoners incarcerated in the Jasenovac concentration camp established in Croatia in August 1941. Jasenovac was the third largest camp in Europe during the Second World War and it remains the least understood. During his trial, Šakić often laughed loudly when witnesses testified about his gruesome activities. The prosecutors, who had worked for years to gather evidence, faced powerful public hostility. For many young Croatians, the sneering old man in the dock was a patriot, a national hero. Every day, hundreds of noisy supporters crowded into court to provide him with 'moral support'. Šakić was eventually convicted and imprisoned – a landmark judgement in Croatia. But his many admirers refused to give up his cause. In 2007, at a huge concert by Croatian singer Marko Perković, young Croatians turned up wearing Ustasha uniforms to honour Šakić – and when this convicted murderer died in prison the following year, he was publicly buried in his Ustasha uniform. His priest eulogised him an 'example to all Croatians'.[2] The Ustasha militias served the Croatian puppet regime set up by the German Third Reich in the spring of 1941 after the destruction of Yugoslavia. Later that year, in August, the Germans authorised the establishment of a camp system on the banks of the Sava River at Jasenovac. Inside, Ustasha men like Šakić tortured, raped and murdered without restraint. At least half a million Jews and Serbs died at Jasenovac. One survivor recalled:

> Victims would wait in the Main Warehouse or in some other building or out in the open ... the Ustasha would strip them naked. Then they would tie their hands behind their backs with a wire ... A victim would be forced to his knees ... they would hit the victim with a mallet, a sledgehammer or with the dull side of an axe on the head. They would often cut their stomachs open with a butcher's knife and dump them into the Sava.[3]

On the morning of 6 April 1941, Palm Sunday, wave after wave of heavily laden Luftwaffe Ju52 heavy bombers and Stuka dive bombers roared into the air from Romanian airfields and set a course for Belgrade, the capital of Yugoslavia, the fractious national kingdom of the Serbs, Croats and Slovenes. This airborne assault was the opening move of Operation Punishment – chastisement for the Serb coup that had scuttled Hitler's plans for a Balkan pact. The bombardment lasted three days and killed between 5,000 and 10,000 undefended people. Hitler's attack caught the Yugoslavian army on the back foot. Lightning raids smashed the air force and shredded lines of communication. Mobilisation had just begun and Yugoslavian troops were still crisscrossing the country to reach their units. Croatian fascists known as Ustasha undermined the resolve of Croatian units. Many soldiers deserted and reservists failed to report. On board his special train *Amerika*, halted in the foothills of the Alps close to the entrance of the Aspangbahn tunnel in case of air assault, Hitler followed the progress of Operation Punishment in the map room attached to his command coach.[4] Three German armies, backed by the SS 'Grossdeutschland' and 'Das Reich' divisions, advanced rapidly towards Belgrade. The Yugoslavs fought back hard, but demoralised Croatian soldiers deserted en masse and turned against Serb regiments. The Germans reached the old Croatian capital of Zagreb on 10 April then followed the Drava and the Danube south towards Belgrade, seizing the Avalla Heights overlooking the city two days later. On the night of 13 April, SS-Hauptsturmführer Fritz Klingenberg led advance units of the SS 'Das Reich' across the Danube – and Belgrade fell into German hands.

It is often remarked that the German onslaught on the Kingdom of Yugoslavia and the Balkans was 'improvised' by an enraged Hitler, and that it delayed the German attack on the Soviet Union, which had been planned for the spring. To be sure, Hitler never contemplated acquiring 'living space' in south-eastern Europe, preferring to exploit Balkan mineral resources instead. But Hitler's invasion may not be quite as impulsive as it appeared. As soon as victory was secure, Hitler shredded the Kingdom of Yugoslavia and set up a compliant regime in Croatia dominated by anti-Semitic, Serb-hating fascists under dictator Ante Pavelić. The conquest of the Balkans allowed German military occupiers and SS administrators to broaden their racial war against Jews and in this case the South Slavic Serbs. The occupation of Yugoslavia and Greece provided a rehearsal for the 'war of annihilation' that would soon engulf the Soviet Union. As they had in the puppet state of Slovakia, the Germans encouraged indigenous native militias to slaughter the ethnic enemies of the Reich. The murderous activities of the Hlinka Guard in Slovakia, the Ustasha in Croatia and the Iron Guard in Romania provided a model of lethal collaborations which would just months later be applied in the Baltic States and Ukraine.

Contingency, fate and luck all played their parts. It was ever thus. Until the end of 1940, Hitler left the Mediterranean and the Balkans to his Axis partner Benito Mussolini. Germany was fixated with Romania's rich Ploieşti oilfields, needed to power and lubricate the planned attack on Russia. And, fearing that either the British or his Soviet allies planned to disrupt the flow of oil to Germany, Hitler ordered the Luftwaffe and Wehrmacht special forces to secure the oilfields at the beginning of October 1940. He informed Mussolini the day after, provoking a tremendous temper tantrum. At a heated summit with his Foreign Minister Ciano, Mussolini bellowed that he had had enough of Hitler's fait accompli and 'would pay him back in his own coin. He will find out from the papers that I have occupied Greece.' It was an eccentric kind of revenge: Ioannis Metaxas, the Greek dictator, regarded himself as a friend of Fascist Italy. But the seizure of the Ploieşti oilfields and Mussolini's angry retort set in motion a chain of events that culminated with German occupation of the Balkans and the dismemberment of Yugoslavia. Mussolini's attack on Greece soon turned into a humiliating catastrophe, forcing Hitler to come to come to the aid of his volatile Axis partner. In the meantime, the British began landing troops in southern Greece and attacked the Italian navy. They occupied Crete and, in North Africa, crushed an Italian army at Sidi Barrani. Once Hitler had digested the scale of the swiftly deteriorating Italian engineered catastrophe in the Mediterranean, he had no doubt that Mussolini's impetuous adventurism had put at risk his plans for Operation Barbarossa, the attack on the Soviet Union, then scheduled for the second half of May. Once the British had access to Greek airbases, they could in theory launch attacks on the precious Romanian oilfields. In broader strategic terms, it would be foolhardy to strike east against Russia if Germany's south-eastern flank was, thanks to Mussolini's ineptness, in shreds and tatters and the Mediterranean wide open to the Royal Navy. The time had come to sort out the mess and rescue the Italians before it was too late.

On New Year's Eve, Hitler wrote to Mussolini promising military support. He sent General Erwin Rommel and his Afrika Korps to Tripoli and began planning Operation Marita, the invasion of Greece. Its success would depend on the co-operation of other Balkan nations like Romania, Bulgaria and, of course, Yugoslavia. At the end of March 1941 German Foreign Minister Ribbentrop bullied Yugoslavian leaders led by Prince Paul to sign up to a Three Power Pact leaving the way clear for Operation Marita. Hitler was convinced that there would be 'no more surprises'. But he had reckoned without the Serbs. Proclaiming that the deal with Hitler was a Croatian plot, anti-Nazi army commander Dušan Simović staged a coup, backed by a wave of widespread public revulsion about the pact. American citizen Ruth Mitchell (the only foreigner who ended up fighting with Chetnik insurgents) recalled: 'Belgrade lay silent in a paralysis of horror, of shame, of slowly kindling fury. Then the storm broke ... university students were demonstrating fiercely, shouting: "Down with the traitors! Better war than the pact!"'[5]

When news of these impertinent mass protests reached Berlin, Hitler exploded with rage. Goebbels commented in his diary, 'The Führer does not let himself be messed around in these matters.' The coup leaders soon fell out and the coup looked increasingly fragile. As the Serbs and Croats bickered in Belgrade, Hitler ranted that Yugoslavia must now be regarded as an enemy. He had after all always regarded the kingdom as an illegitimate 'Versailles state'. His swiftly formulated 'Directive 25' and ordered Wehrmacht (army and air force) commanders 'to smash Yugoslavia militarily and as a state form ... with merciless harshness'. On the afternoon of 29 March, Major General Friedrich Paulus (who less than two years later would surrender to Soviet armies at Stalingrad) presided over detailed planning for a two-pronged ground thrust closely co-ordinated with an aerial assault on Belgrade. Goebbels prophesied that 'the problem of Yugoslavia will not take up too much time ... The big operation then comes later: against R'.[6]

Goebbels was right. In just three days, Operation Punishment had been wrapped up and Hitler's generals could turn their forces against Greece – and the Anglo-Greek forces deployed along its borders. Three days after the destruction of Yugoslavian forces, the German 12th Army rumbled into the northern Greek port of Salonika and then began pushing south towards Athens. When British forces sent a message to the northern city of Jannina they received the cheeky reply: 'The

German army is here.' Operating way ahead of the fast-moving front line, German bombers targeted city after city, stopping only when they closed in on Athens. Hitler had forbidden any bombing of the Greek capital; this was, he believed, the birthplace of Aryan culture. On 27 April, a little after 8 a.m., the German 6th Armoured division rumbled into the ancient city's drab northern suburbs. On the same day, Walther Wrede, a young German archaeologist, recalled: 'A police official ... tells us that German troops are making their way to the Acropolis ... I spring to the lookout post on the upper floor. Correct! From the mast of the Belvedere of the city shines the red of the Reich's flag.' German forces soon chased out the last British troops (in fact Australians and New Zealanders) from the southern tip of the Peloponnese. On 4 May, Axis troops (Italians and Germans) staged a victory parade on Athens. A beaming Wrede delightedly escorted a flood of 'war tourists' led by Field Marshall Walther von Brauchitsch around the Acropolis. Himmler toured the Greek monuments a few weeks later. On 20 May, German airborne forces surprised the complacent British defenders of Crete – and in North Africa, Rommel recaptured territory lost by the Italians.

In the Balkan campaign, the Wehrmacht had smashed military nuts with sledgehammers: Hitler sent twenty-nine divisions against just six weak enemy ones. But as it turned out, just ten German divisions saw action for six days. This new blitzkrieg provided another demonstration that the 'German soldier can do anything'. Now it was high time to make the same brutal point to his Soviet ally, and as soon as the Balkan war had been wrapped up, Hitler ordered the bulk of his forces to rejoin their comrades massing along the Soviet border.

According to Hitler, what one philhellenic German general called the 'lofty culture of Hellas' had once been the ancestral homeland of the Aryan Master Race. But these sentiments did not protect the modern inhabitants of the region whose once noble ancestral bloodlines, many Germans suspected, had been contaminated centuries ago by Semitic Phoenicians and by Slavs. Greece would not become another Poland but such pseudo-scholarly balderdash would have appalling consequences for the people of the Balkans. 'The Germans,' wrote novelist Giorgos Ioannou, 'suddenly introduced ... all the abysmal medieval passions and idiocies of Gothic Europe.'[7]

Hitler's Balkan campaign may well have begun as a way to get Mussolini out of trouble. The destruction of Yugoslavia which has so often been presented as a fit of rage is much less easy to accept as a spontaneous response to an inconvenient coup. Even for the mighty German army, military campaigns required planning time. Enormous numbers of troops had to be diverted to south-eastern Europe. Premeditation surely guaranteed that the Balkan campaign was enacted at breakneck speed and achieved its strategic aims within a few weeks. Although Hitler was forced to delay launching Operation Barbarossa until the summer, the destruction of Yugoslavia and the occupation of Greece very effectively tied up his ragged south-east flank which had been so impetuously weakened by Mussolini. In any case, the occupation of the Balkans provided another opportunity for the Germans to refine the apparatus of occupation. As they had for Operation Tannenberg, Reinhard Heydrich's RSHA cataloguers had diligently compiled 'special lists' of Greeks considered to be 'hostile elements'. Gestapo agents combed Athens seeking out their human prey.[8] In the meantime, Wehrmacht and SS troops swept across Greece, like grey locusts, seizing whatever they could lay their hands on – from goats and olive oil to clocks and even lingerie. While German troops ravenously 'lived off the land', the Greeks went hungry, and soon began sicken and then to die. The numbers of dead overwhelmed the authorities and mass graves had to be dug outside Athens. It is estimated that the final death toll from hunger and disease by the end of 1942 was in the hundreds of thousands. Other species of locust followed. Alfred Rosenberg's Sonderkommando units scoured ancient Salonika in search of cultural treasures and artefacts to stock his new 'Library for Exploration of the Jewish Question'. Robber barons dispatched by the Reich Economics Ministry and German industrial giants like Krupp seized iron, chrome and nickel mines, and dismantled entire factories to send them piece by piece back to the Reich.

To begin with, the SS adapted a dilapidated army barracks not far from Athens in Haidari to use as a holding centre for political prisoners and hostages. The camp commander SS-Sturmbannführer Radomski and his mainly ethnic German guards were drunken brutes. From their offices in Athens, with splendid views of the Acropolis, SS administrators brought bloody terror to Greece and destroyed some of the oldest Jewish communities in the world. At the end of 1942, SS-Sturmbannführer Adolf Eichmann sent trusted aides Dieter Wisliceny and Alois Brunner to organise the deportation of 50,000 Jews from Salonika. When the Greek Jews finally arrived at Auschwitz-Birkenau, camp commander Rudolf Höß recorded that 'they were of such poor quality that they all had to be eliminated'. Anti-Semitism had never contaminated Greek political culture and yet at least 90 per cent of Greek Jews did not survive the war.[9]

The fate of the former Kingdom of Yugoslavia was equally terrible. In the wake of the German armies came savage ethnic cleansing and bloody civil war. Inevitably Jews were the first victims. They had first come to the Balkan region after 1565, when

Sephardi Jews settled in the Miljacka river valley near Sarajevo in central Bosnia. They had been expelled from Spain half a century earlier. Now they had to rebuild their lives in a poor, harsh land ruled by the Ottoman 'Caliphate' in faraway Istanbul. The Sephardi and later Ashkenazi Jews were, by and large, welcomed by their Muslim neighbours (the Bosniaks) and over time became a small but influential Yugoslav community. Then came catastrophe on an unimaginable scale. Between 1941 and 1945, at least a million citizens of the former Kingdom of Yugoslavia died violently; many of the victims were Jews: few survived the German occupation.

In the nineteenth century, Ashkenazi Jews began to settle in Yugoslavia for the first time, mainly in Croatia-Slavonia and the province of Vojvodina. Like the Sephardim, the new arrivals were by custom urban dwellers who toiled in commerce, crafts or the liberal professions. Many Jews contracted 'mixed marriages' to Muslims or Christians and by the end of the nineteenth century, the majority identified themselves as Serbs or Croats and later Yugoslavs. In 1919, Jewish leaders had joined the National Council of Serbs, Croats and Slovenes as 'self conscious nationalist Jews'.[10] It was only when Nazi Germany began meddling in the Balkans that anti-Semitism, as one historian put it, 'crept into the Yugoslav'.[11] Racial chauvinism became especially potent in Croatia, the most fractious region of the kingdom. But even there, by the mid-1930s radical nationalists like the Ustasha movement had been marginalised. Most Ustasha leaders, including Ante Pavelić, had fled to Fascist Italy. Very few Croatian Jews could have foreseen the tragedy that would engulf their community in the spring of 1941.

In Sarajevo, Jews prospered – as metalworkers, tanners, doctors. Sometime in the 1580s, they built their first synagogue, Il Kal Grande. In the mid-1600s Rabbi Samuel Baruch established the beautiful Jewish cemetery in Kovacici. For centuries, these Bosnian Jews survived and prospered. But in 1940, the Yugoslavian government passed 'Numerus Clausus' laws to restrict Jewish enrolment in schools and universities. The intent was to please Hitler's Germany, an aggressive trading partner with a rapacious interest in Yugoslavia's natural resources. Then came the catastrophe of May 1941. On 16 April, German troops tramped through the streets of Sarajevo. That day, they ransacked then demolished every one of the city's eight synagogues. They plundered sacred books, silver, entire libraries and sacred manuscripts. By the summer of 1942, the old Jewish communities in Bosnia had vanished. Any Jews who survived the German onslaught had escaped to the mountains and joined partisan units led by Josip Broz (Tito). One of his closest advisors, Moshe Pijade, was a Bosnian Jew.[12]

On 12 April, even before the Yugoslav army had surrendered, Hitler issued a directive dividing Yugoslavia into German and Italian spheres of influence. He then hacked the Balkan Peninsula into territorial morsels and divided them between Italy, Hungary and Bulgaria. In occupied Serbia, Wehrmacht commander Heinrich Danckelmann appointed a 'quisling', General Milan Nedić, the former minister of the army and navy who had been sacked by Prince Paul in November 1940. Nedić was backed by the Serbian fascist ZBOR movement and its military wing, the Srpski dobrovoljački korpu, which had for some time courted the German Reich. The main beneficiary of the German occupation would be a then little-known Croatian nationalist called Ante Pavelić and his Ustasha movement. The Ustasha militia, modelled on the German SS and Italian Blackshirts, would become the vanguard agents of a spasm of mass murder that in many important respects prefigured the escalation of the German war against the Jews that began in the occupied Soviet Union that summer. The Balkan genocides have rarely been discussed by historians of the German Holocaust. And yet Croatian nationalists had been especially responsive to radical chauvinist ideas that had been hatched and incubated in Germany.

After the creation of Yugoslavia in 1921, many Croatians resented Serbian domination of the new federated kingdom. Tensions between Serbs and Croatians steadily grew more intense. On 20 June 1928, during a parliamentary session, a Serbian deputy of the Radical Party assassinated two representatives of the Croatian Peasant Party and wounded three others. One was Stjepan Radić, the most charismatic of the Croatian leaders who died of his wounds six weeks later. This shocking event, it was said, 'plunged all of Croatia into indescribable agitation' – and led to the foundation of a Croatian terrorist faction the Ustasha (meaning 'uprising') by Ante Pavelić, a former lawyer who represented the minuscule separatist Croatian Party of the Right in the Yugoslav parliament. Described by an American intelligence agent as an 'extremist even in his youth ... quarrelsome ... sulky', Pavelić revered Mussolini and adopted his histrionic mannerisms.[13] In 1929 when the Serb King Alexander abolished the 1921 constitution and turned Yugoslavia into a royal dictatorship, Pavelić fled first to Vienna and then Italy accompanied by a cadre of Ustasha men, dedicated to creating an independent Croatia by any means, including terror. With Mussolini's blessing, the Ustasha exiles set up military-style training camps in Italy and at Janka Puszta in Hungary, where they joined forces with Macedonian and Bulgarian nationalists to plot the downfall of King Alexander and his bastard state. On 9 October 1934 an Ustasha gunman who had been trained at Janka Puszta, assassinated King Alexander and the unlucky French Foreign Minister Louis Barthou during a state visit to Marseilles. Pavelić was naturally delighted by the success of the mission, but the international hue and cry it provoked severely embarrassed Mussolini who had no interest in upsetting either the French or the Yugoslavian governments. He had the Ustasha leaders arrested. While Pavelić was incarcerated in a succession of comfortable villas in Sienna and Florence, the Italians banished Ustasha rank and file to the bleak and windswept Lipari Islands.

As a political force, Pavelić and the Ustasha might then have simply disappeared into political obscurity. In the mid-1930s, Mussolini began to flex his political muscles in the Adriatic and once again began cultivating Croatian radicals like Pavelić, hoping to use them to undermine Yugoslavia. But the quarrelsome Pavelić was his own worst enemy. He squabbled unendingly with other Croatian nationalists, accusing them of being in league with Serbs, Jews and other 'enemies of the Croats'. This habit did not impress his Italian hosts. Like many fissiparous ultranationalist factions in the 1930s, the Ustasha looked increasingly marginal. But as Hitler plunged Europe into war after 1939, Pavelić's fortunes began to change. Hitler's attack on Poland inspired Italian Foreign Minister Ciano to raise the 'Yugoslav question' again. He persuaded Mussolini that Yugoslavia posed the same threat to Italy as Poland allegedly had to Germany. Hitler weighed in too, urging the excessively proud Italian dictator that 'Italy should grasp the first favourable opportunity to dismember Yugoslavia and occupy Croatia and Dalmatia'.[14] Hitler's statement clearly indicates the destruction of Yugoslavia was on his agenda well before he struck in the spring of 1941. With the Balkans thus set out before him as a prize, Mussolini decided once more to renew his acquaintance with his Ustasha friends.

Ciano now held a series of meetings with Pavelić; he later described the temperamental Croatian, rather oddly, as 'an aggressive, calm man'. Naturally Pavelić had no interest in having his homeland occupied by a foreign power. So Ciano mollified him by proposing that the Ustasha, backed by Italian troops, return to Croatia and proclaim an independent state, forcing the break up of Yugoslavia.[15] But Mussolini's Axis partner did not appreciate this Italian meddling in the Balkans. Hitler hoped to avoid providing the British with another excuse to get more deeply involved in the region. So Ribbentrop twisted Italian arms and persuaded Ciano to leave Yugoslavia alone. It was this heavy-handed Axis 'diplomacy' that fuelled Mussolini's intemperate response to the German seizure of the Romanian oilfields that led in a few short months to the German destruction of Yugoslavia.

Now in April, with the Balkans in the German bag, Foreign Minister von Ribbentrop (who had opposed Hitler's plan to invade the Soviet Union and fallen badly from grace) spotted an opportunity to mend his reputation by creating a new state on the Slovakian model. He dispatched a message to the German Consul General in Zagreb ordering him to inform Croatian leaders 'that we would provide for an independent Croatia within the framework of a New Order for Europe'.[16] Ribbentrop began casting around for a suitable puppet to lead his new state. Pavelić was not his first choice. He shared Hitler's disdain for fractious extremists and, unlike other European ultranationalist factions, the Ustasha had hitherto received minimal support from Germany. The pig-headed Pavelić was viewed as a creature of the Italians. Well-informed German diplomats favoured instead the moderate Vladko Maček. But Ribbentrop was aware that his rival Alfred Rosenberg backed Maček. To follow Rosenberg's lead would have been out of the question. Maček, in any case, refused to have anything to do with a fabricated 'Croatian' state and fled to his farm pleading 'incorrigible pacifism'. So Ribbentrop had no choice but to swing behind Pavelić and the Ustasha.

So it was that SS-Brigadeführer Edmund Veesenmayer, representing the 'Dienstelle Ribbentrop' and already an expert 'manufacturer' of client states such as Slovakia, arrived in Zagreb on 10 April. He was accompanied by Ustasha leader and a former Austro-Hungarian lieutenant colonel Slavko Kvaternik. A few hours before the first German troops rolled into the Croatian capital to a tumultuous welcome, Veesenmayer took Kvaternik to the local radio station, sat him down in front of a microphone and jointly proclaimed a 'free, independent Croatia' – '*Nezavisna drzava Hrvatska*' – usually referred to as the NDH. In his first report, Veesenmayer declared:

> the proclamation of a free, independent Croatia was made; this fact called forth tremendous rejoicing and the immediate decorating of [Zagreb] with flags ... The faith and trust of the entire Croatian people in the Führer and his Wehrmacht ... is moving ... I have not committed myself in any way as regards the interpretation of the concept of freedom.[17]

In other words, Veesenmayer and his boss Ribbentrop would be the real *metteurs en scène* of Croatian statehood.[18]

Ribbentrop's machinations pleased Hitler who had previously assumed that Croatia would fall into the hands of the Italians. Now he could add another puppet state to his collection. For his part Mussolini, still smarting from his Greek disgrace, had good reason to hope that Pavelić, who had enjoyed his largesse for so long, could still be of use. He sent him to Trieste, on the border with Croatia, where a few days later he rendezvoused with Ustasha men who had been held on the Lipari Islands. The Italians provided buses and a few rickety cars and the motley Ustasha crew drove south and crossed the border into independent Croatia. As Pavelić and his cronies approached Zagreb, a delegation led by Kvaternik and Veesenmayer waited to greet the new Croatian head of state. In the early hours of 15 April, under cover of darkness, the Ustasha government slipped quietly into Zagreb, at the head of a few hundred paramilitaries kitted out in Italian uniforms.

That same day, the elderly German Plenipotentiary General in Croatia, Edmund Glaise von Horstenau, and the German envoy Siegfried Kasche presented their diplomatic bona fides to the Croatian leader, known as the Poglavnik – a title with the same connotations as Führer. Although the NDH, as Jonathan Steinberg writes, 'lacked everything' including enough cars to drive its Cabinet members to meetings, Pavelić shrewdly inaugurated his rule by appointing a head of propaganda, Vilko Rieger, a journalist who had studied for his doctorate in Berlin. Pavelić told him to 'Consider as friends those I consider friends, and as enemies those I consider enemies'. Dr Rieger lavished his modest departmental budget on conjuring up a conspicuously Catholic leadership cult. The Poglavnik, Rieger proclaimed, was 'the most ideal man of contemporary Croatia, since in the eyes of the people He is their saviour and redeemer'.[19] This was rhetoric all too reminiscent of Josef Goebbels. Although Mussolini had hoped to use Pavelić to dominate 'independent Croatia', Italy would end up Hitler's frustrated junior partner in the Balkans – and the Poglavnik and his henchmen the willing tools of German-inspired terror. Rieger wrote later (on the eleventh anniversary of Hitler's seizure of power), 'The Ustasha movement is the only movement in this part of Europe that according to its programme and activities is so close to German National Socialism'.[20] The national policy of the Ustasha puppet regime would be to build a mono-ethnic state dominated by the Roman Catholic Church but respecting the historic faith of the Bosnian Muslims to secure their allegiance. The message for Serbs, Jews and gypsies was plain. The Ustasha regime spewed out a steady stream of political decrees that embodied a brutally plain maxim: 'only Croats rule always and everywhere.'

In Croatia, radical nationalist ideology had been nourished both by what Michael Burleigh calls a 'crude nativism' that claimed that anyone not descended from a peasant family was 'not Croat at all, but a foreign immigrant', and by academic pseudoscience.[21] In the aftermath of the First World War, the Croatian intelligentsia violently rejected 'Yugoslavism'. 'Yugoslavists' advocated a kind of Balkan melting pot in which the super-heated force of modernity would dissolve the old ethnic and religious barriers. Academics and government officials joined forces to develop a radical social policy that they believed would bring forth the new 'Yugoslav man'. They embraced the full armoury of modernist social engineering from rapid urbanisation to eugenics.

Croatian scholars took an altogether different line. Unlike Serbia, which had thrown off the shackles of empire in the nineteenth century, the Croatian lands remained locked inside the Austro-Hungarian Empire until the end of the First World War. Croatian nationalism emerged against a backdrop of tremendous ethnic diversity and Croatian nationalists looked for any distinctiveness that gave them an edge over their neighbours. The father of Croatian separatism was Ante Starčević, a philosopher and theologian who founded the Croatian Party of Rights. A prolific writer of popular pseudoscientific books, journalist and political agitator, Starčević insisted that Serbs at best disguised Croats or, at worst, an inferior and degenerate race descended from nomadic 'Vlach' shepherds. He promoted what would become a persistent theme in Croat racism: the idea that Bosnian Muslims, the Bosniaks, possessed the purest Croatian bloodlines. This was, even on its own terms, nonsense: the Bosnian Muslims were Slavs just like the Croats and Serbians and had no special claim to ethnic purity. In any case, Starčević's claim was not intended to flatter Muslims, who resented the 'pure Croatian' appellation anyway because it implied that the Bosnian heartlands, to which they claimed a vague kind of right, were, by default, an integral part of Croatia. But for Catholic Croatians, these fantasies of ethnic singularity had insidious popular appeal, at least in part because Starčević and his followers were unaffiliated 'gentlemen scholars' rather than elite university academics. This populist ethnic advocacy soon found its political champions in Josip Frank's Pure Party of Rights and the Croatian Peasant Party, led by future Croatian martyr Stjepan Radić. On the streets of Zagreb, party activists began attacking Serbs, chanting slogans picked up from Starčević's pamphlets and books.

Another advocate of Croatian ethnic singularity was Ćiro Truhelka. He was an archaeologist and anthropologist who welded together anti-Serb rhetoric with the tropes of anti-Semitism. He claimed, for example that the despised Vlachs, descended from the pre-Roman inhabitants of the Balkans, would always be 'recognizable at a hundred paces'. Any intelligent child on meeting a Serb would exclaim 'That's a Vlach!' This peculiar formula was borrowed, of course, from the anti-Semitic maxim that even assimilated Jews could never hide certain telltale physiological traits. Like Hans F.K. Günther, the German *Rassenpapst*, Truhelka relied on a set of allegedly fixed physiological tropes: Serbs were dark skinned, brown eyed and 'pigeon-chested'; Bosniaks and Croats were blonde and blue eyed. Serbs were swarthy, degenerate types, who, unless they were 'removed', threatened to spread their dark blood among true Croatians. The solution naturally was to build national barriers to protect Croatian lands and blood: Croatia for Croatians.

This strand of Croatian nationalism faded somewhat after the creation of Yugoslavia, but as the kingdom began to fracture after 1925, the Croatian cause found a brilliant new advocate. This was Milan Šufflay, a genuinely brilliant scholar in the Anthropology Department of Zagreb University. Although he was respected outside Croatia, Šufflay developed a fixation with the pre-eminence of the white race and the threat of 'Asiatic' that echoed the hysterical theories of the German anthropologists. Like them, Šufflay believed that his own 'Gothic' blood line must become a cordon sanitaire between the west and the 'Asiatic' east. 'The blood of Croatdom means civilization,' he wrote, 'it does not mean simply a nation. Croatdom is a synonym for all that is beatific and good that the European West has created.' In the Balkans, Serbs had distinct and threatening 'Asiatic' characteristics; they did not belong with Croats within the same pseudo-national borders: 'Yugoslavism' was a dangerous delusion.

To proclaim these ideas in royal Yugoslavia was dangerous. In 1931, Serb fanatics ambushed Šufflay in the street outside his home and beat him to death with iron rods. Prominent intellectuals like Heinrich Mann and Albert Einstein, who surely cannot have read Šufflay's nationalist tracts, denounced the Yugoslavs for failing to protect an academic luminary. Šufflay joined a Valhalla of martyred heroes. The murder galvanised nationalists, including the embryonic Ustasha militia. A new convert to the cause was a young law student called Mladen Lorković who, in 1939, published a pamphlet, 'The Nation and Lands of the Croats'. Following Šufflay's lead, Lorković explicitly introduced the language of German racism into Croatian nationalist rhetoric. He stated that Vlach nomads (a pejorative way of referring to the ancestors of Serbs) and Turkish mercenaries had 'stolen Croatian living space'. He resuscitated the old idea that Bosniaks were the purest Croatians – and that Bosnia thus belonged to Croatia, just as German nationalists had claimed that the Sudetenland and other ethnic German strongholds were an integral territory of the German Reich. But Lorković went a lot further even than Šufflay when he proclaimed that Croatians were originally of Persian descent, and were thus Aryans – not South Slavs at all.[22] After 1941, in Ante Pavelić's puppet state, pseudo-history like this had deadly consequences as the Ustasha regime rushed with unseemly haste to bolster the legitimacy of the NDH. The fantastical notion that Bosniaks had distant Aryan ancestors would later be taken up by Muslim leaders when, in 1943, they sought German backing for their own autonomist aspirations.

Now in the early summer of 1941, the new German-backed Ustasha government would incite a crusade against the hated 'Asiatic' Serbs. But to satisfy their sponsors in Berlin, their rage fell first on the Jews. In September 1942 Monsignor Augustin Juretic fled Croatia and submitted a series of reports to the American OSS and the Yugoslav government-in-exile in London. He denounced the genocide as a 'dark blot on the conscience of many Croats'. Croatia, he said, had become 'a real slaughterhouse'. Some 80 per cent of Croatian Jews would perish at the hands of Ustasha murder squads and in Croatian camps like Jasenovac.[23]

In Berlin, as soon as Balkan matters had been settled, the demands of Operation Barbarossa again took precedence. The bulk of German forces were withdrawn from the Balkans and were replaced by garrison units. In Croatia a single division, the 718th headquartered in Banja Luka, was left behind. The German diplomatic corps headquartered in Zagreb proved to be either fanatical Nazis or feeble 'yes men'. The Plenipotentiary General, the Austrian Edmund Glaise von Horstenau, was not in Hitler's view, reliable. Glaise von Horstenau was a dedicated Nazi and had served under the fanatical Arthur Seyß-Inquart. But he was hostile to Himmler's SS and his reports to Berlin show that he was repelled by Ustasha violence, while barely lifting a finger to stop it. The German envoy Siegfried Kasche, on the other hand, a fanatical Nazi who had joined the party in 1926, had no such doubts. He remained a staunch supporter of the NDH to the very end. What troubled Kasche and the German Foreign Office was not Pavelić, but Mussolini. State Secretary Ernst Weizsäcker confusingly warned Kasche that 'the Croats and Italians would not get along well' and that he should in all matters 'spare Italian sensibilities' and let 'Italian hegemony in Croatia prevail'.[24] But 'sensibilities' could not get in the way of German strategic plans. Hitler's solution was to bind Pavelić to the Reich by exploiting what historian Marko Attila Hoare calls 'the Ustasha's genocidal proclivities'.[25]

In other words, Pavelić and Ustasha militias would serve German interests by fully embracing the core Nazi doctrines of state terror and ethnic cleansing. Croatian propaganda soon enshrined both. In the 1930s, a Croatian 'legion' had been trained by German officers in Vienna. In 1941, this became the core of a new Ustasha militia that was loyal to Pavelić and dedicated to 'Croatia for Croatians'. According to an Ustasha propaganda leaflet: 'knife, revolver, bomb, and the infernal machine, these are the means that are going to return to the peasant the fruits of his land.'[26] But who, in Ustasha minds, was guilty of purloining these fruits? The answer was obvious: Orthodox Serbs, Freemasons, Gypsies and Jews. The hard-faced Croatian nationalists and the Catholic 'Clericalists' who had been allotted senior positions in the Pavelić government, including propaganda and mass media, believed ardently that Jews were a toxic 'foreign element' that spread poison through the Croatian body politic. But a parochial hatred of Serbs far outweighed the regime's anti-Semitism.[27] One of the terrible ironies of the Croatian genocide is that Ante Pavelić, the head of state and Slavko Kvaternik, commander-in-chief of the armed forces, had half-Jewish wives, and a few high-ranking Ustasha Cabinet members had married 'full Jews', whom they secretly protected from the attentions of NDH murder squads.

Hoare makes a crucial point: the Croatian genocide was both 'a Nazi-led genocide of Jews and Gypsies and an independent ... genocide of the Serbs'. What drove Ustasha anti-Jewish measures was a desire to fit in with Hitler's New Order.

The Ustasha campaign against Croatian Jews began early in April with a flurry of reactionary legislation. The law for the establishment of the army and navy excluded both Jews and Serbs from military service except in labour battalions. Then on 30 April, the law decree on racial belonging and the law decree on the protection of Aryan blood and honour of the Croatian people (modelled on the German Nuremberg Laws) proscribed marriage between gentiles and Jews and sexual relations between Jewish men and Croatian women. Jews would have to register with the Ustasha authorities and if they had changed surnames after 1918 they were compelled to resume the use of their original names. The Ustasha regime prohibited Jews from working in the liberal professions or frequenting restaurants and hotels, cinemas and theatres. Jewish religious and cultural institutions were plundered, and a new 'Office for the Reconstruction of the National Economy' plundered Jewish and Serbian businesses.

The impact of this legal blitzkrieg was, as Michael Burleigh succinctly puts it, to 'abrogate all constitutional and legal provisions that granted religious equality and freedom of conscience'.[28] NDH anti-Jewish legislation stated explicitly:

> Since Jews spread false reports in order to cause unrest among the people, and since by their speculation they hinder and increase the difficulty of supplying the population, they are considered collectively responsible. Therefore the authorities will act against them and *beyond criminal legal responsibility*; they will be confined in assembly camps under the open sky.[29]

A tiny handful of Jews – if they had rendered noteworthy service to Croatia, voluntarily given up their property or had married a government official – could sometimes evade persecution as 'honorary Aryans'. It was this kind anomaly which in 1942 so infuriated intelligence officer SS-Sturmbannführer Wilhelm Höttl (another Austrian) and provoked direct SS intervention to finish off what the Ustasha had started – and then bungled.

After 1933, the Nazis had at first proceeded cautiously against German Jews, starting with the Nuremberg Laws in 1935 and escalating their onslaught two years later in November 1938. In the new puppet NDH, there was no need for such finesse. The killing must begin immediately. On 6 June, Pavelić travelled to the Berghof for a summit with Hitler. They focused on the future ethnic composition of the new state; Hitler advised Pavelić that 'if the Croatian state was to be truly stable, a nationally intolerant policy had to be pursued for 50 years'.[30] Hitler's lethal arm-twisting must be seen in context with what had already begun to take place in Serbia and the Banat region, which was dominated by radicalised ethnic Germans. As soon as they had crossed the Romanian border into the Banat, German troops and local ethnic Germans squads rounded up and imprisoned about a third of all adult male Jews. Then in the summer of 1941, the German military administration, with minimal SS prompting, began to systematically kill Jews and Serbs. As German troops had withdrawn to the Eastern Front, Yugoslav partisans had begun attacking the German garrison troops. According to brutal German reprisal doctrine, Serb villagers would pay the price in blood. At the same time, just as Himmler did, Wehrmacht generals blamed Jewish 'bandits' for inciting attacks. Anti-Semitism and hatred of Serbs intricately blended in the German military psyche. Many soldiers called Serbs a *Rattenvolk* (rat people), a term borrowed from the lexicon of anti-Semites. Austrian general Franz Böhme egged on his troops: 'Your mission ... lies in the country in which German blood flowed in 1914 through the treachery of the Serbs, women and children. You are the avengers of these dead.' In the Banat, the depleted German military administration could call on the assistance of radicalised ethnic Germans – one reason why the liquidation of both Serbs and Jews was accomplished there with such horrible speed.[31]

So when Pavelić kowtowed to Hitler at the Berghof, the mass murder of both Serbs and Jews was already under way. If the Ustasha failed to act, the German army would take over, exposing the illusory autonomy of the NDH. Pavelić returned to Zagreb with a renewed sense of mission. His Education and Culture Minister, Dr Mile Budak, proclaimed: 'For minorities such as the Serbs, Jews and gypsies, we have three million bullets.' The genocide tore open, one Ustasha renegade confessed, 'a great Croatian wound ... Our faces burn for shame'.

The Germans kept up the pressure. On 22 June, Hitler summoned Pavelić's military henchman Slavko Kvaternik to his new eastern headquarters at Rastenburg. The attack on the Soviet Union had begun at 3.15 that morning. As 3 million German soldiers advanced across the Russian border and the Luftwaffe pounded Soviet cities and airfields, Hitler found time to rant at Kvaternik:

> The Jews are the bane of human kind. If the Jews will be allowed to do as they will, like they are permitted in their Soviet heaven, than they will fulfil their most insane plans ... This sort of people cannot be integrated in the social order or into an organized nation. They are parasites on the body of a healthy society ... There is only one thing to be done with them: to exterminate them ... it would be nothing less than criminal to spare these bastards.[32]

Historians have often debated precisely when Hitler 'gave the order' to inaugurate the destruction of all European Jews. It is hard to conceive of a more direct instruction than the Rastenburg tirade lapped by Slavko Kvaternik.

German diplomats stationed in Zagreb left us with detailed accounts of the Croatian Holocaust. The main perpetrators were the new Ustasha militias, formed with German assistance in April and modelled on the SS. As in the Waffen-SS, service was voluntary. The majority of recruits were young, awash with testosterone, devoted to the Poglavnik and unquestioning believers in the maxim 'Croatia for Croatians'. Their modus operandi resembled murderous Hutu militias like the Interahamwe and Impuzamugambi, who rampaged through Rwanda half a century later in the spring and summer of 1994. Like the Hutu squads, the Ustasha and the Hutu death squads used knives, clubs, hatchets – weapons procured from field and farmyard. The killers customarily herded victims into churches and school buildings, closed spaces where the brute force of hatchets and knives had the most deadly effect. The killing was both grisly and intimate. In July 1941, the German Plenipotentiary General Edmund Glaise von Horstenau reported that the Ustasha campaign incited by his own government was driven by 'blind, bloody fury'. 'Even among the Croatians,' he went on, 'nobody can feel safe. The Croatian revolution is by far the harshest and most brutal of all the different revolutions I have been through at more or less close hand since 1918.'[33]

The killers were almost without exception good Catholic boys. In the 1960s, Italian journalist Carlo Falconi investigated the wartime Croatian massacres to try to shed light on how much the Vatican had known about the slaughter.[34] His report documented in detail what 'blind bloody fury' meant for the Jews and Serbs trapped in Croatia. In the villages, where the targets were mainly Serbs, an attack would usually begin just after dawn. Villagers would be woken by the rumble of truck engines and a blaze of headlights. The trucks would stop in the village square and black-capped Ustasha men would throw open the tail gates to fan out through the village, driving people from their farms and houses. In one case, the Ustasha men herded hundreds of Serbs into a local church to attend a service of 'thanksgiving for Croatian independence'. The Ustasha men thrust their way inside brandishing knives and axes. Serbs were traditionally Orthodox not Catholic, but a minority had converted. Ustasha officers were often loath to murder Catholics and they demanded to see any 'certificates of conversion'. Just two men possessed the correct documents. They alone were released. The Ustasha men now set about butchering everyone else, men, women and children. Falconi discovered that according to many eyewitness accounts, Franciscan priests took a leading part in the slaughter. These men armed themselves with knives and clubs, set fire to Serb homes and sacked villages. One priest performed a celebratory dance around a pile of Serbian corpses. During one especially hideous massacre that took place in Nevesinie, Ustasha militia rounded up 173 Serbs. Wielding hammers, picks, rifle butts and knives, they mutilated and severed ears, noses, genitals and fingers. They gouged out eyes, they ripped off hair, beards and eyebrows and stuffed them into the mouths of victims. The Ustasha beheaded men and severed the breasts of their wives or sisters. Others they tortured to death in front of their families. When a small boy begged for water he was shot in the head.

The Germans were dismayed by this litany of horrors. They deplored the fact that their Croatian killers acted in 'hate and hot frenzy'; this was not the German way. Glaise von Horstenau grumbled that the Ustasha militia had 'gone raging mad'. At a conference of Axis military top brass in Rome in 1942, Generaloberst Alexander Löhr described the situation in Croatia as 'very unsatisfactory': the Croatian army was unreliable; the Ustasha were savages. Mussolini concurred: 'it was madness of the Poglavnik', he complained, 'to think he could exterminate two million Serbs'.[35] There was a steep learning curve on how to use these non-German executioners.

Croatia was the only Axis satellite state that murdered more non-Jewish than Jewish civilians. But this is not to say that Axis occupation did not energise autochthonous anti-Semitic energies. The NDH, as we have seen, passed anti-Jewish legislation that defined Jews in racial terms. Under Edmund Veesenmayer's tutelage, the Ustasha regime immediately Aryanised the capital and removed Jews from public office and professions with ruthless energy. By May 1941 they had begun rounding up Jews in Zagreb and dispatching them to concentration camps built on the German model. The biggest was Jasenovac, an archipelago of destruction that sprawled alongside the Sava River close to the village of Krapje. Here in different sections the Ustasha segregated Serbs, Jews, Gypsies and Croats. At Jasenovac at least 25,000 Jews perished. The numbers of Serbs and Gypsies who died was possibly even higher.[36] Jasenovac was a place of profligate cruelty. Father Petar 'Pero' Brzica, the camp's most notorious guard, wielded a custom-designed blade strapped to his wrist that he called a 'Serb cutter'. Inside Jasenovac, the Ustasha gassed their Jewish and Serb victims or cremated them alive. They hurled their bodies into the Sava River. A Croatian prisoner who survived to describe his experience described 'the screams and wails of despair and extreme suffering, the tortured outcries of the victims, broken by intermittent shooting'.[37] At the trial of the Jasenovac camp commander Dinko Šakić in 1994, a 77-year-old witness recalled speaking to Ustasha Lieutenant Zrinusic. 'He told me (once) he had competed in slaughtering, but lost to Ustashi Lt. Brzica. For him, the genuine Ustashi was the one who had bloodied his hands. The Ustashi who would not kill were punished.'[38]

Hitler sanctioned this hideous regime – and urged its fanatical leaders to be 'nationally intolerant' and to 'exterminate Jews'. The German Plenipotentiary General Glaise von Horstenau may have, on occasion, deplored Ustasha 'excesses', but in 1943 his subordinate Siegfried Kasche proudly reported that Croatia had been successfully cleansed of Jews. On 20 January 1942, at the Wannsee Conference, which was convened to plan the administrative details of the 'Final Solution of the Jewish problem', Himmler's deputy RSHA chief Heydrich informed delegates that: 'In Slovakia and Croatia the matter [of the Final Solution] is no longer so difficult, since the most substantial problems in this respect have already been *brought near a solution*.'[39] In the Balkans, Himmler and the SS leadership learnt important lessons. The Croatian experience showed that native militias like the Ustasha had a role to play in organised mass murder. But while the Ustasha squads had proved themselves to be effective killers, they had 'run amok'. According to German observers, there was a right way to carry out mass murder and the Ustasha militias had failed to conform. On 17 February 1942 an SS intelligence chief sent a memorandum to Himmler blaming the Ustasha for igniting Serbian 'bandit activity' by committing *Greueltaten* (acts of horror). He estimated that 300,000 Orthodox Serbs 'have been massacred and sadistically tortured to death'. This was counterproductive.[40] Worse, by the summer of 1942 the Ustasha killing spree had run out of steam, forcing the Germans to step in to mop up survivors.

In Himmler's mind, these acts of horror were merely teething problems. German values would ultimately prevail. There could be no doubt that for the SS planners of genocide, properly organised non-German execution squads would have a vital role to play. As Ustasha militias rampaged through independent Croatia on the eve of Operation Barbarossa, Heydrich issued orders to his Einsatzgruppe commanders that no obstacle should be placed in the way of what he called *Selbstreinigungsbestrebungen*, meaning autonomous cleansing efforts; 'on the contrary they are to be intensified if necessary and directed into the right channels'.[41]

3

Night of the Vampires

> Romanians! For each Jidan [Jew] that you kill you liquidate a communist. The moment for revenge has come!
>
> Iron Guard proclamation, 1941

On 22 June 1941, along a 1,000-mile front line that stretched from the Finnish border in the north to the Black Sea in the south, three vast German armies waited in the dark; 3 million troops, 3,350 tanks, 7,146 artillery pieces and 2,713 aircraft – the largest invasion force ever assembled. To the north and south, Finnish and Romanian divisions allied to Germany bracketed this stupendous horde. The Polish campaign had been a dress rehearsal for this new 'war of annihilation'. Hitler was supremely confident. 'We have only to kick in the Soviet front door,' he declared, 'and the whole rotten edifice will come tumbling down.' At 3.15 a.m. the storm finally broke. The fire and thunder of a monstrous artillery barrage rippled along the front – and the armies of the Reich and its allies roared into the barbaric east, the lair of the 'Jewish-Bolshevik' enemy. Above, the massed engines of great fleets of Luftwaffe aircraft pulsed through a cloudless sky. 'Such a thing only happens once in the whole world!'[1] The German attack on the Soviet Union had begun.

At about the same time, another act of barbarism began to unfold in the grim Romanian city of Iaşi, situated close to the Soviet border. As the German armies unleashed a storm of destruction on the astonished Russian defenders, paramilitary bands hustled young Jews to the city's Jewish cemetery and ordered them to excavate a long, deep pit. In the city, crucifixes appeared on the walls or doors of Christian houses, marking them off from the homes of their Jewish neighbours. On 27 June, Romanian dictator Marshall Ion Antonescu telephoned the commander of the Iaşi garrison and instructed him to begin 'cleansing Iaşi of its Jewish population'. On the evening of 28 June, as German forces punched ever deeper into Soviet territory, an unidentified aircraft flew over Iaşi and released a blue flare. Immediately afterwards, sporadic gunfire was heard all over the city. Romanian militias and civilians armed with hatchets and firearms laid siege to Jewish homes and businesses. Backed by German military units, the Romanians began marching the city's Jews to the central police station. In the courtyard, Romanians and Germans began beating their captives to death. Others they marched to the railway station and loaded them on board sealed train wagons. By the end of the month, 14,850 Jews had been killed, both in the city and on locked and sealed death trains. It was the first major pogrom of the Second World War – and the beginning of a campaign of violence in which 270,000 Romanian and Ukrainian Jews would perish.[2]

In the 1930s, Himmler had cultivated the Romanian 'Legion of St Michael' and its militia the Iron Guard. He admired the legion's founder Cornelius Zelea Codreanu and, following his murder, backed the new Iron Guard leader Horia Sim. But early in 1941, Marshall Antonescu expelled SS agents from Romania – and most historians have concluded that the SS was not involved in the Iaşi massacre and its aftermath. This terrible event would seem to be an open and shut case of autochthonous anti-Semitic mass murder. The Romanian regime had long ago abandoned any pretence of democracy and was dominated by some of the most vicious anti-Semitic demagogues outside of Germany. Deeply ingrained Romanian bigotry would seem to provide a ready and logical explanation for the mass murder of Romanian Jews that began in January 1941 and reached a hideous climax in Iaşi in June. By the end of the war, the Romanian 'Legionary' state had killed or deported well over half of its Jewish citizens. Antonescu and his henchman called this 'Romanianisation', a polite term that disguised a disease that afflicted all the new nation states thrown up by the tidal wave of autonomist fervour that engulfed Europe in the nineteenth century. In the case of Romania, the strain of chauvinism that infected every state apparatus was especially virulent and was most forcefully directed at Romania's Jews, who had played a dynamic role in the new nation's rapid industrialisation. In Romania, anti-Semitism transcended social class and educational status. Its principal spokesmen emerged from the universities and a powerful intelligentsia wrapped the crude emotions of race hatred in the trappings of high culture.

There had been speakers of the Romanian language in south-eastern Europe for many centuries, but the nation of Romania was the surly child of fanatical nationalism. The first autonomous Romanian state, the so-called *Vechiul Regat* (Old Kingdom), was recognised by the Great Powers in 1881; but since it was hemmed in by the three great empires of Russia, Austro-Hungary and Turkey, it could never satisfy the most fervent nationalists. In the period between 1881 and the First World War, the *Regat* was steadily enlarged and was proclaimed *România Mare* (Greater Romania) in 1912. At the end of the war, with the disintegration of the old empires, Romania became positively bloated with the acquisition of Transylvania, Bukovina and, most importantly, Bessarabia, which extended its reach along the Black Sea coast. Such rapid expansion led to indigestion and chronic discontent. Territorial aggrandisement had inevitably enlarged Romania's pool of ethnic minorities which now included Hungarians, Bulgarians, Germans, Ukrainians, Serbs, Greeks, Russians, Roma and Jews. So-called 'Regateni' Romanians regarded themselves as a Latin not a Slavic people, the descendants of Roman settlers. They resented and feared their new Magyar and Jewish neighbours. Romanian academics began to promote a language of exclusion and cleansing. The powerful Romanian Orthodox Church had traditionally deprecated Jews as Christ killers. Between 1867 and 1918, Hungarian Jews enjoyed a golden age of tolerance and prosperity, but in Romania, official documents record no less that 196 violent anti-Jewish incidents during the same period. In Romanian universities, anti-Semitism was fashionable, even glamorous. Radical, chauvinist professors had a stranglehold on Romanian academic life, and their shrill and venomous voices shaped the hearts and minds of generation after generation of young men and women. The intelligentsia was in thrall to the most radical brand of Romanian nationalism. There was nothing civilised about their crude message. Only violence could solve Romania's 'Jewish problem'.[3]

By the 1930s, two violent anti-Semitic factions dominated Romanian political culture: the League of National Christian Defence (*Liga Apararii National Crestine*) and Legion of St Michael (*Legiunea Arhanghelelui Mihail*). The latter is often referred to as the Iron Guard, which was, strictly speaking, its military wing. Neither was a minority faction. Both wielded enormous influence. The legion's aggressive agitation strategies led directly to the election of an openly anti-Semitic government in 1937 and propelled the rise to power of Ion Antonescu, whose 'National Legionary State' would pass forty-one anti-Jewish decrees between 11 September 1940 and 21 January 1941.

A conclusion might easily be drawn that the mass murder of Romanian Jews, which began in Bessarabia at the end of June 1940 but climaxed the following summer, was simply the expression of indigenous Romanian chauvinism. But this 'Romanian' explanation fails to account for an important factor: timing. Why did the Romanian state unleash this murderous assault on its Jewish citizens in late 1940 and not earlier? Why did the terrible blow of Romanian anti-Semitic violence fall on the Jews of Iaşi at the precise moment German armies roared across the Soviet border? Might the climactic radicalisation of Romanian chauvinism have been somehow entangled with Nazi aggression? This clearly was the case in Croatia – but the NDH was a puppet entity manufactured by the German Foreign Office. Though it was allied to Germany, Romania was an independent nation state. To answer this question we must dig a deeper into the story of the Legion of St Michael and its malevolent founder Corneliu Zelea Codreanu.

According to historian Jean Ancel, 'Hatred of Jews was [the legion's] true faith, a dogma of their Christianity'. Unlike the neo-pagan NSDAP in Germany, the legion made a direct appeal to traditional Orthodox religiosity and enjoyed the backing of the influential and reactionary Patriach Miron Christea. Like one of its most celebrated ideologues, the philosopher Mircea Eliade, the legion despised modernity and exploited the darker rhetoric of the mythical, the primitive and instinctive. Legionaries detested so-called 'Judaic' values such as logic and materialism. It offered both liberation to Romanian peasants and a semi-licit thrill for students and intelligentsia. The Legion was modelled on the notorious 'Black Hundreds', the Russian anti-Semitic terrorist movement that first co-opted the cult of St Michael. But it forged close links with Hitler's agnostic Reich and Himmler's modernist SS. Legion rank and file would be dominated by vigilantes, who proudly rejoiced in practising 'righteous violence' – a decades long orgy of political murder.

On 10 November 1926 a group of Jewish students appeared in a Bucharest court accused of demonstrating against one of their professors who was in the habit of making insulting remarks about Jews. As the students left the court, a young man stepped forward and shot one of the students, David Falik, three times in the stomach. He died in agony forty-eight hours later. The culprit, who had carried out his crime in front of a small crowd that had come to jeer the students, was arrested and brought to trial. He was defended by Paul Iliescu – a member of the Romanian parliament who denounced the victim thus: 'so will die all the country's enemies. By innumerable bullets which will be fired against the filthy beasts.' The jury found the accused not guilty and he was escorted from the courtroom on the shoulders of his supporters. His name was Nicolae Totu and he had travelled all the way from his home city to kill at least one of those insolent Jewish students. That city was Iaşi; it has been said that this rather grim, heavily industrialised northern metropolis was to Romanian anti-Semitism what Munich was to Nazi Jew hatred. It was, in other words, the source, 'steeped in anti-Semitism'. The city was above all the Mecca of the Legion of the Archangel Michael.[4]

The movement's founder was Corneliu Zelea Codreanu, who named the legion to honour the patron saint of Romania's wars against the Turks. The Archangel Michael, he claimed, had visited him while he waited in prison to be tried for murder. This self-proclaimed myth tells us quite a lot about Codreanu. He cultivated a political persona as a charismatic mystic who disdained doctrine. He talked of saving Romanian souls and restoring the nation. One of Codreanu's most fervent admirers, professor Nae Ionescu, explained why cast-iron doctrine was anathema: 'Ideology is the invention of the liberals and the democrats. No one among the theoreticians of totalitarian nationalism creates a doctrine. Doctrine takes shape through the everyday acts of the legion as it evolves out of the decisions of him [Codreanu] whom God placed where he orders.'[5]

Codreanu exploited redemptive nationalist rhetoric to tap deep into the national psyche. 'The spiritual resurrection!' he proclaimed. 'The resurrection of nations in the name of Jesus Christ!' He was born in the small Moldavian village of Husi in 1899. At the time, this part of northern Romania was still under Hapsburg rule. His father Ion Zelinski had travelled here from Bucovina and married Elizabeth Brunner, a young German woman. Corneliu was their first son. When he was 2, Ion changed the family name from the dubious sounding Zelinski to its Romanian form Codreanu. To say that Ion Zelea Codreanu was a violent bigot would not do his memory justice. He liked to strut around Husi dressed in national Romanian costume and brandishing a wooden club, worn smooth, it was said, by frequent use.

By the time Ion's son completed a perfunctory period of military training, Corneliu had matured into a tall, striking young man who had thoroughly absorbed his father's fierce hatreds. Soon after the war ended, he left home to study law at Iaşi University. Like Munich or Berlin, Iaşi, the old capital of Moravia, was in those febrile times a bubbling cauldron of bigotry. Since the mid-nineteenth century rapid industrialisation had transformed a city once dominated by its university into a sprawling version of Manchester. New factories sprang up belching smoke and undermining old certainties and relationships. Students and their powerful professors still dominated the old centre of the city. After classes at noon and then in the evening, young men and women spilled into cafes like the Fundatia or Ceasocornicaria Goldstein, where the odours of strong coffee and *Brenza*, the pungent local cheese blended with *ciorba de pui*, a greasy chicken broth made with vinegar. Since the 1960s, we have become accustomed to associating university students with left-wing radicalism. In Germany and Romania in the 1920s, students formed the vanguard of the ultranationalist right. In the cafes of Iaşi, students talked of little else but Romania's 'Jewish problem' and the menace of that associated political religion, Bolshevism.

From the balcony of the neo-classical Jockey Club, plump and complacent Romanian aristocrats languidly watched little dramas unfolding in the street below where the main thoroughfares the Strada Brocuraru and Strada Carol met. Here a wealthy new middle class, many of them Jewish, rubbed shoulders with peasants who had come in search of work and a new life, and angry-eyed students. Romanian soldiers begged on street corners, bewildered and resentful. Romania was in turmoil.

Iaşi was a city of faith too – it boasted over a hundred Orthodox churches and scores of synagogues. For despite the bitter vinegar of anti-Semitism, Iaşi was one of the most important European centres of Jewish enterprise and learning, where the first ever Yiddish newspaper had been published more than half a century earlier in 1855. By 1900, close to half of Iaşi's population was Jewish. But the confidence and prosperity of many Romanian Jews was intolerable to young chauvinists like Corneliu Codreanu and his fanatical followers. Anti-Semitism bonded gentile factory worker and peasant, student and bureaucrat, soldier and professor. The crucible of reaction was the university. Codreanu fitted well into this reactionary cesspool and soon settled on a mentor: Professor A.C. Cuzu, the university's political godfather and veteran of numerous anti-Semitic parties and campaigns.

By 1927 Codreanu, thanks to Cuzu's enthusiastic patronage, had proved himself a competent and charismatic political organiser. He was not afraid to use violence and a campaign of assassinations led to several appearances in court and a prison sentence. But Codreanu soon returned in triumph to Iaşi – and as his train passed through stations, priests held masses and children threw flowers. He had long outgrown even Cuza's fanatical National Christian Union and his old mentor too. Cuzu's own thinking was brutish, but Codreanu, thrilled by the success of Mussolini and Hitler, his ego inflated by the adulation of student acolytes, demanded a new party of action. We must assume that the archangel had paid his fateful visit to Codreanu's cell in Turnu Severin, for on 24 June 1927 Codreanu issued 'Order No 1' to found the Legion of the Archangel Michael.

He called his handful of followers 'Legionaries' and kitted them out in green shirts, Sam Browne belts and high black boots. The 'Roman salute' was de rigueur. Codreanu appointed himself captain and many Romanians simply referred to him as 'the Captain'. In Iaşi, Codreanu began to recruit cells or nests (*cuiburi*) of converts, and to begin with the legion was unashamedly a kind of devotional sect – and deliberately so. It had all the dark ritualistic attraction of elite university fraternities. By 1929, Codreanu had attracted just 1,000 followers, but in those early years that was enough. For his followers, he conjured up the ideal of the *omul nou* (new man): 'Everything depends on will,' he proclaimed. The Legion, Codreanu said, was a 'form of life' devoted to the 'spiritual resurrection of the nation'. Its task was 'to hand out justice to the righteous and death to the wicked'. Action followed rhetoric. In December 1927, at a Legion conference in Oradea, delegates took time off to set synagogues on fire and incinerate their ancient Torah scrolls. As the legionaries travelled home, they stopped at Huedin, Târgu Ocna and finally at Iaşi, where they gleefully set Jewish households alight and pillaged property.[6]

Cultist ritual reinforced absolute devotion. Codreanu initiated Legionaries into a 'Brotherhood of Christ' at bizarre ceremonies reminiscent of Freemasonic rites. Aspirants slit their arms to fill a communal cup with blood which was passed from hand to hand and eagerly slurped.[7] Codreanu, it was said, resembled the handsome American actor Tyrone Power. An Italian admirer who met him at his headquarters, the Green House, recalled that the 'Captain' had 'an uncommon nobility of expression, frankness and energy imprinted on his face, azure eyes, open forehead, a genuine Roman-Aryan type'. Codreanu showed his visitor his version of Hitler's *Mein Kampf* – a bulging volume called *The Iron Guard*. Page after page, Codreanu assailed 'the filthiest tyranny, the Talmudic, Israelite tyranny'. He denounced what he called Mussolini's 'implicit' anti-Semitism. Romania must be 'de-toxified': Judaism had to be purged.

Codreanu, like many fascists, was a born ham; he often swept into Legion gatherings riding a white horse and clad in peasant garb. But there was nothing rustic about Codreanu's organisation – it was a terrorist cult of death. One of Codreanu's 'Brotherhood', General Zizi Cantacuzino, remarked casually that the only way to solve Romania's 'Jewish problem' was to kill the Jews.[8] A striking photograph of Codreanu, 'a god descended among mortals' according to his disciple Horia Sima, shows the 'Captain' squatting among the relics and remains of Romanian disinterred by Legion archaeologists.[9] Codreanu grasps a worm-eaten regimental banner and gazes solemnly into the empty eye sockets of a dead Romanian warrior. The image is both a macabre tableau and a threatening prophecy. Destruction, Codreanu proclaims, will be the fate of those his mentor Professor Cuzu denounced as the 'dirty beasts and enemies of the country'.

Codreanu cultivated powerful friends. He had studied in Germany in the mid-1920s (his mother's family came from Munich) and in 1928 he contacted Herman Esser, the NSDAP's first head of propaganda. Nothing much came of their meeting, but he soon became well known in German circles. In 1934, Hitler met with his foreign policy advisor Alfred Rosenberg, Rudolf Hess and Himmler at the German Foreign Ministry to discuss 'Romanian internal problems and the Iron Guard'. Soon afterwards, Codreanu and his successor Horia Sima acquired a number of high-ranking German admirers – including Himmler and Goebbels. The SS poured money into Legion coffers. In the short term, however, what would transform Codreanu into a big player was not a flood of Reich marks, but the return of the king.

King Carol II has been judged the most corrupt crowned head in Europe. No other king 'abused to such an extent the sincere love and faith with which the people surrounded him'.[10] He was lascivious, cynical, corrupt and ravenous for power. Carol inspired lists: according to one historian he was also 'profoundly corrupt, unscrupulous, superficially educated, perverse and depraved ... an opportunist'. A scandal had forced him to renounce the throne in 1925 but he had made a comeback in June 1930 and taken power in a constitutional coup. Romanian ministers tried to force him to discard his 'Jewish mistress' Magda Wolf-Lupescu. (In point of fact, Magda had mixed ancestry – her father was Jewish, but her mother was a Catholic Austrian.) But the king refused. The woman anti-Semites called the 'Red Witch' or 'Jewish Wolf' became to all intents and purposes the Romanian queen.

In any case, Carol was as anti-Semitic as he was venal and had become fascinated by the legionary movement and the alluring 'Captain' Codreanu. Once he was securely installed in Bucharest's opulent royal palace, Carol's instinct was to do away with troublesome political parties altogether; in other words, follow the example set by King Alexander in Yugoslavia. But he soon realised that the Yugoslavian strategy would gain little support from Romania's military elite, which had supported his coup. Instead, Carol chose to meddle and manipulate, at the cost of destabilising Romania's fragile democratic institutions. After 1931, thanks in large part to King Carol's devious and divisive tactics, Romania endured a dizzying succession of governments that pulled ever closer to the legionary right. It was the Romanian reprise of the Weimar period in Germany. Chronic volatility suited the revolutionary nationalists like the legion just fine. Many Romanians were drawn to the despotic certainties of the legion. Codreanu took full advantage of the feebleness of the Romanian state and police. He recruited death squads (*echipa mortii*) that carried out a succession of gruesome assassinations of prominent critics and Jews. When a Legionary squad bungled an attack on a Jewish journalist and were arrested, Codreanu remarked, 'What was illegal about trying to put a hole in the head of this snake with a kike rattle?'

The journals of Mihail Sebastian, an ambitious young Jewish writer, provide a chilling sense of the overwrought, frightening atmosphere in Bucharest, where he lived and worked. On 24 June 1936, he anxiously noted that 'We may be heading for an organised pogrom. The evening before last Marcel Abromovici was knocked down in the street by twenty of so students who then dragged him unconscious into the cellar and only released him a couple of hours later.'[11] In his own mind, the liberal, agnostic Sebastian did not doubt that he was a Romanian. But his Christian peers, even those who became close friends and colleagues, could not accept that Sebastian the Jew was truly one of them: a Jew would always be a Jew, not a proper Romanian. Sebastian admired Nae Ionescu, who had helped to get his first books published. He believed he was a friend. But Ionescu, as mentioned before, revered Codreanu and was a prolific Legionary propagandist. When he agreed to publish Sebastian's second book in 1934, he inserted his own preface which admonished his young disciple: 'Remember that you are Jewish!' Many assimilated Romanian Jews had to negotiate such intimate betrayals. The Iron Guard and their allies, the playwright Eugen Ionescu wrote, had created a 'stupid and horrendous reactionary Romania'.

For Codreanu and his Legionaries, 1936 would prove a watershed year. Workers and peasants swelled membership lists. A cult became a mass movement party. Codreanu, who wore a little bag of soil around his neck to honour the sons of toil, responded by forming the 'Legionary Workers Corps' – these would form the 'shock troops' of the pogroms that erupted in 1940. The sudden escalation of support for the legion, now more commonly called the Iron Guard, rattled King Carol who admired Codreanu but feared such a charismatic rival. He tried to counter the corps by forming a rival youth organisation, the Guard of the Nation. But it was swiftly infiltrated by youthful Legionaries. Mihail Sebastian describes how:

> university professors, students, intellectuals were turning Nazi, Iron Guard, one after another ... one of our friends would say: 'Of course I don't agree with them at all, but on certain points, for example the Jews, I must admit ...' Three weeks later, the same man would become a Nazi. He was caught up in the machinery, he accepted everything.

The troublesome Legion haunted the nightmares of King Carol. The Iron Guard had somehow to be neutralised, taken in hand. His first move was to inveigle Codreanu to share power. He turned to a popular Legion sympathiser and military strong man General Ion Antonescu to set up a meeting. Antonescu summoned Codreanu to his villa at Predeal – but the discussion led nowhere. Charismatic demagogues rarely share power. So Carol carried out a purge of known Legionaries in the government and, after yet another chaotic election, ordered the radical right-wing Goga-Cuza Party (which had won a pathetic 9 per cent of the vote) to form a coalition government to siphon off some of Codreanu's support. Antonescu was appointed Minister of War.

In his journal, Sebastian denounced the new regime as a 'typical government of panic'. Coalition leaders A.C. Cuza and the poet Octavian Goga were of course outspoken anti-Semites and worked hard to take on the populist mantle of Legionary radicalism. For the first time in official speeches, Sebastian noted, 'one could hear the vocabulary ... of "Yid", "the Jews", Judah's domination'. Carol tried to rationalise his new government's anti-Semitic measures to a journalist from the British *Daily Herald*; the story was published on 10 January 1938. Carol refused to pull any punches: Romanian Jews must be forced to emigrate. Printed in a populist broadsheet, the king's comments unsettled British public opinion – and the British minister in Bucharest, Sir Reginald Hoare, conveyed to the king his government's concern, as did his French counterpart. Moral outrage provoked a flight of capital. Foreign businesses boycotted Romanian trade. The economy stuttered; economic collapse threatened. The government was too fragile to withstand these tremors. On the night of 10 February 1938, Sebastian recorded, 'the Goga government fell'. It had held on to power for barely a month. For a few days, Sebastian hoped that life might return to what passed for normality. He was wrong. For his friend, Iron Guard enthusiast Mircea Eliade, agreed that the king's strategy had failed dismally but 'three quarters of the state apparatus has been "Legionized"'. In his farewell address, Goga proclaimed, 'Israel, you won'. On 20 February, King Carol threw in the democratic towel, abolished the constitution and imposed a royal dictatorship and swore in a puppet government led by the Patriach Miron Christea. He banned all other political parties. To the king's astonishment, Codreanu made no protest and called for the Iron Guard to disband. Ominously, however, he demanded the return of the 'true king', Carol's son Michael. He proclaimed: 'The hour of our triumph has not come. It is still their hour.'

In March, Hitler's armies marched unopposed into Austria. The rapid expansion of the German Reich thoroughly unnerved Carol. The Nazi elite had never hidden its high regard for Codreanu and the legion and the King feared that the guardist movement might become a German Trojan Horse. On 16 April, the Interior Minister Armand Călinescu ordered the arrest of Legionary leaders, starting with Codreanu, who was accused to begin with of insulting a minister, the historian Nicolae Iorga. The 'Captain' appointed a fanatical 31-year-old lawyer called Horia Sima, who was much admired by Heinrich Himmler, to take over as acting leader. Codreanu was convicted and sentenced to six months' imprisonment. This did not satisfy King Carol. In May, Codreanu was re-arrested and accused of conspiracy: organising terrorist activities and collaborating with a foreign power – namely, Nazi Germany. When Wilhelm Fabricius, the German minister in Bucharest protested, Călinescu cited as evidence the draft of a letter Codreanu may have sent to Hitler in 1935. By now, Hitler had begun to court General Ion Antonescu, who had been Minister of War in the short-lived Goga government, through the good offices of Veturia Goga. Antonescu was much admired in Berlin as a 'strong man' with the right pro-Legionary sentiments. As a favour to his German friends, Antonescu testified at Codreanu's trial in favour of the accused: 'I cannot believe that the accused would be guilty of treason,' he declared as he publicly shook Codreanu's hand. This gesture infuriated Carol, who had the general banished to a villa attached to the Bistriţa Monastery. In the meantime, Codreanu was convicted and sentenced to ten years' forced labour.

As Codreanu began serving a long stretch in the Râmnicu Sărat prison, King Carol was summoned to meet Hitler. At their meeting, Hitler tried hard to persuade the king to release Codreanu and form a 'Guardist Government' which would draw on Legionary support. The meeting confirmed Carol's worst fears. When he returned to Bucharest he resolved to be rid of his rival once and for all. On the night of 30 November, a squad of gendarmes marched into Râmnicu Sărat prison. They seized the keys to Codreanu's cell, dragged him into the prison yard and bundled him into a waiting truck. He was driven to a remote forest road where other kidnapped Legionaries waited, terrified, in the dark. The gendarmes strangled Codreanu and shot the others. On the king's orders, they secretly buried the 'Captain' in the courtyard of Jilava prison and poured concrete over his body. Romanians refer to the bloody events of 30 November 1938 as the 'Night of the Vampires'. As news of the murders spread, Carl Clodius, a German economics specialist, observed that the 'The murder of Codreanu and his followers has changed the situation considerably. Condemnation of this murder is equally strong in almost all circles of the population ... The murder of Codreanu has shaken [Carol's] moral position ... he will recover from it only very slowly'.[12] This was prescient. The king's prestige was further eroded by the onward march of the German Reich. In September 1938 the British and French (Romania's traditional allies) capitulated to Hitler over Czechoslovakia. As Romanians digested the implications of the Munich Agreement, many felt bitterly disillusioned. France and Britain had betrayed the Czech government; Romania might be next in line. It was imperative to mend relations with Germany and 'orientate towards the Axis'.

Martyrdom suited Codreanu. His fanatical disciple and heir, Sima, fled to Germany where he tirelessly promoted the 'Captain's' posthumous political canonisation and sought revenge. On 21 September 1939, he dispatched a guardist squad to the Ministry of the Interior where they shot dead Călinescu, the minister who had ordered Codreanu's murder. The six assassins fled to the local radio station where they announced: 'The Captain has been avenged!' and promptly gave themselves up. The gendarmes took the six men back to the ministry building and executed them, leaving their bodies to rot for several days where Călinescu had been killed.

In the aftermath of the Bucharest bloodbath, King Carol set about tightening his political grip. He ignored his ministers and handed draconian new powers to the secret police. Any potential opponent of the king was put under surveillance. Hidden microphones were installed in private homes and offices. Romania plunged ever deeper into a climate of fear. But the wellspring of terror was the king himself. He resembled a Shakespearean monarch haunted by his past misdeeds and the spirits of his victims. And the most disquieting spectre that stalked the king's nightmares was Corneliu Codreanu. The resilient power of the murdered 'Captain' and his Legionary movement would force King Carol ever closer to the Reich.

As Hitler's armies marched into Poland then began to threaten Belgium and France, Carol conceded the stark truth that to hold the Reich at bay he would need to make peace with the Iron Guard. At the beginning of 1940, the king began releasing Legionary prisoners and in May Codreanu's anointed successor Horia Sima returned to Romania.

In Bucharest, Sima was summoned to meet the head of the Romanian Intelligence Service, Mihai Moruzov, who intimated that the king wanted to strike a deal. On 13 June, Sima issued orders to the guard to co-operate with the king and, after protracted negotiations, signed a 'declaration of obedience'. From this union sprang a new government party, the 'Party of the Nation', led by the pro-German Ion Gigurtu, who appointed Sima Minister of Cults and Arts. Although two other Sima allies also became ministers, the king had an attack of cold feet and vetoed any further guardist appointments. Three days later Sima resigned. On 1 September he called on the king to resign.

The day after Sima joined the government, the king faced a fresh crisis – also engineered by Germany. Hitler's pact with Stalin was, by the summer of 1940, fraying as the Soviets stepped up territorial demands in Eastern Europe. On 26 June, Stalin sent an ultimatum to King Carol insisting that Romania cede Bessarabia and northern Bukovina to the Soviet Union. At the same time, he pressed the Hungarian government to reclaim half of Transylvania. These opportunist machinations racked up tensions between the Reich and the Soviets. Hitler had already made the decision to break the pact and attack the Soviet Union but could not yet afford to show his hand. Even though Romanian oil lubricated the German war machine, Hitler, for now, chose to accede to Stalin's demands and in effect wrench Bessarabia and northern Bukovina out of Romanian hands.

This land grab was a body blow for King Carol, but a death sentence for thousands of Romanian Jews who lived in the ceded territories. As twenty-four Soviet divisions marched into Bessarabia, hard on the heels of the humiliated Romanians, guardists spread tales that Romanian Jews had welcomed the Soviet forces and insulted the Romanian troops. 'The Jews are to blame!' cried the Legionaries. As Romanian troops prepared to cross the Pruth River, now the Soviet border, they began to attack Jews and burn their homes. In the town of Dorohoi, a full-scale pogrom erupted. Romanian soldiers even began shooting their Jewish comrades.

Then it was Hungary's chance to seize a few Romanian morsels. With the connivance of Ribbentrop and the Italian Foreign Minister Ciano, Admiral Miklos Horthy gobbled up vast chunks of Transylvania – the Vienna Award. This was the final straw. On the streets of Bucharest, humiliated Romanians wept openly. The national mood became increasingly volatile. As Romanian divisions retreated to their garrisons, blocking roads and cramming into railway carriages, huge crowds gathered in Bucharest and other cities, demanding that the king, who had caved in so easily to Hitler's demands, abdicate. Guardist thugs began firing shots under the windows of the royal palace.

From his headquarters in Berlin, SS Chief Himmler followed events in Romania closely. On 31 August, he sent Waffen-SS commander 'Sepp' Dietrich to Bucharest, where he had meetings with Sima to thrash out strategy once the king had been forced to abdicate. At the beginning of September, as the crisis deepened, Sima manoeuvred to seize power. Somewhat ineptly, he tried to persuade the German minister Fabricius that the Gigurtu government intended to resist the Vienna Agreement. We have no evidence that Dietrich had suggested this ruse, but Sima clearly believed that he had German backing. But Himmler was playing a devious game by secretly opposing Hitler. In any event, Sima's scam was exposed and he once more fled Bucharest. Himmler's clumsy intervention did not impress Hitler. He had other plans. He needed a strong man, not a fanatic; a strong man with the right ideas. Antonescu was a professed anti-Semite who forged close bonds with Codreanu and the guardist movement. Instead of resenting the loss of Romanian territories, Antonescu looked forward to retrieving them as a military ally of the Reich. He ticked all the boxes.

In Bucharest, at the beginning of a warm and muggy September, King Carol anxiously paced the echoing corridors of the royal palace. He could not shut out the furious chants of the huge crowds that surged along the grand Calea Victoriei into Palace Square. From his window, the king noticed that many demonstrators wore the bright green shirts of the legion – and he called Gigutu and ordered him to execute a few imprisoned Legionaries. He refused.[13] One of his advisors Valer Pop, who unknown to Carol was feeding inside information to the Germans, urged him to call on Antonescu for support. On 3 September, the king gave in and summoned Antonescu to the palace. He refused to offer full powers but agreed to 'take guidance'. Antonescu refused point blank. By now, the palace was to all intents and purposes under siege. Guardist agitators ratcheted up the pressure, and the following day Antonescu was recalled and offered 'all necessary power'. In the meantime

Fabricius cabled Berlin confirming that Antonescu was 'firmly resolved to carry out our important demands here'.[14]

The king hoped that a deal with Antonescu would allow him to remain on the throne. But outside the palace, news of the Carol's offer set off loud volleys of rifle fire. The guardist mob surged close to the palace insisting that the king abdicate. Horia Sima had by now returned to Bucharest. On the evening of 4 September, Antonescu met Sima and a 'Legionary Forum' to discuss ways of ending the stalemate. On 5 September, the guardists returned to the streets in even greater numbers. At 9.30 a.m. Antonescu returned again to the palace to deliver a final ultimatum. It was at last checkmate. At dawn the following morning, King Carol II, that much unloved monarch, handed Romania's poisoned crown to his son Michael. Two weeks later, Antonescu, prompted by the German minister Fabricius, proclaimed himself Conducător (leader) of the Romanian state and chief of the Legionary Party.

The reactionary Legion finally held the reins of power. Romania was ruled by a 'National Legionary State'. Antonescu appointed Horia Sima vice-president of the Council of Ministers – in effect his deputy. On 23 November, the Legionary state of Romania joined the Tripartite Pact with Germany, Italy and Japan. A few days later, Antonescu authorised the exhumation of the remains of Codreanu and other martyrs from beneath the courtyard of Jilava prison and gave them a grand state burial. Peasants and workers journeyed from all over Romania to celebrate the posthumous triumph of their heroes. A vast procession wound its way to the Bellu cemetery, known to many Romanians as the 'Garden of Souls'. Marshall Antonescu and Sima marched side by side. The sight of Codreanu's decayed remains provoked a hysterical reaction from some of the mourners. An enraged Legionary gang broke into Jileva prison and began hacking to death inmates whom they believed to be linked to Codreanu's murder. When he heard about this bloody episode, Himmler sent Sima a congratulatory telegram: he was doing what needed to be done. Legionary propagandists promised that the new Romanian state would 'make the country seem like the holy sun in heaven'.[15]

But as Legionary violence escalated, threatening to topple the new pro-German regime, Hitler faced a dilemma. The planned attack on the Soviet Union, Operation Barbarossa, depended on Romanian oil. Wehrmacht strategy assumed Romanian troops would join the invasion in the south-east. Hitler observed the chaos unfolding in Romania with alarm. He concluded that General Antonescu needed to apply the lessons of the 'Night of the Long Knives' – when Hitler, in league with the SS, had eliminated the leadership of the troublesome Brownshirts. In short, Antonescu must do away with Sima and his volatile Legionaries.

Hitler's backing for Antonescu was either not communicated to Himmler or the SS chief and his paladins chose to ignore it and persisted in stoking up anti-Antonescu factions. Sima, for his part, plainly mimicked SS strategy. On 6 September, he had established a new 'Legionary Police' to defend the regime and take vengeance on its enemies. He reorganised Codreanu's Corps of Legionary Workers as a paramilitary unit – the *Garnizoana*. Sima regarded these new squads as Romanian versions of German Einsatzgruppen. In late October 1940 Himmler sent RSHA representatives to Bucharest to reinforce bonds with the Iron Guard and bolster the new Legionary Police. Sima promised that 'the time of revenge on all opponents of the Iron Guard' was near.[16]

The Legionary terror began on 27 November.[17] Murder squads began assassinating former members of King Carol's administration, including the Prime Minster Nicolae Iorga. Shortly afterwards, Sima's squads began to take 'revenge' on Romanian Jews: the legionary state, which was infested with Iron Guard ministers and officials, imposed illegal fines and taxes, and Legionary police units carried out arbitrary arrests, then torture, rapes and public degradations inspired by German practice in occupied Poland. In rural areas, army units also took part in anti-Jewish actions. While Antonescu was, of course, no friend to Romanian Jews, and would demand the deportation of 'foreign Jews' before the end of the year, he could not afford to let Sima's terror campaign destabilise the legionary state, and at the end of November he ordered the Legionary Police to disarm. This was not a humanitarian gesture. As well as public mayhem, Antonescu feared that the legion was growing rich on the plunder and pillage of Jewish businesses and homes. A Legion with bloated coffers would be a dire threat indeed to his own grip on power. Addressing Legion ministers Antonescu ranted: 'Do you really think we can replace all Yids immediately? Challenges to the state should be addressed one by one, as in a game of chess.'

On 14 January 1941 Antonescu was summoned to meet Hitler. Sima refused to accompany him – rank-and-file Legionaries had never forgiven the Germans for handing over Bessarabia and Bukovina to Stalin. This was a mistake: Antonescu's plan was secretly to get Hitler's backing to crush the legion. At the Berghof, Antonescu demanded: 'What am I supposed to do with the fanatics?' As the two dictators watched storm clouds gather over the mighty flanks of Mt Watzman, Hitler replied without hesitation, 'You have to get rid of them ... revolution is not a condition to be perpetuated'. He reminded Antonescu that in the summer of 1934 he had been forced to quash the troublesome Brownshirts: 'You have to get rid of fanatical militants who think that by destroying everything they are doing their duty.'[18] Later at a conference in the Berghof's great hall, Hitler's counsel was reinforced by Foreign Minister Ribbentrop and OKW Supreme Commander Wilhelm Keitel, who warned Antonescu that he could not afford to let the Iron Guard 'infection' spread to the Romanian army. SS Chief Himmler had not been invited.

As Antonescu bonded with Hitler, Sima agitated against Antonescu, egged on by Himmler. On 21 January he ordered a call-up of all the legionary militias. As thick snow fell on Bucharest, armed workers seized government buildings and the radio station and threw up barricades in the streets. As the legionary uprising gripped Bucharest, Sima and the Iron Guard vented their fury on Jews. On 22 January, the Minister of the Interior ordered the burning of Jewish districts in Bucharest. Legionaries, students, priests, the anti-Semitic intelligentsia and even women and children descended on the Jewish districts.[19] Vigilantes raped Jewish women in front of their families. They beat, tortured and killed rabbis, community leaders and gentile citizens caught by the mobs and denounced as 'Yids'. Sima's squads detailed at least 2,000 Jews, aged from 15 to 85, in police stations, the Prefectura, the legion headquarters, the town hall and the old Codreanu farm. Many were tortured. In his journal, Mihail Sebastian described in detail the most egregious incident. In the Bucharest suburb of Straulesti, a Legionary mob rounded up some 200 Jews and hauled them into a farm abattoir: '[they] hanged [them] by the neck on hooks normally reserved for beef carcasses. A sheet of paper was stuck to each corpse: "Kosher meat"'[20] Later, it was reported, Sima's Legionary killers 'chopped up the bodies'.[21]

The Legionaries rampaged at will through Bucharest's Jewish quarter; the Romanian authorities took no action for at least seventy hours. Rabbi Tzwi Gutman and ninety other Jews were dragged to the Jilava Forest, stripped naked in freezing snow then shot at point-blank range. The Legionaries hacked at their victims' mouths to find gold fillings. Astonishingly Rabbi Gutman, who had been shot twice, survived. His two sons died. The Legionary mobs incinerated one of the most beautiful of all European synagogues; the Cahal Grande was consumed by a raging inferno that 'lit the capital's sky'. Legionaries danced around the flames and pushed three Jewish women into the inferno.[22]

But as Sima's Legionary hordes ran amok, Antonescu had little difficulty reasserting his authority. Hitler backed his ally with force: 'I don't need any fanatics. I need a sound Romanian army.' On Friday 24 January, Sebastian reported that 'long motorized German columns, with machine guns and rifles at the ready' rumbled into Bucharest; 'they certainly made an impression. And it was crystal clear that the German army was on the side of General Antonescu.'[23] On Thursday 23 January, Sebastian reported, Legionary squads gathered outside Sima's headquarters on the Strada Roma. With a deafening roar, German motorised units suddenly appeared. The demonstrators greeted them: '*Heil Hitler! Duce! Duce! Duce! Sieg Heil! Sieg Heil!*' But the German troops ignored the crowds. Instead they took up positions at each of the entrances to the square. More German troops arrived, again to the delight of the Legionaries. 'But then what a stunning blow!' Sebastian went on. Once all the exits had been closed, a German officer ordered everyone to leave the square. 'And everyone left. Just like that.'[24]

The Legionary revolt had backfired, bolstering Antonescu's power and binding Romania closer to the Reich. That bond would endure until Red Army troops smashed Romanian armies at Stalingrad. The news of the fall of Horia Sima was not well received at SS headquarters in Prinz Albrecht Strasse; Himmler's *Außenpolitik* (foreign policy) was applied racial hatred, and it was logical to back murderous anti-Semites like Sima. Antonescu, to be sure, shared the legion's hatred of Jews and Bolshevism but he was Ribbentrop's man. Himmler hesitated to speak out against Hitler, but he would never forgive the Foreign Minister for the humiliation of 'his' Iron Guard. In the aftermath of the revolt, Heydrich's agents found sanctuary for Sima and other Iron Guard leaders in the home of Romanian ethnic German leader Andreas Schmidt (who it will be recalled was Gottlob Berger's son-in-law). Shortly afterwards, Sima fled to Italy, while the SS spirited other Legionaries to a special unit at the Buchenwald concentration camp, near Weimar. Many hundreds of lower ranking Iron Guard men found refuge in Germany.[25]

Goebbels noted: 'The Führer is on Antonescu's side. He wants an agreement with a state not a world view.' He added: 'Still, my heart is with them [the Iron Guard]' and a week later when he heard about Antonescu's triumph: 'the Führer … needs Antonescu … for military reasons. One point of view. But it wasn't necessary to wipe out the Legion.'[26] Hitler had more astute insight than either his SS chief or Propaganda Minister. Antonescu was not only a military strong man. He was, to be sure, a political pragmatist, but he was just as loyal to the murderous spirit of Codreanu as Horia Sima. At the Berghof meeting in January, Hitler and Keitel had informed Antonescu that Germany planned to attack the Soviet Union that spring. In exchange for Romanian military assistance, Hitler promised to return to Romania the lost territories snatched away by Stalin. Scattered across Bukovina and Bessarabia were large Jewish communities. Once the attack on the Soviet Union was under way, these would fall into German and Romanian hands. Both the Germans and Antonescu understood that reclaiming territory would also provide a fresh opportunity to solve the Romanian 'Jewish problem'. Antonescu's pact with Hitler was much more than a strategic alliance – it represented a shared understanding about shared ideological objectives.[27]

Antonescu was indebted to Hitler for reinforcing his rule and taming the legionary movement, at least for now. In the aftermath of the revolt, Antonescu deported representatives of the SS, the RSHA and the Abwehr who had helped

Sima and other Legionary conspirators escape – and may well have helped stir up the revolt. But in March, according to Deputy Prime Minister and Foreign Minister Mihai Antonescu, 'special emissaries of the Reich and of Himmler' led by SS-Hauptsturmführer Gustav Richter, an expert on 'Jewish matters' arrived in Bucharest to discuss the 'handling of Romania's Jews'. Plainly, with 680,000 Germans troops already on Romanian soil, Richter expected that (as Mihai Antonescu reported) 'responsibility for the handling of Romania's Jews be handed over to the Germans exclusively'. General Antonescu refused. The dictator had no interest in protecting Jews, but wanted to retain control of strategy.

The Germans, and in particular Himmler's SS, did not give up. On 21 February, Himmler met with the former SD representative in Bucharest, Otto-Albrecht von Bolschwing, and a handful of other SS bureaucrats.[28] At the end of April, RSHA emissary Richter returned to Bucharest and held more meetings with Mihai Antonescu with 'excellent results', he reported to Ambassador Manfred von Killinger. At last Himmler had an agreement with Antonescu's regime that harmonised SS plans with Romanian strategy. Later that year, Mihai Antonescu made a remarkable statement to his Cabinet: 'I can report to you that I have already conducted intensive negotiations with a high ranking German representative: they understand that the Jewish problem with ultimately require an international solution, and they wish to help us prepare *this international solution.*' (My italics.)[29] Given that when Mihai Antonescu made this statement many thousands of Jews had already been murdered by Romanians and German squads in Eastern Europe, it implies that the idea of a 'Final Solution of the Jewish problem', i.e. liquidation of Jews on an international scale, was already well advanced by the late summer of 1941 – and not months later as many historians assume.

On 12 June 1941 Hitler met Antonescu again, this time in Munich.[30] Antonescu later informed Ambassador von Killinger that Hitler had shown him with a document titled '*Richtlinien für die Behandlung der Judenfrage*' (Guidelines for the Handling of the Jewish Question – in some versions *Ostjuden*). Killinger reported that 'there is no reason to doubt the accuracy of General Antonescu's assertion'.[31] The meeting with Hitler had an immediate impact on Romanian anti-Jewish plans, especially with regard to the 'lost provinces'. Antonescu began promulgating a radical new policy which he called 'Cleansing the Land'. This would require identifying 'all *Jidani* [Jews], communist agents or sympathisers … in order to enact whatever orders I may transmit at a given time'.[32] A few days before 22 June, Antonescu ordered the Romanian *Serviciul Special de Informaţiuni* (the Special Intelligence Service or SSI) to begin forming Escalon Special (Special Echelons) modelled on Heydrich's Einsatzgruppen. On 22 June, the leader of Special Task Force D, Otto Ohlendorf, arrived at Romanian military HQ at Piatra Neamţ in Moldavia – and remained there until the beginning of August, acting as (to borrow an American term) a special advisor. The Romanian Special Echelons were charged with 'defending the army rear area' and, like Heydrich's execution squads, split into teams (*echipe*). These Special Echelons would spearhead the Romanian assault on the 'Jewish enemy'.[33]

General Antonescu had crushed the revolt led by Codreanu's successor Horia Sima, but in spirit he was a Legionary. Antonescu absorbed Iron Guard chauvinism and cruelty into his own 'ethnocratic' state: he and his ministers firmly believed that 'the Jews pose a permanent threat to every nation state'. At the end of June, Mihai Antonescu, who had been a professor of international law at Bucharest University, echoed Hitler's speech to his general before the Polish campaign: 'I beg you to be implacable. Saccharine and foggy humanitarianism has no place here. The Roman Empire performed a series of barbarous acts ... yet is was the greatest political creation ... I take full legal responsibility and tell you there is no law!'[34]

And it would be in Iaşi, the birthplace of the Legion of St Michael, that the SSI Special Echelons would launch a campaign that Mihai Antonescu (eschewing saccharine and foggy humanitarianism) called 'total ethnic liberation'.

At the end of June 1941 Iaşi was a frontier city just 10 miles from the Prut River, which marked the Romanian border with the Soviet Union. Earlier that summer, General Antonescu had pledged to Hitler that Romania would join his crusade 'against Russian Bolshevism, the arch-enemy of European civilization'. The German 11th Army and the Romanian 3rd and 4th Armies waited on the Prut. Crammed into Iaşi's barracks and milling about its streets was a flammable mix of Romanian troops, Romanian SSI agents and gendarmerie units, as well as thousands of Iron Guard Legionaries. Stationed here too were German soldiers from the 198th Division of the 30th Army Corps, and the Todt Organisation. Although the Jewish community had endured more than a decade of persecution and harassment by Iron Guard activists and other Romanian anti-Semites, it remained relatively prosperous. Altogether 100,000 people lived in Iaşi – just over 50,000 were Jewish. By the beginning of July, at least 13,266 Jews had been murdered either in the city itself or on the 'death trains'.[35] This abrupt escalation of violence is firm evidence that Antonescu had fully grasped German intentions with regard to European Jewry – and chose to emulate them.

The full extent of German responsibility for inciting and managing the Romanian pogroms of 1941 has only very recently come to light. In 1996 an affidavit written by Captain Ioan Mihail, who took a leading role in the events in Iaşi, revealed that German soldiers stationed in Iaşi collaborated with the Romanian army and robbed, beat and murdered Jews. Although Einsatzgruppe D men did not directly participate in the Iaşi massacre, Himmler still exerted his baleful influence through the Special Echelons modelled directly on the Heydrich's murder squads. This point can be reinforced by scores of eyewitness accounts that refer to the activities of German troops in Iaşi. SSI Chief Eugen Christescu testified that SS and SD agents had arrived in the city, as well as an Abwehr major, Hermann von Stransky. Although details about the precise role of the SS agents are scant, we know that the SSI Special Echelons went on to collaborate with Einsatzgruppe D elsewhere in Romania. Further evidence of this intertwining of German and Romanian interests comes from their military communications. Both German and Romanian military commanders reported their unease about Jews in the army rear areas and demanded their removal. As German and Romanian troops advanced into Bessarabia later in July, according to historian Matatias Carp, they executed 'almost the entire Jewish population living in villages'.[36]

Carp himself argued that this was a Romanian Holocaust – the culmination of decades of fervent nationalist bigotry and what he called the 'rotting system of Romanian pseudo-democracy'. But would the destruction of Romania's Jews have taken place at all if Hitler had not launched his attack on the Soviet Union? Would the Iron Guard and its Legionary militias have wielded such deadly power had they not been promoted by Himmler and Goebbels? And would the Romanian Holocaust have taken place at all had Hitler not levered Ion Antonescu into power? None of these questions is likely to have a simple answer. Nevertheless, we should hold them in mind as we try to piece together the events that took place in Iaşi at the end of June 1941.

Iaşi was under military jurisdiction and fell within range of Soviet artillery and bombers. Italian journalist Curzio Malaparte arrived in the city not long after 22 June and recorded his experience in his remarkable semi-fictionalised work *Kaputt*. He rented a small room in a building next to an 'abandoned orchard', which was in fact, as he soon discovered, an ancient Orthodox cemetery.[37] He recalled 'The Soviet bombers were hammering hard'. The planes flew back and forth at about 900ft, some approaching low enough to clip roof tops. A Soviet bomber crashed in a field near the city. When Malaparte arrived on the scene, Romanian soldiers were tormenting the female crew – 'two brave girls', one a 'study blonde with a freckled face'. Everywhere in the city, the atmosphere was tense. Rifle fire was frequent and nervous Romanian soldiers sometimes let loose without warning. By 25 June, Soviet forces, dug in along the Sculeni ridge overlooking the Prut

River still stood firm against the German assault. For a while, it seemed as if they might push Axis troops back across the river. As news of this unexpected setback spread, the mood of Romanians darkened. A whispering campaign accused Jews of acting as Soviet agents. No one could be trusted.

General Antonescu stoked the furnace: he proclaimed that '[Barbarossa] is not a struggle with the Slavs but one with the Jews. It is a fight to the death. Either we will win and the world will purify itself, or they will win and we will become their slaves.'[38] Shortly before 22 June, in Bucharest General C.Z.Vasiliu, the general inspector of Romania's gendarmerie, called a meeting with his officers to discuss how the 'Cleansing of the Land' orders would be enacted once Romania's lost provinces had been recaptured. He pointed out that the operation could not begin until Soviets had been pushed out of Bessarabia and Bukovina. So at the end of June, Vasiliu's brigades were transferred to Iaşi where they waited tensely for the signal to move across the border.

Post-war interrogations of Romanian military officials revealed an intricate web of contacts between the German and Romanian army intelligence units, the gendarmerie and the SSI. At the hub of this web was the mysterious Abwehr Major, Hermann von Stransky. He was a nephew of Ribbentrop who had lived in Romania for many years, spoke fluent Romanian and fed information to head of the SSI's German section, Colonel Ionescu-Micandru.[39] According to SSI Chief Christescu, SD, Gestapo and Geheime Feldpolizei (Field Police, the secret military police) agents had also arrived in Iaşi.

Four days before the German attack, on 18 June, the SSI Special Echelon comprising 160 men had set off from SSI headquarters in Bucharest in a convoy of automobiles and trucks. They were heavily armed and, like the German Special Task Forces, had been issued with catalogues listing 'target' Jews and communists. Before leaving Bucharest, the Special Echelon commanders had printed thousands of posters that showed Jewish caricatures, modelled on the Nazi *Der Stürmer*, as spies and saboteurs.

In Iaşi, a police superintendent reported that a number of Iron Guard Legionaries had begun 'taking a sort of course under the tutelage of two uniformed officers'.[40] These officers were SSI men. At least forty Legionaries, still officially 'enemies of the state', assembled for 'lessons' in a rented apartment on Florilor Street in the Păcurari district. Their task would be carry out 'noisy acts of violence' to test the reaction of any authorities who had not been informed of the pogrom, and to stir up Christian citizens to attack Jews and plunder their houses. By 23 June, 'Legionary mercenaries' had been stationed in every city district. They had been issued with side arms, 'Flaubert' guns, blank cartridges (to make a noise), as well as lethal weapons.

Romanian military authorities scarcely troubled to conceal what they had in mind for the Jews of Iaşi. Two weeks before the German invasion began, Jewish forced labourers from a nearby camp were marched to the Jewish cemetery, also located in the Păcurari district, where they began digging trenches; each was 100ft in length and 6ft wide and deep.

The tinder had been laid. All that was required was a spark. At sunset on 24 June, Italian journalist Curzio Malaparte woke to the rising scream of air-raid sirens. As he ran into the street, he heard the roar of aircraft engines, the rattle of anti-aircraft guns, then the thud of bombs and the crash of crumbling masonry. In the city the railway station was on fire and German and Romanian soldiers mustered in the streets, weapons cradled in their arms. Between the roar of anti-aircraft fire, Malaparte heard 'the hoarse voices of German soldiers'. On 26 June, a second raid ratcheted up the tension. This time, Russian pilots scored direct hits on the St Spiridion Hospital, the telephone exchange and the HQ of the Romanian 14th Infantry division. Some 600 people were killed, including thirty-eight Jews.

Soon afterwards, Malaparte heard 'a confused din, a rattle of machine guns, and the dull thud of grenades' from the Jewish districts. By then, the SSI Special Echelon (Esalonul I Operativ) had arrived in Iaşi and set to work plastering walls and meeting places with its stock of posters slandering Jews as enemy agents. Iron Guard Legionaries, dispersed all over the city, fuelled rumour and counter-rumour about Jewish saboteurs. Fear stoked hatred – and shrill calls for revenge. Romanian military authorities issued 'official reports' claiming that among captured Soviet air crew they had found renegade Romanian Jews from Iaşi. Rumours spread that signal lights had been discovered installed in the chimneys of Jewish households. Malaparte sensed that 'something was in the air'; a storm was building. He noticed squads of gendarmes waiting, hidden in doorways, and the streets echoed with the *click click click* of military patrols. 'A strange anguish weighed upon the city. A huge, massive, monstrous disaster, oiled, polished, tuned up like a steel machine.' One evening a group of rabbis visited Malaparte in his rented room next to the old Orthodox cemetery: *please try to stop the pogrom*, they begged. The Italian journalist could do nothing, of course.

On Thursday 26 June, the day of the second raid, Abwehr Major von Stransky arrived in Iaşi accompanied by Colonel Ionescu Micandru, the chief Romanian liaison officer with the Wehrmacht. That afternoon, police superintendant 'Chestor' Constantin Chirilovici ordered Jewish community leaders to attend a meeting at the central police headquarters. He accused them of collaborating with the Soviet air force and ordered them to surrender flashlights, binoculars and cameras within forty-eight hours. Elsewhere in the city, soldiers of the 14th Division arrested three

Jews accused of providing information to the Soviets about the location of military buildings. Although all three men were released after interrogation, a Legionary squad hauled them off to the garrison firing range and shot them. They botched the job – two men escaped.

Back at police headquarters, Chirilovici organised detachments of gendarmes to begin house-to-house searches to find 'saboteurs'. Already, Christian houses had been marked with a C or a cross. This meant that, on the evening of the 26 June, gendarmes swiftly located Jewish 'suspects' and, in the Jewish quarter, arrested anyone who owned a flashlight or a 'suspicious' article of red clothing. That evening the father of Dr Marcu Caufman was shot by a Romanian artillery officer as he walked through the Nicolina quarter. Gendarmes began shooting Jewish 'suspects' arrested the day before.

On 27 June, General Antonescu telephoned Colonel Constantin Lupu, commander of the Iaşi garrison, and ordered him to 'cleanse Iaşi of its Jewish population'. Two days later, at 11 p.m., Antonescu called again. According to Lupu, he made clear that 'The evacuation of the Jewish population from Iaşi is essential, and shall be carried out in full, including women and children. The evacuation shall be implemented *pachete pachete* [batch by batch], first to Roman and later to Târgu Ji … Suitable preparations must be made.'[41]

The final cataclysmic eruption of violence came on Saturday morning. As soon as it was light, Romanian soldiers from the 13th and 24th Artillery regiments began attacking and robbing Jews. In the vicinity of the slaughterhouse, German soldiers went on the rampage. Malaparte tells us what a pogrom sounds like: the frenzied barking of dogs, banging of doors, shattering of glass and china, smothered screams, imploring voices calling *mama mama*, beseeching cries of *nu nu nu*. Everywhere in the city, he heard 'strident, frightful German voices'. In the central Unirii Square, SS men set up a machine gun and fired on a crowd of women huddled by the statue of a Romanian prince. Soldiers threw hand grenades through the windows of Jewish homes. The streets became slick with blood. The night filled the sounds of with weeping, terrible screams – and laughter.[42]

That evening, Police Inspector Gheorghe Leahu issued orders that no one must interfere with 'what the army is carrying out'.[43] An unidentifiable aircraft flew low above the city and fired a number of blue flares. Below, Romanian soldiers had begun marching out from their barracks to begin their journey to the front line. Panic erupted. All over Iaşi volleys of gunfire erupted as armed Legionaries ran amok. Ordinary Romanian citizens brandishing shotguns and metal pipes joined in. One eyewitness described what unfolded in the slaughterhouse district as German soldiers and a 'group of young Christians' smashed down doors and plundered Jewish houses. At 9 p.m. the banshee wail of air-raid sirens added to the mayhem – and soldiers accompanied by 'paramilitary reservists' (almost certainly Legionaries) shot the owner of a textile store. At the Binder Hotel, owned by the Jewish Blau family, Legionaries broke down the door and, after a perfunctory search, announced that they had discovered a machine gun in the attic. They dragged Mr Blau, together with his wife, baby daughter, sister-in-law and mother-in-law, from the hotel and shot them all in the street. The discovery of the weapon was, of course, a put-up job.

That night, a violent storm broke over Iaşi. All over the city, the shootings continued, illuminated by flashes of lightning. On Sunday morning, Malaparte reports, human forms lay scattered in awkward positions about the streets. Yet more had been heaped in gutters, one on top of the other. The police organised work parties of Jews so that the piles of bodies would not block the flow of traffic. Many hundreds of corpses had been dragged into the churchyard close to Lapusneanu Street. Dogs sniffed the air and began gnawing at the dead. German and Romanian trucks rumbled past. A murdered child sat bolt upright on the pavement. Laughing German soldiers and Romanian gendarmes, as well as chattering civilians, set to work mutilating the bodies and stealing clothes and shoes. They discarded the dead where they lay – twisted, naked.

Worse horrors were still to come. At police headquarters on the Saturday evening a decision was made to reinforce patrols and bring in 'suspects'. The following day, Romanian soldiers backed by gangs of Legionary vigilantes began ordering terrified Jewish families to begin assembling in the streets outside their homes. Many still wore their pyjamas. As Iaşi's church bells rang out, Romanian and German soldiers began to herd terrified Jewish families, including very young children, through the streets to the centre of the city. Legionaries and ordinary citizens lined the streets to spit and hurl rocks and bottles. They battered the Jews with iron bars and rifle butts; anyone who fell was shot and the road became lined with the dead. As Mihai Antonescu had demanded, there was no 'foggy humanitarianism' – on one street a small child was killed in front of a Jewish store, then disembowelled. A few Romanians did what they could to rescue Jews, frequently suffering the same fate as those they tried to assist.

Soon more than a thousand Jewish families been incarcerated inside the courtyard of the central police headquarters. By noon, Chirilovic testified, numbers had reached more than 3,000 – by sunset, 5,000. Given that General Antonescu had ordered that 'The evacuation of the Jewish population from Iaşi is essential, and shall be carried out in full, including women and children', what took place at the Iaşi police headquarters is hard to comprehend, especially since Romanian army commanders like General Gheorge Stavrescu made several visits to monitor progress. In official terms, the Jews held at the police headquarters were suspects. Romanian police went through the motions of assessing individuals and in some cases issuing tickets of release bearing the word 'Free'. Many of those set free were women – but very few reached their homes. Legionary patrols remained at large, arbitrarily executing anyone they encountered that they suspected or knew to be Jewish.

As this charade played out, many more Jews arrived at the police headquarters hoping to acquire a ticket of release. At noon, the vice tightened. According to witnesses, SS troops and German soldiers attached to the Organisation-Todt Einsatzgruppe Südost appeared outside the police headquarters. A German professor of Ottoman history at Iaşi University, Dr Franz Babinger, testified after the war that he observed a German infantry unit shooting Jews outside the police headquarters. He protested, but was informed by a German officer that it was the Jews' own fault; later he noticed several more German officers arriving at the police headquarters.[44] As Jews continued to arrive in the courtyard in large numbers, the Germans formed a cordon, apparently to control access. This grey wall was soon reinforced by Romanian gendarmes and Legionaries armed with iron bars or wooden staves. The real purpose of the cordon now became clear. As desperate and bloodied men and women struggled to reach the illusory refuge of the courtyard, the Germans and their Romanian accomplices rained down blows without mercy. Then inside the courtyard at about 2 p.m., as panic spread among the Jews trapped inside, policemen and soldiers suddenly opened fire with machine guns.

Leizer Finkelstein, an eyewitness, recalled these terrible moments:

> The chaos at the Police precinct was indescribable. I was 17 at the time. There were Romanian gendarmes and I think I even saw a few German soldiers wearing helmets with 'SS' written on them who were delivering blows left and right with a baseball bat. Dead people were already lying in the courtyard of the Police Precinct, there was blood and scattered brains everywhere. It was for the first time in my life when I saw dead bodies. I was so terrified.[45]

As this dreadful 'cleansing' continued, a train clanked slowly into Iaşi station hauling a long line of more than thirty closed freight wagons. Soon afterwards, trucks pulled up outside the police headquarters and anyone still alive was crammed inside. Led by two German tanks with motorcyclist outriders, the convoys set off at about 8 p.m. At the station, Romanian troops conducted a head count, forcing their captives to lie face down on the station platform. Travellers stepped, without looking down, between the prone men and women. When the count was finally completed, the Romanians herded the Jews towards the wagons, manhandled them inside then hauled shut the doors. Early the following morning another convoy arrived from police headquarters and a second train steamed into the station.

The loading took all night. The wagons had recently been used for transporting carbide and the rough planks that made up the floors were thickly smeared with an evil-smelling waste. Soon 3,000 men, women and children, many of whom had been terribly injured, were crushed inside thirty-three wagons. Night became a hot summer's day and hundreds of kilometres from Iaşi on the German front line, temperatures reached 40°C. Exultant German soldiers marched forward, bare-chested and tanned by the fierce Russian sun. They cooled off in streams and rivers. But in Iaşi, no water or food had been provided to the families in the freight wagons held at Iaşi station. Bored Romanian soldiers waiting for orders to move occupied their time daubing the sides of the cars with slogans 'killers of German and Romanian soldiers'. They poked bayonets between the wooden slats, laughing if they scored a hit and someone inside screamed. Hours later, the trains let out shrill yelps and unhurriedly began to grind forward. Many had already died inside the wagons. Leizer Finkelstein reported:

> After being taken to the Police precinct on 'that Sunday,' I was boarded on 'the death trains'. It is extremely difficult for me to talk about this. I think no film director will ever be able to depict the experiences on 'the death trains'. To lie with the dead, covered with excrements. We made chairs and benches out of the dead. We stretched the dead bodies and sat on them, stepped on them. Later, on reading about Auschwitz and other concentration camps, I told myself: 'By God, perhaps those people were more fortunate than us. At least they entered the gas chamber and were dead in a matter of minutes.' We stayed inside these train cars which turned into gas chambers and people would die just like that, standing up. Now one, another one 10 minutes later, and so on. Nobody had any hope left of escaping with their lives. There were over 100 people in our train car, of which about 20 survived. When I was among those who stepped off the train cars and were instructed to bury our dead, I still had no hope left of ever returning home. Anyone could kill you, nobody was accountable for their actions. One of my brothers, Leon, who was also on these trains, was taken to the hospital, as he slipped when he got off the train car and a portion of skin from his back was torn off. At first, we didn't even notice that Leon was missing, that's how exhausted and terrified we were.

At police headquarters, municipal workers began to clear away the corpses and hose down the streets that were encrusted with blood and brain matter. A work party of

Jews was compelled to scrub every single stone in the courtyard. An 80-year-old woman recalled: 'I remained there without food for three days. On the third day, a general arrived … admonishing us that whatever happened there was because of the Jews who had fired on the Romanian-German army.'

Months later, the Italian journalist Curzio Malaparte visited Hans Frank, the Governor General of the German General Government, at his headquarters in the Wewel Castle in Kraków. At a sumptuous dinner, Frank asks Malaparte about 'that night in Iaşi'. Mihai Antonescu, he says, 'mentions 500 dead'. Malaparte corrects him: the unofficial figure is 7,000. 'That is a respectable figure,' Frank responds. But he adds: it 'wasn't nice'. Also at Frank's dinner table is the Austrian governor of Kraków, Baron Otto von Wächter. 'It's an uncivilised method,' he says with a tone of disgust. Frank has a ready explanation: 'The Romanians are an uncivilised people … We use the art of surgery, not that of butchery. Has anyone seen a massacre of Jews on the streets of a German town?'

Frank's disdain was widely shared. Otto Ohlendorf, commander of Einsatzgruppe D which, as mentioned before, was active in Romania, grumbled:

> the way in which the Romanians are dealing with the Jews lacks any method. *No objections could be raised against the numerous executions of Jews*, but the technical preparations and the executions themselves were totally inadequate. The Romanians usually left the victims' bodies where they were shot, without trying to bury them.

He concluded: 'The Einsatzkommando has recommended that the Romanian police be more orderly from that standpoint.'[46]

German complaints about inefficient Romanian methods had results. By the end of July, historian Raul Hilberg estimated 10,000 Jews had been murdered in Bessarabia and northern Bukovina by German and Romanian militias. The Iaşi pogrom heralded systematic mass murder that engulfed Bessarabia and Bukovina. The Romanian Holocaust was driven not only by indigenous hatreds, but by a much broader and calculated strategy that had been hatched up in Berlin and then mimicked in Bucharest. From the German point of view, they learnt the bloody events at the central police headquarters was more evidence that what Reinhard Heydrich called 'self-cleansing' had to be properly managed – and that the principal means of doing this would be specialised local militias formed under the auspices of the SS and modelled on SS paramilitary police divisions.

This is not in any sense to exonerate the Romanian terrorists and soldiers who killed and murdered so many thousands of Jews in the summer of 1941 and in the months that followed. In October, Romanian troops burned alive 19,000 Jews in Odessa; at nearby Dalnick they shot another 16,000; and they grossly mistreated Jews who were being pushed east across the Dniester River.[47]

The second lesson that the Germans learnt in Romania was more complex. Romanian hatred of Jews, to be sure, had deep roots. But to the Nazi mind, indigenous anti-Semitism was, as Hans Frank and his dinner guests agreed, a species of barbarism. The task of the Reich was to modernise these rusticated hatreds and to replace the club and the hunting rifle with the scalpel. Modernisation would be driven by the image of a new kind of villain: the 'Jewish-Bolshevik' – a foe that, like the Antichrist of medieval eschatology, menaced the foundations of European civilisation.

The axiom that Marxism was, behind its egalitarian mask, 'Judeo Bolshevism' underpinned Nazi ideology. In an important study, Jeffrey Herf quotes art historian E.H. Gombrich, who monitored German radio broadcasts during the war. He pointed out that what characterised Nazi propaganda was not so much lies, but the imposition of a 'paranoiac pattern on world events'; by this he meant a global, overarching narrative that shaped the chaos of history into a simple story of good and evil. According to this modern fairy story, the duty of virtuous Germans was to wage war on evil 'Asiatic' Jews who had somehow penetrated the political bloodstream of Anglo-Saxon nations like the United States and Great Britain and then compelled them to take up arms against Germany. The global power driving this devious plan was the Soviet Union. According to Nazi ideology, the roots of Bolshevism could be traced to the Jewish culture; it was the Jews' declaration of war against western culture. A Jewish cabal in Moscow, Hitler said in Nuremberg in 1936, intended to exterminate the 'existing blood and organically rooted leadership and replace it with Jewish elements alien to the Aryan peoples'. Bolshevism was the mask of the Jew. A 'Jewish head rested on a communist body'.

Hitler's crusade against Bolshevism was another way of waging war on world Jewry. In his notorious speech made to the Reichstag on 30 January 1939, Hitler used this seamless identification of Jews and Soviet Communism to make a chilling prophecy: 'If international finance Jewry inside and outside Europe should succeed in plunging the nations once more into a world war, the result will not be the Bolshevization of the earth and thereby the victory of Jewry, but the annihilation of the Jewish race in Europe.'

All over Europe, the German crusade against Jewish Bolshevism would soon supply an intoxicating rallying cry for Hitler's foreign executioners.

Part Two: June 1941–February 1944

4

Horror Upon Horror

> The [Sicherheitspolizei und SD] was determined to solve the Jewish question by any means necessary ... It had to appear to the outside world that the native people themselves had reacted naturally to decades of oppression by the Jews ... and carried out these first measures of its own accord.
>
> The Stahlecker Report, 1941

> All nationally conscious Latvians ... who would like to actively participate in the cleansing of our country of destructive elements should report at the administration of the Sicherheitskommando [SD] at Valdemara.
>
> Public announcement in Tevija, 4 July 1941

The small Lithuanian town of Svencioneliai can be found some 50 miles north-east of Vilnius in a region of forests and lakes. In 1941, many Jews lived here alongside Poles and Lithuanians. Today only Lithuanians live in Svencioneliai. Not far from the town, close to the banks of the Zeimena River, a simple memorial stands in a quiet wooded area. Here, we discover, German SS men and Lithuanian partisans murdered 8,000 Jews on 7 and 8 October 1941. A metre or so above the ground, there are curious depressions in the gnarled trunks of some of the trees. In this now silent place, Lithuanian *žydšaudžiai* (shooters of Jews) smashed the heads of Jewish infants against these trees. Traces of mass murder, like these wounded trees, can be found all over Lithuania – though very few Lithuanians choose to look.[1] Some 220,000 Jews or 'Litvaks' lived in Lithuania before 1941. Just 8,000 remained alive in 1945: a 'victimology rate' of 96.4 per cent.[2] The majority of victims were murdered close to their homes, often by their neighbours. The killing began very soon after 22 June, when German troops crossed into the Soviet-occupied Baltic States, followed by Special Task Force murder squads. There is incontrovertible evidence that Lithuanian citizens eagerly participated in the mass murders that followed the German assault at places like Svencioneliai all over Lithuania. Many survivors and some historians of the Holocaust believe that the majority of Lithuanians became willing executioners, murdering their Jewish neighbours with spontaneous zeal. RSHA chief Reinhard Heydrich frequently asserted that his Special Task Forces merely encouraged local pogroms by Jew-hating Lithuanian villagers, a process he euphemistically called 'self-cleansing'. But was the Lithuanian genocide more closely managed than the German occupiers cared to admit? And if so, how?

On the morning of 27 June 1941, a colonel in the German army arrived in the Kaunas (Kovno), the main centre of Jewish life and culture in Lithuania. It was nearly a week after the German invasion of the Soviet Union had begun and both Hitler and his generals were increasingly confident of a quick victory. The colonel had come to Kaunas to arrange suitable accommodation for the commander of Army Group A, Field Marshall Wilhelm Ritter von Leeb. As he passed the small Lietukis Garage opposite the Kovna cemetery on the junction of Greenwald Street and Vytatas Boulevard, he noticed a large, noisy crowd gathered on the forecourt.[3] The colonel appears not to have noticed that a German army photographer was also present and was taking pictures with his Leica camera. Both men provided eye-witness accounts of what took place. In the forecourt, mothers had hoisted children on to their shoulders or stood them on chairs or boxes to see better what was happening. There was a carnivalesque atmosphere. The colonel asked another onlooker to explain what was happening. He was told that the 'death dealer of Kaunas' was at work. This was where 'collaborators and traitors' received their 'rightful punishment'. Pushing forward, the colonel now faced a spectacle of medieval horror. At the centre of the crowd stood a young blonde man of medium height leaning on a 'wooden club', in fact an iron crowbar, which reached as high as his shoulders. It was crusted with blood and other body matter and was 'as thick as [his] arm'. In the photographs taken by the German army photographer, the death dealer looks unashamedly into the lens. He is well turned out in knee length black boots, dark trousers and a dark jacket with a white shirt. At his feet lay between fifteen and twenty dead and dying men, all bleeding copiously from head wounds. Someone had turned on a hose and blood-stained water gushed into a drainage gully. But the death dealer had not yet completed his day's work.

A short distance away armed Lithuanians guarded some twenty other men who awaited execution 'in silent submission'. All were Jews. As the German officer watched, the blonde executioner raised the iron bar, made a 'cursory wave' and another 'collaborator' was 'beaten to death in the most bestial manner'. Each savage blow brought enthusiastic cheers and cries from the crowd. A bystander informed the German colonel that when the Soviets had occupied Lithuania, the parents of the death dealer had been arrested and 'immediately shot'. Now he was taking his revenge. The German colonel made no attempt to intervene and left the scene before the death dealer had completed his repugnant task. The photographer's account, however, tells us what happened next: 'after the entire group [of Jews] had been beaten to death, the young man put the crow bar to one side, fetched an accordion ... stood on the mountain of corpses and played the Lithuanian national anthem.'

We now know that the death dealer was Algirdą Antaną Pavalkį, some of whose family had indeed been deported by the Soviets, as had many thousands of Lithuanian Jews. Later, Pavalkį served in the Gestapo, but changed sides at the end of the war and became a Soviet agent. A photograph of him taken in 1950 shows he was working as a rather well-paid doctor – 2,000 roubles a month.[4]

Scenes of equal barbarity unfolded elsewhere in Kaunas. Many were witnessed by gawping German soldiers. A lance corporal of the 562nd Bakers' Company watched Lithuanian convicts and Freikorps armed with clubs, cudgels and iron crowbars killing Jews in a small, cobbled square. A Baker reported: 'The actions seemed extremely cruel and brutal.' He had turned away: 'I could not watch any longer.' He heard someone say: 'these were Jews who had swindled the Lithuanians before the Germans arrived.' The German colonel who had watched the death dealer at work returned to headquarters and reported what he had seen to his commander-in-chief, Colonel General Busch, who listened impassively. Since the German campaign had begun on 22 June, pogroms like the one he had witnessed had become all too commonplace. Busch explained that these 'cruel excesses' were 'spontaneous action on the part of the Lithuanian population'. They had to be treated as 'internal matters' to be dealt with by the 'Lithuanian state'; the German army could not intervene. Orders to this effect had been received from the highest military authorities. He was, he regretted, 'powerless' to take action. In any case, he had been 'forbidden' to do so.[5]

These eyewitness accounts of apparently spontaneous pogroms in Eastern Europe figure significantly in accounts of the Holocaust. They provide evidence, it is claimed, that autochthonous anti-Semitism fuelled mass murder under German occupation. But a closer examination of many eyewitness statements reveals that these 'spontaneous pogroms' may not have been quite what they appeared. In the photographs taken by the Wehrmacht photographer, the self-proclaimed death dealer of Kaunas unmistakably wears a uniform: black jacket and trousers, high black boots. Armed men stand guarding the Jewish men he will kill. In the case of the town square massacre, the Wehrmacht Baker describes the perpetrators as 'Lithuanian criminals' and 'Freikorps', in other words a militia of some kind. These sometimes overlooked details strongly suggest that this was 'organised spontaneity'. But organised by whom? There was another eyewitness at the Lietukis Garage and his account provides at least part of the answer.

On that terrible day, Julius Vainilavičius had been fishing and passed by the garage on his way home: 'I saw some civilians there. The Germans were treating them roughly.' Vainilavičius noted that the civilians were all Jews. The Germans had, he discovered, ordered the Jews to clear horse dung from the garage forecourt using their bare hands. When this humiliating task was completed, Vainilavičius goes on, 'a great massacre began'. The Germans and ten to fifteen Lithuanians 'swooped down on the Jews, belabouring them with rifle butts, spades, sticks and crow bars'. Soon they lay moaning and crying. Then a hose was turned on and some of the Jews revived. A truck then appeared, with a number of Jewish prisoners already on board. The corpses were loaded into the rear of the truck and 'the Germans dispersed the onlookers'.[6] Who were these mysterious Germans?

The answer is in the German colonel's report. He tells us that by the beginning of July, when Army Group North arrived in Kaunas, 'the squads of [Lithuanian] guards now wore a kind of militia uniform of German origin. Amongst these men were also members of the SD who had, as I subsequently learned, started their activities in [Kaunas] on 24 June.' So German SD agents had arrived in Kaunas three days before the slaughter at the petrol station; plenty of time to organise a 'spontaneous' pogrom. The other account of the petrol station murders by the German photographer confirms that SD men were present: 'an SS officer came up' and tried to confiscate his camera. He refused and produced his official military pass and suggested that the 'SS man' discuss the matter with Colonel General Busch, 'whereupon I was allowed to go on my way unhindered'. Both SS and SD men wore black uniforms.

The evidence strongly suggests that this notorious slaughter was not a spur-of-the-moment pogrom carried out solely by Lithuanians, but a 'joint operation' instigated directly by Heydrich's SD and Lithuanian militias. Lithuanian historian Alfonsas Eidintas has remarked that 'reading the hundreds of memoirs by surviving Jews, including those by Lithuanian Jews ... I sometimes got the impression that it had been only Lithuanians, Latvians, Poles and Ukrainians who had executed mass murders in their own countries, but not Germans'.[7] His remark goes to the very heart of the matter. Leonic Rein, in a paper about local collaboration in Belorussia, asks: 'was the Holocaust in fact, as Goldhagen argues, a purely "German undertaking"?'[8] Take for example this account from a recently discovered diary: 'on

Tuesday the 24th of June, we went out into the street and saw that the town was already full of German soldiers. Already there were "Partisans" – Lithuanian bandits wearing white armbands with swastikas ... They could already do anything they wanted to a Jew.' Shortly afterwards, we discover, systematic killing began, conducted – according to this witness by Lithuanian 'partisans' and villagers – without any German supervision.[9] This short, telling narrative lays out the key elements of this enquiry: the arrival of German troops; the appearance of partisans wearing arm bands; followed by apparently unsupervised mass murder. Many accounts of the Holocaust take for granted that Eastern European peoples, conditioned by centuries of anti-Semitic loathing, impulsively set about murdering their Jewish neighbours. This is an intricate and contentious matter. Historian Dovid Katz, who has devoted his life to the study of the Lithuanian Holocaust, has fought a long battle against Lithuanian apologists – many of whom are active members of the government – who seek to 'sanitise' the historical record. The facts are not in doubt: many Lithuanians took a direct and lethal part in the Holocaust. How and why they did so is the subject of this chapter.

As we have seen, between September 1939 and June 1941, the Germans exploited native antipathies in Slovakia, Croatia and Romania to wage an undeclared war on the racial and ideological enemies of the Reich. The SS and German military intelligence cultivated relations with ultranationalist factions and militias like the Hlinka Guard, the Ustasha and the Romanian Iron Guard. The Germans judged the first results to be unsatisfactory. The Governor General of the General Government, Hans Frank, complained that the Romanian pogroms that began at the end of June had been 'barbaric'; what was required, he told his dinner guests at his headquarters in Kraków, was not butchery but surgery. German Einsatzgruppe commander Otto Ohlendorf came to the same conclusion. He insisted that 'the Romanian police be more orderly'. In short, Hitler's foreign executioners needed expert tuition. So it was that as Hitler's armies and air force fell on their Soviet allies on midsummer's day 1941, Reinhard Heydrich's Special Task Forces set about applying the lessons learnt in Croatia and Romania. Driving this joint operation was a potent mythology that permitted Germans and their allies to share a language of destruction – the myth of the 'Jewish-Bolshevik'.

On 30 January 1941 Hitler told a wildly cheering crowd in the Sportspalast in Berlin that 'if Jewry were to plunge the world into war, the role of Jewry would be finished in Europe'. In German strategic planning, this delusional claim had a

peculiar logic. Hitler understood that America, aligned if not allied to Great Britain, posed a grave threat to his imperial plans. Nazi doctrine and propaganda blamed American Jews for encouraging American rearmament – and inspiring President Roosevelt's increasingly bellicose posture. Seen in this light, Hitler's decisions to strike 'hard and fast' against the Soviet Union before the Americans entered the war made strategic sense.

This logic, it should be emphasised, sprang from racial paranoia. Hitler's war would be a racial struggle directed against a global enemy, namely 'World Jewry'. This chimerical foe had the power to manipulate both London and Washington, and infested the 'Bolshevik' state in Moscow. For this reason, the German attack on the Soviet Union marks a decisive break in European history: it was both the biggest land grab in history and a monstrous chastisement that sought to end for good the baleful influence of Jewry. The security of the new German empire depended on the elimination of their ancient menace.

Heydrich's Special Task Forces would spearhead the assault on the 'Jewish-Bolshevik' stronghold. At regular meetings with his commanders, Heydrich frequently drummed in the bond between Bolshevism and Jewry. In the meantime, Himmler pushed forward plans for the colonisation of the east – plans of staggering ambition that covertly envisaged the physical annihilation of millions.

Since the 1980s the Holocaust has dominated accounts of the Second World War – and rightly so. But in the minds of Hitler and German imperial strategists, mass murder and the forced 'evacuation' of millions was a beginning not an end; a 'cleansing procedure' that would pave the way for a complete ethnic reordering of the east. Historian Adam Tooze has shown that Operation Barbarossa was a first step towards a 'long term programme of demographic engineering', summarised in that ugly word 'Germanisation'.[10] The idea was not an original one. Germanisation already had a long and shabby pedigree. Heinrich von Treitschke, the nineteenth-century advocate of *Drang nach Osten* (the drive to the east), celebrated the 'most stupendous and fruitful occurrence of the later Middle Ages – the northward and eastward rush of the German spirit and the formidable activities of our people as conquerors, teachers, discipliners'.[11] Now in 1941, the 'formidable rush' of Hitler's war machine and Himmler's SS militias promised to fulfil that old German dream. Hitler compared his quest for 'living space' in the east to the American colonisation of the west: the Volga, he proclaimed, would be Germany's Mississippi. In 1939, when Germany invaded the then dismembered Poland, Himmler and his racial experts (*Ostforschung*) had embarked on an ambitious experimental programme to settle German colonists in the new German provinces and deport Polish Jews into the Lublin district in occupied Poland. This fist stab at 'Germanisation' proved a dismal failure. More than half a million ethnic German settlers, uprooted from the Baltic and South Tyrol, abandoned their homes only to end up stranded in sordid transit camps. In occupied Poland, the General Governor Hans Frank stymied SS plans to dump millions of Jews in what he regarded as a personal fiefdom. Clearing this ethnic logjam demanded the most radical of solutions: the conquest of the east and the subjugation of its peoples.

Himmler commissioned an ambitious young agronomist SS-Oberführer Dr Konrad Meyer to begin devising a 'Generalplan Ost'.[12] Between May 1941 and the following spring, Meyer toiled away at his office in the upmarket Berlin suburb of Berlin-Dahlem. As German troops penetrated deep into the Soviet Union and Caucasus, Himmler continuously stepped up pressure on Meyer urging him to consider ever more radical solutions. Meyer began by assuming his schemes would take twenty-five years to complete – Himmler wanted that reduced to five. Finally, in May 1942, Meyer delivered his 'Legal, Economic and Spatial Foundations for Development in the East'. The Generalplan Ost was the high point of a succession of toxic German occupation plans devised after the destruction of Poland. The 'Hunger Plan', developed not by the SS but German army planners, proposed diverting Russian agricultural supplies to Germany, condemning to certain death by starvation 30 million people in Belorussia, northern Russia and the major Soviet cities. German military reversals in the winter of 1941 forced the partial abandonment of this wicked scheme, although the German army chiefs used planned famine as a weapon of war during the 900-day siege of Leningrad and other Soviet cities. After February 1942, the Germans focused more intently on the 'Final Solution of the Jewish problem' promulgated at the Wannsee Conference.[13] Many historians have claimed that the German occupation of the east was chaotic and unplanned. In fact, no other imperial project has generated more occupation plans.

It is also widely assumed that German racial experts believed that the east was occupied by a homogeneous mass of 'Slavs'. Hitler certainly held this opinion, as his 'table talk' frequently demonstrates. But German race science was by no means monolithic and underwent a number of conceptual upheavals, which intensified after 1941 when German anthropologists seized the opportunity to study Russian prisoners of war. For now, we need simply to understand that German race experts increasingly recognised the diversity of eastern peoples – and that Himmler acknowledged this in his grandiose plans for the east. As early as 24 May 1940 Himmler presented Hitler with a short paper: 'Some Thoughts on the Treatment of Foreign Peoples in the East'.[14] He begins by arguing that 'we must endeavour to recognize and foster as many individual groups as possible'; he lists Ukrainians, White Russians, Gorales, Lemkes and Kashubians. From a purely strategic point of view, he argues, it makes sense to 'divide them up into as many parts and splinters as possible'. In other words, divide and rule. By 'dissolving' these ethnic groups 'into countless little splinter groups and particles', any sense of 'unity and greatness', 'national consciousness ad national culture' would be eliminated. Himmler then goes on: 'we will of course use the members of these ethnic groups ... as policemen and mayors.' The planned outcome would be the 'dissolution of this ethnic mishmash [in the east]' so that the most 'racially valuable' people could be 'fished out' and then 'assimilated' in Germany. Himmler then turns to education. Schools for the non-German eastern population would teach the majority only basic maths 'up to 500' and how to 'sign one's name'. All would be taught that 'it is God's commandment to be obedient to the Germans'. But – and here is the crucial point – more ambitious parents could apply to SS authorities to have their children educated to a higher level. On condition that the candidate was 'racially first class', successful applicants would be removed from their families and placed in a German school 'indefinitely'. Himmler assumed that such parental zeal signified the possession of 'good blood'. Himmler then appears to realise that Germanisation could not depend on parental whim alone, however praiseworthy. German teachers would be required to constantly sift their six to ten charges to winnow out 'valuable blood'.

Himmler used the very same metaphors when he discussed recruiting Germanic volunteers for the Waffen-SS: 'I really have the intention to gather Germanic blood from all over the world, to plunder and steal it where I can.' It was a short step, in other words, from the classroom to the parade ground. Although the Generalplan Ost proposed the extinction of at least 80 per cent of indigenous peoples, Himmler also recognised the level and complexity of ethnic diversity in the east and proposed exploiting certain racial 'splinters', as he put it, as 'mayors and policemen'. Recruitment of non-German volunteers thus formed part of a grossly ambitious imperial plan that depended on the physical liquidation of many millions of 'surplus' people.

Between the spring and summer of 1941, occupation experience in the Balkans and close involvement with the Romanian 'National Legionary State' had taught Himmler and his SS planners valuable lessons. In Croatia and Romania two thorny problems had become all too evident. Now they would have to be solved. First, factions like the Hlinka Guard, the Ustasha and the Iron Guard could not be relied upon to perform the task of mass murder in an orderly manner. They had a tendency to run amok or lacked 'staying power'. Their energies needed to be disciplined. That was the German way: a matter of proper organisation and proper training. Surgeons not butchers! The second problem – nationalism – would prove much less tractable.

The birth of nationalism in the old empires in the nineteenth century was from the start wedded to extreme ethnic chauvinism directed mainly at Jews. This union was if anything deepened as new nations stumbled on to the stage of history. When the old empires collapsed at the end of the First World War, a new bout of nation building bonded nationalism ever tighter to chauvinism. Most, if not all, ultranationalist ideologues believed that Jews, either as the agents of international capital or as Bolsheviks intent on spreading a revolutionary message, menaced the fragile new nation states that emerged from beneath the wreckage of the old empires.

Why was this a problem for the Nazi imperial strategists? The reason is simple: anti-Semitism and nationalism came as a package. The Germans wished to exploit the one without satisfying the other. Hitler was no nation builder. In Western Europe and Scandinavia, Nazi administrators, whether military or civilian, soon found to their cost that failed ultranationalist demagogues like Vidkun Quisling in Norway, Anton Mussert in the Netherlands and Léon Degrelle in Belgium assumed that German occupation would provide the fast track to power. This was a delusion: for Hitler, collaboration was a one-way street. Power could flow in one direction only. Hitler ultimately planned to rebuild the Holy Roman Empire, the First Reich, by extending the western borders of Germany as far as the Pyrenees. That was bad news for the conquered peoples of Europe. On 9 April 1940, he proclaimed that 'the Greater German Reich will arise today': Danes, Norwegians, Dutch and Flemings would join together in a new community defined by its racial purity and dominated by Germany. In this 'Germania magna' the old national borders would be dissolved away. In the end, Hitler failed to rebuild the old Reich, but because he expected to he had no interest in promoting nationalists.

In the east, the Germans would encounter the same difficulty, but in an altogether different form. At the end of the First World War, a number of brand new nations had emerged kicking and screaming from the wreckage of Europe's old empires. For two short decades, the peoples of Estonia, Latvia and Lithuania tasted the joys of sovereignty. Then in 1939 Stalin signed the non-aggression pact with Hitler and snatched it away. Soviet armies and the hated agents of the Soviet security service, the NKVD, occupied the Baltic States and eastern Poland. A year later, Hitler's armies drove out the Soviet occupiers. Many greeted the German invaders as liberators – and nationalists looked forward to the revival of their sovereign rights. They had no idea that for Hitler military conquest would mean the extinction of these insolent 'little states'. The only winners would be the minority deemed suitable for 'Germanisation'.

Nazi ideologue and head of the 'Ministry of the Occupied East' (Omi) Alfred Rosenberg flirted with the idea of granting some kind of suzerainty status to a few privileged eastern peoples like the Estonians. Rosenberg, who took the most 'liberal' approach to Eastern European nationalist aspirations, had no doubt that in the long term anything resembling a nation state would be completely digested by the 'Greater German Reich'.

Hitler's contempt for Slavic nationalism was profound. But in the German political tradition, his views were by no means original. Michael Burleigh argues in his essay 'The Knights, Nationalists and Historians' that the idea of *Drang nach Osten*, expansion to the east, was a leitmotif winding through German foreign policy – from Otto the Great through to Frederick the Great, to Bismarck and the Wilhelmine Empire and on to Hindenburg and then Hitler. The nineteenth-century apostle of eastward expansion, von Treitschke, denounced the 'anarchic crudity of the Slavs' which made them incapable of state formation. Only the Germans could be masters, teachers, discipliners and the bringers of civilisation to their crude eastern neighbours. 'In the unhappy clash between races,' Treitschke argued, 'a quick war of annihilation' would sort out 'the brute beasts of the East'.[15]

How then might nationalist eastern collaborators be rewarded if German imperialism demanded the destruction of their nation states? That circle could never be squared. But in the euphoric aftermath of conquest, Himmler would offer nationalist factions an alluring reward: the chance to seek revenge on the Jews they blamed for the 'Bolshevik' occupation of their nation states.

Evidence of this murderous skulduggery can be found in the Einsatzgruppen reports, a huge collection of German documents discovered in RSHA headquarters in Berlin at the end of the war by American lawyer Benjamin Ferencz. In cold, detached language they document how Heydrich's Einsatzgruppen executed 2 million Eastern European Jews in forest clearings and excavated pits between 22 June 1941 and 21 May 1943. The entire document collection weighed 200 tonnes – testimony to German managerial and reporting zeal. It comprises 195 'Morning Reports USSR' and 55 longer 'Weekly Reports'. When the four Special Task Forces crossed the Soviet border in June 1941 they brought with them back-up teams: secretaries and clerks, teletype operators and wireless operators equipped with the most up to date equipment. The duty of these men and women was to send detailed accounts of the previous day's activity to local Task Force headquarters (for example in Tilsit on the Lithuanian border) by wireless or courier. Heydrich's officers filed reports every morning until May 1942, when the 'Reinhard' murder camps and Auschwitz-Birkenau began to take on a bigger role in the genocide. After May, reports had to be filed weekly. This raw data flow from the front line listed execution sites, numbers killed and, crucially for our purposes, 'the mood of the general population'. At the Special Task Force HQ, higher ranked officers collated the raw information and compiled 'meta reports' and dispatched them to Heydrich's offices in Prinz Albrecht Strasse in Berlin. The Einsatzgruppe reports documented the mass murder in chilling, voluminous and meticulous detail. Heydrich distributed the final reports to high-ranking Wehrmacht, police and SS officers, to members of the German Foreign Office, and to Göring and the German industrial magnates.

Himmler and Heydrich both studied the 'Morning Reports' closely and radioed fresh instructions to Task Force officers in the field, invariably urging them to show greater 'harshness'. Each Task Force commander was subordinate to the three Higher SS and Police Leaders (*Höhere SchutzStaffel- und Polizeiführer*, HSSPF) in charge of different regions of the occupied Soviet Union. These SS officers became the managers of genocide on a day-to-day basis – 'little Himmlers' who co-ordinated the work of the Special Task Forces with Order Police battalions and the 20,000-strong Waffen-SS brigades. We will hear a great deal about these men in the chapters that follow.

The charge of the Special Task Forces would be 'carrying out fundamental special measures against the Jews'.[16] Heydrich, it must be emphasised, set out strict *Sprachregelung* (language rules) to camouflage German plans. The methodical large-scale execution of Jews and 'Soviet Commissars' was referred to using a blizzard of code words: action, special action, large-scale action, reprisal action, pacification action, radical action, cleaning up action, overhauling, cleared or cleared of Jews, freeing the area of Jews, special treatment or measures, rendered harmless, handled according to orders, severe measures, treating according to the previous procedure ...[17] It was not considered necessary to provide a crib. As well as the standard kind of Einsatzgruppe report, Ferencz discovered three 'authored' Einsatzgruppe reports that have special significance. Two bear the signature of the commander of Special Task Force A, Franz Walther Stahlecker. The third was written by the Swiss-born leader of Einsatzkommando 3, Karl Jäger (in his own words, a 'person with a heightened sense of duty').[18] These chilling documents tells us a great deal about the management of mass murder in Lithuania and the other Baltic States.

The 40-year-old Stahlecker was a dedicated and proficient *génocidaire*. He fervently believed that 'the East belonged to the SS'. Colleagues noted that Stahlecker was often 'jumpy and unpredictable ... obsessed that they [his superiors] would realise in Berlin that he was absolutely obedient concerning this [Heydrich's] order: not just obedient, but had a special mission to carry it out'.[19] Stahlecker would consider only those recruits who could 'tolerate hardships and burdens of the soul'. In this respect, he himself provided the model. In his second, shorter report, Stahlecker attached a map of the Baltic region and Belorussia on which he or his assistants had inscribed numerous graphic coffins which enumerated how many Jews had been killed in particular regions or places. The third special report was filed by Stahlecker's subordinate, Karl Jäger (b. 1888), who became the commander of the Security Police and SD for Lithuania.[20] Jäger lists with remorseless thoroughness the murder of precisely 137,346 Jews and communists. Jäger documents over a hundred 'special operations' in seventy-one separate locations (he made return visits to the same village if he discovered from informers that Jews had survived). The report demonstrates that the Einsatzkommando could move very fast: in the course of a single day in September, the SD men performed their 'racial duties' in four different villages. Roland Headland notes that 'in no other surviving document do we get as detailed a picture of the steady accumulation of victims'.[21]

There are a number of studies of the SD Einsatzgruppen, both in German and English. Here I will focus on a somewhat neglected aspect of the reports. One of the tasks of the Special Task Force commanders was to provide information about the 'Mood and general Conduct' of civilians and the 'value' of local activists. Most of the reports contain paragraphs that provide very revealing insights into the thinking of both the Einsatzgruppe men and the 'activists' they encounter. One report, for example, makes the following observation: 'All experiences confirm the assertion made before that the Soviet state was a state of Jews of the first order.' For this reason, the report continues: 'the Jewish problem has become a burning problem [sic] for the Ukrainian people.' The SD men and many non-Jewish Eastern Europeans shared the same perception that the agents of Soviet rule were Jews – and that the entire edifice of Bolshevism was a 'Jewish conspiracy'. It was this mythology of 'Jewish Bolshevism' that would sustain the mass recruitment of non-Germans in the service of the Reich. The chimera of the 'Jewish-Bolshevik' forged a shared ideological language and practice that allowed the Germans to continuously refer to mass murder as 'spontaneous actions'. In a letter to Special Task Force commanders, Heydrich emphasised that:

> no obstacle is to be placed in the way of the *Selbstreinigungsbestrebungen* (self-cleansing efforts) of the anti-Communist and anti-Jewish circles in the newly occupied areas. Rather, they are to be intensified, when required, without a trace, and channelled onto the proper path, without giving these local 'self defence circles' any opportunity later to claim that they acted on orders or were given political assurances.[22]

The point about 'political assurances' makes clear that 'self-cleansing' could be a prelude to independence.

Heydrich went on to make a second sometimes overlooked point. He insists that reports must 'make clear that it was the local population that spontaneously took the first steps against the Jews'. Why? Because 'it was preferable, that at least at the beginning, the cruel and unusual means, which might upset even German circles, would not be too conspicuous'. In other words, the inciting role of the German murder squads needed to be as covert as possible. In an especially telling aside, Heydrich recommended that the Special Task Force commanders film or photograph any 'spontaneous pogroms'. This meant that SS propagandists would be able to show the world that Jews and other undesirables had somehow invited their own chastisement at the hands, not of Germans, but of their fellow Lithuanians or Latvians or Ukrainians. The 'spontaneous' slaughter of Jews by native executioners became, in a perverse twist, justification for persecution. Since it was Lithuanians who first carried out these slaughters, the victims surely deserved their fate. Germans merely facilitated natural justice.

Like Himmler, Heydrich was also preoccupied with the 'nationalism problem'. As we have seen, he insisted that 'self-defence circles' must not be provided with any 'political assurances'. In other words, mass murder could not be rewarded with promises of nationhood. This was tricky because authorising any kind of native militia as the Germans planned implied the de facto recognition of statehood. It was a dilemma that would plague SS efforts to exploit eastern peoples until the very end of the war. Stahlecker was well aware of the quandary: 'The security of [Riga] has been organised with the help of 400 [Latvian] *Hilfspolizei* (auxiliary police) ... care has been taken to assure that these troops *would not become a Latvian militia* ... two further independent units have been established for the purpose of carrying out pogroms. All synagogues have been destroyed.' (My italics.)[23]

This then was the German dilemma: how to encourage 'self-cleansing efforts' without igniting nationalist agitation. What of the other side? As we have seen in the Balkans and Romania, the arrival of German armed forces acted like a catalyst on the peculiar mosaic of each national culture and the nature of the ruling elite. The destruction of Yugoslavia offered Croatian fascists the opportunity to strike decisively at Serbs. Alongside this civil war, the Ustasha regime also targeted Jews – partly to satisfy Croatian chauvinism but also to reinforce its bond with the Reich. In Romania, Germany promoted a radical ultranationalist and anti-Semitic regime led by Ion Antonescu to secure vital economic resources and the services of the Romanian army. When Hitler attacked the Soviet Union, Germans and Romanians colluded in the destruction of Romanian Jewry. Hitler made no claim to 'living space' in puppet states like Croatia or Slovakia and he had no wish to undermine the national integrity of Romania so long as Antonescu stayed on side.

The Baltic nations and Ukraine had a very different significance in German imperial plans. The purpose of Operation Barbarossa, the invasion of the Soviet Union, was to seize living space for the German people and to smash the 'Jewish-Bolshevik' state. German radical imperialism, founded on the blood right of Germans to exploit the east as (in Hitler's words) a 'garden of Eden' or territorial *tabula rasa*, had no room for nation states. Since the twelfth century, Germans had sought hegemony in the Baltic region. The Teutonic Knights, armed with the 'cross and the sword' brought Christianity – and serfdom. In Riga, founded by a Bishop of Bremen, and Reval (Tallinn), Hansa merchants dominated commerce and trade. In rural areas, big German baronial estates, which resisted any attempt to abolish feudal relations between master and serf, remained largely intact until the end of the First World War. Serfdom retarded national aspiration, and while Lithuanians, Latvians and Estonians spoke distinct languages and nourished different cultures, it was only the Grand Duchy of Lithuania that had ever achieved genuine statehood. In the seventeenth century, the Polish-Lithuanian Commonwealth swallowed up much of the Baltic region, including what is now modern Latvia and Estonia which were mere duchies. Catholic Lithuania had successfully resisted 'Germanisation' by the knights and their feudal successors while Lutheran Latvia and Estonia had succumbed. But at the end of the eighteenth century, the Commonwealth had been dismembered and Lithuania was split between the Russian empire and Prussia. This turbulent history meant that the Baltic national movements that bubbled up from the wreckage of the Russian and German empires after 1918 had grown the shallowest of roots. The Baltic States suffered bloody and traumatic birth pains. Even in defeat, Germany was unwilling to give up its Livonian fiefdoms, and the Freikorps 'Iron Brigade' and Baltische Landeswehr exploited the threat of a Red Army incursion to make a last ditch attempt to establish a German state. Allied intervention eventually drove out both the Freikorps and the Russians – and in 1920, recognised three new sovereign states.

The rebirth of Lithuania had been especially painful. Lithuanians squabbled with Poland over Vilnius and laid claim to Klaipèda – the German Memel. The new Lithuania was a feeble reiteration of the old grand duchy – a mere buffer state between Germany and the Soviet Union. This inspired a kind of national siege mentality and a succession of authoritarian governments. From 1926, Antanas Smetona ruled Lithuania as a virtual dictator. Latvia and Estonia, the other Baltic nations, proved equally rickety and followed much the same path. In Latvia, Kārlis Augusts

Vilhelms Ulmanis seized power in a coup in 1934; he banned Latvian political parties, locked up his opponents and closed newspapers, including those published in Yiddish. That same year Kontantin Päts introduced martial law in Estonia. These regimes were authoritarian rather than fascist in a strict sense. Ulmanis, who liked to compare himself with Oliver Cromwell, openly rejected any kinship with Italy. At a political congress in 1935, Lithuanian president Smetona denounced what he tartly called Nazi 'zoological nationalism' – and on the surface, there was little overt evidence of 'eastern' anti-Semitism. The Lithuanian Minister of National Defence, Balys Giedraitis, even passed legislation forbidding attacks on Jews.

This fragile tolerance reflected the long history of Jewish settlement. Since the eighteenth century, the Russian Pale of Settlement had included Lithuania (but not the territory of the other more Germanised Baltic States). And while many Jews who lived in the Shtetls of the Pale endured both poverty and frequent pogroms, Jewish social and cultural institutions flourished: Vilnius was celebrated as the Jerusalem of Lithuania, and Jews made up nearly half the city's population. Tolerance was the public face of the regime. Dig deeper and a rather different picture takes shape however. The coup that brought Smetona to power had been engineered by an extremist faction of Lithuanian army officers called *Geležinis vilkas* – the Iron Wolf. This was the guard movement attached to Smetona's Tautininkai Party, headed by Augustinas Voldemaras, who became prime minister after the coup. Voldemaras was a charismatic, brilliant radical nationalist (educated in St Petersburg) who soon fell out with Smetona. The president was honorary head of the Iron Wolf, but feared the fanatical young army officers who gravitated to Voldemaras' extremist camp. The Iron Wolf, for their part, viewed Smetona as too moderate, especially with regard to the alleged 'influence' of Lithuanian Jews. In 1934, Iron Wolf officers tried to oust Smetona and replace him with Voldemaras. But the coup faltered and Smetona had his rival arrested.

The Iron Wolf never became a mass movement as Codreanu's Legion of St Michael did in Romania, but its anti-Jewish agenda reflected the secret views of many Lithuanians. From the mid to late 1920s, organised anti-Semitism became increasingly evident. Gangs defaced Yiddish street signs; attacks on Jewish shops and individual Jews in cinemas, restaurants and other public places began to rise noticeably. As the world depression deepened, attacks on 'Jewish influence' in the press became increasingly venomous. Driving this up-swelling of anti-Semitism was the *Tautos valia* (*Will of the Nation*) newspaper which began appearing in October 1926 and the Union of Lithuanian Business (LVS), whose paper *Verslas* (*Business*) agitated for the Lithuanisation of the national economy.

Smetona had been thoroughly rattled by the Iron Wolf and now tried to buttress his power by tapping into the new chauvinism. He abolished the Ministry of

Jewish Affairs (established in 1918), stripping Jews of effective political representation at a stroke. After conferring with gentile business leaders, Smetona introduced a series of measures designed to smash Jewish enterprise. The new legislation denied Jews access to cheap credit, forcing many businesses to declare bankruptcy. In the countryside (where so many thousands of Jews would be killed in 1941) the hoary myths of Jew hatred revived. In 1935, a rural newspaper reported that in the village of Plunge two Christian children had vanished and insinuated they had been abducted by local Jews. In deeply traditional Lithuanian villages, there was no need to spell out the old 'Blood Libel' – the medieval myth that Jews used the blood of Christian children to bake matzo bread. Soon after the newspaper report appeared, a rash of flyers urged ethnic Lithuanians to take revenge. The children had not yet been found and feelings ran high and ugly. At a *Jomarkas* (open-air market), an angry mob attacked Jewish traders. Another child disappeared a few months later – and again, mobs attacked Jewish homes and vandalised synagogues. A gang of young men ambushed some Jewish travellers who were watering their horses by a stream. The local police, already under attack for failing to find the missing children (or their remains) carried out a few token arrests, but this merely provoked a fresh surge of anti-Semitic leafleting.

This was not an isolated incident. The disappearance of any child provoked the same kind of hysteria in other villages and towns. Anti-Jewish feeling in rural areas was reinforced by state legislation. All over the old Pale, Jews traditionally made a living as agricultural middlemen, trading corn and other produce between country and city. If they did well, gentile peasants and merchants resented their success and accused them of sharp practice. In the mid-1930s, the LVS promoted the idea of rural co-operatives in effect to take over the services performed by Jewish merchants. Smetona became an enthusiastic convert to the co-operative movement and its rapid success bankrupted many rural Jewish businesses. Many middle-class Lithuanian Jews admired Smetona. He appeared to respect the bastions of Jewish culture and faith in Vilnius and Kaunas. But his apparently benevolent dictatorship pushed them to the margins of Lithuanian society, where they would be exposed to terrible peril. A younger generation, deeply impressed by Zionism, felt differently. Lithuania, they rightly suspected, was no longer safe. One young journalist wrote on the eve of war that: 'Jews and Lithuanians lived alongside one another ... on the same street and often in the same building. That should have brought them closer to one another but that never happened.' The Lithuanian liberal elites failed to notice the groundswell of hatred, openly expressed on the streets and in the farms, until it was too late.[24]

In the republic of Latvia, anti-Semitic agitators spoke louder and wielded greater influence. Latvian Jews had made a vital contribution to the independence movement, but President Ulmanis was a fanatical nationalist who promoted 'Latvia for the Latvians' (meaning ethnic Latvians) and privately resented Jewish business expertise. We associate anti-Semitism in Eastern Europe with the pogrom and the mob. But in Latvia, as in Romania, an influential intelligentsia had been radicalised by student fraternities, the Korporacijas. These elitist reactionary student associations slavishly mimicked the German fraternities, the Burschenschaften. Like their German counterparts, the Korporacijas, such as the Lettonia and Selonija, cultivated a broad network of contacts in government and business. Membership provided a fast track to business and state elites. But not for every Latvian citizen. All the Korporacijas refused to admit Jewish students and promoted a heady brand of radical nationalism. Many graduates of the Lettonia, like Arveds Bergs, turned to journalism and he, it was said, educated an entire generation. Stirred by Bergs' rhetoric, Gustavs Celmiņš, another Lettonia graduate, set up a new ultranationalist party, Ugunskrusts – inspired by the Romanian Iron Guard – that proclaimed 'Latvia to the Latvians, bread and work to the Latvians!' Recruits donned quasi military uniform (dark grey shirt, beret, trousers with knee high boots) adorned with swastikas – and staged impressive public drills and organised mass meeting at rural camps.

Celmiņš claimed that by 1933 he had 12,000 members – almost certainly an exaggeration, but he made enough noise to rattle the Ulmanis government and the Ugunskrusts was banned. Undeterred, Celmiņš simply changed the name to Pērkonkrusts (Thunder Cross). His programme was not especially complex: 'The sovereign power in Latvia belongs to the Latvians and not to the people of Latvia.' This slogan alluded to Latvian Jews, but also to Baltic Germans. 'Already now our Germans, anticipating the arrival of their messiah Hitler, feel like half-masters in our house … If today the general struggle is against Jews, it does not mean that we shall not purge Latvia of the pitiful baronial detritus.'[25] By the time Hitler seized power in Germany, Latvia had become infested with nationalist factions like the Pērkonkrusts that promoted radical nationalist agendas and threatened to destabilise the young Latvian republic.

The crisis suited Ulmanis who, like his presidential neighbour in Lithuania, had grown weary of democracy. And like Smetona, the Latvian president had powerful backers. At the end of the First World War, Latvian vigilantes formed Aizsargi (defence units) to fight off incursions by Soviet troops or German Freikorps. By 1922, the Aizsargi had voluntary armed units in every township in Latvia and attracted tens of thousands of recruits. Like the German SA Brownshirts, the Aizsargi unnerved the regular Latvian army. But Ulmanis strenuously cultivated the Aizsargi leadership. In 1934, they duly proclaimed him 'Vadonis' – Führer. To defend his seizure of power, Ulmanis claimed that Latvia was threatened by dangerous nationalist agitators – and Celmiņš organised a few outrages to help him make his case. Once Ulmanis had tightened his grip, he had the Pērkonkrusts banned and deported Celmiņš, who, like legions of other exiled radicals, found his way to Germany, where he offered his services to the Reich. In Latvia, Pērkonkrusts radicals went underground, organising secretive cells inspired by the Iron Guard to continue Celmiņš' crusade. For the next six years, Ulmanis resisted all efforts by the Latvian army to demobilise his Aizsargi benefactors and instead used them as a private militia.[26]

It is tempting to draw a simple causal line between the Baltic radical nationalists and the explosion of mass murder after June 1941. That would be misleading. We cannot ignore the psychological impact of an event that profoundly destabilised civic relations, however fragile, in the Baltic States. On 23 August 1939 Joachim von Ribbentrop, the German Foreign Minister, and his Soviet counterpart Vyacheslav signed a 'Treaty of Non-aggression between Germany and the Soviet Union'. This pact contained 'secret protocols' that divided Eastern Europe between German and Soviet 'spheres of influence'. The protocols sealed the fate of the Baltic States by ceding them to the Soviet Union's sphere of influence. Stalin did not rush to take advantage of the pact. Soviet troops moved up to the borders of Latvia and Lithuania in October 1939, but full-scale occupation did not begin until 15 June 1940.[27] Nevertheless by the middle of July, the Baltic presidents and their governments had been forced to resign, and newly installed puppet regimes voted for incorporation into the Soviet Union. By the end of 1940, the three former states had been digested by the Soviet Union as the 'Pri-Baltic Military District'. Immediately after these rigged elections, Soviet NKVD units rounded up 15,000 'hostile elements' and police tribunals were set up to try 'enemies of the people'. On 21 January 1921 General Ivan Serov, Deputy People's Commissioner of State Security of the Soviet Union (NKVD), signed Order No 001223: 'On the procedure of carrying out the Deportation of anti-Soviet Elements from Lithuania, Latvia and Estonia'. Historian Edgars Dunsdorfs, author of *The Baltic Dilemma*, estimates that, between June 1940 and June 1941, the number of Baltic citizens executed, conscripted or deported after the Soviet annexation was at least 125,000 men, women and children, including heads of state and ministers and allegedly dissident members of the intelligentsia. Stalin's definition of 'enemies of the people' threw a net over both individuals and economic classes (for example, large landlords and factory owners), as well as specific professions, including prostitutes and clerics. According to Order No 001223 fitting punishments included 'confiscation of their property, arrest and incarceration in camps for a term of five to eight years, and after serving their term in camps, to settlement in remote areas of the USSR'. The great forced migrations set in motion by the Nazi Soviet agreements continued in the Baltic. At railway terminals, NKVD battalions herded many thousands into cattle wagons, often left standing for days, before they began the long journey east. Only a few of the deportees ever returned home.[28]

In Latvia and Lithuania, Soviet deportations hit Jewish communities harder, proportionally, than any other ethnic or religious group. It is estimated that between 1939 and 1941 the Soviets arrested and exiled 100,000 Jews – which was about 5 per cent of the Jewish population in the annexed territories. The fact that the Soviets deported Jews in such large numbers had no impact on how other Latvians or Lithuanians perceived the 'terror'. As in Romania, Soviet aggression was instinctively attributed to Jews or some unspecified 'Jewish' agency. In short, the Jews were to blame. Soviet aggression galvanised and refashioned age-old hatreds.

Saul Friedlander argues that a balanced assessment of Jewish involvement in the Soviet occupation is 'quasi impossible'. What nationalists observed was that Jews were well represented in officer schools, mid-rank police appointments, higher education and some administrative positions.[29] Many NKVD officers were Russian Jews – a statistical fact that is still exploited by Holocaust 'deniers'. In his book *The Whisperers*, Orlando Figes confirms that Jews had 'flourished in the Soviet Union'. They recalled, of course, the persecution of Jews under the tsars which had reached such a grisly climax in the 1880s. In the 1920s, the Soviet government energetically promoted Yiddish culture, especially in Moscow. For delusional believers in the anti-Semitic conspiracy theories of 'The Protocols of the Elders of Zion', such apparent favouritism seemed to be hard evidence of a 'Jewish-Bolshevik' conspiracy.

It was, of course, nothing of the sort. Soviet tolerance, in any case, was skin deep. Stalin himself, educated in a seminary, was no friend of the Jews and in the run up to negotiations with Ribbentrop, purged prominent Jews from conspicuous positions to curry favour with Hitler. The Nazi-Soviet Pact in fact traumatised Soviet Jewry and weakened their commitment to the Soviet ideal. To be sure, in the annexed regions of Eastern Europe, many Jews came to see Soviet occupation as the 'lesser of two evils'. But they soon learnt to their cost that while the Russians may have spared them at least temporarily from the attentions of the Germans, the Soviet occupation authorities targeted any sign of Jewish political activism. The Soviets arrested leaders of the World Zionist Organisation and the radical Zionist Betar group, among them Menachem Begin. Zionist Jews flooded into underground resistance organisations.

Jewish opposition to Soviet rule was especially pronounced in Lithuania. During the Smetona years, the city of Vilnius (Vilna) was a vibrant centre of Jewish life and culture. When Stalin agreed to transfer the city to neutral Lithuania, before annexation, many Jews in Soviet-occupied Poland hoped that Vilnius would offer an escape route from an increasingly dangerous Europe. 'Vilna fever' ignited a stampede of desperate Jews on trains, cars and wagons. A year later, when Lithuanian became a Soviet republic, the Vilnius door slammed shut. The old Polish city had become a trap and the NKVD turned its attention to 'counter revolutionary' Zionist Jews. This crackdown had no impact on the entrenched chauvinism of Baltic nationalists. They had noted only Jewish acquiescence or 'collaboration'. They had witnessed Lithuanian Jews welcoming visiting Soviet writers and artists. It is one of the ironies of this troubled period that Soviet Jewry, which had suffered a long religious and cultural decline since 1917, was revitalised through contact with other Jews they now encountered the Soviet 'sphere of influence'. This energetic fraternisation repelled Baltic nationalists and reinforced the mythological bond between Bolshevism and Jewry. This bond was, of course, the foundation of Nazi propaganda. In the Baltic States, nationalists plotted revenge – and German agents would actively promote their simmering resentment.

RSHA chief Heydrich diligently cultivated the prejudices of Lithuanian nationalists. In 1939, as Soviet forces began to occupy border strongholds, many Lithuanians in the army, government police and state security fled to Germany. Many made contact with German military intelligence and the SS. As Hitler began to prepare for Operation Barbarossa, Wehrmacht planners eagerly tapped Lithuanian expertise. On 17 November 1940 one of the most fanatical exiles, Kazys Škirpa, set up the Lithuanian Activist Front (LAF), which energetically pursued contacts with the SS. In Lithuania, the Germans began to arm local activist groups. Encouraged by the devious Heydrich, Škirpa and the Chairman of the LAF Propaganda Commission, Bronys Raila, drafted a proclamation of Lithuanian independence that is riddled with racist slurs and declares that Lithuanian Jews are 'outside the bounds of the law': 'Traitors [collaborators] will be pardoned only if they provide certain proof that every one of them has liquidated at least one Jew. The Jews must be informed immediately that their fate has been decided upon ... The crucial day of reckoning has come for the Jews.'[30] In his book *The Shoah in Lithuaniase*, Joseph Levinson reprints some of the LAF appeals that were widely distributed before the German invasion. They are drenched in ethnic hatred: 'Away with the Jews, Communists and Lithuanian Judases,' shrieks one. 'Let us liberate our Fatherland from the Jews,' demands another. 'We will rectify past mistakes and repay Jewish villainy.' Attacks on

'Jewish perfidy' far outweigh references to Soviet misdeeds.[31] At RSHA headquarters in Berlin, Heydrich and Škirpa jointly hatched up the idea of 'self-cleansing actions' that would provide a rationale for Lithuanian participation in German mass murder. Their secret agreement was then passed through a network of Lithuanian spies and informers attached to the LAF. This ensured that when Hitler's armies crossed the Lithuanian border on 22 June, Lithuanian activists were ready to act. Naturally Škirpa and his LAF friends expected to be rewarded for their zeal – and at this stage Heydrich was careful not to disillusion them.

Hitler's attack on the Soviet Union has been called 'the most appalling, devastating and savage conflict in the history of warfare'. Above all, it was an ideological crusade, a war of irreconcilable world views, a clash of races. 'And the hour will come,' Hitler ranted, 'when the world's most evil enemy of all time will have no further role to play for at least a 1,000 years.'[32] This eschatological logic appealed to Protestant clerics in Germany who dispatched a telegram to Hitler congratulating him for 'summoning our nation' to a 'decisive passage of arms' to 'eradicate the source of this [Bolshevik] pestilence'. In Moscow, a traumatised Stalin fled to his dacha for two days, either in a funk or to test the loyalty of his satraps – and moaned that 'we [Lenin's] heirs have fucked up [his inheritance]'.[33]

On the Baltic front, twenty-nine Soviet infantry divisions, four cavalry divisions, four armoured divisions and armoured brigades faced von Leeb's Army Group North, which included the SS Death's Head division. In the build up to 22 June, German commanders had done little to conceal the masses of German troops crossing the Nemen to reach their assembly points or furious bridge-building activity. This made Soviet commanders on the front line increasingly anxious, but Moscow appeared blithely unconcerned and even ordered the withdrawal of some frontier divisions. German strategy relied on fast flanking movements, spearheaded by panzers that could race deep and fast into enemy territory. Russian forces were thus divided and chopped into pockets to be mopped up by a second German wave.

The stunning surprise of the German attack (which was also a political shock) overwhelmed Soviet forces. During these first weeks, German advance divisions sometimes advanced as much as 50 miles a day. But the strategy of encirclement left in its wake an archipelago of intact Soviet strongholds. Many put up fierce resistance. One such was the city of Gargždai (Garsden) on the Lithuanian border, where German troops fought a protracted and bloody battle to crush fanatical Soviet troops.

When Stahlecker's Einsatzgruppe A men arrived hard on the heels of Wehrmacht troops, this would be site of the first *Judenaktion*. And this time Himmler had made sure that his men could get on with the job without any whining from weak-minded army generals. His confidence that SS shock troops would not be hindered was well founded. In the months leading up to Operation Barbarossa, Himmler and Heydrich had wrung crucial concessions from Wehrmacht Quartermaster-General, Major (General Staff) Hans-Georg Schmidt von Altenstadt, concerning the 'execution of political tasks'. These hard won agreements authorised the SD Special Task Forces to 'carry out on their own responsibility, executive measures concerning the civilian population'. Schmidt von Altenstadt agreed too that the SS commandos could carry out 'special tasks' not only in rear areas but close to the front line. For Himmler, this was a decisive breakthrough. His SS militias, instead of being confined to the rear, had secured a place on what Heydrich called 'the fighting line'. Combat and ideology could be inextricably woven together. Wehrmacht negotiators had not been innocent dupes. Schmidt von Altenstadt accepted that while the army would 'fight the enemy into the ground', it was also necessary to fight 'a political police struggle against the enemy'. Co-operation between the two wings of the German assault would guarantee the 'final liquidation of Bolshevism'. He listed politically dangerous individuals: 'Jews, émigrés, terrorists, political churchmen'. This threat warranted measures of 'extreme hardness and harshness'.

It would be a mistake to view this rapprochement merely as a shotgun wedding. At a meeting with army commanders-in-chief on 30 March Hitler insisted that Operation Barbarossa, like the Polish campaign, must be grasped (according to Halder's notes) as a 'war of extermination'. 'We do not wage war,' he continued 'to preserve the enemy' – otherwise Germany would need to fight the same battles all over again in a few decades times. Since Bolshevism was by definition a criminal regime, the German army must be freed from any legal restraints: 'This is no job for military courts.' The majority of army top brass led by Dr Rudolf Lehman, head of the Wehrmacht's legal department, completely agreed. Between March and June, a stream of decrees and army ordinances created the conditions that would transform traditional Prussian ruthlessness into barbarism. These reached a climax on 6 June when the OKW issued a draft of 'Guidelines on the treatment of political commissars', the so-called 'Commissar Order'. This repellent document sanctioned the execution of 'exponents of the Jewish-Bolshevik system' either at the moment of capture or as soon as possible at POW collection points. It was widely recognised by the OKW as well as Dr Lehmann that the Commissar Order defied international law. This was justified by means of grotesque sophistry: political commissars of all kinds, the drafters argued, were 'originators of barbaric Asiatic fighting methods', and thus had thus had placed themselves beyond the reach of 'the principles of humanity or international law'. We should note that an army education pamphlet described these 'commissars' as 'mostly filthy Jews'. This then was how the German Wehrmacht and the SD militias would fight a 'war of annihilation'.[34]

On the eve of the German invasion, Himmler and Heydrich met with the head of the Order Police, Kurt Daluege, to co-ordinate strategy along the front line. In Tilsit on the East Prussian border, where Army Group North awaited the signal to march into Soviet Lithuania, SS General Hans-Adolf Prützmann, HSSPF for 'Northern Russia' (which included the Baltic States), had overall command of the SS militias. Himmler would soon discover that Prützmann was rather too squeamish for the task in hand – and he came to rely on the hard nosed Stahlecker. Stahlecker arrived in Tilsit on 24 June and was informed that Gargždai had been chosen for the first *Judenaktion*. Stahlecker in turn communicated this decision to his subordinates: SS-Major Dr Martin Sandberger in charge of Sonderkommando Ia and SS-Colonel Karl Jäger heading Einsatzkommando 3.[35]

Stahlecker's Einsatzgruppe A was the biggest of Heydrich's murder squads, totalling 990 personnel. According to Hilberg,Einsatzgruppen A included 340 Waffen-SS, 172 motorcycle rider, 18 administrators, 35 SD men, 41 criminal police, 89 state police, 87 auxiliary police, 133 Order Police, supported by 13 female secretaries and clerks, as well as teletype and radio operators.[36] Each Kommando crossed the Soviet border equipped with trucks and motorcycles, shovels to dig mass graves and an extensive armoury of Lugers, Bergmann machine pistols, hand grenades and Walther P-38 pistols (considered ideal for administering the *coup de grâce* in the back of the neck). Every member of the Task Forces, Heydrich had confided to SS spymaster Walther Schellenberg, 'will have the opportunity to prove himself'. When one SD man realised the meaning of 'special tasks', he stammered '*Du bist ja verrückt!*' ('You must be mad!') His informant replied, '*Ihr werdet ja sehen*' ('Wait and see').[37] The most dedicated became known as 'Dauen-Schützen' (permanent shooters). Extinguishing so many lives required only the most rudimentary skills. Thanks to Heydrich's deal with the LAF, Lithuanian auxiliaries usually dug a pit to German specifications and then helped round up local Jewish men. The Germans then rushed their victims to the edge of the pit: 'Hurry up Isidor [anti-Semitic term]! The faster you go the sooner you will be with your God.' Then: 'Gustav, shoot well!' Later when a few recruits fretted about the day's work, an officer bucked them up: 'For God's sake don't you see? One generation has to go through all of this, so that our children have it better.'[38] So began the catastrophe that would engulf so many tens of thousands of Lithuanian, Latvian and Estonian Jews.

In a few Lithuanian villages and small towns, the general population reacted to news of the German 'liberation' by turning with horrible savagery on their Jewish neighbours. On the roads out of Vilnius, Polish peasants ambushed and killed Jews fleeing the city. Others set fire to synagogues and burnt Torah scrolls. They plundered homes. They killed. Some accounts of the Holocaust give the impression that the majority of ordinary Lithuanians and later Latvians and Estonians, when they had the opportunity, took part in killing sprees. To accept this would be to fall for Heydrich's carefully laid trap that pogroms must be made to appear spontaneous. The SS managed mass murder because they had learnt that 'cleansing' could not be delegated to the mob or unreliable national militias. Surgery not butchery – the 'Stahlecker reports' reveal how this lesson was applied 'in the field'.

In the last week of July, Stahlecker reached Kaunas. He informed Berlin that 'To our surprise, it was not easy at first to set in motion an extensive pogrom against Jews'. What did Stahlecker mean by these puzzling words? In the first place, 'to our surprise', he is implicitly criticising SS experts who almost certainly exaggerated the level of anti-Semitic hostility in the wider Lithuanian community. He makes the same point more than once: 'Native anti-Semitic forces were *induced* to start pogroms … but this inducement *proved to be very difficult*.' But Stahlecker boasted that he quickly came up with a solution: 'every attempt was made from the start to ensure that *reliable elements* in the local population participated in the fight against the pests in their country, that is the Jews and Communists.'[39] Instead of relying on the general population, Stahlecker turned not to any passing Lithuanian but to known activists – who had, as we have seen, already been informed of Heydrich's pact with the LAF.

Before they arrived in Kaunas, Stahlecker and his adjutant SS-Sturmbannführer Horst Eichler made contact with Iron Wolf stalwarts Major Kazys Simkus and Bronius Norkus, who had founded the Voldemaras partisans to harass the Russians. They came to the meeting wearing Lithuanian army uniforms and it was evident that they had assumed that the Germans would assign them to a military division.[40] Stahlecker reported that he saw straightaway that this would be a mistake. If partisans fought alongside German soldiers against a foreign power, it implied that they had been accepted as military allies – and thus as a Lithuanian army, which could be used to legitimate a sovereign Lithuanian state.

Stahlecker had every reason to be cautious. The Germans had been ambushed by Lithuanian nationalists right at the start of Operation Barbarossa. On 23 June, as the Soviet authorities fled Kaunas, Lithuanian LAF gangs seized the radio station and announced that a provisional Lithuanian government had been formed under Juozas Ambrazevičius-Brazaitis and called on Lithuanians to 'extirpate the Soviet regime'. Even in Polish-dominated Vilnius, the German 'liberators', whose tanks arrived in the night, inspired wild enthusiasm. A rash of Lithuanian flags erupted and radios played the old national anthem ad nauseam. A Citizens' Committee was set up to press for independence. Sideswiped by the LAF, the German military authorities reacted with shrill indignation: Lieutenant-General Wilhelm Schubert protested that the Lithuanians had the effrontery to regard themselves as 'equal partners in the territory liberate from the Russians'. They plainly had the impression, he blustered, that Germany had 'only gone to war' with the Bolsheviks to grant Lithuania independence! When Schubert met the former Lithuanian Foreign Minister he made sure he knew who was in charge – and he had the tanks to back him up. The new Lithuanian state was strangled at birth.[41]

Despite this bitter disappointment, Stahlecker had little difficulty harnessing nationalist energies – for his own strategic purposes. He ferreted out 'reliable elements' that would do his bidding. Assisted by ethnic German Richard Schweizer, who spoke fluent Lithuanian, he sidelined Major Simkus and the Iron Wolf army faction and turned instead to radical journalist Algirdas Jonas Klimaitis who was well known as a self-proclaimed radical anti-Semite. Stahlecker authorised Klimaitis and a 'Dr Zigonys' to recruit dependable types as auxiliary policemen. In his report, Stahlecker states: 'Klimatis [sic] succeeded in starting a pogrom *on the basis of advice given to him* … in such a way that no German order or German instigation was noticed from the outside.' (My italics.) Stahlecker set his willing Lithuanian auxiliaries to work with breathtaking speed. Led by Klimaitis, the Lithuanian auxiliaries set fire to synagogues and houses in the old Jewish quarter. They began plundering homes and shooting down Jews caught in the street; on the first night alone, some 1,500 Lithuanian Jews were killed. On the second night, double that number.

Opportunity for plunder provided a powerful motivation (as it did for the German SD men and soldiers). After their owners had been murdered, auxiliaries looted homes and warehouses. They ripped valuables from bodies. On 28 June, Einsatzkommando 1B reported with satisfaction that 'During the last three days Lithuanian partisan groups have already killed several thousand Jews'.[42] Two days after arriving in Kaunas, Stahlecker had moved on to Riga in Latvia. The means of carrying out mass murder had now been settled. It was succinctly described in Einsatzgruppe Report No 21, sent to Berlin from Minsk on 13 July: 'By 8 July in Vilnius, the local Einsatzkommando liquidated 321 Jews. The *Lithuanian Ordnungsdienst* which was placed under the Einsatzkommando … was instructed to take part in the liquidation of the Jews. 150 Lithuanian officials were assigned to this task.'

The Diary of Herman Kruk, who lived in Vilnius, recounts the destruction of Lithuanian Jews day by day, as events unfolded. On 23 June, Kruk records that German bombers roared overhead and pounded the city all night long. A day later, he hears that 'the Germans push forward with dreadful force and thrust with enormous speed'. Many thousands of Jews, including Kruk himself, tried to escape either by following the fleeing Soviet troops or by booking passage across

Siberia to Vladivostok. But the Russians abandoned the trains and every escape route became barred: 'Today has turned me into an old man ... Everything is lost.' On 24 June, the Germans entered the city. Kruk then reports a new development: Jews are ordered to wear armbands, 10cm wide and worn on the right arm. In the streets, Lithuanian gangs and Poles rob and beat Jews. It sometimes seems, Kruk writes, 'as if whole streets scream'. Then in July, Kruk hears stories about Lithuanian 'Snatchers' (Yiddish *Hapunes*). At first these sinister figures appear only at night to roam the Jewish districts, seizing anyone unfortunate enough to run into them. They take people 'wherever they want'. A few days later, Kruk reports, 'snatching at night has become a frequent event'. On 17 July, Kruk fears 'The Snatchers are making progress ... carrying off entire courtyards'. 'Horror upon horror.'[43]

What was really happening in Vilnius? Who were those mysterious 'Snatchers'? The Einsatzgruppen Reports provide the answer. Between 24 June and 2 July two Einsatzkommandos, 9 and 7a, had arrived in Vilnius – and organised Lithuanian 'snatch squads'. These squads had begun kidnapping Jews and holding some in Lukiskiai prison. The rest they took to the forest near Ponary and executed them in shallow pits. The Germans had recruited their 'snatch' squads from members of an ultranationalist faction called the Ypatingas Burys (the special ones).[44] A Lithuanian witness described what took place: 'The Gestapo [i.e. SD] come in cars and stop in front of Jewish houses. They take out males and order them to bring along a towel and soap ... Groups of Polish and Lithuanian youths wearing white armbands appear in the street and snatch the Jews ... People call them *Hapunes*.' The Einsatzkommando leader Dr Alfred Filbert was under pressure; his task was to 'liquidate the Jews of Vilnius' and he was a competitive, driven man. He urged Lithuanians to ramp up their 'productivity' and organise more 'Jew hunts'. It was after this that, as Kruk recalled, the snatch squads appeared in daytime, surprising their victims and even preying on Jewish labour gangs recruited by the German administration. This provoked protests from German army officers who resented losing 'their Jews'. But Dr Filbert was not to be stopped. On 5 July, Kruk reported in his diary that many of those kidnapped had been taken to Lukiskiai. He travelled to the prison to try to find out what had happened to them. 'Many women,' he reported, 'congregate outside the prison.' Kruk soon discovered that large groups of prisoners had been led away 'in the direction of Ponary'.

The journey to Ponary was a death sentence.

Less than 10 miles south of Vilnius, and close to the road and rail links to Grodno in the former Russian zone of Poland, Ponary (now Panariai) was a bucolic patch of pine and beech forest. Before the German attack, Poles and Lithuanians both Christian and Jew had spent happy hours here picnicking and hiking. In the hot, dry summer of 1941 the Ponary forest became a 'Valley of Death'. In 1940, the Russians had dug deep pits in the forest for fuel tanks. These were 20ft deep and up to 150ft in diameter, and ringed by high earthen embankments that were bisected by primitive earthen passageways. In photographs they resemble Palaeolithic earthworks. Now in July 1941, these deep pits had become the fiefdom of SS-Obersturmführer Franz Schauschutz, who would turn the Soviet fuel pits into a mass grave.[45]

It was a Lithuanian 'snatcher' who had informed the Germans about the existence of the Ponary pits – and Ypatingas Burys murder squads had already carried out executions using machine guns. The Germans considered this a profligate waste of ammunition. Schauschutz would now apply proper German *Ordnung*. He forbade the use of machine guns. Only rifles could be used; and he taught his Lithuanian comrades how to site precisely and kill instantly, without wasting ammunition.

Schauschutz reorganised the way Jews arrived at the edge of the pits, and by doing so increased daily kill rates to a hundred Jewish men per hour. He set up a waiting zone where victims undressed and were relieved of their valuables. A German soldier from a motorised division witnessed a typical day's work. He observed a group of approximately 400 Jewish prisoners led from Vilnius to the execution site by Lithuanian civilians armed with carbines and wearing coloured armbands. At the edge of the two large circular pits, whose sides were braced with planks, an elderly man stopped and asked (in good German), 'What do you want with me? I am just a poor composer.' Two Lithuanian guards stepped forward and beat the man with such ferocity that he 'flew into the pit'.[46] A ten-man execution squad waited on the opposite embankment. In less than an hour, they had killed the entire convoy of Jewish prisoners. How can you do this, the German driver asked a Lithuanian: 'After what we've gone through under the domination of the Russian Jewish Commissars [sic] …we no longer find it difficult.' A second 'SD man' stood nearby guarding a landau coach drawn by two horses. Inside sat two elderly Jews; both were shaking violently. The first SD man made the terrified couple walk to the edge of the pit; one carried a towel and soapbox. The SD man shot them both in the head.[47]

We know the identity of at least one of the Lithuanians who served alongside German executioners at the Ponary site. His name was Jonas Barkauskas – though when he was arrested in 1972 he was using the Polish version of his name Jan Borkowski and was first trombonist in the orchestra of the Warsaw Opera.[48] Borkowski was born in 1916 and grew up in Vilnius, then Polish Wilno. He did not speak Lithuanian, and after the destruction of Poland in 1939, this deficiency led to Borkowski having problems finding work. He did speak Russian, however, and an ethnic Russian neighbour introduced him to the Ypatingas Burys. He now became Jonas Barkauskas and was assigned to guard duty at Ponary. To begin with Barkauskas escorted groups of Jews to the edges of the execution pits – but he was soon 'rotated' to join execution squads and began killing men, women and children. He began plundering his victims. He stole boots and a pair of dark green trousers that, together with a Lithuanian army cap, became a rough and ready uniform. He grabbed suits, fur coats, wristwatches, leather jackets – even some children's sheepskin coats that he gave as presents to his nieces. After many weeks of dedicated and profitable service at the execution pits, Barkauskas won a transfer to the Ypatingas Burys headquarters where he processed the goods stolen from murdered Jewish families. At his trial, many decades later, Borkowski explained that he saw Jews as parasites and had no difficulty carrying out his duties.

A more privileged species of collaborator was Antanas Gečas-Gecevičius who would end his life as 'Anthony Gečas' in Edinburgh, Scotland, in 2001. He too was born in 1916 into a family of prosperous landowners and attended the prestigious Lithuanian Military Academy. Gečas was an out-and-out opportunist. When the Soviets occupied Lithuania, he joined the NKVD as an undercover agent and worked in western Lithuania as a police spy. But just days after the Germans arrived, Gečas signed up for the German sponsored 'Battalion for the Defence of National Labour' which soon became a Schutzmannschaft (protective battalion). No doubt fearful that his services to the Soviets would be discovered, he sent an obsequious letter written in German to the local commander claiming that he was descended from old German stock and was dedicated to serving the greater glory of the Reich. His trick worked. By the time the new Lithuanian security police, the Saugumas, issued arrest warrants, Gečas was serving under General Baron Gustav von Bechtholsheim in Belorussia, who was in charge of 'Operation Free of Jews' (*Aktion Judenrein*). The plan was to amalgamate the German 707th Division with Gečas' Lithuanian 2nd Schuma Battalion and the German 11th Reserve Police Battalion to liquidate every Jewish family in western Belorussia – they must 'disappear without trace'. Like many German officers, von Bechtholsheim was convinced that Soviet partisans and Jews were inextricably connected. Since Gečas spoke excellent German, he received orders directly from German officers and took a leading role in *Aktion Judenrein*. At Slutsk, the Lithuanian 2nd Battalion and German police rampaged though the village shooting and beating Jews and local Belarusians. After that first assault, the Lithuanians rounded up Jewish families and herded them to pits on the edge of the village. Bellowing and screaming in German and Lithuanian, Gečas organised the killing, rotating platoons and personally shooting anyone who remained alive after the first volleys. Gečas was later awarded an Iron Cross.

In the former Lithuania, the Saugumas was the main agency that brought together German administrators and local collaborators. A decade older than Gečas, Aleksandras Lileikis had been born into a peasant family but managed to get into university to study law. He was forced to continue his studies part time and joined the Saugumas in 1939. In 1940, when the Soviet occupation began, Lileikis and other Saugumas officers escaped across the border and fled to Berlin where he applied for German citizenship and remained until August 1941. We have no documented records but it is almost certain that Lileikis received detailed briefings about the role of the Saugumas once the Germans had expelled the Russian occupiers. When he returned to Vilnius he immediately assumed responsibility for the city's security and reorganised the Saugumas along Gestapo lines, setting up a special division to deal with Jews and communists – the Komunistų-Žydų Skyrius. His main job was to deal with escapees from the Vilnius ghetto. Records show that Lileikis issued a succession of orders handing over captured Jewish men, women and children to the German security forces. One was a 6-year-old called Fruma Kaplan (b. 1935), who would be 'treated according to orders'. Fruma was shot with her family at Ponary on 22 December 1941.

Himmler's SS closely managed the mass murder of Jews in Lithuania. But this fact does not exonerate. As a Christian doctor Elena Kutorgiene wrote after the war: 'With the exception of a few individuals, all the Lithuanians, especially the intelligentsia, hate the Jews … The coarse Lithuanian mob, as opposed to the total apathy of the intelligentsia, acted with such beastly cruelty that by comparison the Russian pogroms seemed like humanitarian deeds.'[49]

5

Massacre in L'viv

> The Reichsführer-SS is not willing at this stage to take any actions regarding the combat training of these [Ukrainian] men.
>
> Rudolf Brandt, 31 April 1941

On 30 June 1941, at 4.30 in the morning, a Ukrainian battalion sonorously named the 'Nachtigall', recruited by German military intelligence and wearing Wehrmacht grey uniforms, marched into the city of L'viv (then known as Lemberg), once the capital of the old Austrian province of Galicia. They yearned to free their homeland from the Soviet yoke and arrived just hours before the first German units. In the city, chaos reigned. Between parked military vehicles, people swarmed waving blue and yellow Ukrainian flags. They shouted wildly and fired rifles and pistols with abandon. In the old Austrian office buildings, Soviet files lay strewn on the floor or flowing into the streets to be tramped underfoot by men wearing blue and yellow armbands. Officers of Einsatzgruppe B had moved into a building vacated by the Soviet NKVD just days before. Everywhere Ukrainians in old Austrian uniforms attacked Jews – '*Yids! Yids! Kaputt!*' – to the ubiquitous sound of the national anthem played on accordions.

The commander of the Ukrainian battalion, Roman Shukhevych, had family in L'viv. He soon discovered that the body of his brother lay rotting inside Brygidki prison, murdered by the NKVD along with thousands of other Ukrainian victims of the Soviet terror. Many Jews had also been murdered by the retreating Russians. Their corpses too lay in the prison yard and cells. But Shukhevych had no doubt who was responsible for the death of his brother. Jews were the agents of Moscow and must be punished.

A few days later, German soldiers exhumed some of the bodies and publicly displayed the rotting corpses. 'Murdered by Jews!' Soon the 'Nachtigall' men forgot all about liberating their homeland – and began to round up Jews wherever they found them. They murdered men, women and children as a German military cameraman calmly recorded the unfolding bloodbath.

The massacre in L'viv appears to confirm one of the enduring stereotypes of the Second World War – and the destruction of European Jewry – that is summed up by the phrase: 'The Ukrainians were the worst!' All over Eastern Europe, as Jews and other victims of the genocide met their terrible ends, survivors reported that Ukrainians always behaved 'worse than the Germans'. At every murder site and in the extermination camps like Treblinka and Sobibór, it would seem that Ukrainians barked and bellowed the last words heard by victims as they disembarked the death trains. Ukrainian guards and police showed themselves to be uniquely cruel, brutal and merciless. In the chapters that follow, we will encounter many such murderous Ukrainians. But as in the case of the Baltic States, a more complex historical narrative lies beneath the surface. Under German rule, millions of Ukrainians perished or were enslaved. Thousands of Ukrainian partisans fought German soldiers and SS men (as well as Poles and Jews). The barbarism of some Ukrainian guards and police auxiliaries cannot be generalised to all Ukrainians – just as crimes committed by Soviet agents of Jewish origin cannot be blamed on 'the Jew'.[1]

This point should be made more precise. After 1941, the German occupiers recruited auxiliary police battalions from all over Ukraine. But they did not treat this immense region of Eastern Europe as a single homogeneous entity. The ideology of occupation policy, often referred to mistakenly as 'chaotic', reflected German racial speculation that split the Ukrainian lands into two broad enclaves: the semi-Germanised lands west of the Dneiper River and the Slavic east. For this reason, 'elite recruitment' by both the Abwehr and later the SS targeted a specific region of Ukraine region that had once been the 'Kingdom of Galicia-Volhynia' and, many centuries later, an Austrian province. To understand German occupation and recruitment strategy, we need to decode the enigmatic territorial concept of 'Galicia', whose schizophrenic identity remains a contentious matter to this day.[2]

Today Galicia, or in Polish 'Halychnya', has no independent political status. It is a kind of shadow land split between modern Poland and Ukraine. There are no Galicians distinct from Poles and Ukrainians. In the early Middle Ages, a Kingdom of Galicia or Halych-Volhynia flickered intermittently into life. It reached an apogee under King Danylo in the thirteenth century (his son Lev established the Galician capital at L'viv) but was in less than a century being tossed back and forth between more powerful neighbours above all Poland and Kievan Rus'. The latter, broadly speaking, was the ancestral form of the modern state of Ukraine. The history is fiendishly complicated and need not detain us here. By the sixteenth century, Galicia had been absorbed by the Polish-Lithuanian Commonwealth. Throughout this turbulent history, impoverished Prussian Germans had migrated to Galicia to farm its rich, dark soils and happily intermarried with their Slavic neighbours. This rich ethnic mix was frequently stirred and reworked over time. To bolster their new state and mercantile ambitions, Polish rulers encouraged enterprising Yiddish-speaking Ashkenazim Jews who migrated to the Galician region from Silesia, Bohemia, Moravia and other less tolerant 'Germanic' lands. Sephardic Jews that had been expelled from Spain and Portugal settled here too, transforming Galicia into one of the most vibrant centres of Jewish commerce and religious culture. In the mid-eighteenth century, the empires of Prussia, Russia and Austria ripped apart the Commonwealth. The 'Kingdom of Galicia and Lodomeria with the Duchies of Auschwitz and Zator' became the northernmost province of the Austrian Empire. And so Galicia remained until the end of the First World War, when once again the old kingdom became the fulcrum of violent conflict and ethnic upheaval.

For the representatives of the victorious Entente Powers who gathered at the Paris Peace Conference in 1919, these relics of the Russian and Austrian empires presented a mighty challenge. British Prime Minister Lloyd George said that 'Russia was a jungle in which no one could say what was within a few yards of him'. President Wilson lamented that 'Russia ... goes to pieces like quicksilver under my touch'.[3] After the Paris Conference, the three Baltic States won fragile independence. But not one of the Entente Powers was prepared to back an independent Ukraine. They pressed instead for a unified Russia, ruled by an anti-Bolshevik government and with the Ukrainian lands securely bolted to its western border. 'I only met a Ukrainian once,' opined Lloyd George, '... and I am not sure that I want to see any more.' For the lobbyists and negotiators gathered in Paris, western Ukraine threw up the most acute difficulties. Since Galicia was former property of the now defunct Austro-Hungarian Empire, the Entente Powers had to choose a new owner: but which one? Few disputed that western Galicia was Polish in character, but the east was less tractable. While 'Ruthenians' dominated the countryside, the big cities like the capital L'viv and Tarnopol resembled Polish islands in a Ukrainian ocean. Overall, Poles made up just one-third of the East Galician population; Jews just over a tenth – so the majority was indisputably Ukrainian. This ethnic patchwork is still evident today in modern L'viv, 50 or so miles east of the modern Polish-Ukrainian border. The Austrians erected grand state buildings, expansive parks and a handsome opera house – and often referred to 'Lemberg' as 'Little Vienna'. For their part, Polish Catholics erected scores of churches, whose spires still bristle along the skyline. Further out from the Austrian centre, Galician Jews constructed some of the most impressive synagogues in Eastern Europe, like the famous Di Goldene Royz, which was designed by an Italian architect, and the Reform Synagogue near Market Square. After 1941, the Germans destroyed both of these imposing structures; today only ruins remain. Outside the cities, Ukrainian peasants and the majority of rural or Shtetl Jews endured chronic poverty. In the mythology of Ukrainian victim-hood, these same Jews were to blame.

For the new Polish government, the idea of a Ukrainian *state* was a joke. They also coveted the oilfields that lay beneath Galicia, close to L'viv. This is why, as the Entente Powers debated the fate of eastern Galicia, the Poles resolved the matter on the ground – by occupying eastern Galicia and neighbouring western Volhynia. Polish farmers seized Ukrainian farms, harassed peasant farmers and provoked violent responses from Ukrainian factions, equipped with arms pilfered from German stock. The 'Galician question' remained open when the peace negotiators were wrapped up in Paris and the problem was taken up by the new League of Nations. But relentless 'Polonisation' proved hard to resist and so in 1923, de facto annexation was internationally sanctioned by the League's Council of Ambassadors. Ukrainians, along with the Arabs, were the main losers at a succession of post-war settlements. The old empires had collapsed – but the Ukrainian ethnic territories remained divided between foreign powers. East of the Zbruch River, the Bolsheviks created a new state entity, despised by nationalists as a Soviet puppet: the Ukrainian Soviet Socialist Republic (UkSSR). The militant nationalist organisations melted away underground; their leaders fled west to Vienna and Berlin. Here they began to build ties with German and Austrian ultranationalists, including members of Hitler's new NSDAP. Two decades later, Ukrainian nationalists would make a sublimely foolish mistake when they turned for assistance to a resurgent German Reich.

As Sol Littman, one of the first historians to investigate the record of the 'Ukrainian' SS division, puts it, 'Whatever quarrels existed between nationalists and federalists, Ukrainians and Russians, peasants and aristocrats, hatred of the *zhidy* (the Jews) was common coinage'.[4] Ever since the seventeenth century, when Count Bohdan Khelnitsky's Cossacks rampaged through the Russian 'Pale of Settlement' murdering more than 200,000 Jews, violence fell frequently and hard on the Shtetls of Ukraine. Any social upheaval, whatever its cause, provoked pogroms. It was a pattern repeated at the beginning of the twentieth century after the first Russian revolution, when pogroms flared all across Ukraine and Bessarabia. In Ukraine after 1917, civil war brought a fresh wave of attacks. Trotsky's Red Guards, White Russian brigades, Nestor Makhno's anarchists, a fledgling Ukrainian army led by Symon Petliura, mobs led by fickle Ukrainian chieftains or 'Atamans' – all at one time or another took up arms against Ukrainian Jews. Only the Bolshevik officers defended them, inadvertently reinforcing that pernicious axiom that Bolsheviks and Jews were one and the same and that communist Jews planned to enslave the Slavic peoples of the east. It was this chimera of the 'Jewish-Bolshevik', hatched in Paris but nurtured in the court of the Russian tsar, that Alfred Rosenberg brought to Munich and the meagre ideological coffers of the NSDAP. When it came to Jews, Ukrainian nationalists and Hitler's Nazis shared the same currency of hate.

From the mid-1920s, a number of belligerent Ukrainian factions jostled for favour with the German far right. These included Petliura's UNA government based in Warsaw in exile; the so-called 'Hetmanites' in Berlin, who took their name from Pavlo Skoropadsky who had been briefly installed by the Germans as a puppet Hetman (ruler) in 1918; and the rather obscure Organisation of Ukrainian Nationalists, the OUN. Led by Colonel Ievhen Konovalets, the OUN was to begin with a fierce little splinter group devoted to terrorist attacks on Poles and Jews. It was authoritarian, anti-democratic, anti-Bolshevik and anti-Semitic. But in 1928, an event took place that pushed the OUN to centre stage. On 25 May, Samuel Schwarzbart, who had lost family members in the Ukrainian pogroms, approached Petliura as he walked down a street in Paris and shot him dead. He gave himself up immediately to the police. In court, Schwarzbart pleaded that he was taking revenge for the murdered Ukrainian Jews, including his own kin. In a shock verdict, he was acquitted.[5] The news delighted Ukrainian nationalists. It seemed to authenticate both the perfidious power of Jewry and the clout of the Soviet regime. It was widely assumed that the Jewish Schwarzbart was a Russian agent and that his French legal advisors were puppet actors in a global conspiracy. The radical OUN was uniquely placed to take advantage of this surge of anti-Jewish and anti-Russian revulsion that swept through the Ukrainian nationalist movement. 'The Jews are guilty,' ranted OUN spokesman Dmytro Dontsov, 'horribly guilty because they were the ones who helped secure Russian rule in the Ukraine.'[6]

Colonel Konovalets convened the First Great Congress of Ukrainian Nationalists in Vienna that same year, and the OUN noisily unfurled its ideological colours. The OUN was unashamedly a terrorist organisation: violence dominated its vaguely defined ideology, legitimated by the image of the 'Apostle of Battle' with a sword firmly grasped in his hands.[7] Leaders of the OUN's youth movement published 'Ten Commandments' to guide new recruits. The first commandment was 'Attain a Ukrainian state or die in battle for it'. Others were: 'Regard the enemies of your

Nation with hate and perfidy'; 'Do not hesitate to commit the greatest crime'. This was murderous occultism rather than ideology. Its source was Dmytro Dontsov, an inspiration still for Ukrainian neo-Nazi skinheads, who advocated strategies that were 'irreconcilable, uncompromisable, brutal, fanatic and amoral'. He believed on the basis of unspecified scientific 'measurements' that Ukrainians were biologically superior to other Slavs and thus destined to rule 'inferior races'.

In 1920, Galicia was absorbed by Poland and suffered an especially vicious bout of 'colonisation'. The OUN unleashed a terror campaign that would be resumed with ever greater savagery after 1943. Poles bore the brunt of Ukrainian national terrorism but according to Betty Einstein-Keshev, who is quoted by Littman, OUN leaders 'wanted to finish off all their enemies at once'. Their method of waging war 'was murder and destruction with no quarter shown. The independent Ukraine promised death and destruction to Jews.' Hatred of Jews dominated OUN thinking; according to OUN ideologue Professor Mycjuk, in a polemic published in 1932, Jews were dangerous because they 'did business and made children'.[8] OUN style nationalists regarded Jews, Poles and Bolsheviks as forming a hydra-headed monster that would be forcefully denied a place in a 'Ukrainian Nation'. In German, OUN ideology was summarised in a single word: *Pogrompolitik*.

For many OUN activists forced into exile, Vienna and Berlin became sanctuaries of choice. After 1933, German foreign affairs and intelligence experts began to cultivate suitable Ukrainians like the OUN leader Konovalets, who shared German loathing of Poland, and the '*mosko-jüdischer apparat*'. Konovalets had served in the short-lived 'Ukrainian Army' in 1918, and remained on good terms with officers in the Reichswehr. They provided him with contacts inside the army's military intelligence wing, the Abwehr. Abwehr head Admiral Wilhelm Canaris continues to puzzle historians. He was close to Himmler and a personal friend of Reinhard Heydrich. But Canaris loathed Hitler and allegedly passed secret information to the Allies. Certainly Hitler believed so and in 1944 had him arrested and executed – in the same prison, in the same brutal manner and on the same day as Pastor Dietrich Bonhoeffer. But Canaris, unlike Pastor Bonhoeffer, was neither honourable nor righteous. Like many other 'resistors', he was a radical – if not rabid – conservative, and enjoyed fraternising with the nastier species of émigré nationalist. It was said that he liked the 'smell of reaction' and the OUN clearly emitted just the right kind of odour. Canaris liberally poured Abwehr cash into OUN coffers in exchange for intelligence about the Polish army. In 1938, Konovalets was assassinated by a Russian NKVD agent – but the flame was swiftly passed to war veteran Andreii Melnyk and his younger rival Stepan Bandera, who was then locked up in a Polish prison cell. Canaris would remain the OUN's staunchest backer. The rest of the Nazi elite proved more fickle. Alfred Rosenberg funded anti-Bolshevik factions through his sham foreign policy think tank, the Foreign Political Office (*Außenpolitische Amt der NSDAP*), but he distrusted the OUN on the grounds that it was too rooted in 'Austrian' Galicia. For his part, Hitler had no scruples about trampling over Ukrainian nationalist sensibilities even though, as an Austrian, he was well versed in Galician history. In 1939, his betrayal of the semi-autonomous province of Carpatho-Ukraine made plain his aversion to any kind of Slavic nationalism.[9]

At the end of the First World War, this splinter of land on the Hungarian border sometimes called 'Ruthenia' had been granted to Czechoslovakia, but promised autonomy. Carpatho-Ukraine was viewed by the OUN as an embryonic or proto Ukrainian state, but it was also coveted by the Hungarian government, which had close ties to Germany. On 14 March 1939, after months of agitation, the Soym or Diet of Carpatho-Ukraine declared total independence. Before dawn on the following day, Hungarian troops crossed into Carpatho-Ukraine and were met with fierce resistance from the poorly equipped 'Carpathian Sich' and OUN volunteers. Later that same day, a gleeful Hitler entered Prague at the head of German forces. Meanwhile, the Soym urgently requested protection from the Reich. Hitler refused outright and ordered the Ukrainians not to oppose the Hungarian troops. After five days, the independent state of Carpatho-Ukraine was completely erased.

A handful of OUN fighters later denounced Hitler 'the well known carnivore ... and sworn enemy of the Slavic race', but the Germanophile OUN leadership still refused to properly learn their German lesson. As we saw in Chapter 1, as Hitler prepared to invade Poland, Colonel Roman Sushko, encouraged by Canaris, volunteered the services of a Ukrainian brigade of approximately 200 men. Sushko and the OUN leaders hoped that once Poland had been defeated, the Germans would hand over Galicia. The Abwehr secured the release of Ukrainian POWs held in Hungary, then welded them together with OUN exiles based in Germany to form a 'Nationalist Military Detachment' (*Viis'kovi viddily natsionalistiv*) – known to the Germans as the 'Bergbauenhilfe'.

On 15 August, training began in Slovakia. Then three days later, OUN leader Andrii Melnyk received a summons from an Abwehr officer to 'hold himself in readiness in case the political situation would demand' his presence. Melnyk's reaction is not documented – but if he assumed that he was about to be made leader of an independent Ukraine his hopes would have been swiftly dashed. For on 23 August, Ribbentrop and Molotov signed the Non-Aggression Pact in Moscow.

The Ukrainians' German sponsors had made common cause with the hated Soviet foe. This was stunning news – and German intelligence agencies received orders to closely monitor the reaction of Ukrainian émigrés, who were forbidden to leave Germany. At the German Foreign Office, Melnyk was informed that the Reich could 'make no promises'. At their training camp in Slovakia, Sushko's volunteers found themselves being 'Germanised'. On 1 September, they crossed the Polish border, led by Abwehr colonel Erwin Stolze rather than Sushko. Once they had crossed the old Czech border in September, the Ukrainian legion fought not on the front line but as partisans, knocking out Polish communications and making unpleasant mischief behind enemy lines.

On 12 September, the Wehrmacht high command and Canaris met on Hitler's train (which was stationed at Ilnau in Silesia) to debate the fate of the Polish lands. They discussed a number of alternatives including declaring 'Galician and Polish Ukraine' independent.[10] A few days later, Canaris met with OUN leader Melnyk and informed him that a western, i.e. Galician Ukrainian state, was back on the German agenda. Melnyk rushed away to prepare a list of 'West Ukrainian' government officials, but two days later, the Soviet army advanced into eastern Poland, trampling into the mud all hope of an 'independent Galicia'. On 27 September, the Polish government capitulated and Ribbentrop dashed back to Moscow to agree additional protocol, establishing a definitive new border: the Narva-Buh-Bug-San line.

Once again, Hitler had dashed OUN hopes. The 'secret protocols' of the Nazi-Soviet Pact meant that western Ukraine and Belorussia now fell into Soviet hands. Stalin, in fact, unified all the Ukrainian lands as a Soviet republic. The 'Bergbauenhilfe' was hastily disbanded and the Ukrainian volunteers deported to German-occupied Poland. The whole episode was clumsily executed but set a clear pattern. If the Germans had a strategy to deal with military collaborators it went something like this: Armed non-Germans might be temporarily functional but, in the long run, would prove dangerous. The trick was to allow a little growth then apply the pruning shears. As we will see, even that limited relationship was a step too far for Hitler.

The slippery Canaris did what he could to mollify the disheartened Ukrainians. Once the Wehrmacht had secured Warsaw, he had Bandera released from prison. In Kraków, the Germans discovered a highly organised community of some 3,000 Ukrainian émigrés, most with strong German sympathies. As soon as Hitler had decided on the division of Poland, the German administrators in the General Government made sure that their Ukrainian subjects received a stream of privileges and favours at the expense of Poles and Jews. They set up a Ukrainian Central Committee (UTsK), to promote Ukrainian welfare. The UTsK was headed by Volodymyr Kubiiovych, a professor of geography who had strong ties to the OUN and would, in 1943, play a crucial role recruiting Ukrainians for Himmler's Waffen-SS. The UTsK foreshadowed later SS strategy by authorising Roman Sushko to turn his disbanded legion into a Ukrainian police force, which was eagerly seized on as a move towards national autonomy. As local administration posts in the General Government fell into Ukrainian hands, Sushko and Kubiiovych urged their German masters to transfer agricultural land from 'Jewish hands' to Ukrainian co-operatives, which grew significantly in number after 1939. In the short period before Hitler invaded the Soviet Union, German administration in occupied Poland effectively reversed two decades of 'Polonisation'. At the same time, Ukrainians seized Jewish businesses and helped build new forced labour camps and ghettos. Even this was not enough. In April, 1941, the UTsK head Kubiiovych approached General Governor Hans Frank to urge him to completely purge all 'Polish and Jewish elements'.[11]

But in the pressure cooker of the General Government, the OUN ruptured. Like every nationalist movement in history, it split between irreconcilable moderates and radicals. On one side, Andrii Melnyk, Konovalets' successor, advocated a gradualist approach. He had served in the Austrian army and was accustomed (like the Bosnian Muslims) to that long-vanished Hapsburgian munificence. A sovereign Ukraine, Melnyk believed, would be the reward bestowed for long service to Hitler's empire. But in 1940, his rival Stepan Bandera, who had made his mark assassinating Poles, impatiently rejected such abject kowtowing to a fickle and opportunist foreign despot.

The Bandera-Melnyk 'split', bitter though it was, was tactical rather than ideological. Melnyk and Bandera were both committed integral nationalists and anti-Semites who, in some form or other, wanted German National Socialist backing. In 1940, the firebrand Bandera set up camp as OUN-B while Melnyk's conservative supporters regrouped as OUN-M. In March 1941, at a congress in Kraków, the split became public. German intelligence followed events closely. A NSDAP foreign policy expert, Arno Schickendanz, warned both Canaris and SD chief Heydrich that the OUN was a 'purely terrorist organisation', with a 'Galician colouration' (meaning that its influence was confined to the western Ukraine) that had forfeited any influence over other Ukrainian nationalists after the Soviet cession of western Ukraine. The OUN, he recommended, should be banned.[12] Canaris stoutly defended the OUN: it was 'too early', he argued, to take a drastic measure which would have disastrous consequences for German relations with potentially useful Ukrainian émigrés. At the RSHA, Schickendanz's warning was noted, but thanks to Canaris, ignored. A split OUN might even be of greater use than a unified one. Melnyk and Bandera bickered and skirmished but neither abandoned core OUN doctrines. The glue that kept them both in the German camp was the lingua franca of the radical right: hatred of Jews. One OUN writer put it concisely: 'Long live greater independent Ukraine without Jews, Poles and Germans. Poles behind the San, Germans to Berlin, Jews to the gallows.'[13]

By the summer of 1941, Alfred Rosenberg, Hitler's self-appointed 'eastern expert', had begun hatching up convoluted plans for the administration of the European East. He proposed preserving 'national units' such as in the Ukraine – but, unknown to either OUN leader, these pseudo-nations would soon be swallowed whole by immense German-controlled administrative blocks called Reich Commissariats. In instructions issued to future commissars (the Germans preserved the old Soviet titles) Rosenberg referred to establishing a 'free Ukrainian state *closely linked* to Germany'. These viper words disguised the shabbiest window dressing. Hitler, as Rosenberg understood very well, would never recognise any kind of non-German national sovereignty in the east. All Rosenberg could offer was a kind of wishy-washy status as vassal nations under German suzerainty. These temporary 'national units' would eventually vanish. In a few decades, any national identity would have been dissolved in the acid bath of German occupation. For all his posturing as a defender of anti-Bolshevik national identities, Rosenberg never doubted the Nazi maxim that the outcome of conquest would be 'the total destruction of the Judeo-Bolshevik administration' and the 'vast exploitation' of former Soviet lands – above all the rich Ukrainian farms.[14]

Rosenberg did not, of course, share these plans with Ukrainian émigrés. In ignorance of German intentions, in the spring of 1941 as Hitler became increasingly bellicose towards Stalin, OUN leaders began to plot ways of exploiting a future German attack on the Soviet Union. Both Melnyk and Bandera sent pleading memoranda to the Reich Chancellery urging the formation of Ukrainian military units to be trained by the Wehrmacht. OUN-B representative Colonel Riko Vary met with Canaris and General Walther von Brauchitsch, who was also broadly sympathetic to Ukrainian aspirations.[15] After weeks of haggling, Vary secured an informal agreement with OKW 'eastern experts' Professor Hans Koch and Theodor Oberländer to begin training two battalions, mustering in total some 700 men. Oberländer we will encounter again. He was, like Himmler, a trained agronomist and had a long record of reactionary agitation. In 1935 he had been appointed assistant to Erich Koch, then Gauleiter for East Prussia, and charged with 'investigating' ethnic minorities on the Polish border. He was therefore considered an 'eastern expert'.

Canaris did not get or even seek Hitler's approval to begin negotiations with the Ukrainians, and he and the other Abwehr officers involved dealt exclusively with Bandera's more radical faction, OUN-B. The German officers insisted that the agreement be kept secret and Hans Koch warned Vary that, realistically, still developing German policy in the east might well end up thwarting their political goals.[16] There were risks – but 'nothing ventured, nothing gained'. It was imperative that Canaris keep the Russians in the dark, for the Non-Aggression Pact remained in force. He feared that Ribbentrop, the German architect of the pact, might betray the Abwehr plot. His protracted negotiations with Molotov had been the crowning achievement as Hitler's Foreign Minister and he had no desire to see his great scheme unravel so swiftly. It is almost certain that Heydrich and Himmler both knew what was afoot and it is not inconceivable that Canaris intended that the two Ukrainian brigades take the same 'self-cleansing' role as the LAF militias recruited in Lithuania.

The Ukrainians called the battalions '*Druzhyny ukraïnskykh nationalistiv*' (Units of Ukrainian Nations) or DUN. The Abwehr awarded them romantic sounding codenames: Organisation Roland (for the medieval French knight who died in battle against the Saracens) and Sonderformation Nachtigall (apparently the recruits enjoyed singing). Both were disguised as 'labour divisions' and trained in Austria and Silesia by Abwehr officers. Canaris assigned the 'Nachtigall' recruits to the 'z.b.V800 Brandenburg' – Special Task Forces or K troops, first deployed in Poland. Many 'Brandenburg' officers were Sudeten Germans or Polish Volksdeutsche. At the end of the Polish campaign, Canaris reformed the K troops as 1st Training Company (German company for special missions) based in Brandenburg-am-Havel – hence 'Brandenburgers'. The 'Brandenburg' resembled a German foreign legion or the British SAS and has acquired a spuriously romantic post-war reputation. Canaris recruited lower ranks from the Baltic, Romania, the South Tyrol, Africa, Palestine and even Australia. 'Brandenburg' commander Theodor von Hippel had fought with General Paul Emil von Lettow-Vorbeck in East Africa during the First World War. In 1914, the Germans had recruited tens of thousands of African mercenaries known as 'Askaris' (Arabic for soldier) – a term that would now be applied to Ukrainian and other Eastern European recruits. Hippel boasted that the 'Brandenburgs' 'could snatch the devil from hell'.[17]

Thousands of Ukrainian exiles had washed up in Vienna. It was in the old capital of the Austrian Empire that Vary found his 'Roland' recruits. Abwehr officers then transferred recruits to Saubersdorf in Austria for training. In Kraków, Roman Shukhevych, head of Bandera's military section, rounded up some 300 Ukrainians for training at Neuhammer in Silesia. According to the *Aufgaben für Ukrainer-Organisationen*, the task of the two battalions was to aid in establishing 'the marching security for German troops on grounds not occupied by the German military, especially by disarming Russians'.[18] The Ukrainians had been encouraged to believe that if they fought well, the two battalions might be amalgamated. The Abwehr issued the 'Roland' men with Czech uniforms that resembled those worn by Ukrainian soldiers in 1918. The 'Nachtigall' received Wehrmacht *feldgrau* uniforms with a blue and yellow shoulder badge. Vary, when he negotiated with the Wehrmacht, had insisted that Ukrainians recruits would swear allegiance not to Hitler, but to Ukraine and the OUN. Canaris had secretly recognised Ukrainian autonomy, in the service of the Reich.

Although Canaris' plans remained secret, Gottlob Berger, Himmler's opportunist recruitment chief, had also spotted an opening. By the end of April 1941 Berger had recruited ethnic Germans, Flemish, Dutch, Danish and Norwegian volunteers. Why not, he thought, grab pro-German Ukrainians too? On 28 April 1941 he wrote to Himmler proposing that a few hundred Ukrainian émigrés who spoke both German and Ukrainian might be recruited by the Waffen-SS. Rudolf Brandt, Himmler's secretary, replied with a terse note a week later: 'The Reichsführer is not willing at this stage to take any action regarding the combat training of these men.'[19] Himmler's response has been interpreted to imply that he was repelled by the idea of recruiting 'Slavic sub humans'. This is mistaken. Himmler was a keen student of history so it can hardly have escaped his attention that Galicia had been an Austrian *Kronland* and that some 'Ruthenians' must therefore have acquired an infusion of Nordic blood. Germans called the principal Galician city Lemberg. Modern-day L'viv was popularly known as 'Little Vienna of the East' and was home to more than 50,000 proud German language speakers. We can be certain that Himmler did not turn down Berger's proposal on racial grounds. But like Rosenberg, he was suspicious of the OUN – above all, Bandera's radical wing, though he sympathised with its rabid chauvinism. Brandt's qualification 'at this stage' is frequently overlooked. Less than two years after he rejected Berger's proposal in 1941, Himmler would authorise an SS recruitment drive in Galicia.

Far from playing the part of passive collaborators, the leadership of OUN-B harassed the German government throughout the period leading up to 22 June with memoranda that insisted on the primacy of Ukrainian interests and warned of the consequences if these were ignored.[20] Himmler's priorities had to do with police operations and pacification; as Heydrich made clear in a letter to the HSSPF, the extermination of 'undesirable' elements had to be pursued with 'ruthless vigour'.[21] In central and northern Ukraine, this would fall to Einsatzgruppe C. Like the commanders of Einsatzgruppe A in the Baltic, the SD commanders assigned to Ukraine had received Heydrich's instructions to incite local pogroms and to use local activists 'to attain our goals'. It was on this point that German and Ukrainian interests converged. At an OUN gathering in Kraków, Bandera proposed:

> The Jews in the USSR constitute the most faithful support of the ruling Bolshevik regime, and the vanguard of Muscovite imperialism in Ukraine. The Muscovite-Bolshevik government exploits the anti-Jewish sentiments of the Ukrainian masses to divert their attention from the true cause of their misfortune and channel them in time of frustration into pogroms on Jews. The OUN combats the Jews as the prop of the Musovite-Bolshevik regime and simultaneously it renders the masses conscious of the fact that the principal foe is Moscow.[22]

Then two weeks before the scheduled date for the invasion of the Soviet Union, Bandera's 'moderate' rival Melnyk sent a telegram to Hitler, asserting that he alone could best represent German interests in Ukraine. Like so many other nationalist supplicants who banged on the door of Hitler's Reichs Chancellery, Melnyk received no reply. On the eve of Operation Barbarossa, he made a second appeal demanding to march shoulder to shoulder with the Wehrmacht, to build a new Europe 'free of Jews, Bolsheviks and plutocrats'.[23] An opportunity to begin fulfilling that rabid dream would come soon enough. On 18 June, its training completed, the 'Nachtigall' battalion crossed the territory of the General Government and arrived in Przemyśl, which, before the German invasion, was split between the Reich and the Soviet Union by the San River, which flowed through the middle of the city. That summer, the waters ran low and sluggish between muddy banks that glittered in the mid-summer sunlight.

On the night of 22 June, the Ukrainians crossed the San into Soviet territory, meeting no opposition, and began marching towards L'viv. The 'Nachtigall' was commanded by OUN leader Roman Shukhevych – but his orders came from German Abwehr Oberleutnant Dr Hanz-Albrecht Herzner Oberländer. Shukhevych said later that he felt 'like an ordinary recruit' not a commander-in-chief. In the 'Nachtigall's' wake came *Pochidne hrupy* (in German *Marschketten*, marching groups) – paramilitary units that Bandera had recruited from fanatical young nationalists to take over police, press and administrative tasks in 'liberated Ukraine'. Also moving fast in the same direction were two Einsatzgruppe C commandos. Bandera had every reason to be confident that within days he would be acting president of a sovereign Ukrainian state, albeit confined to Galicia. The 330 Ukrainian Abwehr recruits pounding their way towards L'viv, desperate to stay ahead of German armies bearing down behind, carried a heavy burden of expectation.

In the early morning of 22 June, dense waves of Luftwaffe bombers and Stuka dive bombers throbbed in the cloudless skies above the western Ukraine. German intelligence had reported that cities like L'viv and Ternopol had been fortified and might hold up the German advance. The Luftwaffe attack was designed to soften them up.[24] The thunderous sound of the German attack struck terror into the hearts of the 160,000 Jews who lived in L'viv, many of them refugees from occupied Poland. Inside the city, the Soviet authorities were taken by surprise and began to organise evacuation. They had grossly inadequate resources, and German bombing raids disrupted road and rail links. The Russians prioritised Communist Party officials and technical experts. In order to make it easier for Russians and Poles to escape, the panic-stricken Soviet authorities issued orders forbidding Jews to leave their homes. Any Jewish refugees who fled the city were turned back. The Jewish Council estimated that Russian obstruction trapped 150,000 Jews in L'viv as the German forces and their Ukrainian battalions bore down from the west.

As the 'Nachtigall' crossed the Soviet border, Bandera had sent couriers ahead armed with flyers: 'Destroy the enemy, people! Know this! Moscow, Poland, the Hungarians, the Jews – these are your enemies – destroy them.' According to Geheime Feldpolizei (Secret Military Police) reports, their Ukrainian interpreters accepted without question that 'every Jew must be killed'. Another OUN-B leader informed Bandera that 'We are setting up a militia that will help remove the Jews'.[25] As the German front line rolled east, a 'pogrom mood' raced like bushfire through the villages and cities of east Galicia. And thanks to Stalin and his NKVD apparatchiks, this conflagration would become a furnace. For, as they fled, the Russians left in their wake the gruesome relics of political terror.

During the Soviet occupation, the NKVD had crammed the prisons of east Galicia, including Brygidki and Zolochiv in L'viv, with Jews, Poles and Ukrainians – all accused of being 'enemies of the people' or saboteurs. On 24 June, as fast-moving German forces swung round to encircle Major-General Vlasov's 4th Mechanised Corps which had been hastily assigned to defend this section of front, Lavrenti Beria, head of the Soviet secret service, sent instructions to his regional chiefs to shoot all 'political prisoners': those arrested for 'counter revolutionary activities', economic and political sabotage and 'anti-Soviet activities'. With some justification, Stalin feared that a German-sponsored 'Fifth Column' was about to wreak havoc in the western Ukraine. Not every prisoner in the Soviet prisons was an OUN activist by any means; many were Jews. But Bandera's plan was indeed to launch uprisings against the Soviet oppressor with German backing and to aid German forces.[26] As soon as they received Beria's orders, NKVD men rushed to murder thousands of prisoners, shooting some, bludgeoning others with hammers or throwing hand grenades into cells. The dead piled up; a handful saved themselves by smearing their faces with blood and hiding beneath their dead comrades. The Russians buried some of their victims but piled many others in the prison yard. Among them was the 'Nachtigall' commander Roman Shukhevych's brother Yuri. In Ukraine, the summer of 1941 was hot; Brygidki became a stinking charnel house.

In 1975, American lawyer Alfred-Maurice de Zayas stumbled on the unexamined records of the WUSt 'Wehrmacht Untersuchungsstelle' (War Crimes Bureau), a specialised department of the OKH which investigated breaches of the law and customs of war by the Allies. In 1941, WUSt reported on massacres in L'viv and other locations in the east by agents of the NKVD.[27] De Zayas' Wehrmacht 'sources' are by definition grossly biased – and, writing in 2000, he refers to the 'Jewish dominated NKVD'. This anti-Semitic slander, which according to de Zayas explains the ferocity of the L'viv pogrom, profoundly compromises the value of his research. His findings have predictably been exploited by anti-Semites and Holocaust deniers.[28] Nevertheless, many of the eyewitness statements collected by WUSt agents provide an unadorned account of the horrors encountered in the NKVD prisons:

> We discovered ... in the first four cellars a considerable number of bodies, the upper layer being relatively fresh and the lower layers in the pile already in advanced decomposition. In the fourth cellar the bodies were covered by a thin layer of sand. In the first courtyard we found several stretchers stained with blood. On one of the stretchers I saw the body of a male who had been killed by a bullet through the back of the head ... I ordered that the cellars should be immediately cleared, and in the course of the next three days 423 corpses were brought out to the courtyard for identification. Among the bodies there were young boys aged 10, 12, and 14 and young women aged 18, 20, and 22, besides old men and women.

'The NKVD men rushed from cell to cell and shot down the detainees ... then I heard "Come quickly to the courtyard, the cars are ready to go".'

During the night of 29 June the last NKVD detachments fled L'viv, harassed by fanatical OUN-B insurgents. It need hardly be pointed out that the Jewish citizens of L'viv bore no responsibility for these hideous atrocities.

At 4.30 the following morning, the 'Nachtigall' battalion marched singing into L'viv and, following orders, seized strategic sites, including the radio station on Vysoky Hill in the centre of the city. The 'Brandenburg' regiment and other German 1st Mountain Division and the 49th Army Corps arrived soon afterwards. It did not take long for the Germans to find in Brygidky prison the remains of the NKVD prisoners. One German eyewitness, cited by de Zayas, describes 'bringing out' 423 corpses from Brygidky and hundreds more from the former OGPU Samarstinov prison. Many of the bodies had badly decomposed and, without proper masks or oxygen, the stench was unspeakable. As the German 'Wochenschau newsreel' (filmed by an army cameraman) shows, the German troops then rounded up Jews and forced them to continue excavating the bodies and bringing corpses loaded on wagons into the city.[29] It was the typical German trick of guilt by association – a theatrical sleight of hand that compelled Jews to display and stand beside the bloated and mutilated corpses brought from the charnel houses of the prisons. It was a crude but horribly effective way to pin blame on every Jew in L'viv for the crimes of the NKVD. At some point, Einsatzgruppe C units drove into L'viv led by SS-Brigadeführer Dr Otto Rasch. They later reported that 'the Russians, before withdrawing, shot 3000 inhabitants. The corpses piled up and buried at the GPU prisons are dreadfully mutilated. The population is greatly excited: 1000 Jews have already been driven together.' In other words, righteous retribution would now be meted out to any Jew who fell into the hands of the vengeful people of L'viv. There was talk that OUN activists and relatives of men in the 'Nachtigall' battalion had also been murdered. The Germans exhibited photographs of the mutilated dead in shop windows and attached descriptions that blamed their murder on Jews.

The myth that all Bolsheviks were Jews, Golczewski points out, was thus turned on its head: all Jews were Bolsheviks, murderous agents of Moscow. According to the Einsatzgruppe report, Rasch 'formed a local Ukrainian militia' (presumably he co-opted Bandera's men), and between 30 June and 3 July, the Germans and Ukrainians rampaged through L'viv. They beat Jews to death or dragged them into Brygidki prison to remove more bodies and wash down the walls and floors. As soon as the 'purification of the prisons' had been at least partially completed, the Ukrainians locked Jews inside. Philip Friedman, who survived the Ukrainian genocide, remembered: 'the newly organized Ukrainian militia began to roam through Jewish houses to remove men – and frequently women also … Eyewitnesses relate that the courtyard and walls of the Brigidky prison were spattered with flesh blood up to the second floor and with human brains.' Dr Rasch reported to RSHA in Berlin that 'the Ukrainian population took praiseworthy action against the Jews', not only in L'viv but further east in Dobromil, Tarnopol and Sambor. One OUN-B man made his intentions plain: 'We don't want the Polish and Jewish landowners and bankers to return to Ukraine. Death to the "Moskales" [Muscovites], the Jews and other enemies of the Ukraine.'[30]

Some Ukrainian historians continue to deny that the 'Nachtigall' battalion took part in the L'viv pogrom, or to defend the L'viv attacks as a 'reaction' to the discovery Soviet atrocities. At the end of the war, a representative of the OUN issued a statement:

> While withdrawing ... the Bolsheviks killed 383 citizens. Mainly Jews took part in the extermination. The reaction of *the population* [my italics] after the flight of the Reds was very firm. *All Jews were slaughtered* [my italics] in the city and after the arrival of the Germans the people refused to bury their bodies.[31]

The statement acknowledges that Ukrainians accepted German propaganda that the atrocities had been carried out 'mainly by Jews'. The statement further accepts that 'all Jews were slaughtered' and that 'the population' took 'firm action'. 'The population' is a very vague term indeed. Did the Abwehr recruits join in the slaughter? This has proved a tricky question to resolve because witness statements, unreliable at the best of times, do not clearly identify who took part in the slaughter, except as Ukrainians wearing some kind uniform. This difficulty is compounded by the fact that the Abwehr issued the 'Nachtigall' recruits with German *feldgrau* uniforms. The 'Roland' men, who wore old Czech uniforms, would have been easier to identify.

Evidence has recently come to light that conclusively implicates the 'Nachtigall' men and has solved many of the puzzles associated with the L'viv pogrom. In the mid-1950s, the East German *Die Vereinigung der Verfolgten des Naziregimes – Bund der Antifaschistinnen und Antifaschisten* (VVN-BdA e.V. or simply VVN) began to investigate the wartime activities of Dr Theodor Oberländer, who had by then become a prominent political figure in the West German Adenauer government. The VVN investigation and trial was, to be sure, politically motivated but it does not necessarily follow that the evidence it gathered has no historical import. It will be recalled that Oberländer had played a leading part in negotiations with OUN-B representatives concerning the formation of a Ukrainian battalion and was then appointed an 'advising officer' to the 'Nachtigall'. In 1960, the East German court sentenced Oberländer to life imprisonment in absentia; the sentence was rescinded after German unification in 1993.[32]

In his review of the evidence, the DDR public prosecutor concluded definitively that: 'individual Ukrainian soldiers of the 'Nachtigall' took part in the raids against Jews in the city centre. To sum up, it can be concluded that a Ukrainian unit of 2nd Company of the 'Nachtigall' battalion committed violent acts against Jews rounded up in the NKVD prison.' One former 'Nachtigall' recruit, interviewed by the German investigators, admitted 'During our march we saw victims of the Jewish-Bolshevik terror [sic] with our own eyes, which increased our hatred against Jews. And after that we shot all the Jews we could find in two villages.'[33] His confession makes it unlikely that the 'Nachtigall' men did not join in the slaughter of Jews in L'viv, where the Germans had displayed the victims of 'Jewish-Bolshevik terror'.

Terrible pogroms engulfed many villages and towns in East Galicia. On the night of 1 July, Złoczów suffered a massive Luftwaffe bombardment – and early in the morning the German 9th Tank Division rolled into town. Hordes of young OUN members wearing blue and yellow armbands flooded into the streets, waving Ukrainian flags and embracing the German soldiers. It was reported that the 'the Ukrainians were ecstatic ... They came from the villages, dressed in Ukrainian national costumes, singing their Ukrainian songs'.[34] The Germans soon discovered the heaped-up remains of individuals executed by the NKVD. As in L'viv, local Jews were rounded up, forced at gunpoint to excavate the corpses, many of them blackened and bloated, and lay them out in orderly rows. The Germans invited villagers view the bodies to identify their friends and kin. Ten Ukrainian policemen and two SS men took up positions nearby. Although the real perpetrators had long fled east, revenge was demanded. The killing erupted with terrifying speed. The SS men began machine-gunning the Jewish men who had been forced to stand next to the row of corpses. By then, units of the Waffen-SS 'Viking' division had arrived in Złoczów and they eagerly joined the OUN fanatics and other SS units to murder Jewish men, women and children. The frenzy raged for four days; Wehrmacht General Karl Heinrich von Stülpnagel, commander of the 17th Army observed the unfolding catastrophe impassively and made no effort to intervene.

On 30 June, Bandera's henchman Yaroslav Stetsko arrived in L'viv. The son of a Greek Catholic priest, Stetsko (b. 1912) was a hyperactive and ruthless young man. He had risen to the top of the OUN thanks to his fanatical dedication; he became a member of the Homeland Executive when he was 20 and was asked to join the Provid, or Leadership, seven years later. When Melnyk and Bandera split, Stetsko sided with Bandera. He had taken a lead role setting up Bandera's network of cells and special action groups in Soviet-occupied Ukraine.

At the end of the day on 30 June, Stetsko summoned the local OUN action group and informed them that he would shortly proclaim 'the restoration [sic] of the Ukrainian state'. He ordered OUN men who had seized the radio station to stand by. At 8 p.m. that evening, Stetsko called hundreds of Ukrainians to a public meeting at the Prosvita Society building in the old city. Wearing an army trench coat, Stetsko arrived late, accompanied by Abwehr officer Professor Koch and another officer. When Stetsko rose to speak, his spectacles glittering in the candlelight that provided the only illumination, he astonished his audience by proclaiming, as a representative of OUN-B, the birth of a 'sovereign and united' Ukrainian state. Stetsko then dashed to the radio station where he read out the proclamation again. His broadcast was picked up by the OUN in Kraków and soon afterwards in Berlin. Bandera and Stetsko caught Rosenberg and Abwehr chief

Canaris, who had recruited the Ukrainian battalions, on the back foot. Preoccupied with the Wehrmacht's exhilarating eastward rush, they had missed the nationalist plot that had been hatched up under their noses.

On 3 July, Stetsko sent a letter to Hitler, it began by congratulating 'His Excellency the Führer and Reichschancellor', prematurely as it would turn out, for defeating 'Muscovite Bolshevism'. Stetsko informed Hitler that the Ukrainian people had a vital role to play as Germany extended 'the construction of a New Europe to its eastern part'. The 'sovereign state' of Ukraine would take its place as a 'fully fledged, free member of the European family of nations'.[35] Stetsko's latter did not reach the Reich Chancellery until two weeks later – by then Bandera's free Ukraine was no more.

Stetsko's rash declaration was a bolt from the blue. The Wehrmacht was then fighting its way towards Kiev and, for a week, there was little the German authorities could do to damp down the chaos in L'viv. In Kraków, however, the General Government under Secretary Ernst Kundt and other high officials of the General Government reacted swiftly. They summoned Bandera and other OUN-B leaders to a conference. Kundt informed Bandera in the most direct terms that Germany was a conquering nation not a liberating power. The Reich was not an ally of the Ukrainians because there was no such national entity as Ukraine. Only Hitler and the Wehrmacht high command had the natural right, as conquerors, to form a new government. When Hitler described Ukraine as a 'Garden of Eden', it meant that he viewed it as a tabula rasa, to be cultivated by German settlers. 'In twenty years,' he prophesied, 'the Ukraine will already be a home for twenty million inhabitants beside the natives. In three hundred years, our country will be one of the loveliest gardens in the world. As for the natives, we'll have to screen them carefully.'

It is of course not difficult to understand why the Ukrainians believed they would get away with proclaiming an independent Ukraine. Canaris had appeared to endorse their demands for a free Ukraine. Rosenberg's fickle ideas about a 'free Ukraine' had only been mooted in secret documents. But in the Nazi 'Chaos State', in which foreign policy was a many-headed monster, it is not difficult to understand that Bandera believed he could seize the initiative. At that humiliating meeting in Kraków, Kundt brutally disabused him. 'Only Adolf Hitler,' he bellowed, 'can determine what will happen [in the Ukraine].'[36] On 5 July, the Germans arrested Bandera and other members of the Ukrainian National Committee. Although the OUN-B refused to withdraw the proclamation of independence, Ukraine had gone the way of the Carpatho-Ukraine.

As Stetsko read out his proclamation on the evening of 30 June, Germans and Ukrainians had begun killing Jews in the streets of L'viv and other cities in western Ukraine. This was not a coincidence. The proclamation and pogroms had the same ideological source. On 9 July, German security police dragged Stetsko off to Berlin, but then released him under house arrest. He spent his time composing a six-page apologia, or *zhyttiepys*, that he had translated into German. It would be hard to exaggerate its historical significance. It reveals OUN strategy and thinking in 1941 – without benefit of post-war rationalisation. Stetsko assumed that the Germans would grant to Ukraine the same 'independent' status as Slovakia or Croatia. In other words, client regimes committed to radical domestic policies and aggressive handling of the 'Jewish problem'. Stetsko did not regard the proclamation as hostile to Germany. He emphasises that Hans Koch and some German army officers had attended his meeting – and one of them had read out a 'greeting'.

In short, while Stetsko insisted on sovereignty, he worked hard to reassure the Germans that any Ukrainian state would provide military and economic support to the Reich. His *zhyttiepys* conveys an even more sinister promise: 'I fully appreciate,' he wrote, 'the undeniably harmful and hostile role of the Jews, who are helping Moscow to enslave Ukraine.' He went on: 'I therefore support the destruction of the Jews and the expedience of bringing German methods of exterminating Jewry to Ukraine, barring their assimilation and the like.' In Berlin, the German translator removed the words 'destruction' and 'extermination' and retained only the reference to 'German methods'.[37]

OUN apologists point out that at its April 1941 conference in Kraków, Andrii Melnyk denounced pogroms (while insisting that the OUN 'combats the Jews as the prop of the Muscovite-Bolshevik regime'). They stress that Stetsko wrote his *zhyttiepys* under house arrest: everything he said and wrote was read by German officials. But many other OUN briefing papers express identical sentiments. The OUN-B 'Guidelines', for example, published in May 1941, state that liquidation of Polish, Muscovite and Jewish activists 'is permitted'. Jews are characterised as the prop of the Soviet regime and the NKVD 'both individually and as a national group'. 'We will adopt any methods', declared another OUN leader that lead to the 'destruction' of the Jews. One OUN policy document states that since Jews welcomed the Russians with flowers, the Ukrainians must greet the Germans 'with Jewish heads'.

As soon as Hitler learned of events in L'viv, he acted promptly. Following Bandera's humiliation in Kraków, the German military authorities withdrew the 'Nachtigall' and 'Roland' battalions from the front. Some recruits would end up in an SS Schuma battalion – No 201. Hitler had Bandera and Stetsko, the architects of the July declaration of independence, arrested and deported to the Sachsenhausen camp near Berlin, where they waited out the rest of the war. A furious Hitler reminded anyone who would listen that in 1918 the perfidious Ukrainians had murdered German General Hermann von Eichhorn in Kiev. OUN treachery reinforced Hitler's conviction that all eastern peoples had to be treated with extreme caution. He ordered that only Germans be permitted to 'bear arms' – an edict that SS Chief Himmler would openly disobey.

Following a conference convened at his Rastenburg military headquarters on 16 July, Hitler exacted 'territorial' revenge on the OUN upstarts by attaching East Galicia to the General Government. Hitler assigned the Ukrainian rump, Volhynia and 'Right Bank Ukraine' to the Reich Commissariat Ukraine ruled by Commissar Erich Koch – who once remarked that 'If I find a Ukrainian who is worthy of sitting at the same table with me, I must have him shot'. German strategy would gut the Ukrainian lands of their mineral and agricultural riches; its peoples would be reduced to slavery. German administrators would be imported to form a new ruling elite as mayors, farm leaders, school directors and above all militia chiefs.

Although Hitler called the east 'our India', he would deprive Slavs of their schools and universities, even books, so that they would not become semi-educated nuisances. He had no time for Rosenberg's proposal that suitable Ukrainians form an anti-Bolshevik national administration or that they be encouraged to take up arms against Stalin. 'Even when it might seem expedient to summon foreign peoples to arms, one day it would prove our absolute and irretrievable undoing.'[38] According to one of Rosenberg's officials, Otto Bräutigam, Hitler explained to Rosenberg:

> If I allow those people to take part in the abolition of Bolshevism with their own blood, they will one day want me to pay for it and I won't be free in the political setup of the European East territories anymore. If they want to help the Germans, they should go and work in the German factories so that German soldiers are free to fight at the front.[39]

More than a million Ukrainians lost their lives under German rule between 1941, when Hitler invaded the Soviet Union, and 1944, when the Soviet army evicted the German occupiers.[40] The victims included Ukrainian Jews and Roma prisoners of war, and many tens of thousands deported to Germany for brutal and merciless 'labour service'. Karel Berkhoff writes that the Reichskommissariat Ukraine resembled, on a massively enlarged scale, a German concentration camp like Dachau. Terror, public beatings and executions, abuse, humiliation: these were the basic instruments of German rule. And yet many Ukrainians when they held positions of power zealously implemented Nazi policy – and rejoiced when their Germans conquerors deported and murdered Jews.

6

Himmler's Shadow War

> Explicit order by the Reichsführer-SS: All Jews must be shot. Drive the female Jews into the swamps.
>
> Order to SS cavalry brigade, 31 July 1941

As Reinhard Heydrich's Special Task Forces swept into Lithuania and Ukraine, unleashing 'a Holocaust by bullets', Heinrich Himmler prepared his own SS onslaught. This shadow war began in early August 1941, in the vast Pripet Marshes (sometimes referred to as the Pinsk Marshes) that straddled northern Ukraine and southern Belorussia – a no-man's-land that lay slap-bang across the middle of the fast-moving German front line and defied conventional military attack. This shadow war would be fought by specialised military divisions, the SS brigades, and their objectives and tactics would shape SS strategy for the duration of the war. Many non-German recruits served in the three SS brigades, including ethnic Germans and Danes. But these little-known SS brigades possess an even greater significance. Himmler's secret war decisively influenced the part all non-German militias would play in this 'Crusade against Bolshevism'. They would become the vanguard troops for a new kind of combat that Himmler called 'bandit warfare'.

The Pripet campaign was assigned to SS-Gruppenführer Hermann Fegelein, Himmler's golden boy, and the SS cavalry brigade 'Florian Geyer'. Himmler instructed his favourite: 'hold fast to the great ideas of the Führer' with 'uncompromising severity, drastic action'. The Pripet action heralded a decisive escalation of the war against the 'Jewish-Bolshevik' enemies of the Reich. Himmler ordered Fegelein to treat all male Jews 'as plunderers' (meaning they could be summarily executed) and to drown their womenfolk in the marshes. For the first time,

Himmler's orders referred to all Jews, not only men. Himmler disguised this *schwere Aufgabe* (grave task) as an anti-partisan special action – and until the very end of the war, the SS 'bandit war' against partisans would be covert genocide.

Hitler had proposed using this deception at the summit meeting that took place on 16 July. 'This partisan warfare,' he told the Nazi leaders gathered at the 'Wolf's Lair' in Rastenburg, 'gives us an advantage by enabling us to destroy everything in our path. In this vast area, peace must be imposed as quickly as possible, and to achieve this, it is necessary to execute anyone who doesn't give us a straight look.'[1] Hitler did not refer explicitly to Jews, although in anti-Semitic literature they are often slandered as 'shifty'. In any case, the German military doctrine of Bandenbekämpfung (bandit warfare) offered the Germans both a code of conduct and a cover story. The mythic figure of the deceitful bandit ('who doesn't give us a straight look') was entwined with the equally mendacious Jew, who was in turn the representative of Bolshevism. In military reports, any use of the nobler term 'partisan' was forbidden. Himmler took up the idea with enthusiasm. After another meeting with Hitler, this time in Berlin, he made a note: 'Jewish question / exterminate as partisans.'[2] This lethal equation that conflated the figure of the Jew with the 'Bolshevik bandit' dominated German strategic thinking. On the Eastern Front, army general Max Schenckendorff invited his colleague SS-Obergruppenführer Erich von dem Bach-Zelewski and Einsatzgruppe commander Arthur Nebe to design a course of lectures to 'exchange experience' between the army and the SS; their core maxim was 'Where the partisan is, the Jew is, and where the Jew is, the partisan is'.[3]

The Pripet action had another sometimes overlooked significance. Himmler could not deploy the Special Task Forces in the marshes because they were officially attached to the German army groups which skirted the marshes, punching into the Soviet Union to north and south of this huge trackless waste. Himmler turned instead to the elite Waffen-SS cavalry brigade, the 'Florian Geyer', which he believed was better suited to penetrate this marshy terrain and root out 'bandits'. Himmler designed these SS brigades to rival Heydrich's SD Special Task Forces – and they would prove to be equally as murderous. The SS brigades, like the Special Task Forces, became the vanguard troops of the 'Final Solution'.[4]

Beginning in the mid-1930s, Himmler had adroitly engineered a succession of decrees that were designed to weld together German police forces and the armed SS. Himmler thus erased the distinction between combat and security; between SS policeman and SS warrior. Both would be devoted to fighting a National Socialist war. In May 1941 Himmler put the finishing touch to this plan by forming an executive body that would, in theory, yoke together the many different roles the SS would need to take on in the occupied east. Himmler assigned to this 'Command Staff of the Reichsführer' (Kommandostab Reichsführer-SS or KSRFSS) a lynchpin role co-ordinating mass murder by police battalions, the SS brigades and different Waffen-SS units. To head the Command Staff, Himmler appointed a career army officer, the 56-year-old Kurt Knoblauch, who had joined the SS in 1935. Himmler hoped that this hardnosed 'professional' would smooth co-operation with the German high command, the OKH, that had proved so troublesome during the Polish campaign.[5]

Throughout the spring and early summer of 1941, while SD Chief Heydrich busied himself with his Special Task Forces, Himmler used the KSRFSS to recruit vanguard SS combat units. This 'private army' comprised just two SS infantry brigades and an SS cavalry brigade. The brisk growth of the KSRFSS in the spring of 1941 meant that when Hitler launched Operation Barbarossa at the end of June, Himmler had under his direct command elite combat units, equipped to Wehrmacht standards and ready to wage war against the 'Jewish-Marxist enemy'. To manage the deployment of these brigades, Himmler would rely on his Higher SS and Police Leaders (HSSPF) like SS General Bach-Zelewski who would, in July 1942, become Himmler's chief 'bandit hunter'. After the war, Bach-Zelewski confessed that 'the fight against partisans was used as an excuse to carry out other measures, such as the extermination of Jews and gypsies, the systematic reduction of the Slavic peoples'.[6] The Pripet Marshes action enabled the rise to power of this highly proficient killer, who was often called on to lecture Wehrmacht officers about 'The Jewish Question, with Special Regard to the Bandit Movement'.[7]

SS General Erich von dem Bach-Zelewski was a fanatical Nazi shackled to a Polish surname. Soon after June 1941, Himmler appointed him HSSPF Central Russia, later promoting him Chef der Bandenkampfeverbände (head of the war on bandits). Wherever Himmler posted Bach-Zelewski, death and destruction followed. He commented in his war diary that when he flew over burning villages 'his trigger finger itched'.[8] In 1945, the Allied prosecutors possessed only fragmentary information about Bach-Zelewski's activities in the east and called him as a witness for the prosecution in return for immunity. Hermann Göring famously called him a *Schweinhund*. Bach-Zelewski later claimed that it was he who had provided the former Reichsmarschall with a cyanide capsule to cheat the Allied hangman. But in 1960, German investigators prosecuted Bach-Zelewski for 'multiple illegal killings' committed in Warsaw in 1944. Thanks to these zealous prosecutors, we know a great deal about Bach-Zelewski's career in the SS and his activities as a dedicated *génocidaire*.[9]

As a Higher SS and Police Leader, HSSPF Bach-Zelewski wielded enormous power. He and his fellow HSSPF, like Friedrich Jeckeln, were often referred to as 'little Himmlers'. As Hitler's forces crushed Poland then occupied much of Western Europe and Scandinavia, Himmler sent forth his loyal SS emissaries. After 1941, the

HSSPFs would become the bureaucratic backbone of SS strategy in the east. In their respective domains, they had a free hand to do Himmler's bidding and the authority to call on all the SS and police agencies, including Heydrich's Special Task Forces and Daluege's Order Police, as well as the specialised SS brigades and Waffen-SS battalions. Every HSSPF was also an SS-Gruppenführer (General). The KSRFSS was the SS equivalent of the Wehrmacht general staff, and it provided Himmler with a means to promote competitive initiative and reward the best performers. Men like Bach-Zelewski and Jeckeln would become the front-line managers of genocide.

The power invested in the HSSPF also reflected Himmler's constant fear of ambitious rivals. In Berlin, he could just about control, as both Reichsführer-SS and chief of the German police, the fast expanding SS Main Offices (Hauptämter) and the sprawling police apparatus. But as the SS empire bloated, Himmler needed loyal placemen on whom he could rely to do his bidding. The HSSPF would be, as it were, 'family' – steadfast members of the SS 'Sippenorden' (kinship order). Bach-Zelewski testified after the war that a few weeks before the invasion of the Soviet Union, Himmler brought his SS-Gruppenführers together at the SS 'Order Castle' at Wewelsburg, near Paderborn. Here Bach-Zelewski, Friedrich Jeckeln and SS notables gathered together in a gloomy, subterranean crypt decorated with Teutonic insignia. In his address, Himmler solemnly revealed that Germany was on the threshold of greatness or the edge of destruction. Hitler had demanded a solution to the problem of living space, and to 'make room', it would be necessary to remove all the Jews of Europe and diminish the Slavic population by up to 30 million people.[10] These monumental tasks demanded resolute harshness. Two years later, in his infamous speech to SS officers at Posen, he reiterated the same creed:

> Most of you will know what it means when 100 bodies lie together, when there are 500, or when there are 1000. And to have seen this through, and – with the exception of human weaknesses – to have remained decent, has made us hard and is a page of glory never mentioned and never to be mentioned.[11]

These Wewelsburg pep talks were just one way that Himmler groomed his top brass. He understood the value of men conditioned (we would say brutalised) by past combat experience. Bach-Zelewski and most of the other higher SS commanders had been profoundly shaped by their experience of the First World War 'Storm of Steel'. This was the SS esprit de corps.

In 1939, a German 'racial geographer' described the Pripet Marshes as 'one of the least developed and primeval areas of Europe', inhabited by people 'vegetating in hopeless apathy'. In the main city Pinsk, sometimes called the 'Jerusalem of the

Marshes', lived (this geographer reported) 'greasy unkempt [Jewish] women whose forms ooze with fat' and who dangled their adolescent brats over street ditches 'to do their morning business'. Jews, the geographer went on, were 'parasites', 'alien to the landscape' 'foreign bodies', 'beneficiaries of work done by others'. Like swamp water, the Jews of the marshes would need to be drained away. Himmler was determined to master this watery gateway to the east; he too viewed the Pripet as a miasmic breeding ground for people 'hostile to the German Reich in heart and soul'.[12]

On 27 July, Kommandostab chief Knoblauch sent Himmler's first set of orders concerning the deployment of the SS cavalry brigade to HSSPF Bach-Zelewski and commander of the 1st Cavalry Regiment, 'Florian Geyer', Hermann Fegelein. The plump SS general and the dashing cavalry officer met in Baranowicze in the Brest province, where Bach-Zelewski had set up mobile headquarters, to plan how to implement Himmler's order for 'Scouring the Marshes by Cavalry'. Most of the time, Himmler relied on word-of-mouth instructions and avoided sending any orders by radio – hence his frequent journeys by train or aircraft across the rear areas of the German front line. 'Scouring the Marshes ...' is a rare document that makes no explicit reference to Jews. That would have been superfluous; Himmler assumed that Bach-Zelewski would 'read between the lines'. His language was distinctly quaint:

> If the local population is hostile, or racially inferior or even, as it seems to be the case quite often in marsh areas, made up of criminals, those suspected of providing support to partisans must be executed. Women and children must be removed, cattle and food seized and taken into security.

He ended: 'Either the [local villagers] beat to death any partisan or marauder by themselves and let us know about it – or they will cease to exist.'[13]

As Himmler intended, such vague instructions forced both Bach-Zelewski and Fegelein to issue plainer orders to the Reitende Abteilung (mounted troops) of the 'Florian Geyer'. Fegelein added a few details about 'Soviet marauders', but sharpened the racial implications of Himmler's instructions: Jews had to be treated 'for the most part as plunderers'. This meant they would be shot on sight.

As Bach-Zelewski and Fegelein prepared their assault on the Pripet Marshes, HSSPF Jeckeln and the 1st SS Brigade began another 'cleansing action' to the south. According to his office diary, Himmler busily sped back and forth behind the German front line, usually in a Junkers 88, co-ordinating the deployment of his SS brigades and meeting with German army commanders to make sure they would raise no objection to his 'security measures'. Few ever did. In Riga, he met HSSPF Hans-Adolf Prützmann, who later informed a subordinate that he had been instructed to 'resettle Jews'. 'Where to?' Prützmann replied: 'in the next world.'[14] From Riga, Himmler flew on to SS headquarters in Baranowicze, where the SS cavalry brigade had mustered, to meet HSSPF Bach-Zelewski and Fegelein. Soon afterwards, Fegelein issued another clarification to his commanders:

> The Reichsführer has ordered me to remind [you] that only unyielding harshness, fierce determination and obedience to the Führer's vision will prevail against the Bolsheviks. It is up to the leadership to compensate for all those irrelevant personal weaknesses shown by individuals. The Reichsführer-SS will no longer accept any excuses in this matter and will make the harshest decisions regarding those who break ranks.[15]

In the case of the SS cavalry brigade, Himmler had reason to be concerned that they might let him down. The SS cavalrymen who rode into the Pripet Marshes were a multinational military unit, recruited from every corner of the German diaspora: from Romania, Poland, Czechoslovakia, Hungary and the Tyrol.[16] Most had yet to be tested in combat, but all had been heavily indoctrinated with SS ideology.

On 30 July 1941 the two SS cavalry regiments rode pell-mell into the Pripet Marshes. Fegelein had appointed a 46-year-old former car salesman Gustav Lombard to lead the 1st Regiment along a northerly route in the direction of Pinsk. Franz Magill, a professional riding instructor with a drink problem, led the second. In this mysterious new world, the air was stifling and humid. Legions of mosquitoes rose up to assault the SS riders as they blundered through entangling low brush and copses of beech and pine. The horizon stretched away beneath a hot, white dome of sky. Horses and men sweated and cursed as they battled through sucking, swampy ground. The Pripet was a labyrinth of islands, separated by marsh and streams. During the next weeks, the SS men would have to ford no less than thirty-five rivers and streams to reach their prey.[17]

Magill and Gustav Lombard had written orders to make wireless reports three times daily so that Himmler could assess the progress of the operation; he insisted that they list murdered Jews as 'looters' or 'partisans'.[18] Magill was a well-known drunk. From the start of the Pripet action, he appeared to be underperforming and recorded just a handful of kills. His reports exasperated Himmler; he rebuked Magill for being 'soft' and Fegelein dispatched another message: 'explicit orders of Reichsführer: all Jews must be shot. Drive Jewish women into the marshes.' Magill's counterpart Lombard was 'on message'. Soon after receiving Fegelein's message, he ordered his men: 'Not one male Jew is to remain alive, not one remnant family in the villages.' That phrase 'not one remnant family' (*keine Restfamilie*) was a death warrant for women and children as well as male Jews.[19] Still Magill had difficulties.

Fegelein nagged him to perform. Magill responded with this macabre information: 'The driving of women and children into the marshes did not have the expected success, because the marshes were not so deep that one could sink. After a depth of about a metre there was in most cases solid ground (probably sand) preventing complete sinking.'[20] This pedantry grated on Himmler. Lombard had few problems executing instructions. He efficiently drowned women and children by simply using the deeper village ponds.

Magill had another chance to impress Himmler when he reached the first large towns. On 5 August, SS-Hauptsturmführer Walther Dunsch led the leading squadrons of Magill's 2nd SS Brigade along the Brest–Pinsk road into Janów. Dunsch ordered all Jewish males to assemble in the marketplace for 'labour assignments'. As the SS men fanned out through the town, local villagers helpfully pointed out the streets and houses where Jewish families lived. Magill reported:

> The Ukrainian clergy were very cooperative and made themselves available for every *Aktion*. It was also conspicuous that, in general, the population was on good terms with the Jewish sector of the population. Nevertheless they helped energetically in rounding up the Jews. The locally recruited guards, who consisted in part of Polish police and former Polish soldiers, made a good impression. They operated energetically and took part in the fight against looters.[21]

At the end of the afternoon, the SS marched Jewish men a few miles out of the town and into a wooded area. Here they were all executed. When the SS men left, villagers descended on the execution site and took clothes and shoes. Magill's SS brigade now turned their attention to the towns and villages strung out along the Pinsk Road. Himmler was finally getting the numbers he demanded.

From the mid-sixteenth century, Pinsk had been a vibrant Jewish religious and cultural centre. Like any city with a large Jewish population, Pinsk had suffered a succession of pogroms, but in 1941 20,000 Jews lived there and 10,000 more, fleeing the German advance, had taken shelter. On 4 July, a month before Magill's men arrived, the Wehrmacht and an SD squad had entered Pinsk. They harassed the Jews, murdered some and ordered them to form a Judenrat (Jewish council). The Wehrmacht troops then moved east, leaving behind SD Chief Hermann Worthoff to keep order. Magill, shamed by Himmler's rebukes, realised Pinsk, with its large captive Jewish community, would offer him a fresh opportunity to please the SS chief.[22] He sent SS-Hauptsturmführer Stefan Charwat to confer with Worthoff – and together they began planning a large-scale 'Action'. On 5 August, Charwat ordered Jewish men between 16 and 60 to assemble at Pinsk station for a three-day 'work assignment'. That night, SS men began assaulting Jews to persuade the 'Judenrat' to do as it was told. The following morning, several thousand men reported 'for work' at the station. Many carried parcels of food. In the meantime, SS men again led a sweep, assisted by Poles and Ukrainians, searching for any Jews who had managed to avoid the round-up. By midday, some 8,000 men and young boys stood waiting in searing temperatures at the station; SS men walked through the crowd, separating out any doctors and craftsmen. They then confiscated identification papers and valuables like wristwatches from the majority of assembled Jews who had been left behind. The SS cavalrymen herded their captives into columns and marched them out of Pinsk in the direction of a neighbouring village. The Germans shot anyone who tried to escape or fell exhausted by the road. Then a halt was called close to a break of birches and alders. The Jews of Pinsk had, in fact, reached their final destination. In the soft sandy ground between the trees, a Polish work brigade had already dug pits. The SS men and their Polish accomplices organised the Jews into groups of twenty and ordered them to remove their clothes and shoes, and to wait in line. The SS men then executed them using their carbines. At 1.21 p.m., Magill reported to Bach-Zelewski that 2,461 'bandits' had been shot so far. Soon afterwards, Bach-Zelewski himself arrived at the killing site and, after hearing Magill's report, congratulated him and returned to his headquarters 'satisfied'. By 6 p.m., Magill reported by radio, another 2,300 Jewish men had been executed. As the sun began to set, Charwat, faced with more than a thousand Jews who still remained alive, panicked, and ordered his men to start firing at will. His victims prayed and sang. Magill chose not to report these messier executions – but it is believed that by the end of that first day, Magill's men murdered 6,500 Jewish men in the woods on the edge of Pinsk.

It was Lombard, however, who won the lion's share of praise from Himmler. On 14 August, he was invited to a high-level lunch with Fegelein, Bach-Zelewski, Prützmann and Himmler. He was promoted shortly afterwards. Himmler's fleet of black Mercedes then raced on towards Minsk. An SS man who overheard some of the conversation between the SS top brass commented, 'Now things are getting going, the Jews really are going to have their arses torn out.'[23] This casual remark provides a chilling glimpse of the racial triumph that percolated through SS ranks in the weeks and months after the German attack on the Soviet Union. The SS brigades, managed by Himmler's Kommandostab and comprising both Germans and foreign volunteers, had proved to be proficient mass murderers. By the end of 1941, Himmler's 'private army' may have killed at least 100,000 Jews, as well anyone else judged to have 'abetted the Soviet system'. Equally as savage as the SS cavalry was the 1st SS Brigade, commanded by HSSPF Friedrich Jeckeln.

Like Bach-Zelewski, Jeckeln was a decorated war veteran. He had joined the NSDAP in 1929 and rapidly climbed the slippery ladder of SS promotion and became police chief in Brunswick. His hatred of Jews was both personal and extreme. Immediately after the war, Jeckeln had married Charlotte Hirsch whose father was a wealthy landowner. His son-in-law administered the estates. The couple had three children but Friedrich's overbearing manner and occasional violence led to protracted divorce proceedings. In the aftermath, Charlotte's father made sure her former husband coughed up his substantial alimony payments on time. Hirsch was not Jewish, but Jeckeln developed a consuming hatred of him because of his 'typically Jewish characteristics'. This grotesque personal vendetta would inspire some of the very worst German pogroms of the Second World War.[24]

On 22 July, the brigade passed through L'viv, then began 'purification' operations in northern Ukraine on the southern rim of the Pripet Marshes. These tasks were characterised as 'encircling and annihilating the enemy' and 'encircling and annihilating bands in the forests'. What this really meant becomes evident from Jeckeln's report that 'the Brigade faced no resistance … and the Brigade suffered no losses'. In other words, his men were murdering not bandits but unarmed civilians. This is confirmed by another more explicit report sent by Jeckeln informing Himmler that his SS brigade men had killed 'around 800 Jews and Jewesses between the age of 16 and 60' close to Novohrad-Volynsky. Evidently authorisation to murder women and children as well as male Jews was seeping down through the various SS murder squads. But Himmler was not impressed; he complained to Jeckeln that the SS brigade was not 'active enough'.[25] Nevertheless, for a short period, he allocated the 1st SS Brigade to Field Marshall Walther von Reichenau's 6th Army to help pacify rear areas; according to reports 'The Jews who abetted the bands [of partisans] were executed' and 'The territory is pacified: there are no Jews or Bolsheviks there'.[26] Evidently, the German army top brass endorsed the SS doctrine that 'Jews' and 'bandits' were one and the same.

Then on 26 August, Jeckeln flew to the Ukrainian town of Kamianets-Podilsky, located in western Ukraine near the old Soviet-Polish border and now under German army administration. At the end of July, many thousands of Jews had fled here, the majority of them expelled by the anti-Semitic Hungarian Horthy regime from the Marmaros district in disputed Carpatho-Ukraine. The German administrators, both civil and military, warned Berlin that unless the Hungarians took the Jewish refugees back, they would be forced to deal with the problem in more radical ways. Since the SD Special Task Force C had already moved further east, Himmler assigned Jeckeln to the task of dealing with the Hungarian Jews. Neither he nor Himmler nor the German army administrators believed for a moment that the Hungarians would ever change their minds concerning these 'undesirable' Jews.[27]

Jeckeln's task was daunting, and the ghastly events that took place in Kanianets-Podilsky at the end of August 1941 marked another step change in SS methodology. Jeckeln had at his immediate disposal a small personal staff and a few inexperienced German police. He desperately needed reinforcements and called in Order Police Battalion 320 – which had been bolstered with ethnic Germans transferred from the Baltic region. Jeckeln next turned to the Hungarian military authorities, who agreed to provide army and Field Police units. But Jeckeln still needed to find a way to maximise the efforts of this relatively modest force. After the First World War, before he joined the Nazi Party and the SS, Jeckeln had trained as an engineer. Now he devised a technical solution to the demands of mass murder that he called 'Sardine Packing'; this would revolutionise SS strategy in the east. Sardine packing was, in crude terms, a means of ramping up the productivity of the execution squad. Tens of thousands of defenceless people could be extinguished in the most gruesome and debasing manner. The key innovation in Jeckeln's system was the systematic excavation of deep, vertical-sided pits. These simple receptacles permitted successive waves of mass executions and then the layering or 'packing' of victims in their tens of thousands inside the pit. Jeckeln's other improvement was to set up a kind of production line using auxiliary policemen who would progressively strip their victims of their possessions and clothes as they moved in stages to the hidden edge of the pit where the shooters waited. The only tools required for this diabolical system were spades and rifles. The result was mass murder on an industrial scale.

The first new method execution pit was dug a short distance from Kanianets-Podilsky. Jeckeln stood on a nearby elevation, with German army observers, to watch his new system in action. There could be no doubting its terrible efficacy: on the second day alone, SS execution squads 'processed' more than 11,000 victims. When the shooting was finally halted, an exultant Jeckeln radioed SS headquarters in Berlin to report that 23,600 Polish and Hungarian Jews (14,000 from Carpatho-Ukraine) had been liquidated so that, Jeckeln proclaimed, 'we Germans can survive'. In almost every case, the smooth running of this killing spree depended on close co-operation between the SS, the German Wehrmacht and local Ukrainian auxiliaries. The ideological imperative was the same for every executioner: the eradication of Jews as the 'bacteriological carriers' of Bolshevism. To ram this mythology home, Jeckeln forced a Jew to wave a red flag over the execution pit before shooting him dead.[28]

On 5 November 1941 Himmler transferred HSSPF Friedrich Jeckeln to Riga, the capital city of Latvia. Now Himmler ordered him 'on the express wish of the Führer' to liquidate the Riga ghetto. To accomplish this, Jeckeln would turn to one of Hitler's most notorious foreign executioners.

7

The Blue Buses

… after exerting appropriate influence on the Latvian Auxiliary Police, it was possible to initiate a Jewish pogrom in Riga.

Franz Stahlecker, Consolidated Report

With Germans it is thus: if they get hold of your finger, then the whole of you is lost, because soon enough one is forced to do things that one would never do if one could get out of it.

Viktors Arājs, Commander Arājs Commando

On 20 November 1941, HSSPF and SS-Obergruppenführer Friedrich Jeckeln ordered Ernst Hemicker, a German construction expert, to begin designing execution pits at a place called Rumbula a few miles outside Riga.[1] When he was informed of the numbers of Latvian citizens that would need to be dispatched on what the Germans liked to call the 'road to heaven', an astonished Hemicker decided to construct six pits, each one the size of small house. A day or so later, a Latvian auxiliary police 'commando', led by a hard-drinking young man called Viktors Arājs, drove 300 Russian POWs to Rumbula in blue buses leased from the Riga transit authority to begin excavation. These blue buses were already feared in many parts of Latvia. In towns and small villages, their arrival, crammed with armed and intoxicated Latvian auxiliary policemen, heralded the beginning of mass executions of Jews and other 'hostile elements' like gypsies and the mentally handicapped. Their fate had been sealed not in Riga but in faraway Berlin.

Heinrich Himmler's *Dienstkalender* (office diary) 1941/42 reveals a great deal about the SS chief's hectic schedule during that scorching summer of 1941.[2] As

German army groups smashed demoralised Soviet defences and pushed into the Baltic and Ukraine, Himmler refined his master plan to dominate the east. In the jungle of Hitler's court, he knew that he would need to quash fierce competition from Alfred Rosenberg and the Reichsmarschall Hermann Göring, both of whom hoped to become Hitler's most influential eastern potentates. At 12.30 p.m. on 24 June (postponed from noon), the SS chief met SS-Standartenführer Professor Dr Konrad Meyer at his headquarters on Prinz Albrecht Strasse to discuss the 'General Plan East'. As he leafed through Meyer's first draft, Himmler was bitterly disappointed. The plan was timid and lacklustre. With German and SS troops penetrating deep into the lair of the Bolshevik enemy, Meyer's ideas had been rendered obsolete. Himmler lifted the phone and cancelled his regular appointment with his masseur Felix Kersten, the Baltic German crank who treated the SS chief for persistent excruciating stomach pains. Like a frustrated schoolmaster, Himmler took his pen and stabbed and scratched at the offending document. Meyer rushed back to his splendid offices in Berlin-Dahlem to start, as Himmler instructed, 'thinking bigger'.

At 11.30 a.m. the following day, Himmler's special train *Heinrich* steamed out of the Lehrter Bahnhof in Berlin. His destination was Hitler's military headquarters near Rastenburg – the Wolfsschanze (Wolf's Lair). Here Todt Organisation engineers had constructed, in a mosquito infested swamp, a vast concrete city of camouflaged bunkers and huts. From inside Security Zone One, Hitler directed his 'war of annihilation'. Himmler's train halted close to the lake at Angerburg, a short drive from the Wolf's Lair. It was from here that he would supervise the escalating slaughter in the east over the coming months. Onboard the *Heinrich*, sophisticated equipment sucked in the daily reports from the Special Task Force commanders and the SS brigades – as well as Waffen-SS divisions in action on the front line. Assisted by his loyal adjutants Joachim Peiper and Werner Grothmann, Himmler kept in close touch with his Higher SS and police officers Hans-Adolf Prützmann, Erich von dem Bach-Zelewski and Friedrich Jeckeln – the front-line managers of mass murder.

On Monday 30 June, the *Heinrich* sped east toward the old Polish border with Lithuania. In Grodno, he met RSHA chief Reinhard Heydrich. As they toured Grodno, a famous centre of Jewish culture, it was apparent that Special Task Force A had somehow neglected to deal with the city's Jewish district. An embarrassed Heydrich made an urgent call to the SD operational office in Tilsit; a few days later SD and security police units arrived to mop up in Grodno and neighbouring towns like Augustowo.[3] After this regrettable operational lapse, Himmler began to make unscheduled visits to Vilnius, Kaunas, Riga and Minsk to cajole SD commanders into stepping up their efforts. He insisted that no mercy could be shown. 'Hardness' was all: the mark of a true SS warrior.

On the evening of 15 August, Himmler, accompanied by his adjutant Karl Wolff, flew into Minsk, the main city of Belorussia. At the splendid old Leninhaus, they met the hatchet-faced commander of Special Task Force B, Arthur Nebe, with the ubiquitous HSSPF Erich von dem Bach-Zelewski and arranged to observe the execution of a number of '*Partisanen und Juden*'.[4] Wolff revered Himmler as an 'unparalleled man' of 'extraordinary qualities', but he was positive that Himmler had never before seen a man shot. A party of German police drove the two men from their hotel in the centre of Minsk to an athletics field, where two pits had already been dug. Himmler's little party did not have long to wait. A big Mercedes truck rumbled into the field, and SD men threw open the tailgate and hauled out their catch of 'Jewish spies and saboteurs'. They pushed and shoved these terrified, weeping men to the edge of the pit. Some begged for mercy. Wolff noted that the Jewish 'saboteurs' had been stripped of their clothes, but wore 'rags' to spare the Reichsführer-SS the sight of completely naked 'sub-humans'. The SD executioners forced their victims to lie face down and began loading their weapons. At this point, Himmler moved to a higher vantage point directly overlooking the execution pits. A nod from Himmler and the SD officer gave the order to fire. Instantly, there was an eruption of blood and brain matter. The SD 'shooters' had taken up positions much too close to their victims. Wolff glanced at his boss. Brain matter was visibly smeared across Himmler's jacket. The SS chief was sweating profusely, his face turning a distinct shade of green. A second volley – Himmler swayed. Wolff caught his arm. Himmler turned away and vomited.

As the SD men began covering the corpses, Wolff took his shaky boss back to Minsk to recover. In an interview conducted after the war, Bach-Zelewski claimed that he had taken advantage of Himmler's reaction to admonish him about what his men had to endure. According to Wolff, Himmler merely used his usual catechism that they must all be 'hard', without mercy. They were all burdened by a tremendous responsibility to clean up the east. After a short lunch, taken in the Leninhaus, Himmler recovered sufficiently to enjoy a tour of the Minsk ghetto and a hospital for the 'retarded' at Novinki.[5] The sight of these human vermin, 'lives not worthy of life', no doubt reassured him that his great mission was right.

Hitler liked to describe the Soviet Union as possessing a 'Slavic-Tartar body' and 'Jewish head'. In his mind, decapitation was essential – and in July, as Hitler's armies pushed back the Soviet armies along a 1,000-mile front line and gobbled up vast new territories, anything seemed possible. On 3 July, Colonel General Franz Halder wrote in his diary that it was no overstatement to say that 'the Russian campaign has been won in the space of two weeks'.[6] Halder rejoiced prematurely; but inside the Wolf's Lair, an exultant Hitler proclaimed that a new German empire would soon reach out as far as the Ural Mountains: an iron wall to hold back the Slavic 'rabbit family' (*Kaninchenfamilie*). Every big Soviet city would be levelled. German soldier-farmers would be trained to govern the Slavic helot class, and industrious German peasants would till the rich, black soils of the new Germania. Hitler imagined prosperous Teutonic families taking tours of their vast new domain, speeding along grand autobahns in brand new 'Folk Cars' (Volkswagens) and soaking up the sun on the Caucasus Riviera. And of course, any Jewish 'bacilli' that might have spoilt this idyll would have been banished long before. In human history, Hitler concluded 'there's always killing'.

It was with this vision in mind that Hitler called his paladins to Rastenburg on 16 July to plan a glorious future for the conquered east. There was a new world to be won. Unlike the French or British empires, which had evolved gradually and piecemeal, this new German Reich would be thrown together in months not centuries.[7] Himmler would miss the conference: Stalin's son Yakov had been captured near Smolensk and, as we learn from his diary, the SS chief hurriedly left Rastenburg to gaze on this trophy captive. But Himmler could afford to be relaxed: only Alfred Rosenberg posed any possible threat to SS dominance of the east. Hitler had appointed him 'Reich Minister for the Occupied Eastern Territories', complete with a new ministry – the 'Reichsministerium für die besetzten Ostgebiete', or RMfdbO. Himmler could rely on Hermann Göring to slap down any attempt by the new minister to impose his feeble will. He could meet young Yakov Stalin with a light heart.

Inside the main conference room at Rastenburg, Hitler began speaking at 3 p.m. and the meeting dragged on late into the evening, with a single break for coffee. It was, as everyone understood, Hitler's show. Hitler forbade what he called *Schaukelpolitik* (indecisive 'back and forth' strategy). He conjured up a vision of the east as a 'Garden of Eden', a tabula rasa to be planted with purely German stock. When Rosenberg dared suggest that the Ukrainian lands, whose people had endured famine under Stalin's rule and hated the Soviets, might be granted limited independence, Hitler unhesitatingly rejected the idea. He forbade any deal-making with hopeful nationalists; he would not tolerate recruiting foreign militias: 'We must never permit anybody but the Germans to carry arms! Only the German may carry arms; not the Slav, not the Czech, not the Cossack, not the Ukrainians!'[8]

The day after the Rastenburg conference, Party Chancellery head Martin Bormann forwarded minutes to Himmler. He noted that Hitler had reiterated the plan to exploit anti-German partisan attacks to mask the mass murder of civilians; because the Russians 'have now given out the order for a partisan war behind our front', 'it gives us the possibility of exterminating anything opposing us'. In a nutshell, this was SS military doctrine. But on one vital matter, the 'loyal Heinrich' chose to ignore Hitler's orders. He understood that the task of 'cleansing' the east would require enormous human resources that could not be supplied by Germany alone. This meant that non-Germans must be armed. At the very moment that Hitler proscribed arming non-Germans, SD commanders had begun recruiting Lithuanians and Latvians and other eastern peoples to do the dirty work of ethnic cleansing. On 25 July Himmler authorised the recruitment of auxiliary police units in the occupied east – from 'suitable elements' in the local population. These Schutzmannschaft or Schuma battalions would soon become a vital instrument of genocide, killing at least 150,000 so-called 'superfluous eaters' (mainly Jews) between July and December 1941 alone.[9]

In almost every other respect Hitler strengthened Himmler's hand. The day after the Rastenburg conference, on 17 July, he issued 'An Order of the Führer on the Administration of the New Territories in the East'. This specified that where the army had crushed enemy resistance, German military administrators would hand over power to civilian bodies formed ostensibly under the aegis of Reichsminister Rosenberg. The Führer Order set out a rudimentary organisational structure: the conquered territories would be broken up into Reichskommissariate, each headed by a Reichskommissar, subordinate, in theory, to the Reichsminister. Later on the same day, Hitler published a supplementary order establishing the two first Reichskommissariate: Reichskommissariat Ostland – which comprised Estonia, Latvia, Lithuania and Belorussia – and Reichskommissariat Ukraine, which would become the fiefdom of Gau Leader Erich Koch who had no intention of kowtowing to Alfred Rosenberg. Not content with unleashing Koch on German-occupied Ukraine, Hitler issued a third order on 17 July that stripped Rosenberg's ministry (the RMfdbO) of any security powers and handed them lock, stock and barrel to Himmler and the SS. All matters concerning the policing of the eastern Reich Commissariats would be handled by the SS – and through directives issued by Himmler to the Reich Commissars. The HSSPF like Bach-Zelewski and Jeckeln, the so-called 'little Himmlers', took their orders from the Reich Commissars, but only in theory. In practice, all directives passed through SS channels. But Himmler did not win outright. Hitler's topsy-turvy distribution of powers, giving with the one hand and taking away with the other, guaranteed that the SS and Reich Commissar Koch would now wage a succession of brutal turf wars.[10]

In the Baltic, which was absorbed into the Ostland Commissariat, Rosenberg had, to begin with, the upper hand. He had been born in Reval (now Tallinn)

in Estonia, then part of the Russian Empire. He had been educated in Riga and Moscow where he witnessed Bolshevik Revolution at first hand. A rabid anti-Bolshevist and professed Jew hater, Rosenberg had escaped to the west bringing with him the anti-Semitic 'Protocols of the Elders of Zion' – a poisonous tome that he introduced to Hitler and his circle. As a Baltic German, Rosenberg had an obsessive interest in restoring the old Teutonic Order in the Baltic. The man he appointed as commissar, Hinrich Lohse, shared his vision. A fanatical Nazi consumed by the same venal obsessions as other Nazi potentates Hans Frank and Erich Koch, the 'gross, vain and silly' Lohse boasted that he would reclaim the lost lands of the medieval Teutonic Knights and Hansa merchants. He was, he claimed, 'treading the fateful path of the great political legacy from West to East' to 'replace chaos with a system of European order, and in place of destruction reconstruction and culture'. His subordinates called him 'Duke Lohse' after he insisted that he was 'working for my own good' but so that 'my newly born son will be able to place the crown of "Herzog" [duke] onto his head'.[11]

Regardless of their grandiose ambitions, Rosenberg and Lohse faced fierce competition. By August 1941 three cutthroat and competitive Reich agencies had turned the Baltic states into rival fiefdoms. Black-clad SD men led by Special Task Force commander Franz Stahlecker had been the first to arrive, followed by Wehrmacht administrators and finally, months later, by Rosenberg's 'Eastern Ministry' men in their yellow, hand-me-down so-called 'pheasant' uniforms. 'To the objective observer,' Stahlecker complained in his Consolidated Report, 'a picture of disunity emerges, where guidelines are totally absent and where German administrative offices and their staff greatly lack preparation for their duties.' This was disingenuous. From the moment German troops entered the Baltic, it was the SD, Himmler's giant security force, that called the shots. Stahlecker explained: 'the security police was well ahead of everyone else … it was the only office that established a certain stability.' The SD took its orders not from Commissar Lohse but the regional HSSPF, Hans Prützmann – and thus directly from Himmler in Berlin.[12]

This jostling for power should not be taken to imply that there was any fundamental doctrinal divergence among Hitler's paladins. In a letter, written shortly after the 16 July meeting at Rastenburg, Rosenberg set out his own version of National Socialist occupation strategy: 'The aim of the Commissars for [Ostland] must be to establish a protectorate of the Reich, and then by winning over the racially valuable elements and by a policy of resettlement measures this region must be made one with the German Reich.' He then explained that in the Baltic region, these 'racially valuable elements' were hierarchically distributed in a gradient running north-east to south-west. At the 'top', in Estonia, Rosenberg argued,

50 per cent of the population had been strongly 'Germanised' through infusions of German and Swedish blood. Estonians, he concluded, were 'people akin to us'. Lithuanians, however, occupied a position at the 'low' end of the scale because they had been so thoroughly 'Polonised'. Latvians fell somewhere between these poles. Himmler concurred with Rosenberg's analysis. This Baltic racial gradient would have a powerful impact on German occupation policy and SS recruitment of non-German Schuma battalions and Waffen-SS divisions.

In the long run, the rival Nazi bosses all assumed that the Reich would eventually completely 'digest' the Baltic nations. In the short term, that is until the war had been won, the German occupiers needed to buy time. To do this they set up so-called self-administrations to do the donkey work of government. This meant that the Germans would have to square a circle in the sense that they would need to recruit, say, credible Latvian officials while stifling Latvian nationalism. The Germans had already had their fingers burnt in Lithuania when nationalists had declared independence. Stahlecker's solution, hatched up with the army group commander General Franz von Rocques, was sly. According to Baltic German Harijs Marnics, who worked closely with the Germans occupiers, 'it was recommended that the terms 'Latvia' and 'Latvian people' should not be used, so as to get the Latvians to forget about their nation'.[13] From the German point of view, this made historical sense. Before 1920, Latvia and Estonia had never existed as sovereign nations. When Latvians proclaimed independence in 1918, they had to set out their borders using the thirteenth-century defensive frontier drawn up by the Teutonic Knights as a bulwark against Russian incursion. Since the idea that Latvia was a nation state at all was a figment of deluded Latvian imaginations, the German occupiers assumed that it would evaporate along with its name. A German official put a mollifying gloss on this scam: 'the times of political independence are in the past … they have been exchanged for times of peace and prosperity under the protection of the German Reich.'[14] Ignorant of the long-term German master plan, many Latvians continued to bask in the glow of deliverance from Soviet tyranny – and there would be no shortage of eager collaborators.

On the afternoon of 1 July an assembly of Latvian nationalists gathered at the 'Latvian Club' in Riga. Led by Col Ernests Kreišmanis and former partisan fighter Bernhards Einbergs, the group comprised army officers and former government ministers who had served under former President Ulmanis. They called themselves the Latvian Organisation Centre (LOC) and set about forming a provisional government.[15] Kreišmanis forcefully argued that the most effective way of pleasing the Germans would be to appeal to the Pērkonkrusts, the ultranationalist faction previously banned by the Latvian government. The virulent brand of chauvinism espoused by the Pērkonkrusts would, he assumed, strongly appeal to the Germans. Kreišmanis had strong grounds for making this argument.

In 1940, Pērkonkrusts' leader Gustavs Celmiņš had fled to Berlin, where he made influential contacts. At the end of June 1941, he returned to Latvia as a Sonderführer (special leader) with a German army brigade. But Celmiņš was much too self-important to throw in his lot with any provisional government not led by him – and in any case his influence with his German friends proved to be expedient and short lived. On 3 July, Stahlecker, who had set up his headquarters in Riga's imposing medieval Ritterhaus, threw out LOC demands and insisted that there would be no 'Latvian government' of any kind until the war had been won. He may have sent an even more direct message. On 18 July, assassins killed Lt Col Viktors Deglavs, a protégé of German military intelligence, the Abwehr, who had been recruiting a Latvian national army modelled on the Ukrainian Nachtigall. A witness testified: 'Around his head there was a puddle of brains and blood. Down the stairs ... stood two lower ranking SD men, apparently German.'

The purpose of Stahlecker's machinations soon became clear. The Germans imposed a 'Land Self-Administration' (*Landselbstverwaltung*) that, as its name implies, had no national status of any kind and ranked lower in the German pecking order even than the puppet governments of Slovakia and Croatia. To fill the top post, the Germans chose an opportunist Latvian general, Oskars Dankers, on the grounds that he was married to a German Balt and was on excellent terms with his German chauffeur. As for the other 'directors general' who would serve under Dankers, the Germans set some pretty basic criteria: candidates would need to be 'popular with Latvians', but 'obedient and submissive' to the German Reichskommissar. Notwithstanding its chronic impotence, the Latvian SA rapidly turned into a forum for frustrated squabbling. Stahlecker and his SD successors let the various factions slug it out. The conservative Ulmaniesi, who had served under President Ulmanis, went many rounds with a second faction, the anti-Semitic fascist Perkonkrustiesi. They competed in turn with the so-called Repatrianti who had been thoroughly 'Nazified' in Berlin and enjoyed even closer ties with the Germans – to say nothing of the Kalpakiesi, grizzled 'War of Liberation' veterans, and a host of nationalist intellectuals and fraternity old boys. For the Germans, the SA provided gladiatorial entertainment. The general directors wielded not a shred of power, and they all knew it. They were, one admitted, 'seven bleached out pillars'.[16] In 1943, the Latvian diplomat Alfreds Bīlmanis, who fled to the United States, provided the Americans with an excoriating analysis of the Latvian 'self-administration'. Dankers' principal task, Bīlmanis explained, was to 'advertise for the Reich Labour Service' and SS Schuma battalions. The general directors 'are no ministers as in the cabinet of an independent state', but 'subjects' of the German commissars. The SA General Directorate, in brief, was a 'German Nazi stooge organisation to deceive the population'.[17]

It is especially telling that Minister Rosenberg regarded Riga as a 'German city' and appointed his father-in-law, Hugo Wittrock, a Baltic German, to serve as its mayor.

With the Baltic nations extinguished to most intents and purposes as sovereign nations, the way was clear for Stahlecker to begin recruiting native militias. The danger of inadvertently forming a national army was in theory neutralised. As partisans, Latvians or Lithuanians or Estonians fought for national liberation from the Soviet Union. But now that the Baltic nation states had been dissolved away in the acid bath of the Reich Commissariat Ostland, Himmler had carte blanche to recruit the 'foreigners' Hitler feared and who could no longer be distracted by nation-building. The SD led the way. From offices at Reimersa iela 1, which occupied an entire city block, Department IV of the 'resident SD' (to distinguish it from the mobile Special Task Forces) began to plot the destruction of Latvia's Jews. Just hours after arriving in Riga, Stahlecker began recruiting native auxiliaries called, at this stage, Hilfspolizei. He informed Berlin:

> Lieutenant Colonel Voldemārs Veiss [a Latvian] has been made leader of the Hilfspolizei: care has been taken to ensure that these troops would not become a Latvian militia … two further independent units have been established for the purpose of carrying out pogroms. All synagogues have been destroyed; so far 400 Jews have been liquidated.[18]

To lure Latvian young men to join SD auxiliary units, the Germans promoted a perverted brand of patriotism. German propaganda ceaselessly reminded Latvians that 'Jewish NKVD' men had organised the deportation of tens of thousands of Latvians – forgetting to mention that the Soviets had also arrested many thousands of Jews. As in Ukraine, Wehrmacht and SS units employed Jewish forced labourers to excavate the corpses of NKVD victims and then *exhibit* the battered and decaying corpses. This macabre theatre definitively attached blame for the Soviet liquidations to Jews who it was claimed had secret knowledge about where the bodies had been buried. German and Latvian propaganda made much of the fact that Semjon Sustin, a high-ranking NKVD officer posted to Latvia, was Jewish. Latvians who had returned from Nazi Germany with the Wehrmacht or the SD, as well as Baltic Germans who spoke Latvian, played a pernicious role convincing young, often unemployed Latvian youths that Latvian Jews had been responsible for the calamities of Soviet rule. As the German occupiers set about destroying the Latvian nation, they exploited frustrated patriotism of young Latvian men who hated the former president Ulmanis for refusing to fight the Soviets. Now they would willingly take up arms against Jews.

Latvian newspapers played an especially corrosive part, reinforcing the power of the SD and spreading the lethal poison of blame. The fascist Perkonkrustiesi and the 'Germanised' Repatrianti, 'savage Jew haters', all poured forth a stream of stomach turning bile. Martinš Vagulans, a fêted journalist, a former agronomist and Pērkonkrusts die-hard, penned some of the nastiest diatribes directed at 'Muscovite Kremlin degenerates and Jews' and 'ghastly Jewish and Bolshevik bondage'.[19] Vagulans was very likely well known to the SD before 22 June 1941, for en route to Riga Stahlecker's Special Task Force halted briefly in his home city of Jelgava. Stahlecker immediately sought out Vagulans, appointed him commander of an SD auxiliary commando and authorised him to begin publishing a newspaper which was to all intents and purposes an SD front. In the following months, Vagulans and his writers set new standards of vileness with a stream of Jew-baiting headlines and cartoons. As soon as Stahlecker moved on to Riga, Vagulans' new militia the Nacionālā Zamgale signalled their intentions by setting fire to the main Jelgava synagogue. This was, of course, a warning of worse atrocities to come. The offices of Vagulans' newspaper doubled as the Nacionālā Zamgale command centre. Stahlecker dispatched (unsigned) orders from Riga and Vagulans then issued orders and instructions to his men through his paper.

Vagulans was plainly ambitious. He promoted himself from commander Latvian SD Jelgava to Latvian SD Jelgava district commander. He made deals with Aizsargi leaders (the old paramilitary civil defence forces), Latvian police and stray partisan units (who had fought the Russians). He organised teams of cyclists, who sped from one village to the next proclaiming the great victory of the Reich and denouncing Jews, 'the cause of all pestilence and misfortune'. By the end of July, Vagulans lorded it over 300 auxiliaries, organised into guard units, criminal sections, a department in charge of sorting appropriated valuables and a second responsible for reallocating flats formerly occupied by Jewish families. Vagulans ordered building supervisors to register their tenants, and separate Jews and ethnic Latvians. He threw Jews out of local government jobs and banned them from using parks and cinemas. In the meantime, he stepped up the pressure through his newspaper: 'No pity and no compromise must be shown. No Jewish tribe of adders must be allowed to rise again in the renewed Latvia.'[20]

Vagulans' trap soon snapped shut: by 18 July his goons had press-ganged large numbers of Jewish men into labour groups and separated them from their families. They evicted Jewish women and children, threw them into the street then force-marched them to a hastily improvised camp inside a derelict warehouse. Then on 1 August, Vagulans ordered that all Jews must leave Jelgava city and district 'or be punished in accordance with the laws of war'. It was a sham. Vagulans' 'order' was the signal to begin a final assault with the full backing of the German SD authorities in Riga. Three days after Vagulans issued his depuration order very few Jews remained alive in Jelgava. The evidence that would tell us in detail how events unfolded in Jelgava on 2–3 August 1941 is fragmentary. What information we do have comes from the post-war trial of Alfred Becu, an officer in Vagulans' commando, that was held in Cologne in 1968; and from a pre-trial deposition made by an eyewitness Arturs Tobiass.[21] Becu confessed that he had led a German SD unit to Jelgava; Tobiass, who lived close to the synagogue, observed the German SD men arrive in the city either in late July or at the beginning of August. Soon afterwards, he heard a loud explosion and ran into the street to see what was happening. He watched German soldiers pouring gasoline through the main door of the local synagogue and then throwing hand grenades through a side window. The rabbi refused to leave. As flames ripped through the synagogue, German soldiers drove a column of Jews along the street towards the fish market, where they were corralled by Vagulans' men. Many Latvians, Tobiass claimed, commiserated with the Jews as they were driven past about the loss of their 'church'.

Sometime later, the SD men herded their captives from the market to a nearby firing range. According to another eyewitness, one Wilhelm Adelt, Latvian SD men with bolt-action rifles and white armbands then began shooting. Adelt testified: 'Every day a new pit was dug … The offenders [sic] had to take off their over clothes. Becu [the officer in charge] said that the Jews were killed because they did not fit into the Nazi regime, and that Jews in general would be rooted out.'[22]

On 15 August, Vagulans declared on the pages of his newspaper: 'Jelgava is "free of Jews" … the shapers of the new life – German soldiers – displayed exemplary trust in supporting and assuring the success, ejecting that race of people who are creators of all the world's misery.'[23] This proved to be Vagulans' undoing. Three days later, Stahlecker sacked him as Jelgava district police chief and editor. He had said too much. The fall of Martinš Vagulans spelt out a simple message: lackeys are dispensable. And when Stahlecker set up his temporary headquarters in Riga, he found there was no shortage of Latvian fanatics eager to do Germany's bidding.

In 1975, an elderly Latvian man who called himself 'Viktors Zeibots' stood before a German court in Hamburg accused of the vilest war crimes. The 78-year-old Herr

Zeibots had been born in Latvia, but spoke fluent, barely accented German. He had greying blonde hair and striking pale blue eyes. Otherwise his appearance was unremarkable. As Herr Zeibots sat impassively in the dock, the court was shown a photograph taken in 1942 at the German SD school in Fürstenberg. It shows a posed group of officers. In the front row there is a young man sitting erect alongside two German officers, Dr Rudolf Lange and Arnold Kirste. A black SD cap throws a shadow over the upper half of his face, concealing his eyes; we can see that he is smiling. Three decades separate the SD officer in the picture from the accused sitting silently in the Hamburg dock. But the resemblance of the young man in the photograph to the elderly Viktors Zeibots is striking. For the man who called himself Viktors Zeibots had once been SD recruit Viktors Arājs: one of Hitler's most infamous foreign executioners.

A young Jewish mechanic who had once worked for Arājs recalled that he had a 'sympathetic face'. 'He had no particular characteristic that stayed in one's mind,' recalled another Latvian recruit. 'To describe his facial features is difficult,' admitted another. But others could recall meeting Viktors Arājs with terrible clarity. A guest at a dinner party in Riga heard this young man with a 'sympathetic face' boasting how he had murdered Jewish infants by hurling them in the air then shooting them as they fell. Had he executed them on the ground, he explained casually, bullets ricocheting from the concrete floor might have injured his men. It was said that Viktors Arājs once presented the German SD commander Rudolf Lange with a Christmas tree festooned with rings and diamonds pilfered from Jewish homes he had plundered. Frida Michelson, a survivor of the Rumbula massacre, remembered seeing 'Arājs, heavily drunk … working close to the execution pits'. A young woman walking along a bombed out Berlin street in 1945 remembered an encounter with a young Latvian man who introduced himself as 'Arājs, the Latvian Jew-killer'.

The Hamburg court sentenced Arājs to life imprisonment. He died in a German prison cell thirteen years later in 1988. He must have appreciated the irony of this because the German judge Dr Wagner acknowledged that Arājs 'acted on orders of Dr. Lange'. It is estimated that the 'Arājs Commando' murdered at least 26,000 Latvian Jews between July 1941 and the summer of 1942. When Arājs 'ran out work', the Germans assigned his commando men to the Minsk region in Belorussia where they applied their unrivalled experience as mass murderers in other large-scale actions. After 1943, Arājs and many of his commandos served in the SS 'Latvian Legion'.

Vicktors Bernhards Arājs provided at least three very different biographical sketches. We know that he was born in January 1910, in a small town near Riga. According to Ezergailis' account, his mother Berta Anna Burkevics descended from wealthy German-Latvian farming stock – but like many Latvians, Teodors (his father) had ambivalent feelings about German Balts and refused to let his children speak German at home. Viktors himself insisted that he had been 'raised as a Latvian' and when he met Stahlecker in 1941 spoke very poor German. At the beginning of the First World War, Teodors enlisted in the Imperial Army of the Tsar – and vanished. His son was four. When Russian soldiers destroyed the Arājs' family home, Anna fled to Riga and found work in a factory. As the family sank into poverty, Viktors ran amok and his mother sent him to work on her parents' farm. He recalled sleeping in a stable and 'making his own toys'. After the Latvian War of Liberation, Teodors reappeared in Latvia sporting a Chinese wife. By then, Anna had moved to Jelgava and life was looking up. She had inherited money and bought a rooming house. When he turned 16, Victors left home to become a farmhand. He recalled that one day, his father unexpectedly approached him as he worked in a field and they had a brief conversation.

Viktors would never see Teodors again and grew up without a father. He went to school only in the winters. He joined a wandering band of carpenters, building houses, farms and saunas. Viktors was a diligent worker, and after he lost his job as a carpenter, won a place at the Jelgava Gymnasium. Here he excelled, despite taking on a number of manual jobs to make ends meet while he was studying. After graduating, he joined the Vidzemē artillery regiment and in 1932, still serving as a corporal, enrolled at Riga University to study law. It would take him eight years, plagued by interruptions, to complete his legal studies and graduate. He was plainly both intelligent and driven. He was shrewd enough to join the prestigious Lettonia fraternity – no mean feat for this former farm boy. The Latvian fraternities, as mentioned before, bred generations of Latvian chauvinists. When Arājs began recruiting for the German SD in 1941, he would turn to former 'Lettonians' to fill the officer ranks in the Arājs Commando.

In 1937, Arājs married Zelma Zeibots and was forced to drop his university studies once again to make ends meet. For the second time, he joined the police. In 1939, he returned to university, dropped out again – then, as the Soviet occupation began, made one last sortie. He took an obligatory course in Marxism and finally graduated as a Soviet Jurist in March 1941. At his trial, many decades later, Arājs claimed that he had in good faith come to believe that 'Bolshevism was the best of systems'. Historian Andrew Ezergailis offers an intriguing suggestion to explain this anomaly. Although Arājs had inveigled his way into the Lettonia, his exposure to more privileged lives and presumably elitist attitudes of brother students may well have led him to welcome the Soviet occupation and a promise of greater social equality. 'Indubitably, I was then a communist,' he claimed after the war. After the occupation began, Arājs found work with a Latvian lawyer, but, not long afterwards he was arrested and deported by the NKVD. Arājs escaped and fled to the countryside. 'My communism vanished.' At this point his account becomes remarkably confusing. By the time the main Soviet deportations began, Arājs may have joined a Latvian army brigade attached to the Red Army.

As the German juggernaut rolled on north through the Baltic, the Russians began withdrawing their Latvian divisions to the east. The retreat degenerated into panicked flight. Chaos and mayhem overwhelmed the retreating Soviet divisions. Many Latvian soldiers deserted, often throwing themselves from the transport trains and, if they survived, joining partisan units. In his final plea to the Hamburg court Arājs said, 'I too was in one of the [partisan] units.' On 1 July, Arājs returned to Riga 'with my partisans' and made for the police prefecture. On the same day, Stahlecker led his Special Task Force across the Daugava River. At 1 p.m., the Germans too arrived at the police prefecture. A Baltic German officer called Hans Dressler who served with Stahlecker had known Arājs at his gymnasium and then served under him in the Latvian army. He introduced Arājs to Stahlecker because 'he had a favourable memory of Arājs from the days of his military service'. That afternoon, Stahlecker met a number of other Latvian police commanders who were eager to join up. Lt Col Voldemārs Veiss, an army veteran, had already recruited a 400-strong Latvian 'self-defence' unit that had on its own initiative begun to round up Latvian Jews and communists. Veiss was eventually appointed Chief of Auxiliary Police – and in 1944 died serving with the 'Latvian Legion'. He remains a national hero in Latvia. The following day, Stahlecker appointed Arājs as a 'Sonderkommando leader' and, he later told the court in Hamburg, 'My assignments began on July 3 or 4'.

As a newly appointed commando leader, Arājs had one tremendous advantage: he was a former Lettonia fraternity 'brother'. As he raced to organise his 'Special Commando' and impress Stahlecker, Arājs cleverly called on his old university comrades and transferred his activities to the Lettonia fraternity house, where he set up tables in the street outside to attract recruits. The brothers flocked to join – in such large numbers that Latvian Jews referred to the commando as '*Arājsen Burschen*', from the German word for fraternity, *Burschenschaft*. As the Arājs Commando expanded, leadership positions went to Lettonia old boys like Feliks Dibietis, Arājs' second-in-command who later committed suicide in Minsk. This was a battalion of murderous frat boys.

To proclaim the elite status of his unit, Arājs soon moved his operational base to a splendid town house at 19 Krišjāņa Valdemāra iela, close to the Latvia National Art Museum. No 19 was one of the best addresses in town; it had previously been occupied by A. Schmuljanš, a Jewish banker who had been deported by the Russians.

Arājs shared No 19 with the Pērkonkrusts, who used the basement to 'question' Jews in improvised cells – many were tortured or murdered. In a grand room on the first floor, Arājs arranged German lessons for his recruits, though he himself would continue to need a translator for some time to come. In the leafy Esplanade Park just across the street, Arājs organised rifle drill.

Arājs seemed to have fallen on his feet. But though he was ambitious, to begin with Stahlecker and the SD commanders who took his place remained wary. An Einsatzgruppe report sent to Berlin on 20 July even refers to the imminent 'dissolution of the Security [i.e. Arājs] commando'.[24] And yet the Arājs Commando would prove to be the most resilient of all the auxiliary units formed by German occupiers in the east. It survived in different forms and under a variety of names until 1944, when it was absorbed by the Latvian SS divisions. It will be recalled that following Heydrich's instructions, Stahlecker warned that any auxiliary units must not be allowed to become 'a Latvian militia'. Now Arājs had only the most tenuous connections to Latvian nationalist circles. As a student, he flirted, to be sure, with a fringe fascist faction but had never joined the Pērkonkrusts. To finally complete his legal studies, he had joined the Soviet legal apparatus and then taken up arms to fight the retreating Russians at the last possible moment. In short, Arājs made no bones about seizing the main chance. Stahlecker and the SD powerbrokers welcomed such energetic 'hard men' who made it clear that they believed that heartfelt patriotism was for suckers. Arājs swiftly proved himself dedicated and proficient *génocidaire*. And loyalty, obedience and devoted service brought, as Himmler promised it would, rewards.

On 2 July, Stahlecker ordered the Latvian police auxiliaries to begin rounding up Jews in Riga and then bring them to square in front of the police prefecture. Stahlecker assumed that ethnic Latvians, attracted by the hubbub, would spontaneously turn on the Jews. But this did not happen – and both Stahlecker's consolidated report and the Einsatzgruppe morning reports sent from Riga described the Latvians as 'absolutely passive in their anti-Semitic attitudes'.[25] Stahlecker turned for help to Arājs who took immediate action. He published an appeal in a semi-official newspaper *Tevija* for 'all nationally conscious Latvians – Pērkonkrusts members, students, veterans and others to participate in the cleansing of our country of destructive elements'.[26] Stahlecker had supplied the words. The response was 'overwhelming'; thousands of young Latvians volunteered. Many were no doubt lured by the thought of plunder. German propaganda that associated the Soviet terror with 'the Jews' had also done its insidious job. Many Latvians, like the death dealer of Kaunas in Lithuania, now sought revenge. A former policeman, whose family had been arrested by the NKVD, promised 'I will kill every Jew in sight'.

On a warm evening at the beginning of July, Arājs unleashed his pent-up paramilitaries. Armed bands rampaged along streets with known Jewish residents. Like the Lithuanian snatcher squads, they kidnapped anyone they believed was a Jew. Arājs and his men dragged their captives, both men and women, back to their headquarters, where they were beaten in basement rooms and left to die. They raped and tortured Jewish women. Arājs' headquarters soon acquired a terrible notoriety. No 19 had become a house of horrors.

This first action was much too tentative for Stahlecker. A week later, he ordered Arājs to begin attacking Riga's synagogues; this new campaign would be murder by arson. All over the east, images of burning synagogues would provide potent emblems of German power over life and death – a signal that Jews had forfeited any protection from the law and their fellow Latvian citizens. Arājs commenced operations with the Gogolu iela Synagogue (the Gogol Street Choral Synagogue). He sent Jewish prisoners into the building to remove valuable Torah scrolls and candelabras. Then the Arājs men herded hundreds of men, women and children inside the synagogue, emptied cans of gasoline around the high alter and made their getaway. Arājs himself gleefully fired the shots that ignited the fuel and rapidly set the entire building on fire. His men then formed lines to prevent any firefighters from trying to damp down the inferno. The Arājs men incinerated other synagogues that same evening – as well as the prayer house in the Smerli cemetery. As the synagogues burnt, the killing escalated. By the end of the second week of the occupation, 8,000 Jews had been murdered in central Riga.[27] Arājs and his Pērkonkrusts friends began hoarding looted treasures at No 19, eventually acquiring enough artefacts to set up an 'Anti-Semitic Institute' modelled on the German museums that displayed Jewish books and artefacts as evidence of their eternal perfidy.

Stahlecker was under pressure to report the most impressive figures to SD headquarters in Berlin. This meant that large-scale 'special actions' needed to be moved outside the centre of Riga so that victims could be easily disposed of in the soft Baltic sands. In the centre of Riga there was no single 'Jewish Quarter' and before the establishment of the ghetto in October, attacks on Jews had necessarily been patchy. Arājs or one of his Latvian officers suggested transferring operations to the Bikernieki Forest, a thickly wooded area 5 miles to the north of Riga. The forest encircled a garden city called the Mežaparks, which German residents called the Kaiserwald. Affluent Latvians had built scores of luxury villas dotted about in bucolic forest glades, but Arājs tracked down a secluded spot close to the Riga road. Here he would carry out Stahlecker's instructions. Over a period of several weeks, the Arājs men arrested Jewish men and alleged communists and held them 'for questioning' at their headquarters. They then drove them in batches (usually about seventy individuals) to the execution site. During the first phase of operations, between 6 and 7 July, the Arājs men murdered at least 2,000 Jewish Latvian men in Bikernieki Forest. As they refined the operation, the numbers rose steadily through July, August and September.

By mid-July, Stahlecker had moved on (he stayed in Riga just two weeks) and the commander of Einsatzkommando 2, Dr Rudolf Lange and his assistant Arnold Kirste, assumed overall command of the Latvian auxiliaries. Under Lange, the SD became the dominant occupation agency in Latvia. The 'intense and dedicated' Lange hated Jews so much that, according to Joseph Berman, a Holocaust survivor, 'he could not look at them'. A favourite of Heydrich's, Lange was a fervent ideological anti-Semite. In January 1942, Heydrich invited him to attend the Wannsee Conference – called to plan the 'Final Solution of the Jewish problem' – and he was congratulated on his 'success' in Latvia. Short and dark complexioned, Lange was a vain man who toured his fiefdom decked out in capacious military greatcoats with fur-lined collars. He rejoiced in his demonic legend as the 'Bloodhound of Latvia'.[28] Lange was a German 'officer and a gentleman' – and a killer. Under Lange's management, the SD reached into every nook and cranny of Latvian society. His SD men, with their field grey uniforms, black ties, yellow shirts and peaked caps bearing the death's head motif, became all too ubiquitous.

At the end of July, Himmler returned to Riga. He must have been impressed by the activities of the Latvian auxiliaries; as we have seen, just a few weeks before Himmler had authorised the formation of Schuma battalions in the east, disregarding Hitler's insistence that 'it must never be tolerated that a person who is not a German carries weapons'. Now he was beginning to mull over deploying the Schuma outside their national territories as bandit hunters.[29] In Riga, Lange's SD recruited scores of different auxiliary Latvian police brigades. The most feared were the 'Schutzmannschaft in geschlossenen Einheiten' (the closed brigades).[30] Lange usually referred to his Latvian murder squads as the 'Arājs people'. The Arājs Commando (now officially the 'Latvian Security Police and SD Security Force') had proved its worth to the SD and Lange soon reduced the numbers of German SD officers responsible for supervising operations, though he personally liked to attend large-scale executions. One of the commando recruits, Genadijs Mūrnieks, described a typical *Judenaktion*: 'the actions began at 2 or 3 am ... From the place where the victims were let off [the truck], they were driven through a 'corridor' that consisted of Arājs commando policemen, and they were shot by the pit.' The Germans always provided generous quantities of alcohol, usually schnapps or vodka to inspire their executioners. Most such operations were completed in time for a hearty breakfast.[31]

The SD squads faced formidable logistical problems. Outside Riga, many Latvian Jews lived scattered in small market towns, often situated at crossroads. In bigger towns, the SD could make use of local Latvian police – but Lange needed some means of throwing his net deeper into the countryside. For the people of Riga, a fleet of sturdy, cheerful-looking blue buses imported from Sweden had become a part of everyday life. Arājs had commandeered a few of the familiar blue buses to transport prisoners to the Bikernieki Forest. Now at the end of July 1941, he commandeered an entire fleet of the blue buses to convey the 'Arājs boys' to every corner of Latvia in style and comfort. Each bus could carry about forty commandos, equipped with rifles and shovels – as well as supplies of vodka, cigarettes and sausages. When the buses rumbled into some remote Latvian hamlet, local police officers had often already rounded up Jewish men from nearby villages. Arājs himself always arrived in a chauffeur-driven car. He was frequently inebriated after the long, vodka-refreshed drive. The job swiftly became routine. The Latvians disembarked from the bus and began excavating a shallow pit. The local police pushed their captives to the edge and the Arājs men began firing. As the pit was covered, the shooters leant casually against the sides of their blue buses, slurped vodka and wolfed down steaming hot sausages. For Arājs' SD bosses, these blue bus actions proved most satisfactory: the Latvian auxiliaries, fuelled by nicotine, sausage and vodka, efficiently carried out mass executions and burnt synagogues in every corner of Latvia.

Post-war Soviet investigations and German legal proceedings have provided a great deal of information about the men who served with Arājs. His deputy commander Herberts Cukurs was a celebrated long-distance pilot, sometimes called the 'Latvian Lindbergh'. After 1941, the 'Eagle of the Baltic' acquired a second reputation as the 'Hangman of Riga'. After the war, Cukurs fled to Rio de Janeiro, where he lived openly under his own name. The Brazilian government turned down a succession of requests by the Soviets to extradite Cukurs for war crimes. On 23 February 1965, Mossad agents, posing as agents of an aviation business, lured Cukurs to Uruguay. In a beach house called the Casa Cubertini in Montevideo he was tortured and then killed. His decomposed body was found, weeks later, locked in a trunk. It is regrettable that Herberts Cukurs was never brought before a court.[32] When Latvia became independent, the Cukurs family and supporters launched a campaign to restore his name. In 2005, an exhibition opened in his home town of Liepaja proclaiming 'Herberts Cukurs – Presumed Innocent'. Even Latvian historian Andrew Ezergailis has made statements backing the rehabilitation of Cukurs. Ezergailis is alleged to have said that 'MOSAAD killed an innocent man'.[33] But Ephraim Zuroff, Director of the Israel Office of the Simon Wiesenthal Center in

Jerusalem, has discovered eyewitness accounts in the archives of Yad Vashem that prove beyond any reasonable doubt that Herberts Cukurs was an ardent anti-Semite and killer. This 'Latvian Lindbergh' burned to death the family of a Riga synagogue sexton. According to one eyewitness, Cukurs molested and tortured a young Jewish girl; another observed Cukurs shooting and torturing numerous Jews at Arājs commando headquarters and asserts that he tried to force an elderly Jewish man to rape a 20-year-old Jewess. At the end of November, Arājs and Cukurs took a leading part in the Rumbula massacre – as we shall see shortly.[34]

The trial records also tell us a great deal about the commando rank and file – who did the digging and shooting while Arājs and Cukurs sipped vodka. Most recruits were aged between 16 and 21 and the majority had completed secondary level education; 21 per cent had gone on to university. Latvian historians describe the educational practice of the Ulmanis period as 'conformist'. These were bright, relatively ambitious young men, used to following orders – and awash with testosterone. At their post-war trials, some of these young men tried to explain their motivation for joining Arājs: one 'did not wish to do manual work, was eager to advance in life'; a second 'was hostile towards Jews because they had arrested many Latvians [sic]'; another 'wished to enter the University of Latvia, for which he needed a background of 1 year of service in police, German army or RAD'; many 'yielded to the influence of German propaganda'. Other recruits had more mundane reasons: one had 'lost his warm clothes in a card game'. These statements make it quite clear that anti-Semitic propaganda, spewed from SD backed newspapers, had a significant impact on Latvian ears and minds. Many of the volunteers parrot German-inspired mythology that linked the hated Soviets with all Jews – and a few refer to Jews as 'parasites that must be eliminated'. When Arājs led the attack on the Gogol Street synagogue, he is reported to have bellowed: 'Since the people of Riga hate Jews, we must demonstrate our position by setting fire to the synagogue so that nothing of Jewish culture remains.' Anti-Jewish rhetoric rather than opportunism inspired many of the young men who joined the Arājs Commando; these men were not just thrill-seeking hooligans.[35]

In September 1941 Himmler made another visit to Latvia and this time toured the port city of Liepāja (Libau) in the Kurzeme, which had been turned into a German naval base. Since early July, as elsewhere in Latvia, SD and Latvian Schuma brigades had been carrying out executions of Jews, gypsies and suspected Soviet agents, most notably in the Rainis Park shootings. In Liepāja, these actions usually took place in open public places rather than nearby forests. At the end of the month, the Arājs Commando had arrived in Liepāja and carried out a number of much bigger-scale mass executions. But in September, Himmler was appalled to discover that many thousands of Jews remained alive in Liepāja – many of them in the Liepāja ghetto, where they worked as forced labourers for German companies. Himmler vented his spleen on the HSSPF Prützmann, who had plainly not shown quite the same zeal as Stahlecker and Lange. But the bigger stumbling block was the Commissar 'Herzog' Lohse and his boss, the despised Alfred Rosenberg. Their plan for the east depended on extracting maximum profit from Jews incarcerated in ghettoes. So far Lohse had resisted Himmler's importunate demands that every Jew must be liquidated; when profits could be made, murder was simply wasteful. At the end of October, Himmler's patience ran out. He replaced the allegedly 'soft' HSSPF Prützmann with a dedicated and efficient SS killer who could be relied on to ignore any bleating protests from Hinrich Lohse.

The new HSSPF was SS-Obergruppenführer Friedrich Jeckeln.

Himmler admired Jeckeln a great deal. Sometime between 11 and 12 November, he summoned the inventor of the 'sardine packing' method of mass murder to Berlin. The time had come, he informed the new HSSPF, that every Jew in the Ostland Commissariat must be liquidated, '*bis zum letzten Mann*' ('to the last man'). 'Tell Lohse that it is my order, and also the express wish of the Führer.'[36] Jeckeln returned to Riga. On 25 October, he ordered that the Jewish ghetto in Riga be sealed – and so too was the fate of some 33,000 Jews who 'lived' behind its walls.

The 'wish of the Führer' was no rhetorical flourish. In the autumn of 1941 Hitler had begun racking up anti-Jewish rhetoric with a barrage of public speeches and private harangues; it was, says historian Saul Friedländer, 'an explosion of the vilest anti-Semitic invectives and threats'. By September 1941 the German armies had pushed over 600 miles into the Soviet Union along a front line that extended 1,000 miles from north to south; they had occupied the most heavily industrialised Soviet regions – home to over half the Russian population and extending over an area the size of Britain, Spain, Italy and France rolled together. Millions of Soviet soldiers had fallen into German hands, along with many tens of thousands of tanks, guns and artillery pieces. But by the end of September, the German military behemoth was beginning to run out of steam – and Hitler and his generals squabbled bitterly about future strategy, in particular when, and in Hitler's mind *if*, to attack Moscow. Goebbels' diaries provide many insights into Hitler's fluctuating state of mind, and that autumn it would appear that his moods were extremely volatile. At the end of September, Hitler finally resolved to go all out for Moscow. In the first weeks of October, Operation Typhoon achieved spectacular success. Hitler was, it was reported, euphoric. Panic gripped the Soviet capital, and the communist elite and other big wigs began fleeing the city. A special train was put at Stalin's disposal and after considerable vacillation, he decided not to join the flight east. It was the most important decision the Soviet dictator ever made. Sometime after 17 November, as it reached the outer suburbs of Moscow, the German attack unexpectedly faltered. Hitler's war had exhausted his troops. As the German divisions rumbled ever closer towards the great prize, the capital of Jewish-Bolshevism, abysmal roads, chronic food, fuel shortages, and, worst of all, rapidly deteriorating weather sapped morale. As his war machine stumbled, Hitler's public invective against the Jewish 'World Enemy' intensified. The records show that he referred obsessively to the 'extermination of the Jews' on 19 October, 25 October, 12 December, 17 December and 18 December.[37] This poisonous flood of invective seeped down through the German high command. Field Marshall Walther von Reichenau urged his men to exact 'just atonement from the Jewish sub humans'. General Hermann Hoth preached the extermination of the 'spiritual supporters of Bolshevism' and Erich von Manstein urged German soldiers to avenge all 'atrocities' perpetrated by 'Jewry, the spiritual bearer of Bolshevism'.[38]

On 14 November 1941 HSSPF Friedrich Jeckeln descended through thick fog towards a landing strip on the Gulf of Riga. He was preoccupied not with Operation Typhoon, but the Führer's order to escalate the war against the Jews. As soon as Jeckeln had settled into Stahlecker's old headquarters, the Ritterhaus (where he liked to sit fondling purloined jewellery), Jeckeln began searching for a suitable location where he could begin carrying out Himmler's new orders.

German execution sites pockmark the great belt of sand that stretches from East Prussia to the Urals.[39] In Riga, Jeckeln had one special requirement. He knew that in the autumn water seepage could wreak havoc with his 'sardine packing' methods. He would need to find and select a site situated on raised ground. Shortly after his arrival Jeckeln and his aides drove out of Riga in their gleaming Mercedes, along the right bank of the Daugava where the Jewish ghetto was situated, to inspect a new concentration camp under construction at Salaspils. They had just left the city boundary when, a few hundred metres to the left side, Jeckeln spotted a few low, rounded tumuli, dotted with birches sandwiched between the highway and the main railway line that linked Riga to Daugavpils. The place had a name – Rumbula Pines. Although it was not especially secluded, the site had many advantages since it was located next to both the main road and railway. Himmler had ordered Jeckeln to liquidate the ghetto – and that meant he needed to 'process' more than 30,000 'pieces' in a very short time. He was also aware that transports of 'Reich Jews' had already been dispatched from Germany to Riga for his attention. Jeckeln made a decision: the site of the 'great action' would be Rumbula.

Jeckeln commenced detailed planning. Like modern German management practice, his system depended on breaking down every operation into manageable segments, run by different specialist teams. Jeckeln assigned SS-Untersturmführer Ernst Hemicker, who had been trained as an engineer, to supervise the excavation of an appropriate number of pits. Hemicker testified that he was 'shocked' by the number of people Jeckeln planned to do away with, but 'chose not to protest'. Jeckeln drafted in 300 Russian prisoners to dig the pits, each one the size of a small house. Temperatures had fallen below zero so the sandy ground was frozen hard. Work was back breaking. Jeckeln often visited Rumbula Pines to check that construction work was proceeding according to his precise instructions. He was often observed staring down intently into the pits.

In Riga, Lange and Kirste liaised with Arājs and other Latvian SD recruits. On 27 November, Jeckeln invited all units to police headquarters in Riga to finalise the liquidation schedule. This may have been the first time he met Arājs and the other Latvian commanders. So far he had avoided having any contact. It was essential, Jeckeln declared, that everyone 'obtain and maintain a German character': the action had to be carried out in an orderly manner. He had calculated that with only seven hours of daylight available, his men would have to march the ghetto Jews in columns of 1,000 and 'process' at least 12,000 people every single day. Two days later, Jeckeln called another meeting at the Ritterhaus. He insisted that all Germans stationed in Riga must come to Rumbula to observe the action; it was a patriotic obligation. Anyone who refused would be considered a deserter.[40]

On 28 November, the Germans issued a 'resettlement order' and dispatched Latvian police to enter the ghetto to begin preparations. The news of 'resettlement' 'hit like a thunderclap', survivor Frida Michelson recalled.[41] For Latvian Jews, the Riga ghetto had become a fragile refuge. Now they would be moved out the next day for an unknown destination. That night snow fell on the ghetto and on the Ritterhaus, where Jeckeln and his officers made last-minute refinements to their plans. The snow fell too on trains pulling into Skirotava Station near Rumbula and bringing 'Reich Jews' from Germany. German SD squads killed them all while it was still dark.

At 4 a.m. on 30 November, a hundred Arājs men led by Herberts Cukurs and Arājs himself entered the ghetto accompanied by a small, unsuspecting Jewish guard unit. The Latvians were already intoxicated. Arājs and his men strutted through the ghetto streets, ordering the terrified Jews to be ready to leave in half an hour. Some refused. Shooting broke out in houses and on stairwells. Soon enough, Frida Michelson watched an 'unending column' filing through the ghetto; she could see young women, women with infants in their arms, handicapped people assisted by their neighbours, young children – 'all marching, marching'. A German guard began firing an automatic weapon into the crowd. As confusion spread, Latvian guards cried 'Faster! Faster!' People began trotting, running, stumbling … falling.

By noon, the Latvian commandos had already killed between 600 and 1,000 Jews inside the ghetto walls. The frozen ground was streaked with blood. Corpses, suitcases, toys, furniture and prams lay scattered. The road that led from the ghetto to Rumbula led past small wooden houses in which ethnic Latvians were enjoying a quiet Sunday morning. A few 'noble' citizens dashed into the road and began beating Jews. Those who walked too slowly or collapsed were shot. Many abandoned suitcases in front gardens. The road to Rumbula was fast becoming a scene of carnage. 'We were no longer people, only shadows,' Ella Medale wrote. 'Everything around us reminded us of a butchery.' Herberts Cukurs appeared on horseback. He reached down from the saddle to seize small children and kill them on the spot. Michelson went on: 'a German SS man started firing with an automatic gun point blank into the crowd. People were mowed down by the shots, and fell on the cobblestones … The Latvian policemen were shouting "Faster, faster" and lashing whips over the heads of the crowd.'

At 9 a.m. the head of the first column reached the execution site. 'At the entrance,' Frida Michelson recalled, 'stood a large wooden box. An SS man armed with a club stood next to it and shouted over and over "Drop all your valuables and money in this box!"' Now the Jews filtered through a succession of human funnels that stripped them of their clothes and finally their lives. Standing at the highest point overlooking the pits, Jeckeln watched, impassive. Alongside him stood his officers and honoured guests, including the Commissar Hinrich Lohse and Franz Stahlecker who had returned to see the Jeckeln's 'sardine packing' system in action. Ella Medale remembered that people 'overcome by a sense of irreversibility and inescapability rushed forward in a quick stream without protestation'. It was this stripping of both dignity and human will that was most cunning about Jeckeln's industrialised killing system. 'Mechanically I took off my coat,' Ella wrote. When all hope seemed to be lost 'the head executioner Arājs fastened his eyes on me. His face was disfigured, beast like, and he swayed back and forth, horribly drunk. A shriek broke out of me: "I am not a Jew!" … Arājs waved me away. "Here are only Jews! Today Jewish blood must flow".'

Jeckeln halted the slaughter on 8 December. He reported to Berlin that his men had used up over 22,000 rounds of ammunition. If we add to that figure the Jews killed inside the ghetto and on the road to the execution site, as well as the numbers of 'Reich Jews' killed on the first morning, the Germans and their Latvian auxiliaries murdered at least 24,000 people. A report issued by the RSHA in Berlin stated flatly: 'The number of Jews who remained in Riga – 29,500 – was reduced to 2,500 as a result of the *Aktion* carried out by Higher SS and Police Leader Ostland.'[42]

On 12 December, as heavy snow fell steadily on the mutilated earth of the Rumbula Pines, Himmler issued secret instructions to his SS commanders. He had become concerned that the 'heavy duty' of mass murder might 'brutalise' SS men. He recommended organising 'comradely evenings': hard-working SS killers should 'sit around the table and eat in the best German domestic tradition'. Officers must organise music and special lectures to 'introduce our men into the beautiful domains of German spiritual and emotional life'.

As Jeckeln and his guests enjoyed the gruesome spectacle unfolding at Rumbula, Japanese aircraft attacked the American Pacific Fleet anchored in Pearl Harbor. On 11 December, Hitler declared war on the United States – and the following day reiterated a warning first uttered in 1939, that a global war would provoke the destruction of world Jewry. Pitted against the most powerful capitalist nation, Germany now faced certain defeat. Few of the men fighting Hitler's war understood the catastrophic implications of his decision – least of all SS Chief Heinrich Himmler and his busy collaborators in the east.

By the end of 1941 the Arājs Commando had, as Ezergailis puts it, 'run out of work'. A stunning frenzy of killing had extinguished the lives of all but a few thousand Jews in the former Baltic nations. Jeckeln may have briefly contemplated 'sardine packing' Arājs and the Latvian Schuma leaders to silence them. Instead, they made Arājs an SS-Hauptsturmführer and transferred his core staff, now called the Latvian Security Police and SD Auxiliary Force, Riga, to the Military Academy of Latvia. Soon afterwards, Himmler agreed to send small groups of Schuma veterans to the SD training school in Fürstenberg. Early in 1942, the head of the SD school arrived at the Military Academy to select the first class of Latvians; about 300 men travelled to Fürstenberg for a two-month course in 'intelligence, counter intelligence and national socialist ideology'. When the first graduates returned to Riga, Dr Lange welcomed them back: 'After the liberation of Riga, you fought shoulder to shoulder with the German security police for further liberation [sic] and cleansing the land from the remnants of Bolshevism.'

Now Hitler's foreign executioners would be moulded into true SS warriors.

8

Western Crusaders

> A Jew in a greasy Kafkan walks up to beg some bread, a couple of comrades get a hold of him and drag him behind a building and a moment later he comes to an end. There isn't any room for Jews in the new Europe, they've brought too much misery to the European people.
>
> Danish Waffen-SS recruit[1]

Historians of the Second World War tend to split German occupation strategy into eastern and western modes. Each mode of occupation produced different species of collaborator. There are good reasons to make this distinction between east and west. The Germans deliberately situated the apparatuses of mass murder in the east and, as we have seen with the Rumbula massacre, deported Jews from Western Europe to the killing grounds of the east. Security forces that collaborated with the German occupiers in France and the Netherlands, for example, rounded up and arrested French Jews for deportation. They did not, in strategic terms, murder them in situ as their Lithuanian and Latvian counterparts did. In other respects, German occupation plans in both Eastern and Western Europe shared fundamental common grounds. In the long term, the German architects of the New Order looked forward to abolishing the older European national entities just as they planned to do with more recent manifestations of nationalism in the east. When it came to the ethnic restructuring of Europe, German planners simply regarded Norwegians or Netherlanders as having a racial head start over say their Estonian counterparts. In the long term, all Europeans would be remoulded as Germanics. This meant that, as we will see in this chapter, the recruitment of Western Europeans by the SS was driven by the same racial imperative.

In Western Europe, German foreign recruitment had been launched in 1940 with high hopes but negligible results. On 20 April, the SS 'Nordland' regiment was established 'from Danish and Norwegian volunteers', and on 25 May the 'Westland' began recruiting Dutch and Belgian-Flemish volunteers. Waffen-SS recruitment chief Gottlob Berger set up offices in Oslo, The Hague, Copenhagen and Antwerp – and turned an old French barracks in Sennheim, Alsace, into a training centre. Himmler had high hopes. He was fascinated by the Nordic races and had cultivated the far right in the Netherlands and Flanders. But on 6 June 1941 an abject Berger reported to Himmler that only about 2,000 foreign volunteers had so far been 'harvested' by the Waffen-SS.[2]

The German attack on the Soviet Union changed everything. Hitler's war was recast as an international crusade against Bolshevism. After 22 June, Goebbels' propaganda machine began churning out a torrent of newsreels and newspaper stories about the 'pan-European war of liberation'. Hitler could muster tremendous forces to throw against Stalin's armies, but he still needed coalition allies and the veneer of legitimacy. Operation Barbarossa would appear to be a joint European action against the 'World Enemy'; a war waged by civilised peoples against barbarians. *Krieg als Kreuzzug*, war as crusade, was for Hitler mere window dressing, a hoax designed to entice unwilling allies. But Hitler's public statements did not completely mask German intentions. As historian Jürgen Förster points out, 'the propaganda attempted to exploit feelings of being under threat, the desire for revenge, ethnic prejudices, and ideological resentment'.[3] Hitler described Bolshevism as 'the bestial degeneration of humanity' – and in 1941 there were many other Europeans who fervently agreed with the German Chancellor. They understood too that when Hitler referred to Bolshevism he implicitly meant a Jewish political ideology. The equation that added together 'Jews' and 'Bolshevism' provided the lingua franca for European ultranationalists.

The crusade swindle worked. After 22 June, as German troops and armoured divisions threw back Stalin's armies, German consuls and other agencies reported a flood of requests to join Hitler's assault on Bolshevism. Berger had already established recruitment offices in occupied Europe and Scandinavia – and the post-invasion flood of eager western crusaders caught the German Foreign Office and the Wehrmacht on the back foot. They feared, rightly, that Himmler would exploit the foreign volunteers to massively expand the Waffen-SS. On 30 June, representatives of the Foreign Office, the foreign department of the German High Command (OKW) and Berger's SS met to thrash out guidelines for what one delegate termed 'crusade foreign legion-gathering'.[4] A few days later, the OKW issued 'Guidelines for the employment of foreign volunteers in the struggle against the Soviet Union'.[5]

This first effort to categorise and regulate foreign volunteers distributed recruits crudely between 'Germanics' like the Danes, Flemings and Dutch, who would be allocated to the SS, and 'non-Germanics' like the French, Croats and Spaniards, who would be allocated to the German army. Although, as we have seen, Himmler was already intrigued by ethnic diversity in the east, there is no suggestion yet that Eastern Europeans would be recruited by either the German army or Waffen-SS.

Pragmatic German army generals viewed and treated their foreign recruits simply as mercenaries. They took no interest at all in their ethnic origin and blithely rejected any kind of cultural sensitivity. They had no respect for their recruits' political convictions, if they had any, and despised them as opportunists, adventurers or simply men who liked 'to be looked after'.[6] Instinctively, the German recruiters recognised that an entire generation had been psychologically blighted by the First World War and its aftermath. This lost generation was ripe for exploitation. The new campaign proved a modest success. By the end of 1941, 24,000 Frenchmen, Croats, Spaniards and Walloonian Belgians had signed up to join the Wehrmacht. Less than half that number of 'Germanic volunteers of non-German nationality' had joined the Waffen-SS: 2,399 Danes, 1,180 Finns, 1,571 Flemings, 4,814 Dutchmen, 1,883 Norwegians, 39 Swedes, 135 Swiss and Liechtensteiners, as well as 6,200 ethnic Germans from Alsace, Romania, Serbia and Hungary.[7] It is tempting to interpret the division of human spoils as Himmler's last word on racial types he regarded as suitable Waffen-SS material. But we should bear in mind that the SS was already recruiting Schuma battalions in the east; and as we will shortly discover, German race scientists were about to embark on a fundamental rethinking of the geographical extent of 'Germanic' ethnicity.

The national elites that sanctioned German recruitment expected rewards for their sacrifice. They bargained honour for power. The men they offered to the Reich would become, they hoped, the currency of influence in Hitler's New Order. The largest foreign Wehrmacht unit was the Spanish División Azul (Blue Division). On the day German forces crossed the Soviet borders, the anti-communist Spanish Foreign Minister Ramón Serrano Súñer persuaded Francisco Franco to permit the formation of Falange volunteer units to join Hitler's crusade against Bolshevism. Serrano Súñer had made the offer to the German ambassador Eberhard von Storrer some weeks before invasion began. Hitler had eagerly agreed. Relations between the German dictator and the 'Caudillo' Franco had been soured by Spain's refusal to take part in a proposed joint attack on Gibraltar, but Foreign Minister Ribbentrop still hoped to bind the hesitant Franco closer to the Axis. For his part, the Spanish dictator assumed that he could placate Hitler and avoid joining the German war in the east by offering up a Spanish volunteer division. At the same time, he could rid Spain of a few thousand discontented radical Falangists who, conversely, hoped that service in Hitler's 'Grand Armée' would give them political leverage when they returned victorious to Spain. Many thousands of Spanish Falangist volunteers enlisted. By the beginning of September 1941 the Azul men were fighting and dying near Minsk in Belorussia. German officers despised these 'gypsy' Spanish soldiers. They disparaged them as lazy and naturally undisciplined. Although Hitler often belittled the Azul, calling it a 'dilapidated' division, he had to acknowledge that they fought 'pluckily'.[8]

Likewise, French far-right ideologues like Jacques Doriot and Marcel Déat saw the crusade against Bolshevism as an opportunity to press for closer relations with Nazi Germany. Said Doriot: 'If there is a war to which I am sympathetic, it is this.' He urged Frenchmen to join a volunteer legion to fight Bolshevism – and found an ally in the Francophile German ambassador Otto Abetz, who saw one of his diplomatic tasks as bolstering pro-German cliques in occupied France. But the Wehrmacht rejected French overtures, as well as an appeal from the Belgian demagogue, the Walloonian Léon Degrelle. According to German race science, the French and the Francophone Walloons were not Germanic peoples like the Dutch or Flemish. Hitler had more pragmatic reasons for spurning Doriot's proposal: he feared that the Vichy puppet government would exploit a 'French division' to elevate its status as a German ally. Abetz realised that Hitler's objection did not apply to occupied France. He also firmly believed that the 'nation of the Franks' had deep historical ties to Germany through the medieval empire of Charlemagne. He sympathised with Doriot and his followers, who believed that active collaboration could restore France to its former glory, albeit as a German satellite. It helped that both parties spoke the language of Jew hatred. Since the Dreyfus Affair, the French right had represented Jews as Bolshevik enemies of France and blamed them for the catastrophe of 1940. For Doriot, the German attack on the Soviet Union was a God-given opportunity to extinguish the Jewish threat once and for all so that France might once again become the 'most Christian Kingdom of Europe'. Ambassador Abetz made a direct appeal on Doriot's behalf to his friend Foreign Minister Ribbentrop, and on 5 July the Wehrmacht finally agreed to form a French volunteer unit. They insisted that French Vichy government support was 'not wanted'.

When Abetz got the green light from Berlin, he appointed Déat and Doriot to organise a recruitment campaign. He set them up in an office at the German Embassy in Paris, and they issued an appeal for volunteers on 8 July. Only Frenchman of proven Aryan descent would be accepted. On 27 August 1941 a call-up ceremony for the 'Légion des Volontaires Français contre le Bolshevisme' (LVF) was staged at the Borgnis-Desborde barracks in Versailles. On the podium, reviewing the rather disappointing turn out, stood the main actors of collaboration: Prime Minister Pierre Laval, Wehrmacht liaison officer Marquis Fernand de Brinon and, of course, Déat and Doriot. German officers noted that a number of dark-skinned French North African troops, as well as elderly White Russians domiciled in France, had somehow slipped into the ranks. As the ceremony reached a climax, one of the volunteers began firing at the assembled dignitaries, wounding both Laval and Déat. The shooter turned out to be a dissident pro-German. The LVF was a fractious coalition of bigots and fantasists. As they began the long journey to the Eastern Front, LVF commander Colonel Roger Labonne rallied the troops; he compared himself to French crusader Godfrey de Bouillon, excoriated Stalin as 'Attila, the scourge of God' and urged his men to fight for a new, healthy France 'free from the yoke of the ghettoes, the lodges, Bolshevism, and British gold'.[9] Their campaign would be less than glorious. As soon as the exhausted French recruits reached Radom in the General Government, German officers purged the ranks of 'coloured troops', former Foreign Legionaries and those Russian émigrés. In November, as Hitler launched Operation Typhoon and all along the Eastern Front temperatures dropped below zero, the Germans sent 2452 LVF (638th Infantry Regiment) men into action on the isthmus of Kubinka, located just 85 miles from Moscow. According to the German report, 'Some [French] units have still not arrived. During moving off, more signs of disintegration became apparent … the men are inadequately trained.' Only a few hundred French volunteers returned alive. In the spring of 1942, the relics of the LVF joined anti-partisan forces in Belorussia.[10]

Many prominent collaborators like Anton Mussert in Holland, Frits Clausen in Denmark, Vidkun Quisling in Norway and Staf de Clerq in Flemish Belgium all hoped, even assumed, that offering up their followers as cannon fodder would open the door to political power. They were wrong. In 1941, Hitler and German administrators in occupied Europe could afford to be discriminating. Foreign volunteers, however 'Germanic', would never be true 'brothers in arms'. When a Vichy newspaper referred to the war on the Soviet Union as a European campaign, Hitler dismissed the description as insolence.[11] His contempt seeped downwards through the ranks. German soldiers called Spanish Azul volunteers 'gypsies' whilst the French generally lacked 'German thoroughness'. In the spring of 1942, the Flemish SS leader Staf de Clerq complained to Berger that German officers habitually insulted his Flemish comrades by calling them 'filthy people' and a 'nation of idiots'.[12] For the Germans, this crusade could never be regarded as a coalition of equals.

For Wehrmacht recruiters, patriotic sentiments threatened to undermine battlefield effectiveness. Himmler's position was even more radical. As we have seen, SD commanders like Franz Stahlecker resisted any attempt to form 'national militia'. In Western Europe, by the same token, a foreign legion must take up arms for the Reich, not any kind of 'national state interpretation'. Himmler's racial doctrines insisted that blood, not nation, would be the foundation of recruitment. Blood kin not national allies would wage war on the 'Jewish-Bolshevik' enemy. In a lecture delivered in June 1942, he discussed deploying 'Germanic' Waffen-SS as a defensive wall along the new eastern frontier of the Reich. As soon as the conquest of the east had been completed, the Waffen-SS would be 'rehabilitated' and absorbed by the 'General SS'; victory would permit the 'bringing home and fusion of the Germanic nations with us'. It was imperative, he went on, not to stop at the creation of a 'Greater Germanic Reich'; the 'road upwards' led to a 'Gothic-Frankish-Carolingian Empire'.[13] In other words, foreign recruitment would become the foundation of a vast SS empire that far exceeded even Hitler's ambitions.

As 'Germanic' SS recruitment at last gathered pace, Himmler relied on two powerful SS agencies. The most significant was the Race and Settlement Office or RuSHA, which had been set up to begin with as the SS Race Office in 1931 under Richard Darré.[14] Its task was to promote ideological indoctrination and screen German SS recruits and their spouses. The Family Office (Sippenamt) meticulously checked the medical history, religious and political affiliations and, above all, the family background of every applicant. As Himmler's racial gatekeepers, the handful of RuSHA experts acquired enormous power. Now ten years later, their expertise would be applied to sift the new 'Germanic' volunteers. The RuSHA allocated SS-Eignungsprüfen (racial acceptability checkers) to SS recruitment offices. In Belgium, for example, Flemish but not Francophone Walloonian volunteers were channelled through the Ergänzungsstelle Flanden. To entice volunteers in greater numbers, a second SS agency, the 'Germanische Leitstelle' (GL), launched a Europe-wide propaganda campaign to convince suitable individuals to join Hitler's crusade. The GL had separate offices in each of the nominated 'Germanic' capitals, like Oslo and The Hague, but worked closely with the RuSHA experts. Their main task was writing and publishing *Germanische Leithefte* (Germanic guides) in Flemish, Dutch, Danish and Norwegian. RuSHA experts wrote many of these booklets – and their work was co-ordinated by Dr Rudolf Jacobsen, an ardent anti-Semite, with a small staff of specialists and translators.

Berger took a close interest in the Germanic propaganda that poured from Jacobsen's little empire. He published his own collection of essays, *Auf dem Wege zum*

Germanischen Reich (*The Road to the Germanic Reich*), which presented the Reich as the saviour of the 'Germanic ideal'. 'The north of Europe was the homeland of the Nordic race,' he wrote. 'The Germans have their roots in the Nordic race.' In the Middle Ages, Berger went on, the Holy Roman Empire (the First Reich) had defended the west from the menace of 'Asiatic' tyrants. Now Hitler's new Reich had taken on the same tremendous task. The wicked Anglo-Saxon English had betrayed the 'Germanic family of nations' but thanks to the SS, mainland European peoples could properly fulfil their obligations as members of the Nordic family. Berger's thinking owed much to a German cultural historian called Christoph Steding (1903–38). A rising star in German academia, Steding had won a grant from the American Rockefeller Foundation which he used to study humanities and social sciences in Switzerland, Holland, Denmark and Sweden. His encounters with the 'tepid democrats' and 'mercenary capitalists' of the 'old Europe' convinced him that only the new Germany 'forged by Bismarck and Hitler' had the answer to the decomposition (*Zersetzung*) of the old Nordic values. He railed against 'wild, demonic, ruthless, bestially savage primordial forces' that allowed Jews, the vanguard of Bolshevism, to manipulate the destiny of the European community. Everywhere he travelled, Steding noted the infiltration of 'inner Semitization' (*Verjudung*) – 'thus the enemy stands right at the heart of Europe'. By the time Steding returned to the beloved Fatherland, he had completed a magnum opus that advocated German hegemony as the only answer to this European sickness: '*Europas Krankheit*'.

Steding succumbed to an aggressive kidney infection soon after he returned to Germany and *Das Reich und die Krankheit der Europäischen Kultur* (*The Reich and the Sickness of European Culture*) was published posthumously. (The edition in the State Library in Berlin displays Steding's rugged features as frontispiece and his handwritten inscription of 10 May 1936 '*Alles für das Reich*'.) Despite its daunting length and ponderous style, Steding's tome soon won over some powerful admirers. One was SD Chief Reinhard Heydrich, who sent a synopsis to Himmler.[15] Although Steding's posthumous tome never acquired the iconic notoriety of Hitler's *Mein Kampf* or Rosenberg's *Myth of the 20th Century*, thanks to Heydrich's backing, Steding's ideas became common pseudo-intellectual currency in SS circles and, a decade later, had a powerful impact on the German founders of the modern European Economic Community.[16]

Steding had tried to explain why other Europeans appeared to begrudge 'natural' German ascendancy. Why did the rest of Europe not do as the virtuous Germans did? The cultural price paid for spurning German hegemony, he argued, had been calamitous: Europe was dominated either by the Roman Church or by its antithesis, Bolshevism. European civilisation had become rootless and cosmopolitan – in other words, 'dominated by Jews'. European institutions, like the League of Nations and the World Court, had become impotent and corrupt. The only way forward, Steding concluded, was that the new Reich sweep aside liberal, decadent Europe and replace it with a Prussian, Protestant empire. After 1941, Steding's book was adopted as the bible of the SS 'Germanische Leitstelle'. SS propagandists seized on Steding's favourite metaphor (which was hardly original) to proclaim that the new Germany would restore decadent European civilisation to rude health. A new Europe would arise defended by a new 'European army' – a proposal that echoed Himmler's idea of a 'Germanic' Waffen-SS, which would become, as he put it, 'a National Socialist order of soldiers infused with a Nordic sensibility, into a distant future'.[17]

The German Jewish diarist Victor Klemperer, in his study of the *Language of the Third Reich*, pointed out that this National Socialist European Union embodied tactical concepts of attack and retreat. Betrayed by the 'English', squeezed between the United States and the Soviet Union, the task of the new Germany would be to defend 'Festung Europa'.[18] The idea was powerful and pervasive. A Dutch SS volunteer wrote to his mother, 'Our great leader Adolf Hitler will construct a new Europe and will lead us towards freedom. Our gracious lord will let Germany be victorious, and I am proud to have marched with the German comrades on the path towards freedom.' But what kind of freedom? Hitler had no doubts that German hegemony would be absolute. Europe would no longer exist as a patchwork of sovereign states. Himmler developed even more radical plans, which he kept secret for obvious reasons. His future 'SS Europa' would be one dominated by Germany, but in the singular shape of the SS. Belgium and Norway, for example, would become SS Gaue (provinces), ruled from SS Main Offices in Copenhagen and Oslo. In Himmler's vision of a New European Order, there was no room for a national party, not even the NSDAP – or, for that matter, its leader Adolf Hitler. Heresy, to be sure – and one that Himmler would certainly have publicly disowned. But it was a high-ranking SS officer who dared imagine this future without Hitler. In a discussion document, 'On the Leadership and Administration of a German European Empire' ('*Über Führung und Verwaltung des europäischen Reiches der Deutschen*'), SS Bewerber Wolfram Heinze set out a vision of a 'Greater German Reich of the Germanic Nation' that Hitler would not have recognised. The attempt to weld party and state, Heinze argued, had failed. Hitler tolerated a wasteful profusion of offices and rival competitors for power. Heinze proposed instead that the SS Main Offices replace both party and state offices, not just in Germany but right across the Greater German Reich. Holland and Flanders, for example, would as a first step become Reichlands, ruled by an SS 'regent'. Heinze's proposal applied to the Reich as a whole, including Western Europe, the administrative structures already imposed on Eastern Europe – the Reich Commissariats that welded, for example, the Baltic nations and Belorussia into a single territorial entity, the Ostland. Although Berlin would continue to have a pivotal governing role, daily administrative tasks would be devolved to this cadre of SS regents, so that the SS alone would bind together the entire Reich. Heinze's paper was never officially published; it was quietly circulated among like-minded SS men.[19]

Himmler understood this future SS hegemony in racial terms. In 1943 he lectured Waffen-SS officers:

> The result, the end of this war will be this: that the Reich, the Greater German Reich of the Germanic Reich of the German nation [sic], will with just title find confirmation of its evolution, that we will have an outlet and a way open to us in the East and that centuries later a Germanic World Empire will be formed.[20]

From his European GL offices, Dr Jakobsen energetically promoted this mythic ideal tricked out as a 'European crusade' with a flood of pamphlets and periodicals like *Der Aufbruch* (*The Uprising*) and *Der Germanische Gemeinschaft* (*The Germanic Community*). Translated into every European language (except Yiddish), these SS pamphlets sold hundreds of thousands of copies. From very early on, the GL emphasised that the task of the Germanic people, including Waffen-SS recruits, was to safeguard Europe. According to one widely read pamphlet, that provided raw material for an eight-week training module: 'Germany will never abandon the task of racially and politically defending Europe. The most valuable races of Europe shall never again be spoiled by alien blood and ideologies of alien races.' The Waffen-SS was an 'assault force for the new Europe' that would claim the east for the west. The task of the SS was to ensure that only people of Germanic blood could occupy the Reich. Himmler's 'European Union' was a kind of blood reservoir whose contents would be poured into the vast expanses of Eurasia, from the Pyrenees to the Urals.

In *Hitler's Willing Executioners*, Daniel Goldhagen singles out a remarkable story of wartime heroism to bolster his case that the Holocaust was a German phenomenon.[21] In 1943, Danish government officials and ordinary civilians saved the lives of 7,000 Jews who were about to be deported by the German occupation authorities. According to Goldhagen, ordinary Germans willingly participated in 'exterminatory anti-Semitism' or, at best, exhibited cruel indifference to the fate of European Jews. Danes behaved in an exemplary, heroic way; the majority of Germans did not.

But analysis by a younger generation of Danish historians has shed a less flattering light on the Danish record. Denmark, as well as other Scandinavian countries, contributed many thousands of 'Germanic' Waffen-SS volunteers, who eagerly participated in the German crusade against Bolshevism. Many hundreds of these SS volunteers worked for the police and terror units that helped to round up Danish Jews who had not been able to escape in 1943.

It was not, of course, the fault of northern Europeans that the Nazi elite so fervently believed that Scandinavians were blood brethren. National socialist ideology embraced Danes and Norwegians, Dutch and Flemish as Germanic peoples, related by blood. Hitler said in 1934: 'the Nordic countries ... as well as Holland and Belgium [belong] to Germany.' Alfred Rosenberg believed in 'the big common destiny of Scandinavia and the people of the Baltic Sea'; the necessity of 'one entire awareness of the Nordic countries'. According to Rosenberg, this 'Nordic ideal' made it imperative to form a 'German-Scandinavian Block' charged with defending Northern Europe against the Bolshevik menace. Himmler warned that 'all good blood on this world, all Germanic blood, which is not German ... could be our ruin'. What he meant by this was that because Nordic blood possessed such potency, it might be used against Germany if it was purloined by enemy nations. That's why a 'Germanic empire has to be created ... as a home for Nordic blood', as 'his will be the strongest magnet, which can attract this blood'. Himmler declared that he had been 'assigned by the *Führer*, to advance the Germanic idea of the empire'. Hitler, he emphasised, did not intend 'to let one Germanic go to America. We must integrate all the Norwegians, Swedes, Danes and Dutch'.[22] And in the Nordic nations, Himmler's 'pan-Germanism' was reciprocated. There were many Danes, as well as Norwegians and Swedes, who longed to be integrated into Hitler's European union.

In his autobiography, the late Swedish film and theatre director Ingmar Bergman confessed not only that his father held ultra-right-wing opinions but that he had himself been deeply impressed by Hitler when he visited Germany on an exchange visit in 1936 and lived with a family of ardent national socialists. They proudly took Bergman to hear Hitler speak at a Party rally – 'unbelievably charismatic', he recalled. When he returned to Sweden, Bergman, his brothers and some friends attacked the house of a Jewish family and daubed it with swastikas. Bergman's confession provides yet another reminder of the centrality of anti-Semitism in National Socialist ideology, and how easily it contaminated other national cultures. Clergy of the Church of Sweden applied the German Nuremberg Laws when Swedish couples applied for permission to marry and prohibited marriage between Swedes and partners of Jewish descent. A mythology of racial superiority infected the national cultures of the Nordic nations – and when Himmler and Berger set up recruiting offices in Copenhagen and Oslo, many young men eagerly signed up. By the end of the war, 13,000 Danish citizens had volunteered to serve in the armed forces of the Third Reich; just over half ended up enlisting. The majority, about 12,000, volunteered for service in the Waffen-SS and ended up serving in three different formations: the Frikorps Danmark (Danish legion), the SS Division 'Wiking' and, following the disbandment of the Danish legions, the SS Division 'Nordland'. About 1,500 SS volunteers came from the German minority in southern Jutland; they ended up serving in the SS Totenkopfdivision and the 1st SS Brigade – both units had a direct involvement in 'special actions' against Jews in the east.

Many Danish volunteers embraced German racial doctrines. And, crucially, these beliefs shaped how they acted as SS volunteers. Here are two extracts from letters written by Danish SS volunteers:

> A Jew in a greasy Kafkan walks up to beg some bread, a couple of comrades get a hold of him and drag him behind a building and a moment later he comes to an end. There isn't any room for Jews in the new Europe, they've brought too much misery to the European people.

> The other day we visited a large lunatic asylum near Munich and attended a lecture on racial science. It was fantastic to watch the mob of human wrecks they'd gathered there, I just wonder why they keep them alive ... Afterwards we visited the famous concentration camp Dachau and saw it from one end to the other. It was a great experience; you all know what one hears about concentration camps in Denmark, like the rest, it's lies from end to end [sic].[23]

One of the most revealing documents discovered by the Danish historians is the wartime diary of an SS volunteer called Harald.[24]

Harald was recruited on 30 July 1941 and discharged on 17 January 1944. His diary records the history of the Frikorps Danmark from its beginnings in 1941 to 1943, when it was dissolved and replaced by the SS 'Nordland'. When he volunteered Harald was 40 and because of his age was appointed quartermaster in charge of supplies. He did not have front-line experience. Nevertheless, Harald served with the Frikorps in Demjansk, Neval and Zagreb and had daily contact with Danish SS recruits who 'did the real fighting'. He provides a grim picture of everyday service in the Danish SS and we hear a lot about the boredom, complaining and drinking; the longing for home and family:

> When one just returns from leave, one can really see much of the stupidity there is in the 'Frikorps', and then one just wants to be alone with his thoughts. I think that everything is idiotic and straight to hell today, but it will probably be better in a couple of days.[25]

But the diary offers a lot more than just a record of often banal experience; it is the testimony of a believer. For Harald was a member of the Danish Nazi Party (DNSAP) and among the first wave of Danish SS recruits.

We know very little about Harald before he began writing his diary in 1941. He was an illegitimate child and seems to have endured an unstable childhood. When he was just 13 he was accused of fraud, but escaped punishment when he was adopted as a ward of court. Later he was in trouble again for assault, probably a pub brawl. Harald eventually married although very little is known about his partner. By the time he volunteered to join the Waffen-SS in 1941, the couple had a grown-up daughter. Harald often refers to family, and when his periods of leave ended he often suffered depression. In 1940, Harald joined the Danish National Socialist Party, which had been established in 1930, and was led by a doctor, Frits Clausen. The DNSAP, like most of the European far right, had modest electoral success and was plagued by schisms. Many DNSAP members grumbled that Clausen was not radical enough about Denmark's 'Jewish problem'. The Danish Anti-Jewish League split away from the DNSAP and tried to forge closer links with the German NSDAP. Its leader Aage Andersen published a newspaper *Kamptegenet*, modelled on Julius Streicher's poisonous *Der Stürmer.*

After the German occupation began in 1940, Clausen hoped to be appointed Danish national leader. Instead the German Plenipotentiary in Copenhagen, Cecil von Renthe-Fink, marginalised the DNSAP. Fanatics made Hitler nervous. Later in the war, Himmler cultivated Clausen, who served on the Eastern Front, and later had him sent to a German sanatorium to cure his chronic alcoholism. For Harald, the DNSAP was a family affair: his wife also joined up and later worked as a cleaner for German staff stationed at Kastrup Airport.

In June 1941, following Hitler's authorisation of national legions, the Germans ordered the Danish government to launch a recruitment drive for the Waffen-SS. For Clausen and the failing DNSAP, the German invasion was a godsend. On 23 June, he made a speech calling on Danes to enlist in the war against *Der Weltfeind* (World Enemy), meaning of course 'Jewish-Bolshevism'. The Danish government protested. Clausen turned for help to a Danish army officer and nationalist Lt Col C.P. Kryssing, who persuaded the Danish War Ministry to permit foreign military service. Danes had already enlisted in the SS 'Nordland' regiment, now

Clausen and Kryssing offered enough volunteers to form a separate battalion, the Frikorps Danmark. On 19 July, 435 officers and men, led by Kryssing, whose two sons also enlisted, staged a ceremonial passing out parade in Copenhagen, attended by Danish officials, to the sound of a German marching band.[26] Recruitment offices staffed by DNSAP men were set up all over Denmark. On a cloudy summer's day Harald walked into the Copenhagen SS office. Soon afterwards he and 600 other SS volunteers took an oath of loyalty to Adolf Hitler and pledged to 'struggle against Bolshevism'.[27] Many officers came from regular and reserve ranks in the Danish army. To be accepted, volunteers like Harald had to prove Aryan ancestry, financial solvency and that they had no criminal record.

Over the next two and a half years, Harald's diary was his loyal companion. He made daily notes – and there is rarely more than a week between full entries. In Copenhagen, he boarded a train for Langenhorn near Hamburg, where he was placed in a mixed SS regiment, the SS 'Standarte Nordwest', which comprised Danes, Dutchmen and Flemish volunteers. After just one week, the Danish volunteers transferred to the Frikorps Danmark. As training began, the German officers became increasingly impatient with the patriotic Kryssing. Although he had co-operated with Clausen, he was not at all a convinced ideological National Socialist and openly objected to any ideological training of his recruits. In October 1941 the Germans transferred the Frikorps Danmark to barracks in Treskau. It was here that serious tensions escalated between Kryssing and the SS, and then between Kryssing and the fanatical DNSAP volunteers. A German Hauptsturmführer, Masell, recommended that the Frikorps relocate to another camp closer to the SS 'Junkerschule' at Posen-Treskau. Here, he argued, the Danish recruits could be properly inducted in SS discipline and doctrine. Kryssing strenuously resisted. In November 1941, Harald made this entry in his diary:

> I guess the Germans will loose their patience soon with the 'Frikorps': it is not the rank and file with which there is something wrong – it's the officers. The Danes are used to having it their way, and Kryssing is too old for an undertaking like this: it should instead have been a younger man with initiative, like Hauptsturmführer von Schalburg. It could be that the day is not far away when changes are going to happen within the command of the Frikorps.

This was prescient. At the beginning of 1942, Himmler, who had followed the disputes with Kryssing, was informed that neither the Danes nor the Norwegians were 'combat ready'. He sent SS-Gruppenführer Krüger to find out why. His report concluded that, as Harald had observed, the Danish officers in the Frikorps were either incompetent or covertly hostile to Germany. He claimed that Kryssing and his deputy Sturmbannführer Jörgensen often expressed anti-Nazi sentiments and were incapable or unwilling to impose discipline. Feuds between rival Danish factions unsettled both officers and men. There was an obvious solution: get rid of the Danish officers and replace them all with Germans. But the cautious Krüger persuaded Himmler to replace Kryssing with another Danish officer, SS-Hauptsturmführer Christian von Schalburg, then serving with the SS 'Wiking' division. Schalburg, Krüger believed, was a 'reliable national socialist'. That was an understatement.[28]

Schalburg was a monster. He had been born in St Petersburg; his father was Danish and his mother an aristocratic. When Christian was 11, the family became caught up in the Russian Revolution and fled to Denmark. From his parents, especially his Russian mother, Schalburg inherited a visceral hatred of communists and Jews – in his mind, one and the same. In Denmark, the Schalburgs gravitated towards the DNSAP. Christian was devoted to DNSAP leader Clausen, who appointed him leader of the youth wing. After leaving school, Schalburg joined the elite Danish Royal Guards (Den Kongelige Livgarde); in an official report he was described as 'dangerous and unstable' – mere Jewish slander, responded Schalburg. When the Winter War broke out between the Finns and the Soviet Union, Schalburg took a brigade of DNSAP youth members, the 'Blood Brothers', to fight alongside the Finns. It was on the Finnish front line that Schalburg and his comrades heard that Denmark had capitulated to the Germans. He said later that he was ashamed. But in September 1940 Schalburg volunteered to join the Waffen-SS and, by the time the German invasion of the Soviet Union began the following summer, he had been promoted to SS-Hauptsturmführer. As Harald had intuited, the no-nonsense, politicised Schalburg proved to be the right man to whip the Danish SS into shape. This he proceeded to do with characteristic ruthlessness.

Harald's diary entries, which inevitably became shorter and less frequent, show that he was increasingly confused about serving with the Frikorps Danmark. From the spring of 1943, Harald began to submit requests for a discharge. His diary reveals that he was frustrated by the harsh, sometimes frightening life of a soldier – but there were also other discontents. Harald remained an ardent National Socialist. The second wave of SS recruiting after June 1941 had attracted a different class – more bourgeois, less fanatical. Here is one vitriolic diary entry from September, 1941:

> Not all of the people who have come down here to the Frikorps are entirely good. It is said that the corps is apolitical – therefore there are people from all camps, amongst the ordinary ranks as well as between the föhrers and the unterföhrers [officers]. There

> are some bad examples in particular amongst the two latter categories: father's sons, who only have on their minds to become decorated as Christmas trees or Shrovetide birch twigs, with as many stars and cords as possible. But let us see what the gentlemen are worth once we get to the front.

Harald had also become disillusioned with Clausen: 'When I joined the Waffen-SS, I was an ardent Fritz Clausen supporter, today I am an ardent Frits Clausen opponent.' Harald explains why he had turned against Clausen:

> Why should we who are a pure white Aryan nation, have a Southern Jutlander as a leader? Those people are a minority: we might as well have an Icelander – a native from Bornholm [a small Danish island in the Baltic Sea] or an Eskimo. We demand a pure white free-born Danish man!

Harald's diary provides some evidence that he knew a good deal about German racial 'eliminationism'. The 'Final Solution of the Jewish problem' had its roots in the treatment of mentally ill patients in German clinics and hospitals who were systematically murdered by euthanasia specialists both in Germany and later on the Eastern Front.[29] This escalation is reflected in Harald's experience. In Treskau, the Danish Frikorps were allocated a barracks that had formerly been a psychiatric hospital. In November 1939, German SS soldiers removed some 900 of the 1,000 remaining patients, marched them into a nearby forest and executed every one. The Waffen-SS then occupied the empty building.

Here is Harald's comment on those events:

> We are moving to the barracks tomorrow. The barracks have previously been used as an asylum for the mentally ill. The story goes that when the Germans arrived at Treskau, during the war against Poland, the mentally ill were provided with knives and sent against the Germans. But the SS occupied these barracks, and the around 1200 patients marched out of here under SS-command, and nobody has seen them ever since. But this doesn't matter, tomorrow 'Frikorps Danmark' will take the barracks.

It is hard to believe that Harald did not know about the fate of those 1,200 patients. But in any case he has no interest: 'it doesn't matter'. Per Sørensen, another Danish SS volunteer, certainly knew all about those executions: 'The barracks [in Treskau] was once a "loony bin" and apart from that a former monastery: when the Germans moved in here they shot all the idiots and converted the asylum into a very nice SS barracks.'[30]

Harald noted other casual brutalities:

> We have Russian deserters and prisoners to work for us to remedy the roads. One of them stole 3 packages of cigarettes today from one of our vehicles. A strm [abbreviation of Sturmmann] Marius was supposed to take him back to the prison camp, on the way he stopped the prisoners in the woods and let the one with the cigarettes dig a hole, after which he shot him. The other prisoners fought over his clothes. I think it is a little rough [sic], if I had been a prisoner of war, I would probably also have stolen cigarettes if there was an opportunity.

Sørensen had a more visceral response: 'The other day I saw a column of Russian war captives, they looked disgusting, such fanatical criminal types that one defends oneself against believing that these are white people, well, they also looked more Asian than European.'[31]

Like the Germans, the Danish SS volunteers regarded Russians as '*untermenschen*'. Here another volunteer writes home in June 1942:

> You should see some of the faces of the prisoners we have here presently. It is not a reproduction of Raphael's angels, but the most horrendous Mongolian face one could imagine ... they are horrendous, but hypocritical as well, cunning as hell, but one knows what one is dealing with and takes them for what they are. They are not one bit too pretty to shoot down every one of them. My opinion of most of the Russians is that they are not humans, but animals.

In a letter written to his young son at the end of 1941, SS Commander von Schalburg passed on his poisonous thinking: 'Your parents have fought the Tartars, the yellow cross-eyed ones. The holy [Tsar] Alexander beat them and kept them away from the Russians, so they could not destroy their houses. We fight today against the Jews, who took the houses and churches and the bread from the Russians.'[32]

In the autumn of 1941, when Harald was in Treskau, he witnessed a German SS officer beating Jewish forced labourers:

> In the last couple of days half a score of Jews with the Star of David on their bags, have been working in the garden of the commander's house, under supervision of a German Sturmmann. They are being appropriately beaten with a stick, and, according to orders, they report to work in the garden with bare legs. The temperature is minus 10 degrees. Cannot sanction that, do not think it is worthy of the German

> spirit. Even though I don't like Jews, I don't think they should be exposed to mistreatment by a bloody Sturmmann, even though they say that his parents were killed and mistreated by Jews.

He went on: 'One with a Star of David died tonight, and another one walked past the barracks this morning with a bloody and bruised face. The Sturmmann says he will kill them all before New Year.'

Read that passage again. It is appropriate to beat Jews – but it is not right to expose them to sub-zero temperatures wearing inadequate clothing. Punishment is acceptable, but *negligence* – the bare legs – is wrong. Providing inadequate clothing is not worthy of SS men. Harald, who 'doesn't like Jews', appears to at least entertain the idea that the German officer can be excused because he said his parents had been 'killed by Jews'; Harald appears to believe this outlandish claim. He reports from a world in which bruised, beaten and frozen Jews pass by his barracks – their fate the whim of a German officer who will 'kill them all before the New Year'.

Harald was not himself a beater of Jews. But in October 1943, he happened to be on leave in Copenhagen. By then Denmark was under German direct rule and its government no longer able to protect Danish Jews. Himmler insisted that the 'Final Solution' be fully applied, overriding the feeble objections of Ribbentrop, and ordered the new Reich Plenipotentiary Karl Werner Best to get on with job. A deportation order was issued to begin on Rosh Hashanah, the Jewish New Year at the beginning of October. Himmler dispatched Rolf Gunther, who was attached to Adolf Eichmann's office, to Copenhagen with a special commando of SS officers. The Germans had lists of addresses where Danish Jews lived, but they needed assistance from the Danish police and SS men to find them. Many Danish police officers refused to assist the Germans. But SS volunteer Harald had no such scruples:

> My team did not have any winnings [meaning arresting Jews]. Out of 4 teams in the car, one team had a winning with 2 old Jewish madams … The Jews were allowed to bring 2 blankets, food for 3–4 days along with the valuables they could carry in one suitcase. It was not a job that interested me, but an order has to be obeyed as long as you are in uniform. According to rumours the Jews had sailed to Danzig.

The Germans and their collaborators arrested about 450 Danish Jews, most of whom were transported to Theresienstadt concentration camp. Three days after the round-up, on 5 November, Harald reports: 'Enjoyable days at home.'

His final entry reads: '[But] now everything is over with, and I am once again a free man. And I don't regret the two and half years I have spent in the SS. I have seen and experienced much during this time.' Then he writes: 'and, first and foremost, I got away with it, to date.'

After this, Harald stopped writing his diary, but he remained in German service working as a 'sabotage guard' at the weapons factory Nordværk, which manufactured parts for German fighter planes. On 5 May 1945 around 8 p.m., Harald surrendered to Danish resistance fighters. Earlier that day they had visited his home looking for him. In September 1945 a Danish court sentenced Harald to five years in prison for his wartime service in the SS, and his later employment by the German armaments company. In August 1947 he was released on parole and 'Harald' vanished from the historical record. He is unlikely to be still alive.

There are Frikorps Danmark veterans still living in Copenhagen. One man was prepared to discuss his experience. Kaj (his real name has been witheld) is 85 – and from his neighbours' point of view is just a rather talkative old man with a fondness for beer. He used to work for the Tüborg Company. His neat and tidy, shared ownership flat is typical accommodation for a Danish pensioner. There is a lot of ageing furniture, cushions strewn everywhere and the accumulated matter of a lifetime. Kaj likes listening to very loud music – mainly German marching songs and Danish folk music. He was happily married for twenty years but, he confesses, left his wife for another woman. His second partner died of cancer years ago, but Kaj only keeps a picture of his first wife. He says he 'made a mistake'. There were no children. But Kaj knows he does have children, three of them, including twins, who live somewhere in Germany. They were the outcome of casual liaisons with nurses in German hospitals; he has never made any effort to locate his offspring. He shows no emotion at all when he talks about this lost German family. The women he recalls only with cynicism. Kaj is not very likeable.

There is another room here, usually locked. Inside is a silent and gloomy shrine. Kaj unlocks the door. Inside, he points out photographs: there he is, arm in arm, smiling, with other Danish SS volunteers. He has only copies of his German medals – he sold the originals to a Danish collector. Most remarkable though is a photograph of a painting of Kaj in full SS uniform commissioned by a German officer. The painting itself no longer exists; it must have resembled one of those heroic propaganda images of noble SS volunteers marching against the Bolshevik foe. Kaj has collected a huge library of books about Hitler and the Third Reich, many in German, which Kaj says he reads. There is a recent picture of Kaj with some Hell's Angels. Another photograph shows him and a few friends standing in front of a controversial memorial that was built to commemorate the Estonian SS division. Kaj is angry that Denmark refuses to do the same service for Danish SS veterans. As he settles down in his sitting room, Kaj rants about the new 'Holocaust

Memorial' in Berlin, designed by Jewish-American architect Peter Eisenmann. It is very ugly, Kaj declares – although he says he has never travelled to Berlin. If he was able to make a visit, he declares that he would urinate on the big grey blocks. When Kaj talks about the Nazi genocide, he claims to feel sorry for the Jews – although he personally 'disliked them'. His one regret, it seems, is that he 'played for the wrong side'.

Kaj (b. 1922) like Harald was born into a working-class family – one of nine children. He was not cherished. He fled home when he was barely a teenager and, failing to get an apprenticeship, ran away to sea. He served on transatlantic merchant ships and still has 'USA' tattooed on his arm. He saw his family on rare occasions: 'No, we didn't speak much with one another.'[33] It was only much later that he had any contact: 'They were notified when I was wounded, when I was in the hospital ... that was in '41.'

Men find different ways of coping with abandonment. Many deny pain and find solace in 'hard' occupations that provide the company of other equally hard men. They acquire a kind of internal crust – protective yet brittle. The job becomes the family. Comrades become brothers. Such men wait for wars. And for Kaj, it came soon enough. In 1939, he got a job on a ship that plied the North Sea importing coal to Poland. In 1940, the German navy blockaded Danish ports and Kaj found himself out of work. After an idle summer, Kaj and his friends heard that Danes could get work in Germany: 'One could go to Germany to work ... they advertised with that, and then with some friends, we signed up. And then we travelled to Germany, we travelled with the first team that ever was sent to Germany.' By the autumn, Kaj had found a job in Hamburg. He claims he took little interest in German life; he had heard only about the 1936 Olympics and then the invasion of Poland: 'I didn't know anything. I knew that there was a guy called Hitler. But otherwise, I didn't have a clue about National Socialism.'

Kaj says that he was just not interested. In Hamburg, he says people shouted 'Heil Hitler' and he shouted back 'Heil – what was his name again?' This is one of Kaj's many jokes – he had what might, generously, be described as a special sense of humour. One day, Kaj's friends noticed posters urging young men to join the Waffen-SS. He appears to have volunteered almost by accident – thanks to his friend Hansen, who could speak German:

> And we are going out of course, and I am with one who – he could speak perfect German – and then he sees that they are looking for people for the Waffen SS. Then we go in there [the recruiting office]. And he knew German, I didn't know a word. Every time they asked something I just said 'ehmm'. But then we went ... we were

> told to go to the changing room, undress and come back in again. Then we were brought in front of different doctors right, all the way around just like the [tests] … And then we were approved.

Seven decades after the war, it appears to be very important for Kaj to make us believe he did it for the adventure: 'It was adventure – all that shit.'

Did he know much about the SS? What it stood for? 'No no no no. I didn't have any idea of what it was. I thought it said 44, when it actually said SS.'

I am not inclined to believe him. Memory is fallible but many SS veterans peddle this kind of front story. In any case, Kaj and his German-speaking friend began their SS careers at the SS 'Westland' barracks in Langenhorn, but were soon transferred to Klagenfurt in Austria. SS training was tough even for someone who had served on Atlantic cargo ships; Kaj uses a rather colourful phrase to describe the rigours of SS training: '[my] tongue hung out like a red tie.' He had the impression that the German officers regarded him as an equal, but that is almost certainly because his German was so poor. For the same reason, he cannot recall any ideological training; there was, he says, 'no propaganda'. Kaj disliked men like Harald, DNSAP members who got privileged treatment, and, he says, even sent on leave whenever their division was sent to the front line. Kaj ended up on the Eastern Front:

> I can remember we were lying in a forest when the Russian artillery came down on us. The entire forest was bombarded within an hour. … God damn it many fell there. They didn't just fall, there were legs and heads and arms that had been ripped of … Bloody hell, there were things flying through the air. It was bloody tough.

Interviewing Kaj is not a comfortable experience. He wants us to believe that he has no regrets – and that he volunteered only for the adventure. Research carried out immediately after the war into the motivations of SS volunteers from the Netherlands implies that Kaj may be telling a partial truth. Dutch psychologist A.F.G. van Hoesel investigated 450 Dutch volunteers and uncovered a significant diversity of motivation.[34] In the case of the Danish volunteers, one of the Frikorps commanders carried out his own study. He came up with the following estimate:

> A. Professional military interest 2–5%
> B. War-adventurer 5–10%
> C. Dissatisfied with home life 3–5%
> D. Anticommunist beliefs 20–25%
> E. Conservative or nationalist beliefs 10–15%

F. Favoured new European political order 15–20%

G. National-Socialist family or member 30–35%[35]

These figures are striking. Ideology motivated at least a quarter of the Dutch volunteers – and we should bear in mind that 'anti-communism' implied anti-Jewish sentiments. Kaj certainly fits within A, B or C; but was he immune to more abstract reasoning? He would like us to think that. To be sure, he was a poorly educated young man. Put him in a time machine and he might be a neo-Nazi in Dresden, a nightclub bouncer in Solihull or a US troop in Baghdad. Politics, the calamitous events of the 1930s, the rise of the dictators – they appear to have passed him by. Kaj had almost certainly heard about the Berlin Olympics because they were reported on the sports pages.

But Kaj remains an actor in history. I am not convinced that he was the simple-minded thrill-seeker he plays. For somehow – and we may never know how and why – Kaj grasped the rudiments of SS doctrine. He claims that he knew nothing about the Holocaust. He denies all knowledge of the famous Danish rescue of the Jews. But pushed to say more, he raises his voice:

> *Kaj:* Not a God damned thing! We first found out about it [the rescue] once we came home, about the Danish Jews who had been taken, and many other places too. But there are some who forget to tell how the Jews were, and how they still are. You can see in Europe today, they direct the whole thing.
>
> *Question:* In America they have a lot of power?
>
> *K:* Yes, and also in Germany and in Denmark. Many people don't know that.
>
> [Kaj refers again to the Holocaust Memorial in Berlin.]
>
> *K:* It is completely wrong that it is down there now. The big memorial they have built in Berlin for the Jews – it's completely wrong.
>
> *Q:* It's completely wrong?
>
> *K:* Yes, it's completely wrong. Every time I pass it down there [sic] I say 'I just have to take a piss on those rocks' because it is completely wrong that they have build something like that. We couldn't even raise a stone for our fallen comrades at home …
>
> *Q:* So you do know Berlin?
>
> *K:* Oh sure …

Danish police arrested Kaj in 1945 and he served a few years in prison. How does he feel today about volunteering? 'I feel bloody fine about it.'

The majority of the Danish volunteers like Harald and Kaj fought on the Eastern Front and some served in the German concentration camps. The letters and diaries written by the Danish volunteers provide evidence that the recruits received the same ideological training as German recruits, mainly at the Bad Tölz Junkerschule in Bavaria, which had close connections with the concentration camp at Dachau. Camp inmates lived in cells constructed beneath the school and carried out maintenance and other menial tasks. In the Junkerschulen, SS officers rammed home the cult values of the SS and made sure the new recruits understood that their job was to master and then destroy the *untermenschen* they would combat in the east. Bayonet practice was carried out using 'Jewish' caricatures. Some Danes joined the Death's Head SS units which were headquartered in German camps. Not only that, but Himmler authorised a number of lethal medical experiments that were carried out at Auschwitz and Buchenwald by Danish doctors.[36]

The Danish rescue of 1943, a central plank in Goldhagen's thesis about German 'exceptionalism', is just one thread in a more complex weave. Hatred of Jews is a recurrent theme in many of the service diaries and letters collected by Christensen and his colleagues. The Danish SS volunteers accepted without question that the campaign in Russia would be fought against a 'Jewish enemy'. Here is von Schalburg again, writing to his wife, in August 1941: 'The Jewish rule [in the USSR] was far greater than even I believed.' The Russians, Schalburg complains, were 'too damned passive'. If the Jews had been 'cut down', 'many lives would have been saved'; he concludes: 'I think that will come.'[37] His poisonous sentiments were echoed by lower ranks: 'Yes we'll eradicate these Jews from the surface of the earth, because while there are Jews there is also war. Now I can imagine that some who would say that the Jews are humans too. My answer would be that rats are also animals.'[38]

These remarkable testimonies by Danish SS volunteers underline the value of documentary evidence, some recorded without benefit of hindsight. The war ended seven decades ago, and the men and women who collaborated with the Third Reich and remain alive have had plenty of time to prepare cases for the defence. In 1945, no one talked about the Holocaust and many of those who served in SS police battalions and Waffen-SS units slipped through the judicial net and went unpunished. Since the emergence of a special historiography devoted to the German destruction of European Jewry, many former collaborators have trimmed their personal stories to suit new times. Many who served the Reich have refashioned themselves as prescient anti-communists. They fought the Bolsheviks – and should surely be judged now in the light of what historians have revealed about crimes of the Soviet Union. Whatever these veterans testify now must be treated with caution; we must read between the lines.

At the end of 2007, I flew to Norway to interview a veteran of the SS 'Norske' Legion, Bjørn Østring (b. 1917). Mr Østring is prepared to talk openly about his service in the SS – in fact, he relishes publicity. He and his supporters have been campaigning for the Norwegian government to recognise his former comrades who died fighting on the Leningrad front as national heroes. Østring runs the Kaprolat Committee to identify and return the remains of Norwegian soldiers that still lie in the hills of Russian Karelia using DNA samples from their living relatives. Mr Østring (who is married to the daughter of Gerhard von Mende, who served in Alfred Rosenberg's wartime Ostministerium) is an alert nonagenarian who lives very comfortably in an Oslo suburb. But make no mistake – Bjørn Østring is a propagandist. He wants us to believe that the Norwegian SS volunteers were 'soldiers like any other'. Early in our interview, both the Østrings made it very clear how much they resent the new Oslo Holocaust study centre opened in 2006; the researchers there want Østring to hand over his records of Norwegians who served in the SS. The Østrings have refused to provide any assistance.

The new Centre for the Studies of Holocaust and Religious Minorities is based in the Villa Grande on the Bygdøy – a peninsula on the western side of Oslo. For most Norwegians, the villa is a shameful reminder of the German occupation. It was the wartime residence of Vidkun Quisling, who founded the Norwegian Nasjonal Samling Party and was appointed the puppet ruler of occupied Norway. During the war, Mr Østring knew the building, then called 'Gimle' after a character in Norse mythology, well: he spent a few years serving in Quisling's personal bodyguard, the Føregarden. Østring is a staunch admirer of Quisling, and even suggests that I photograph him standing next to a portrait of his hero.

Mr Østring and his collaborators insist that Norwegians heroically served the cause of anti-communism on the Leningrad front. But the 900-day siege of Leningrad, commemorated as the 'Blokada', was an act of military barbarism and fitted with German genocidal plans in the east. Hitler ordered the Wehrmacht to 'erase the city from the face of the Earth'.[39] When I meet the Østrings, they are preoccupied with a recent news story: 'Someone', he tells me, has recalled that during the war the Østrings took over an apartment in Dunkers Street that was owned and formerly occupied by the family of Håkon Laksov, a lawyer deported to Auschwitz in 1942.[40] In fact, many Norwegians who had volunteered to serve in the Waffen-SS received property as a reward for services rendered left empty by deported Norwegian Jews. Østring denies that he knew anything about the former occupants of his new home, but the Oslo National Archives show that he was himself active in the 'Liquidation Board' set up under the Quisling regime to distribute Jewish property and chattels to Aryan Norwegians. It was, in short, state-sanctioned looting.

This plunder reflected Himmler's ambitious plans for Norway, which had surprising connections with SS strategy in the Baltic. Himmler was obsessed by Norway. He admired the Viking tradition and liked inspecting restored longboats. Norway had an especially prominent place in Himmler's vision of a Greater Germanic Reich. He hoped that Norwegians, with their pure Nordic blood, would play a leading role colonising the east. It is noteworthy that Dr Konrad Meyer, the author of the Generalplan Ost, was chosen to launch the SS recruitment drive in Oslo. Many Scandinavians had embraced the idea of Nordic racial superiority long before 1933. The Swedish count Eric von Rosen, who became Hermann Göring's brother-in-law, was using the swastika as a personal emblem years before it was adopted by the German NSDAP. In the 1930s, von Rosen became a leading figure in Sweden's National Socialist movement, the National Socialist Bloc. In Norway, Vidkun Quisling's moderately successful Nasjonal Samling Party embraced the notion of Nordic superiority. Alfred Rosenberg championed Quisling's cause in Germany, paving the way for his wartime collaborationist regime.

Himmler's plan to transform Norway into a kind of northern fortress of SS values was directly linked to the destruction of Jews in Eastern Europe.[41] As we have seen, Danes and other Scandinavian volunteers served in the murderous SS brigades that, under the leadership of SS generals like Friedrich Jeckeln and Erich von dem Bach-Zelewski, participated in the mass shootings in Ukraine and elsewhere. Following the occupation of Norway, Himmler began to infiltrate trusted SS emissaries into the German occupation apparatus – including Franz Walther Stahlecker. Stahlecker backed the Norwegian police chief Jonas Lie, who had already served in the Balkans with the Waffen-SS. Hitler had other plans – and in September 1940 appointed Lie's rival, Quisling, to head a puppet Norwegian government. Hitler often stymied Himmler's foreign policy initiatives. SS ambitions in Norway were further frustrated by the ambitious Commissar Josef Terboven. Stahlecker returned to Berlin, where Heydrich assigned him to take command of Special Task Force A in the Baltic.

Himmler, however, did not give up – the Nordic lands were too precious a prize – and in January 1941 he made the first of a series of visits to Oslo to review new SS recruits. In Norway, the SS formed a bewildering number of SS police and combat units: it is estimated that some 15,000 Norwegians had volunteered by the end of the war. To educate these men, Himmler appointed SS veterans who had served on the Eastern Front, murdering Jews in Lithuania, Latvia and Belorussia. In Himmler's plan, the Norwegians, like other western volunteers, would form a Staatsschutzkorps (Corps for State Protection) – politicised soldier-policemen inculcated with SS values. In Norway, Himmler set up a training centre in Kongsvinger on the Swedish border. Its task was to manufacture a new, elite police force, the Ordnenspolitiet, based on the German Order Police, which had, as we have seen, been transformed into a militarised corps of soldier-policemen. They would be deployed to fight partisans on the Eastern Front – and a number of Norwegian recruits were assigned to the SS Kampfgruppe Jeckeln. Later in 1942, the Frikorps Danmark, as well as Flemish and Dutch volunteers, joined their Norwegian comrades serving in 1st SS Brigade, to fight 'partisans' near Minsk. A Norwegian SS officer oversaw the evacuation of the Minsk ghetto.[42] We know too that Norwegians served with the German Polizei-Gebirgsjäger-Regiment 18, which took part in the deportation of Jews from Athens.

These are the Nordic warriors that Mr Østring hopes to commemorate.

9

The Führer's Son

In 1939 National Socialist Germany ... had rebuilt itself in the midst of such lightning bolts, in the thundering and blinding flashes of such cataclysms, that all Europe and all the world felt the tremors.

Léon Degrelle, *Campaign in Russia*

At the beginning of 1944, the German Propaganda Ministry in Berlin battled a relentless blizzard of bad news. On the Eastern Front, Stalin's resurgent armies battered the once mighty forces of the Reich, and the German Empire shrank by the day. But Propaganda Minister Goebbels knew that even if the war was lost, the battle for German hearts and minds might still be won. And on the Eastern Front, an unlikely hero emerged from the blood-drenched ice and snow in the unlikely shape of a Walloonian Belgian called Léon Degrelle. He and a few hundred survivors of the battered SS 'Sturmbrigade Wallonien' had become trapped by the relentless advance of the Soviet army. Degrelle had fought his way through Russian lines to rejoin the retreating German army. Most of his men had been killed, left behind in the snow and ice. But turning desperate flight into an uplifting epic story was meat and drink to Dr Goebbels. The little Walloonian SS man with the cheeky smile who was, some believe, the model for Hergé's Tin Tin would join the pantheon of Germanic war heroes. Degrelle would be sent to meet Hitler in person.[1]

Degrelle devoted many pages to this semi-mythical encounter in his memoir *Campaign in Russia*. The story begins with Degrelle resting with the 'Wallonien' men who had survived his reckless adventure, all of them shaggy and caked with thick, black mud. A German corporal races up with a summons. 'The Führer has telephoned three times. He is waiting for you. We've been looking for you everywhere for two days!' A Fieseler Storch putters out of the clouds to fly Degrelle to Hitler's Wolf's Lair headquarters. These little aircraft had been designed for aerial surveillance and from the air Degrelle has a panoramic view of the seemingly endless black ribbons of German troops trudging west towards Kiev, starkly outlined against the all-encompassing white snow – tiny as flies. In the far distance, giant oil wells loom against a blue and silver sky.

When the Fiesler lands at an airfield near Pinsk, Degrelle is transferred to one of Hitler's Fokker Condors. As the big tri-motor throbs into a thick layer of cloud, the endless Russian steppe slowly recedes. Degrelle would never return to the front. After an hour or so, the Fokker crosses the vast Pripet Marshes and soon afterwards begins descending towards the gloomy Masurian woods near Rastenburg that hide Hitler's military headquarters. Degrelle is in no fit state to meet the Führer. For the purposes of basic hygiene, he is taken first to Himmler's headquarters, Hochwald, which lay hidden in thick pine forest some 20 miles to the east. In the SS chief's personal shower Degrelle washes away layers of grime and legions of lice. Himmler presents him with a clean shirt. SS orderlies remove his mud-encrusted SS uniform. Finally Degrelle is presentable and Himmler drives him to Hitler's headquarters.

By early 1944, Allied air forces dominate German air space. Massive raids have become routine; terror and destruction fall nightly on German cities. As Degrelle is driven into Hitler's headquarters, workers from the Todt Organisation are busy reinforcing the massive concrete bunkers. Security is tight. Rumour has it that a high-ranking Wehrmacht officer plans to assassinate the Führer. Himmler drives on through a succession of gates and barriers, deeper into his master's lair. Hitler spends most of his time in a modest wooden barracks situated at the northern end of the inner compound. The windows face north so that its solitary occupant, who prefers to work through the night, will never be tormented by direct sunlight. Life at the Wolf's Lair was described by General Jodl as 'between a monastery and a concentration camp'.[2] A small party of journalists has been invited to watch as Hitler awards the Iron Cross to heroes of the Reich. As they wait for the Führer, Himmler tries out his execrable French on Degrelle.

A pair of double doors swings open. There is a flickering barrage of magnesium flashlights. Film cameras whir. As Hitler enters, Degrelle is conscious of nothing but his eyes, and then the warmth of his handshake. The voice is hoarse: 'I've been worried about you,' says Hitler. As he moves away to confer with his SS chief, Degrelle has a chance to study the man whose cause he has followed to the ends of the Earth. He is stunned. 'The Führer of before the war had disappeared,' Degrelle confessed later. Gone was 'the fiery Führer with the chestnut hair, the trim body, the back as straight as an Alpine pine'. Now Hitler is stooped from 'bearing the weight of the world'. His hair is white. He is an old man. Grasping a pair of tortoise-shell glasses, Hitler remains silent for some time, his jaw grinding. Suddenly aware of the cameras, Hitler rediscovers some inner reserve of energy and begins to quiz Degrelle about the great breakthrough at Cherkassy. Enthralled, he takes Degrelle by the elbow and leads him into the adjoining map room so that he can demonstrate precisely how he had fled the Russian 'kettle'. The Führer nods sagely: 'If I had a son I would wish him to be like you.'

Hitler was, as Degrelle feared, gravely ill. He suffered severe intestinal problems. His left leg trembled uncontrollably. His physician, the loathsome Dr Morell, had treated him with amphetamine pills and cocaine eye drops. Hitler was, in short, a wreck. But when Degrelle wrote about his triumphal visit to the Wolf's Lair in his memoirs, he chose to recall 'a life of simplicity and order'. Hitler, he imagined, 'worked through the night in profound contemplation', pacing slowly, bent and grey, 'ripening his worries and his dreams'. On that 'night of great emotion', Hitler presented Degrelle with the Ritterkreuz (Knight's Cross). Hitler, in Degrelle's besotted eyes, remained 'a genius at the height of his power'. He was the architect of a New Order that would bring glory to the heirs of Charlemagne and the medieval Dukes of Burgundy. As the meeting ended, Degrelle said his 'soul was singing'.

Collaboration was the faith of men who had briefly tasted power but had had it snatched away. In many countries where fascism had failed to take root, native would-be führers snatched at Hitler's boots as he strutted by, as if some of his tawdry magic would rub off on them. Most saw too late that they had boarded a ship of fools. But few acknowledged guilt or expressed shame. One such was Léon Joseph Marie Degrelle. In the mid-1930s, Degrelle had risen to spectacular heights as leader of the Belgian far-right Rex Party but then fallen back to earth like a spent firework. The self-proclaimed 'Chef de Rex' was a chronic narcissist – and to be sure, his vulpine good looks photographed well even on the front line. He was a brilliant orator, master of spicy turn of phrase and pithy metaphor. Like British fascist leader Oswald Mosley, Degrelle was a champagne fascist – and a seducer of other men's wives.[3] He was erratic, short tempered, a maladroit spinner of plots and schemes, and yet by 1944, this fallen Walloonian demagogue had become a hero of the German Reich.

Degrelle was a consummate survivor. As the Reich disintegrated, he fled to Spain. Here the former Chef de Rex amassed a small fortune throwing up stuccoed villas to blight the Costa del Sol. From a sumptuous, ochre-coloured villa in the mountains north of Malaga, he devoted his old age to spewing out books about the greatness of Hitler and the 'myth of the Holocaust'. He became an icon of a resurgent European fascism and a hero for radical Muslims. He was eulogised thus by Radio Islam:

> The work of Léon Degrelle has always been epic and poetic. As he walks in the environment of his home one feels the greatness of Rome with its marbles, its bronzes, its translucent glass [sic]; one feels the elegant Arabian architecture, the gravity of the Gothic form and the sumptuousness of Renaissance and Baroque art. One feels the glory of his flags. In this atmosphere of beauty and greatness: the last and most important living witness of World War Two.[4]

Inside his Spanish fortress, Degrelle's servants kept the shutters locked tight. Even in daytime, the light was sepulchral. The former Chef de Rex, who in old age had run to fat, received admirers sitting stiffly behind a massive desk modelled on the one Hitler installed in the Reich Chancellery. On the wall behind, Degrelle had hung a Burgundian banner and a Waffen-SS pennant. On a polished table, in solitary splendour, stood a small bust of Hitler. He delivered to order sour denunciations of Jews – and, a Catholic to his last breath, published a morally obscene 'Letter to Pope John Paul II', denouncing the papal visit to Auschwitz in 1979. 'The Holocaust is a myth,' he insisted. History, he told all comers, had proved Hitler right. It was the Bolsheviks all along who had been the real criminals. Léon Degrelle died in 1994, a grand old man of the far right. His life had been a long, morally fetid journey through the putrid landscapes of European fascism. And yet as a Francophone Walloonian Belgian, Degrelle, when he first offered his services to the German occupation authorities, had been dismissed as not much better than an *untermensch*. His Flemish rival Staf de Clercq was Himmler and Berger's choice of collaborator. How did this mercurial Catholic dissident became a hero of the elite SS?

To answer that question, we must begin with Degrelle's troubled and divisive homeland. Starting in 2007, gloomy reports appeared in the European press concerning the dire state of Belgium. Following disputed election results in June that year, Belgians lived without an elected government for six months. A Flemish flower seller was quoted in the London *Guardian*: '"Belgium!" he splutters. "That's something that doesn't exist. The national anthem? Nobody knows it ... The King? A parvenu! A dysfunctional family. We are not going to take it anymore."'[5] Three years after, in 2010, the same bitter wrangling between Dutch-speaking Flanders to the north and French-speaking Wallonia to the south brought down the Belgian government for the third time. Belgium has always been a battlefield.

According to one journalist, 'a whiff of the Balkans' can be detected in the capital of the European Union. 'Long live Flanders, may Belgium die!' According to Filip Dewinter, leader of far-right Flemish nationalists: 'There's no Belgian language. There's no Belgian nation. There's no Belgian anything.' The German *Der Spiegel* speculated: 'Is Belgium Falling Apart?' A Belgian school teacher called Gerrit Six put Belgium up for sale on eBay: 'Belgium: a Kingdom in Three Parts', with free delivery. Bids eventually reached 10 million euro before eBay closed the auction. Today, it is language that separates the Belgian citizens of Flanders and Wallonia. But for the new German masters of Europe in 1940, the differences between Walloons and the Flemish ran deep.[6]

German race scientists viewed the nation of Belgium as a deviant fusion of two distinct European peoples: the French-speaking Walloons and the Germanic Flemish. They naturally favoured the latter and many Flemish Belgians urged union with Hitler's Reich. In the aftermath of Hitler's blitzkrieg, Nazi planners hatched up a now forgotten scheme to remove Walloonian Belgians and hand most of northern Europe to its rightful racial owners, the Flemish and the Dutch under the protection of the Reich. Léon Degrelle passionately admired Hitler. But he had a problem – he was a Walloon. From the German point of view, Degrelle was the wrong kind of Belgian. To serve Hitler's Reich, as he so ardently wished, Degrelle had to somehow overturn the preconceptions of Himmler's race experts.

For the Romans, modern Belgium was merely the 'land of the Belgae'. Once they had mastered these proto-Belgians in the first century AD, they renamed this flat, northern province Gallia Belgica. Here Gallo-Romans called 'Walha' lived cheek by jowl with Germanic tribes in the north. Four centuries later, long after the fall of Rome, it was the turn of the German Franks to rule the descendants of the Belgae. By the Middle Ages, the Low Countries resembled a jigsaw puzzle of fragile feudal states, including Walloonia, that were briefly yoked together as the Kingdom of Burgundy and then broken up again, under Spanish Hapsburg rule, as the 'seventeen provinces'. At the end of the eighteenth century, the ancient land of the Belgae had become the Austrian Netherlands and was ruled from Vienna. After the French Revolution, Napoleon threw out the old Austrian rulers. This turbulent history with its frequent territorial adjustments opened up the deep and fractious fault lines that still divide Germanic and Francophone Belgians. In 1815, after the final defeat of Napoleon, the Congress of Vienna stitched together a few Napoleonic leftovers as the United Netherlands – a kind of buffer state that they hoped would keep the northern lid jammed firmly down on the French. But the United Netherlands soon split between French-speaking Belgians who resented the ascendancy of Dutch-speaking Netherlanders. After a violent revolutionary upheaval in the 1830s, yet another international conference dragged the independent kingdom of Belgium kicking and screaming from the womb of the no longer United Netherlands. Although the new kingdom soon acquired all the necessary trappings of state – a parliament, an army and a constitutional monarch – Europe's newborn had a hard time of it growing up. Belgium remained two nations yoked together by treaty, and by the Roman Catholic Church, the glue that held Belgians fractiously together in a single nation under God.

As European newcomers, Belgian patriots soon demanded wealth and empire to rival their older brethren in the Netherlands and Great Britain. King Léopold III was an aggressive imperialist, and the Belgian Congo became a synonym for the worst excesses of colonial rule. On the home front, rapid industrialisation led to brutal class conflict between militant workers and nouveau riche elite. This freshly opened fault line reflected, albeit unevenly, much older ones. Walloons tended to be urban and wealthy; the old Flemish peasantry flocked to work in the new factories and mines and formed the militant bulk of the Belgian working class. In the twentieth century, Germany would twice violate Belgian neutrality – and, each time, trample over the fragile unities of Belgian society.

Born in 1906, Léon Degrelle grew up on the French border in the little village of Bouillon in the Ardennes – Shakespeare's Forest of Arden. This was the Walloon heartland. Léon's father Edouard was a prosperous brewer and a Catholic Party official. Degrelle could recall playing as a child in the shadow of the ruined fortress of Godfrey de Bouillon, the Burgundian leader of the First Crusade who energetically massacred Jews and Muslims as he marched towards Jerusalem. The Ardennes region of Francophone Belgium had once been part of Burgundy, a long-vanished medieval empire that fascinated Degrelle all his life. In the minds of patriotic Walloons, Burgundy took on the allure of a semi-mythic lost kingdom that might one day be restored as 'greater Belgium'. For very different reasons, Hitler was also fascinated by the idea of creating a 'Burgundian province' in a future Reich.

By the mid-1930s, Degrelle had become one of the most notorious rightist leaders in Europe. He had studied law at the Catholic university in Louvain but failed to graduate. He tried his hand at journalism and was offered a job by a radical conservative journal called *Christus Rex*, which was founded to honour the 1925 Quas Primas Encyclical on the Feast of Christ the King, issued by Pope Pius XI. The Pope weighed in against 'the plague which now infects society. We refer to the plague of anti-clericalism, its errors and impious activities. This evil spirit, as you are well aware, Venerable Brethren, has not come into being in one day; it has long lurked beneath the surface.'[7] That evil spirit was, of course, Bolshevism. This new brand of evangelical anti-communist Catholicism shaped Degrelle's thinking as much as the rise of European fascism. In 1927, the editors of *Christus Rex* sent him to report on the bloody Christero War that had erupted in Mexico, sparked by anti-clerical laws passed by President Plutarco Elias Calles. The Christeros were Catholic terrorist gangs who roamed the Jalisco province led by priests and armed with ancient muskets. They attacked and terrorised villages. The Mexican army responded in kind and began murdering Catholics. Degrelle was inspired by the Christero revolt with their battle cry of 'Long live Christ the King!' and on his return to Belgium began to use *Christus Rex* to build his own radical Catholic political movement: 'Rex'. Degrelle's movement was rabidly anti-communist, but also preached a hazy kind of social equality. Degrelle was a natural orator and noisily attacked corrupt Belgian politicians and denounced 'Banksters'. That term was a coded reference to Jewish financiers and, as Rex grew and expanded, Degrelle added to his ideological arsenal the ideas of other far-right radicals like the Spanish Falangist José Primo de Rivera and the Romanian Corneliu Codreanu. He admired Hitler and the dynamism of the new Germany. But the Nazi ideologues had little time for this Walloonian demagogue. They favoured Degrelle's Flemish rival Staf de Clercq who led the fascist Vlaamsch Nationall Verbond, which, backed by Germany, campaigned for a pan-Dutch state, the 'Dietsland', that would unite Flanders and Holland and eject the Wallonian provinces. Both Degrelle and de Clercq adopted the usual sartorial trappings of European fascist parties with dark or black uniforms and macabre insignia.

By 1936, Rex appeared to be on the brink of electoral success. Degrelle staged huge rallies modelled on the German 'Party Days' in Nuremberg. The youthful, photogenic and relatively glamorous Degrelle appealed to many disenchanted young Francophone Belgians and he was featured in the American newsreel 'The March of Time' along with other rising stars of the far-right European firmament. In the 1936 elections, Rex garnered a decent share of votes and began to look like a serious player. But Degrelle had unwisely attacked and alienated the conservative wing of the Catholic establishment and, added to the fact that he preached a negative message about corruption and the excessive influence of Belgian Jews, support for Rex began to drain away. But by the time Hitler attacked Poland in September 1939 Rex was a spent force. The solipsistic Degrelle became increasingly belligerent. He had glimpsed power and abruptly lost his way. His public attacks on 'Banksters' now modulated into overt anti-Semitism and he made a succession of hopeful pilgrimages to Berlin. Hitler's Germany promised the brightest future for divided Belgium –and for the Chef de Rex, Léon Degrelle.

In the spring of 1940, Hitler's armies gobbled up nations as if they were so many breakfast *Brötchen*. On 10 May, fast-moving Wehrmacht ground forces, paratroops and glider troops swept across the Belgian border backed by screaming Luftwaffe

1 SS Chief Heinrich Himmler in traditional Lederhosen. (USHMM)

2 Governor of the German General Government, occupied Poland, Dr Hans Frank in 1939/40. Frank recommended 'surgery not butchery' when it came to solving Europe's 'Jewish problem'. (Bundesarchiv, Bild 121-0270. Photographer: o.Ang)

3 Danish SS volunteer 'Kaj' who when interviewed said wanted to urinate on the Holocaust Memorial in Berlin. permitted use)

4 The 'Death Dealer of Kovno' Algirdą Antaną Pavalkį, photographed by German observers as he murdered Lith Jews with an iron bar in the Lietukis Garage in Kovno. (Untraceable)

5 Recruitment poster for the Flemish SS 'Langemark'. The Waffen-SS volunteer thrusts his bayonet at a caricatured Jewish figure wearing a Union Jack and apparently dominating Great Britain. German anti-Semitism attributed British resistance to the influence of Jews. (AKG-images)

6 Adolf Hitler, Ribbentrop and other German dignitaries with Romanian 'Legionary' dictator Ion Antonescu, Munich, 10 June 1941. The Germans urged Antonescu to eliminate Romania's Jewish population. (Bundesarchiv, Bild 183-B03212, Scherl agency)

7 Corneliu Zelea Codreanu, founder of the Romanian Legion of St Michael, the Iron Guard. Known as 'the Captain', Codreanu was said to resemble Hollywood actor Tyrone Power. (AKG-images)

8 The Iași pogrom, July 1941, carried out by Romanian army and police units with German connivance. Here Jewish victims lie by the side on Vasile Conta Street. (USHMM/ Serviciul Roman De Informatii)

9 The Grand Mufti of Jerusalem, Haj Amin el-Husseini, who fled to Berlin in 1940. (AKG-images)

10 Himmler greets the Grand Mufti, 1943. (Bundesarchiv, Bild 101III-Alber-164-18A. Photographer: Kurt Alber)

1 The Grand Mufti with Bosnian Muslim SS recruit, November 1943. He urged the Bosniaks to 'kill all Jews'. (Bundesarchiv, Bild 146-1978-070-05A. Photographer: Mielke)

2 Bosnian Muslim recruits serving in the SS 'Handschar'. (AKG-images)

13 German SD Einsatzgruppe (Special Task Force) member murdering Ukrainian Jews, near Vinnitsa in 1941. Waffen-SS troops and Reich Labour Service recruits look on. (USHMM, YIVO Institute for Jewish Research, courtesy of Sharon Paquette)

14 A Ukrainian SS recruit attacking a Bolshevik caricature with bloody hands and knife. (Bundesarchiv, Plak 003-025-061)

15 Propaganda poster enticing Ukrainians to join the SS 'Galizien', 1943. Notice the SS inspector measuring the height of the hopeful volunteer. Height was a critical biological standard for the SS. (Bundesarchiv, Plak 003-025-059)

16 The first Ukrainian Waffen-SS volunteers assemble outside the opera house in L'viv (Lemberg), 18 July 1943. (Michael Melnyk Collection)

17 Over 80,000 Ukrainian men volunteered to join the SS 'Galizien'. That number was whittled down to less than 10,000 after rigorous inspection by SS experts. (Michael Melnyk Collection)

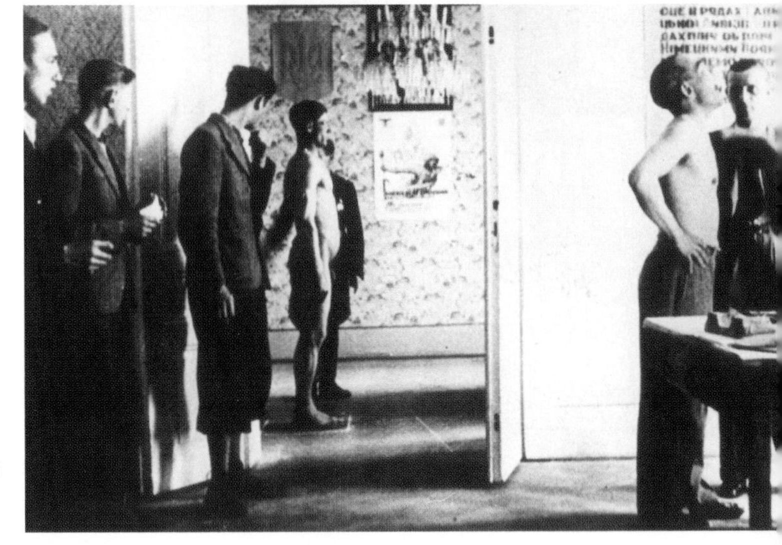

18 SS inspectors examine Ukrainian recruits. It is not the case that Himmler abandoned recruitment standards after 1942. (Michael Melnyk Collection)

19 Himmler and SS-Brigadeführer Wächter inspect the SS 'Galizien' at the Neuhammer training camp, May 1944. Himmler congratulated the Ukrainians for purging their beautiful country of Jews. (Michael Melnyk Collection)

20 Wächter with Ukrainians including Professor Kubijovych, the leading Ukrainian collaborator with the German occupiers. (Michael Melnyk Collection)

21 SS General Otto von Wächter, the main architect of SS recruitment in Galicia. (Michael Melnyk Collection)

22 SS General Erich von dem Bach-Zelewski, the leading SS 'bandit hunter' who followed Himmler's instructions to 'kill all Jews as partisans'. (Bundesarchiv, Bild 101III-Alber-096-32. Photographer: Kurt Alber)

23 Himmler's chief 'bandit hunter' Erich von dem Bach-Zelewski planning an operation in the occupied Soviet Union, March 1944. German anti-partisan operations continued, in some areas, to involve the mass killing of Jews. (Bundesarchiv, Bild 101III-AhrensA-020-31A. Photographer: Anton Ahrens)

24 Belgian Walloon SS volunteer Léon Degralle, photographed by Hitler's personal photographer Heinrich Hoffmann as a hero of the Reich. (Bayerische Staatsbibliotek)

25 Belgian Walloon collaborator with SS recruits in Pomerania, 1944. (Bayerische Staatsbibliotek)

26 Belgian collaborator and SS volunteer Léon Degrelle is decorated by Hitler at his Rastenburg military headquarters, the 'Wolf's Lair'. In the background is SS General Felix Steiner. (Bayerische Staatsbibliotek)

27 Croatian leader (Poglavnik) Ante Pavelić with German Foreign Minister Joachim von Ribbentrop in Salzburg, 6 June 1941. The Germans insisted that the Ustasha regime 'solve' its ethnic problems in the puppet regime of Croatia. (Bundesarchiv, Bild 183-2008-0612-500. Photographer: Henkel)

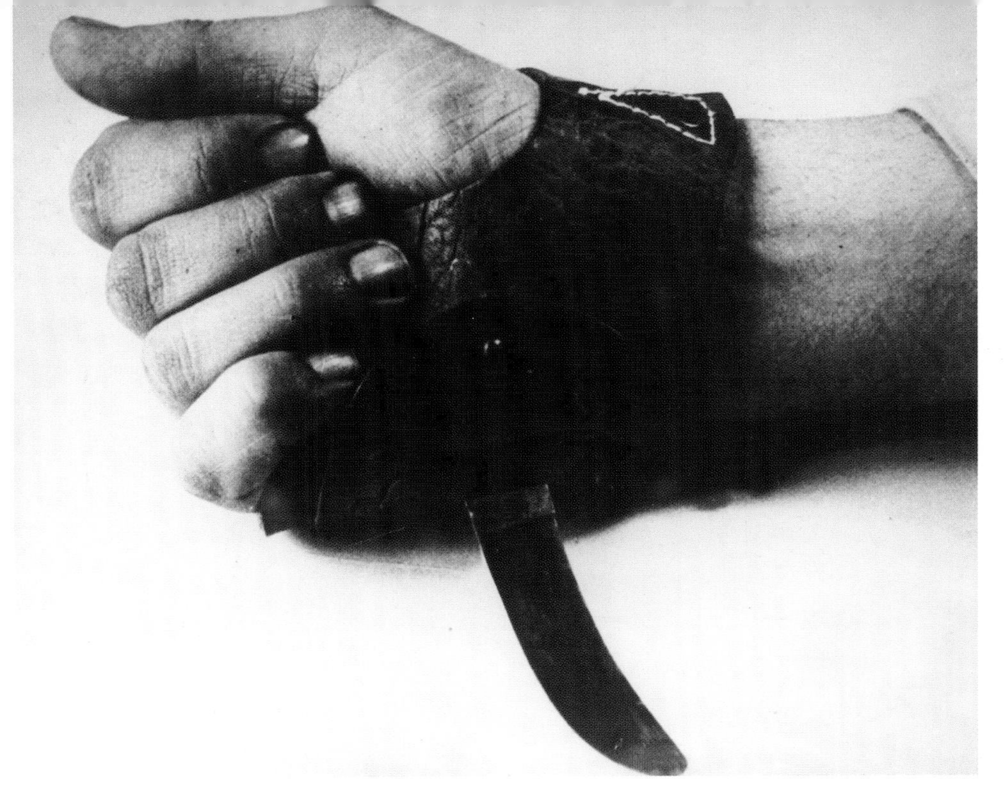

28 Special weapon made by Ustasha guards to slaughter Jewish and Serbian inmates held in Croatian camps. (USHMM/Muzej Revolucije Narodnosti Jugoslavij)

29 An Ustasha murder squad with victims. These squads rampaged through the German puppet state of Croatia murdering Jews and Serbs. (USHMM/Memorijalni muzej Jasenovac)

30 The site of the Rumbula mass killings today. Between 30 November and 8 December 1941, SS General Friedrich Jeckeln directed the murder of 27,800 Latvian Jews, assisted by local auxiliaries. (Author's own image)

31 SS-Obergruppenführer Friedrich Jeckeln, the 'Butcher of Riga' who masterminded mass slaughter in Ukraine and Latvia. (Bundesarchiv, Bild 183-S45466)

32 Viktors Arājs, Latvian commander of SD murder squad the 'Arājs Commando', in British custody after the war. (UK National Archives)

33 The Legionaries begin their march to the Freedom Monument. Young Latvian nationalists head the procession. (Author's own image)

34 Latvian nationalist guarding the Legion veterans. (Author's own image)

35 SS chief Heinrich Himmler inspecting Norwegian SS recruits. (USHMM)

36 Østring with a portrait of his hero, the Norwegian puppet ruler Vidkun Quisling. (Author's own image)

dive bombers. For the second time in a century, German troops incinerated the famous library in the university town of Louvain.[8] The Belgian army fought back – but the national government led by the Catholic Prime Minister Hubert Pierlot was in disarray. Further west, Allied forces had been overwhelmed by the Wehrmacht's surprise advance through the allegedly impenetrable Ardennes. On 24 May, King Léopold III, who despised Pierlot and his ministers, assumed command of the Belgian army and prepared to make symbolic last stand on the River Lys. Pierlot fled to London but the king refused to escape, claiming that he would be regarded as a deserter. Encircled at Dunkirk, the British Expeditionary Force fled across the Channel. And on 27 May, Léopold surrendered his forces to the Germans. He retreated to his castle at Laeken and refused further entreaties to follow the example of King Hakon VII of Norway and the Dutch royal family to escape and form a government in exile. In London, Churchill denounced Léopold for betraying the Allied armies and the French Prime Minister Paul Reynaud accused him of treason. This was unfair. The French fought on but when Reynaud appointed the ageing Philippe Pétain as Minister of State, the die was cast. Pétain urged the French to throw in the towel. On 22 June, he signed an armistice and ordered Renaud's arrest. Hitler's blitzkrieg was over and the reactionaries soon came to terms with the new status quo. Léopold was, like the English Edward VIII, a rabid opponent of democratic government and, as Pierlot feared, he hoped to make terms with Hitler. His hopes were frustrated, however. He had one brief and unproductive meeting with Hitler at the Berghof in November then sulked inside Laeken castle until the end of the occupation.

When news of the rapid German advance had reached Brussels, the Pierlot government, fearing attack by a 'Fifth Column', arrested thousands of suspected German sympathisers, including Léon Degrelle. Most were quickly released, but the Rexists and a few Flemish nationalists remained in custody. To keep Degrelle and the other remaining potential Quislings out of German hands, the Belgian police transported twenty Belgians and fifty-eight foreigners on so-called 'Phantom Trains' across the French border to Abbéville. Here the police hauled Degrelle and a few other Rexists from the train and locked them in a vault underneath a bandstand. They were lucky: French soldiers shot twenty-one prisoners, including the Flemish national leader Joris van Severen, but the rest, including a now heavily bearded Degrelle, ended up an internment camp at Le Vernet in the south of France, close to the Spanish border. In a short pamphlet, 'La Guerre en Prison', written a year later 1941, Degrelle made much of his 'martyrdom'.[9]

In Belgium, it was widely believed that the Chef de Rex had been killed and news of his apparent demise even reached Hitler, who made a reference to the rumour in a letter to Mussolini.[10] In Brussels, the Rexist leaders who had survived the purge leadership now took stock. As news arrived of the German victories over the French and British forces, many of their supporters were euphoric. Surely the Chef had been right and corrupt 'Banksters' and feeble politicians had led Belgium down the road to humiliation and defeat: '*Degrelle avait raison!*' insisted the Rexist newspaper *Le Pays Réel*. In July, Vichy officials released a handful of Rexists, who reached Brussels with the glad tidings that the Chef was alive and well. In July, a small expedition of Degrelle's closest allies travelled to Le Vernet and managed to get him released. By 22 July, Degrelle and his party had reached Paris, now the German administrative centre of occupied France. Degrelle was eager to offer his services to the victorious, clearly unbeatable Reich. Soon, he assumed, he would be making a triumphant return to Brussels as a German-appointed national leader. But as he would soon find out, even the most fervently expressed craving to serve the German occupiers rarely led to a role in Hitler's New Order. This may seem surprising. After all, an occupying enemy power has few friends and would surely welcome the craven overtures of aspiring collaborators. But in 1940, Léon Degrelle had only the faintest idea how Hitler's Reich worked and what stood in his way.

In most of the occupied nations, like Norway and Denmark, civilian commissars closely bound to the NSDAP and the SS soon replaced military administrations. But events in Belgium took a different course. Here a German military administration (Militärverwaltung) held on to power, kept the SS at bay and ruled through the Belgian civil service until July 1944. This unusual state of affairs reflected Hitler's chronic indecisiveness as well as the usual squabbling between his fractious subordinates. In the winter of 1939/40, Hitler was eager to maintain harmonious relations with his generals and reward them for the astonishing success of military operations in Poland and Western Europe. So he played along with the OKW's complacent assumption that when military operations had been wrapped up, new army administrations would assume executive functions in occupied territories, like Belgium, Holland, Luxembourg and France, and began selecting future administrators who played 'management games' to prepare them for their tasks.

When the Dutch army commander-in-chief General Henri Gerard Winkelman surrendered on 15 May, the OKW appointed the elderly General Alexander Baron von Falkenhausen, recently recalled to active service, as the military commander of the Netherlands. But two days later, the High Command received disquieting news: Hitler had overruled their decision and decided to install a civil administration in the Netherlands under Reich Minister Artur Seyß-Inquart, who was then serving as Hans Frank's deputy in the General Government. The German commander-in-chief of the Wehrmacht, Walther von Brauchitsch, now had the embarrassing task of telling Falkenhausen that his services were no longer required – at least for the time being. As it turned out, Seyß-Inquart and the SS dominated the German occupation of the Netherlands until the end of the war – with calamitous consequences for Dutch Jews.

The humiliation of the OKW and Falkenhausen led to grumbling about the 'utter dishonesty of our top leaders'.[11] But shortly after King Léopold, as commander-in-chief of the Belgian army, had surrendered, Hitler shrewdly dispatched Falkenshausen to Brussels to head up a military administration. Himmler lobbied Hitler to appoint a civilian commissar – but this time Hitler fobbed him off. In the Netherlands, the brutal Seyß-Inquart resisted Himmler's efforts to have his fiefdom immediately incorporated into the 'Greater German Reich'.

Hitler, however, never considered the Belgian military solution to be final. In the summer of 1940, he floated the idea of appointing a civilian commissar to manage Flanders and reducing the army's sphere of command to the Francophone Walloon provinces and northern France. The hand of Himmler is clear. His preference would always be for a 'Greater Netherlands'. His Waffen-SS recruitment chief Gottlob Berger, who had long had his finger deep inside the Flemish pie, proposed transferring another notorious bully, Josef Terboven, who had just been appointed Commissar of Norway, to apply his ruthless style to new Reich Gaue 'Flanders' and 'Wallonia'.[12] And so it went on ... But by the autumn of 1940, Hitler's attention had turned decisively to the Soviet Union and he simply lost interest in Belgium. In Belgium, Falkenhausen and his staff still clung precariously to power – and their anxieties about Hitler's intentions would have a profound impact on their treatment of impetuous collaborators like Léon Degrelle.

The rather decrepit, hard-drinking Falkenhausen, like many Prussian aristocrats and career officers, detested Hitler and his movement. As Hitler and the OKW planned their attack on Western Europe, Falkenhausen had secretly warned the Belgian government. But by 1940, he had entered his twilight years and lacked the will to resist the Nazi juggernaut or, for that matter, to take much interest in Belgium.[13] He delegated most of his responsibilities to his deputy Eggert Reeder, formerly the council leader (Regierungspräsident) of Aachen. Like his boss, Reeder was no bleeding heart German liberal, but he was a punctilious and hard-working bureaucrat. Before Hitler's seizure of power in 1933, he had been a member of the moderate German People's Party (DVP), and like many other opportunist 'March Violets' had joined the NSDAP in 1933 to keep his job. According to Elmar Gasten's history of Aachen, Reeder 'remained true to his policy, protecting the administration from encroachments by the [NSDAP]'.[14] Himmler later persuaded the highly competent Reeder to join the SS, but he was never a convinced National Socialist.

Once installed in Brussels, Reeder would successfully keep Himmler at bay for some time. He had a crucial advantage. In compliant Belgium, security remained a low priority until 1943 and gave Himmler few excuses to impose SS control. Reeder's occupation resembled the Danish case, where SS-Brigadeführer Werner Best developed a cheap and efficient method of 'indirect rule'. So long as the subject peoples of 'protectorate' Denmark and Belgium kept supplying Germany with cheap labour, minerals, butter, meat and fish, purged their administrations of Jews and communists, and generally kept their heads down, they would be rewarded with light touch 'supervisory' administrations.[15] Hitler favoured the Best doctrine. According to the 'Table Talk', recorded by his factotum Martin Bormann, Hitler often praised the British Raj, which, he believed, employed a handful of sahibs to rule over millions of Indians.

As soon as Reeder took up his post in Brussels, he had to deal with the importunate Chef de Rex Léon Degrelle. In his *Tätigkeitsberichte* (activities reports), Reeder gives us a vivid portrait of the troublesome Walloon, and shows how he and the Rexist leadership refashioned their faded party as a pro-German faction. In a book published in the 1970s, *Die Verlorene Legion*, Degrelle explained how he hatched up a plan for a 'Fascist Greater Belgium' that could take its rightful place in Hitler's New Order, under his leadership. This new nation would be shaped by what he called the 'hard, pure and revolutionary' doctrines of National Socialism and would 'eradicate pitilessly' 'old democratic, plutocratic, Masonic and even Jewish cliques'.[16] Reeder was no 'righteous gentile'. He did not hesitate to authorise the deportation of 'foreign' Jews. But he may have spared occupied Belgium an even worse fate by shutting the door on Degrelle.

It took Degrelle, a chronic fantasist, some time to understand that neither Hitler nor Reeder and the Militärverwaltung would do many favours for tiresome upstarts like himself who might destabilise the delicate checks and balances of occupation. They had nothing to gain by sponsoring the careers of loud-mouthed petty dictators. They could not, of course, share the Francophone Degrelle's passion for a 'greater Belgium', even as a vassal state. Successful collaboration was a difficult trick to master, as any number of aspiring 'quislings' (including Vidkun Quisling) found to their cost. And in Belgium, Degrelle faced another thorny obstruction. He was, in the German view, the wrong sort of Belgian.

Like any colonial regime, the Germans deepened the fractured Belgian society by turning to a single favoured ethnic group – in this case the Flemish. They viewed the Flemish as a 'Germanic' people, as blood kin. Hitler insisted that the administrative donkey work be carried out by Belgian civil servants, under a small German staff, and ordered Reeder to deal exclusively with the Flemish. Staf de Clercq, the head of the Flemish far-right Vlaams National Verbond (VNV), was soon being courted by the Germans and by June, the plum jobs in the Belgian administration had been grabbed by VNV men. The Germans quickly released Flemish POWs, but Walloons they left to rot behind barbed wire. Himmler's loathing of the inferior Walloons was especially intense.

Hitler had frustrated Himmler's ambitions in both the Netherlands and Belgium. In the late summer of 1940 he tried another tactic. He authorised Berger to recruit Flemish and Dutch volunteers for service in the SS 'Standarte Westland', which recruited volunteers from both the Netherlands and Scandinavia. Then in September, Hitler proposed the formation of a purely Flemish SS legion which, he informed Reeder, 'should serve as a rallying point for all Flemings who are willing to serve in the army'.[17] In his activities report, Reeder noted with alarm that this decision signalled a new plan 'to build a supra-party organisation which could take on all the Völkisch forces in Flanders'. In other words, he feared that Himmler would use the Flemish SS as a Belgian Trojan horse. The news of this Flemish initiative also dismayed Degrelle and his Rexist comrades. Himmler's racial doctrines barred their way to power.

The study of eastern peoples (*Ostforschung*) dominated German race science and especially fascinated Himmler. But it would be wrong to conclude that the race experts had no interest in ethnic diversity in Western Europe. Himmler also energetically promoted *Westforschung*, the study of European peoples. Race science in the Third Reich would never be a merely scholarly pursuit. It had an instrumental political purpose: to facilitate 'Germanisation'. The task of Hitler's experts in both Eastern and Western Europe was to measure the quantity of Nordic blood possessed by different ethnic groups. On that basis, some would be selected for future assimilation as 'Germanics', the rest would be discarded. After 1939, the lion's share of *Westforshung* fell into the hands of Himmler's think tank the SS-Ahnenerbe (ancestral heritage), and for the Ahnenerbe's race experts, the people of the Netherlands, like the Nordic Scandinavians, had a special status.

In October 1940, the Ahnenerbe took over Der Vadaren Erfdeel (DVE), a right-wing Dutch institute linked to the National Socialist Party (NSB). The DVE had been set up in 1937 to study the Germanic ancestry of the Dutch. In the occupied Netherlands, SS race experts and their obliging Dutch counterparts collaborated on what became known as the 'Holland Plan' – the western version of the Generalplan Ost. The core idea had first been proposed by Bonn university professor

Dr Otto Plaßmann, who headed the Forschungsstätte für Germanenkunde, germanische Kulturwissenschaft und Landschaftskunde (Research Facility on Germanic Ancestry, Cultural and Geographical studies). In a letter to the Director of the Ahnenerbe, Plaßmann outlined a wildly ambitious scheme to create a 'Greater Holland' carved out from the Netherlands and parts of Belgium. As in the case of the Ostplan, implementing the Holland Plan meant that any non-Germanic ethnic groups must be removed or even liquidated; and 'non-Germanic', of course, meant the Francophone Belgians, the Walloons. German race scientists already knew a great deal about the Walloons.

The doyenne of Walloonian studies was Franz Petri.[18] In the mid-1920s, Petri and like-minded scholars had become fascinated by different ethnic cultures that lay scattered along German borderlands. These included Petri's own special study, the Flemish population of the Netherlands. In common with many German anthropologists, Petri welcomed the coming of Hitler's Reich since leading Nazis like Himmler and Rosenberg favoured the systematic study of race. In 1936, Petri joined the SA and the NSDAP and began cultivating connections inside the SS. He also formed close ties with Flemish nationalists based in Cologne and was a founder of the Deutsch-Vlämische Arbeitsgemeinschaft (DeVlag), which before too long would fall into the hands of Gottlob Berger. Petri too was a passionate advocate of a 'Greater Holland'.

In October 1939 a Belgian journalist called Maurice Wilmotte read Petri's somewhat obscure papers concerning the 'Flanders-Germanic borderland' and raised the alarm: Petri's ideas, he revealed to the readers of *Le Soir*, were nothing less than an invasion plan. Wilmotte was right to be concerned. After the German invasion of the Netherlands and Belgium, Hitler approved Petri's appointment as Kulturpapst (literally, culture pope) – a key position within the Belgian military administration. His task would be to manage the 'Germanisisation' of Belgium 'in harmony with the methods used by German kulturpolitik in the Danube and Balkan countries'. In these regions, the Germans had used local ethnic Germans as the vanguard of a Germanisation campaign.

Petri shared Himmler's fascination with the Middle Ages and the emperors of the Holy Roman Empire (the First Reich), like Heinrich the Fowler and Barbarossa. He believed that the ancestral Nordic heartlands – the so-called Mittelreich or Niederrhein – remained biologically extant in a region marked out by the Emperor Charlemagne's strongholds in Aix-la-Chapelle, Cologne and Nijmegen. Reminding his SS masters that these Germanic emperors had led crusades against 'Asiatic' and Slavic peoples in the east, Petri argued that *Westforschung* must therefore be the foundation of *Ostforschung*: you had to study the west in order to master the east. This too was Himmler's conviction. It followed that the Germans, of course, but also the Dutch and the Flemish, directly descended from the crusading peoples of the First Reich. They provided the ethnic 'key' to the Germanisation of Western Europe and by the same token must become the vanguard troops for the conquest of the east.

As Kulturpapst in occupied Belgium, Petri had considerable influence. His small but energetic staff took an active interest in everything from education to film and theatre, religion, art and, of course, the vexed question of language. Petri purged the Belgian universities of all Jewish staff and began to restock the library at Louvain, which had been vandalised by German troops, with 'Germanic' volumes. For Walloons like Degrelle, Petri's reign of cultural terror was a catastrophe. In German eyes, only the sturdy, blonde Flemish farmers had 'the right stuff'. Walloons, Petri sneered, tended to shun the countryside, favoured socialistic ideas and spoke an odd French dialect. But he remained undecided about their precise racial origin. Were the Walloons *untermenschen* like the Slavs? At best they might be *Mischlinge*, ethnic hybrids. In any case, the Walloonian provinces simply didn't fit. The Holland plan that Petri and other SS academics developed in 1940–42 proposed an ethnic restructuring of the entire region that reunited all Germanic peoples. The unfortunate Walloons would first be deported to the Dutch province of Limburg. After that 'a few million people' would be moved east into the General Government.[19]

Léon Degrelle had only the vaguest idea that the Germans regarded him as not much different from a Slav, and that Himmler was mulling over a plan to deport Francophone Belgians like himself. He simply wanted to retrieve the power that had slipped through his fingers in 1936. He had met Hitler only fleetingly at the height of Rexist fever, but was convinced that his fame as Chef de Rex would win him influence with the new masters of Europe. After his escape from captivity in the south of France, Degrelle immediately sought out the German ambassador in Paris, Otto Abetz. They had met in Berlin in 1936, and Degrelle had closely followed the career of this silky young star of the German Foreign Service. It was no hindrance to Degrelle's cause that Frau Abetz, *née* Suzanne de Bruyker, had been a childhood friend of Mme Degrelle. Abetz, an ardent Hitlerjugend veteran, owed his present ambassadorial position to his former status as the French expert in the Dienstelle Ribbentrop – the 'shadow foreign office' that Hitler created to counter the more conservative Foreign Office in the Wilhelmstrasse. Before 1940, Abetz enjoyed taking semi-official tours of France where he liked to visit art galleries and cultivate the nastier political factions. When Hitler engineered Ribbentrop's appointment as Foreign Minister, he dispatched the ambitious Abetz to Paris where this oleaginous plunderer set about ransacking the art collections owned by French Jews.[20]

When he sat down to dinner with SS-Brigadeführer and Ambassador Abetz at the German Embassy in the rue de l'Isle, Degrelle, riding on a wave of inflated self-confidence, regaled his astonished host with his plans for a new Burgundian empire – a Greater Belgium that would take its place as part of the German New Order. This phantasmagorical plan obsessed Degrelle, and in 1940, he may have believed it fitted with German plans. It is true that Hitler had discussed breaking France into provinces based on the old medieval kingdoms like Burgundy. Degrelle may well have got wind of this plan and concluded that his own Burgundian vision could be harmonised with Hitler's. He was wrong. A note in Goebbels' diary tells us why: Hitler's new Burgundy would have no place for a Degrelle or any other Francophone Belgian. He planned to settle the province with ethnic Germans from South Tyrol – a region that Germany and her Axis ally Italy frequently squabbled over.[21] On 10 July 1940 Himmler, who, it will be recalled, had been appointed 'Reichkommissar for the Strengthening of Germandom' (Reichkommissar für die Festigung des deutschens Volkstums, RKVD), toured the old Burgundian lands to assess their suitability for 'Germanisation'. He concluded that it would be necessary to import not less than 1 million Tyrolean Germans.

At their meetings in Paris, the suave Abetz listened politely to Degrelle's schemes – and passed on his 'suggestions' to his Foreign Office colleagues in Berlin. It is unclear whether at this stage the ambassador meant to help or harm Degrelle. Abetz was certainly more sympathetic to the collaborators' cause than Hitler. Goebbels noted 'Only Herr Abetz collaborates. I do not. The only collaboration I am willing to consider on the part of our French friends is the following: if they deliver the goods and do it voluntarily ... that I will call collaboration.' In any event, when Reeder and Falkenhausen read Abetz's report, all hell broke loose. They had yet to meet Degrelle but they knew his demagogic reputation. Quite apart from his racial disadvantages, Degrelle had made it all too evident that he was a fantasist with a ravenous appetite for power. Eggert and Falkenhausen refused to even consider a deal with Degrelle. He should have sunk without trace, a *collaborateur manqué*. But it was not in the nature of the Chef de Rex to give up without a fight.

It is natural for a certain kind of unrequited suitor to redouble their efforts rather than try to imagine why they might have been rejected. Now the spurned Chef would transform himself into the most ardent of National Socialists. He began by using the Rexist press to attack and ridicule the British and to denounce Jews in language borrowed from Goebbels' newspapers. Anti-Semitism had never been as potent a vote winner in Belgium as it had been in the Netherlands.[22] To be sure, Degrelle's infamous attacks on 'Banksters' may have implied anti-Jewish sentiment but his speeches contained very few direct pejorative references to Belgian Jews. From the winter of 1940, Degrelle re-engineered Rex as a Jew-baiting party. The party rag *Le Pays Réel* ran regular stories highlighting alleged Jewish deceit and corruption, and Rexist thugs began to attack Jews in the street or in their homes. Degrelle organised a new Rexist militia, the black-clad, SS-inspired Formations de Combat, which rampaged through Belgian cities and towns setting alight Jewish stores. Rexist propaganda portrayed Degrelle as the spurned leader who, alone, knew what was best for Belgium as it faced up to occupation. Fired up by a captive audience of Rexist acolytes, he became, in his own mind, the visionary statesman who would single-handedly lead a renewed Belgium to its proper place in Hitler's New Order.

On 6 January 1941, at a huge Rally in Liège, the Chef de Rex addressed 5,000 Rexists, ringed by his Formations de Combat. For two hours, Degrelle harangued his exultant audience, banging the same drum: only the German Reich could guarantee the future glory of Belgium. He finished with a passionate 'Heil Hitler!' In the days and weeks after the rally, Degrelle noted that this provocative 'Heil Hitler!' had final broken through the barrier of German indifference. His efforts to act out the role of a German backed national Belgian leader forced Reeder to reassure the Belgian government that he continued to back them, and not the Rexists. Degrelle enjoyed Eggert's discomfort, strutting around Brussels in his black uniform: the very model of a strong leader in waiting. General von Falkenhausen shut his door on the persistent Walloon – but Reeder found him increasingly difficult to avoid.

For the bumptious Degrelle, this wooing of the Reich took its toll. At the end of April, Reeder reported to Berlin that the Rexists seemed to be in disarray. He attributed this apparent breakdown to the volatile antics of the Chef: 'a perpetual, not always happy, improvisation'. He reported increased levels of negative opinion concerning Degrelle and his party, even in the Walloonian heartlands. As his fortunes sagged once again, Degrelle (as most frustrated wartime collaborators eventually did) wrote a begging letter to Hitler: 'After six years of violent struggle … I am immobile and sterile.' He did not receive a reply; after all, the German leader was not an agony aunt.[23] For Degrelle, collaboration had come to look a lot like the mythical labours of Sisyphus; he would roll the boulder of sycophancy up one hill only to have it rolled down another.

Degrelle was not stupid. He began to get an idea of how the ponderous and confusing German occupation ticked. He understood at last that his ultrapatriotic talk of 'Burgundian empires' did him no favours. He realised that in German minds only true 'Germanic' peoples could seek favoured status. Degrelle would have to take on the might of German race science. He was a mediocre theoretician but it took only a little digging to find out that even experts like the esteemed Petri had yet to completely make up their mind about the racial status of Walloons.

This uncertainty could be exploited. Perhaps Walloons too might be reclassified as 'Germanics'? This would mean abandoning 'Belgian' nationalism and accepting complete integration with the Reich. But to the power-obsessed Degrelle, the dream of a united Belgium was a political dead end. In the summer of 1941, Degrelle and his closest allies in Rex began to promulgate the idea that Walloons were not at all a French left over, as the Dutch and Flemish experts argued. On the contrary, he and the Walloonian people had long ago descended from a Germanic frontier people.[24] At first, Degrelle's campaign to join the Aryan club, conducted through the party paper and a few obscure journals, made little headway. His new plan would take time to work. But then, in the course of a single night, his political fortunes changed utterly.

In his memoir, Degrelle tells us that 22 June 1941 began like 'all the beautiful Sundays of summer'. He was idly turning the dials on a radio when he picked up the astonishing news that German armies had crossed the Soviet border. He was exultant: 'The real war … had just begun. This was a war of religions.'

For Degrelle and the Rexists, Hitler's crusade against Bolshevism would sanctify their shabby and discredited cause. At the end of June, Degrelle returned to Paris seeking another meeting with Abetz. He found his old friend busy discussing the new Légion des Volontaires Français with Jacques Doriot and Marcel Déat. Abetz urged Degrelle to propose forming a Walloonian legion. This time, Degrelle's timing was a lot better. On 29 May 1941 Werner von Bargen, a German official in Brussels, who later helped organise the deportation of foreign Jews, wrote to the Foreign Office in Berlin: 'It is important for us above all to win Belgium over to the new order of Europe.'[25] He urged that the German administration back all Nazi style organisations, not just Flemish ones. Rex, thanks to Degrelle's talent for mimicry, had just the right Nazi style trappings. On 27 July, Hitler approved the raising of national legions from each country of occupied Europe to join the struggle against Bolshevism. In Belgium, two legions, the Flemish SS Vlaam Legioen and the Légion Wallonie, would be split between the Waffen-SS and the German army respectively. Degrelle had yet to win over Himmler and the SS but the stage was fast being set for the second coming of Léon Degrelle.

So it was that on the morning of the 8 August 1941, Léon Degrelle smartly attired in the grey uniform of a German army private joined a motley crew of 860 men of all ages who had volunteered to join Hitler's 'war of annihilation'. The lowly rank of the Chef de Rex was a consequence of his utter lack of military experience. He had at first insisted on being immediately promoted to lieutenant but the Germans turned him down. He would serve as a humble private (Schütze) in the 1st Group of the 1st Platoon. The fact that Degrelle ended up enlisting at all reflected his chronic insecurity about political rivals like his former deputy Fernand Rouleau, who had first proposed the idea of a Walloonian legion to the Germans. Degrelle could not afford to put his feet up on the domestic front while competitors won glory on the battlefield.[26] In front of the Palais des Beaux Arts in Brussels, callow young thugs waited for orders alongside elderly veterans in badly fitting new boots. The new 373rd ('Wallonia') Infantry Battalion (Légion Wallonie) would be commanded by veteran Captain Georges Jacobs and was firmly attached to the Wehrmacht – so far, only Staf de Clercq's Flemish volunteers had been permitted to join the SS.

Regardless of his humble rank, Degrelle had no intention of abandoning his status as Chef. As Captain Jacobs ordered the Walloonian volunteers to fall in, Private Degrelle appeared standing on the Palais balcony to harangue the recruits about 'the struggle against Bolshevism'. He addressed a motley crew of Rexist die-hards and veterans of the Formations de Combat, and members of a radical 'League' of Belgian anti-Semites. The latter at least had a clear idea about what a 'crusade against Bolshevism' would mean on the front line. As the Chef de Rex returned to the ranks, it began to rain. Private Degrelle fell in and the Légion Wallonie marched off towards the Gare du Nord to begin their journey to the east. An uncertain and distinctly hazardous future awaited every one of them. Only a handful of ordinary Belgians took much interest in the passing show. No one cheered or waved a flag. Most recognised Degrelle and many had come to despise Rex and its preening Chef. Now he had turned traitor. As Degrelle marched into the huge, noisy station and joined the Walloon volunteers struggling with their bulky and unfamiliar kit, he realised that 'there was no going back, there was only ahead'. Only by shaking the dust of political failure could he ever hope to win German backing and the power he craved. In his memoir, written after the war, Degrelle tried to rationalise his decision: if the Reich triumphed:

> it would be the master in the East of a tremendous area for expansion ... The Greater German Reich ... enriched by those fabulous lands, extending in one block from the North Sea to the Volga ... would offer to the twenty peoples crowded onto the old continent [of Europe] such possibilities for progress that those territories would constitute the point of departure for the indispensible European foundation.[27]

A German victory, which he did not question, would allow him to return in triumph to his ungrateful homeland as a Walloonian conquistador. Many of the men who marched as volunteers in Hitler's war machine would never see their homes again. They would end their lives in the blood, mud and ice of the steppe, their deaths the blood price of collaboration.

Campaign in Russia (*Front de l'Est*), Degrelle's memoir about his experience on the eastern front in both a Wehrmacht battalion and later the Waffen-SS, must be classified under 'memoirs, unreliable'. On the first page he boasts that: 'By 1936, I'd already shaken my country to its very core … I could have been a minister in the government: I had only to say one word to enter into the game of politics.' Published by the right-wing Institute for Historical Review and introduced by a Third Reich apologist, who describes Degrelle as 'one of the great men of the twentieth or any other century', *Campaign in Russia* must be treated with caution. It is laced with barefaced lies and misrepresentations; its author is blindly infatuated with Hitler; the writing is grossly solipsistic. Its faults, in short, are legion. But hidden behind the flimflam, the preposterous boasting, the ridiculous self regard is a brutally explicit memoir of Hitler's war.

Degrelle had promised the men who had enlisted alongside him that they would be home by Christmas. He insisted that they were defending 2,000 years of the highest civilisation – and would be fighting alongside innumerable other dedicated young men who had made the same decision to fight for a New Europe: blonde giants from Scandinavia, Hungarian dreamers, whimsical Italians, bantering Frenchmen, swarthy Romanians. On 12 August, the legion arrived in Meseritz, where recruits swore an oath of allegiance to the commander-in-chief of German armed forces, Adolf Hitler. A few days later, the legion men boarded trains for Brest, where they transferred to wider gauge lines and crossed the old Soviet border and began steaming into vast expanse of Ukraine under glorious blue skies.

Few of the men who now gazed on the endless, flat landscape had ever travelled outside Belgium. For hours, days, then weeks they gazed, bewildered, at an astonishing alien new world that had been torn and twisted by the German war machine. Wrecked tanks and armoured cars stretched to the horizon. Every few hours, the long line of wagons and carriages would grind to halt for long, nerve-sapping hours next to ruined, smoking villages. Degrelle and his men watched in amazement as a seemingly endless chain of cattle trucks rumbled in the opposite direction bearing a miserable cargo of tens of thousands of Russian prisoners. More than a million and half Soviet soldiers perished in German camps. The Russian captives stood 80 to 100 in each wagon; 'hairy giants', Degrelle called them, many saffron coloured with tiny 'Asiatic' eyes. At night, he claims, the Russian captives fought over human flesh uttering brutish terrifying cries. They used tin cans to slice up the body of a 'dead Mongol'. The German guards halted these packed trains crammed with starving POWs for weeks at a time. The Belgians watched as prisoners leapt on to the tracks where they plucked long red worms from the glutinous Ukrainian mud which they swallowed immediately, their gullets rising and falling. Degrelle shows no compassion for these 'Asiatic' victims of Hitler's war. They were observing the beginning of a shameful forgotten holocaust that troubled Hitler's foreign volunteers not a jot.

Ever since the German blitzkrieg of May 1940, Degrelle had been strenuously wooing Hitler's Reich. Now, at last he would fight shoulder to shoulder alongside the brutalised heroes of the Wehrmacht. At the start of their campaign, Degrelle and the Walloon leaders worshipped the German High Command. But for their part, the Germans had only the vaguest idea what to do with the legion. By the time Degrelle reached the city of Pervomaisk on the southern Bug River, he and his men began to suspect that all was not well. It was late October and temperatures had started to fall. The German front line had reached a place many hundreds of miles from the Polish border and the German supply lines had become massively overstretched. After the hot, dry summer, the drenching rains of the autumn months had turned roads to mud. Ukrainians call the rainy season *Rasputiza*: the time without roads. Here the German invaders had already discovered Stalin's secret weapon: the mud front.

Before the winter freeze sets in, it is said that mud becomes 'Tsar of the steppe' – and in 1941 *Rasputiza* had come early. The famous black soil of Ukraine, which German troops had plundered and sent west on freight trains in 1918, is impregnated with oil. It has a uniquely viscous quality and resembles black glue rather than the hospitable mud of an English riverbank. Hitler and his generals had struck a very hard blow against Stalin and sent the Soviet Army reeling back towards Moscow. But Wehrmacht planners had not foreseen that this oily black sponge that seized hold of German boots, hooves, wheels and tank tracks could foil their best laid plans.

Along the Bug, the Soviets had destroyed all the bridges. German offices ordered Degrelle and the Walloon legionaries to disembark. They stumbled down to the muddy river edge, waded waist high through the surging torrent, then had to climb for hours up through a viscous wall of sludge to board another train that waited hissing and sighing on the opposite bank. They now steamed ponderously on by night – and for the first time rifle fire crackled and bullets pinged against the side of the wagons. It was colder by the day. In the mornings, the legionaries awoke from troubled dreams to discover that the track had been encased by thick ice which they broke up and heated to make drinking water. Frequently they saw dead Russian soldiers entombed in great icy slabs. The legionaries now began to descend towards the Dneiper, now a broad blue flood nearly a mile wide. A few days later, the Walloonian legionaries had their first experience of real war.

They made camp near the city of Dnipropetrovsk, south-east of the Ukrainian capital city of Kiev. Stalin had built giant apartment buildings here for miners who laboured in the coal fields of the Donets Basin, the Donbass. A savage battle had been fought to secure the city – and Special Task Force commandos had murdered many thousands of Ukrainian Jews here as they had in Kiev. Karl Marx Prospect had been renamed Adolf Hitler Avenue. Stalin's new modernist blocks had become dilapidated and, Degrelle writes, were awash with human waste. The German bombardment had wrought havoc with basic services, but Degrelle concluded that this excremental horror exposed the lie of the Bolshevik dream.

In his memoir, Degrelle says nothing about the bitter quarrels that had erupted in the legionary ranks between Rexists and other Belgian factions. These feuds steadily worsened as the rain poured down and morale slithered downwards. The divisional commander General Maximilian Fretter-Pico concluded that Walloon battalion was 'worthless in military terms', but wanted to 'avoid a row' for political reasons.[28] The operations section of the German 17th Army reported:

> Difficulties with the Walloon Battalion. On one hand, the battalion complains about unfair treatment by the German command to OKW, yet on the other extreme, reports of Group 'von Schwedler' (IV Corps) on behaviour of troops bordering on treason ... Use of the Walloon Battalion remains restricted depending upon its cohesion [*inneren Festigung*].[29]

After three days, the legion men finally crossed the mighty Dneiper River that slices Ukraine into two vast chunks. The advance began at midnight. The men had to cross a long wooden bridge that creaked and wobbled above a rushing torrent. Flak began to explode on every side. Huge, groaning icebergs glided slowly past in the dark, scraping loudly against the hulks of sunken vessels. On the other side of the Dneiper was the Front. Tremendous artillery barrages rumbled. Shells wailed overhead like banshees. 'We had dreamed of dazzling battles.' Degrelle wrote later, 'Now we were to know the real war, the war against weariness, the war of the treacherous, sucking mire, of sickening living conditions, of endless marches, of driving rain and howling winds.' Cold and sickness, rather than combat, wore down the Walloon ranks to just 650 men. The unsympathetic Germans further humiliated the Walloons by confiscating mortars and heavy machine guns. The Germans reassigned the legion to anti-partisan duty close to Dnipropetrovsk along the Samara River sector. Here they would discover some brutal truths about Hitler's war.

As the driving advance of the Wehrmacht slowed, Russian stragglers began to harass the overstretched German troops. These were the first badly organised partisan fighters. 'Partisan', or more correctly 'bandit', was, as we have seen, an elastic term that more often than not referred to Jews as well as Soviet guerrillas fighters. SS-Gruppenführer Bach-Zelewski insisted, 'Where the partisan is, the Jew is'. As Degrelle and the Walloonian volunteers followed the Samara River east, they received orders to target a group of 'bandits' who had taken refuge in a grove of firs a few hundred yards from the line of advance. 'These cunning assailants,' Degrelle wrote, 'had to be caught and wiped out.' It is telling that Degrelle provides few details about what happened next. After a deluge of words, his account abruptly falls silent. No veteran of the Légion Wallonie ever confessed to taking part in anti-Jewish actions. But those 'cunning assailants who had taken refuge in a grove of firs' would have included women and children, and any Jews who had so far survived the German advance. Degrelle's crusaders had become Hitler's 'bandit hunters'. And for Degrelle, the German 'war of annihilation' soon brought rich rewards as he had hoped. When the Walloons 'cleared' the village of Gromovayabalka, he was wounded – and promoted to sergeant. Soon he would be recommended for the Iron Cross. German reports began to take note of Degrelle's 'special personal bravery'. He had already witnessed the fanaticism of Waffen-SS troops and made a friend of SS General Felix Steiner. Soon he would begin lobbying Himmler to embrace the Légion Wallonie as part of the elite Waffen-SS.

10

The First Eastern SS Legions

> You may feel sorry for the Balts because they are a nice reliable people who are frightened of Russia, but when you work for them over here, you realize that they are at least 90% collaborationist. They all worked for the Germans.
>
> Charity Grant, UNRRA, 20 January 1946

In October 1941, when German victory still seemed certain, Professor Wolfgang Abel of the Kaiser Wilhelm Institute for Anthropology Human Heredity and Genetics led a team of race examiners (*Eignungsprüfer*) lent by the SS Race and Settlement Office (RuSHA) to occupied Poland to conduct studies of some of the millions of Soviet POWs held in sprawling, open-air German camps. It was a journey into hell. Historians now believe that the German army killed 2.8 million prisoners through starvation, gross neglect and execution. This barely remembered slaughter has been called the Forgotten Holocaust. Historian Karel Berkhoff argues:

> I submit that the shootings of the Red Army commissars and other Soviet POWs, along with the starvation of millions more, constituted a single process. It was a process that started in the middle of 1941 and lasted until at least the end of 1942. I propose to call it a genocidal massacre. It was a massacre because it was 'an instance of killing of a considerable number of human beings under circumstances of atrocity or cruelty.'[1]

This genocidal massacre was also a turning point in the evolution of German racial pseudoscience.

After the German invasion of the Soviet Union, tens of thousands of Soviet soldiers surrendered to the Germans. Any identified as Jews or 'Bolshevik Commissars' were immediately executed according to Hitler's notorious Commissar Order. They also killed Muslims and 'Asiatics' who were discovered to be circumcised and mistaken for Jews. Completely indiscriminate killing ended in September, when Nazi officials ordered that North Caucasians, Armenians and Turkic peoples, as well as Ukrainians and Belorussians, should be spared. After this spasm of killing, German troops and SS units began marching the Soviet captives to temporary camps known as 'Dulag' and then on to permanent 'Stalag' camps. During these forced marches, prisoners received minimal rations or none at all; guards often shot dead civilians who tried to supply food as the pitiful columns of starving, brutalised men passed through villages and towns. The Germans executed any stragglers who fell behind, even by a few metres. The survivors finally ended up penned inside an archipelago of vast, windswept camps enclosed by rudimentary barbed wire fences. Inside this cruel world, chaos ruled. Or seemed to: German policy was perfectly clear. In the words of Field Marshall Keitel, the purpose of this murderous internment was the 'destruction of a Weltanschauung' – meaning the Bolshevik world view that allegedly infested the minds of the prisoners.

According to the ethos of the German camp system, providing more than a few ladles of watery lentil soup was theft from the German people. Starvation was camp policy. Quartermaster General Eduard Wagner (who had negotiated the 'Einsatzgruppe agreement' with RSHA chief Reinhard Heydrich) insisted that the prisoners 'should starve'. Provision of food, according to Keitel, was 'wrongheaded humanity'. This German army policy reflected a radical ministerial strategy that had been formulated by SS-Obergruppenführer Herbert Backe which assumed that 'the war can only be continued if the entire Wehrmacht is fed from Russia'. As a consequence, 'there can be no doubt that tens of millions of people will die of starvation'.[2] One Ukrainian official was told bluntly: 'The Führer has decided to exterminate Bolshevism, including the people spoiled by it.' Mortality rates varied from camp to camp, but, taken as a whole, were shockingly high. In some camps, over 2,500 prisoners died every day. This was the realm of hunger. To live a few days longer, starving, lice-tormented prisoners would eat anything, including bark. Some resorted, inevitably, to cannibalism. Alexander Solzhenitsyn provided this account of a German camp in *The Gulag Archipelago*: 'around the bonfires, beings who had once been Russian officers but had now become beastlike creatures who gnawed the bones of dead horses, who baked patties from potato rinds, who smoked manure and were all swarming with lice. Not all these two-legged creatures had died as yet.'[3] There was just one way out: to be selected for service in the auxiliary police or for labour service, digging mass graves or rebuilding roads and bridges in the most gruelling conditions. Few Germans who discovered what was taking place in the camps protested – with one surprising exception. The German 'eastern expert' Alfred Rosenberg sent letter after letter to Keitel complaining about the murderous treatment of Soviet POWs. He recognised that Germany was squandering a reservoir of potential good will since many Soviet minorities hated Stalin. Now they were dying like flies in German camps. Rosenberg's appeals fell on deaf ears.[4]

Now in October, the prisoners who remained alive in the hellish German camps would be preyed on by German scientists led by anthropologist and SS officer Wolfgang Abel. Although the camp administrators referred to the prisoners as 'Russians', they came from every corner of the Soviet Empire; for Abel, the gulag was a tainted human treasure trove. The 'Abel mission' examined more than 42,000 prisoners from many different ethnic groups, which included Russians, Turkic peoples, Mongolians and various Caucasians. Abel's team measured, photographed and blood tested their subjects. Then they returned to their spacious offices in Berlin. When they processed their data, Abel was astonished. Their captive subjects revealed that the 'Slavic *untermenschen*' of the east exhibited a markedly higher level of 'Germanic' characteristics than he and his colleagues had anticipated. The new findings troubled Abel and other RuSHA race experts. His findings provided powerful evidence that 'Asiatic peoples' had, during periods of German expansion, been 'strengthened by Germanic blood'; the colonisers, to put it another way, had enjoyed sexual congress with the colonised. History, as geneticist Steve Jones puts it, 'is made in bed' – or the wheat field. The troubling consequence, Abel realised, was a kind of biological theft: German blood had been stolen from its rightful bearers.[5]

The findings of the Abel mission echoed Himmler's remarks about 'harvesting Germanic blood wherever it might be found'. Now he had scientific backing. Traditionally many German anthropologists had regarded the mixing of races or miscegenation as a weakening process. That was certainly the view of Adolf Hitler. But a number of German race experts came to more nuanced conclusions. One was Alfred Ploetz, who argued that racial mixing of peoples 'not too far apart' was a means of 'increasing fitness': he cited the Japanese as an example. Head of the Kaiser Wilhelm Institute of Anthropology, Human Heredity and Eugenics, Professor Eugen Fischer had come to similar conclusions when he had studied the so-called 'Rehobother Bastards'. Fischer recommended that the offspring of unions between Aryans and Jews or Africans should be compulsorily sterilised. But in cases where the two parents had closer ethnic bonds, then their offspring might be treated more leniently. This implied that, as Himmler put it, Germanic blood lines in non-Aryan peoples were a resource that might be 'harvested'. When the Abel mission published its conclusions, the existence of far flung Germanic blood reservoirs had scientific backing. The time had come to exploit these prized corpuscles. The Abel mission to the German gulag would soon have a decisive impact on Waffen-SS recruitment strategy. For Himmler and the SS recruitment experts the question was where to start.[6]

In 2004, in the Estonian town of Lihula, Mayor Tiit Madisson dedicated a memorial statue to Estonians who had served in the Waffen-SS. The memorial depicts an Estonian in German uniform holding a machine gun; the Estonians who had enlisted in the Waffen-SS had, said Madisson, 'chosen the lesser of two evils. They had experienced the Soviet occupation and did not want to return to it.' Madisson, who heads the Eesti Rahvususlikliit (ER, Estonian National Union) has authored a book called *The New World Order* that argues, with some originality, that Hitler was brought to power by Jews and Freemasons – and that the Holocaust never happened. When the Estonian government ordered the removal of the memorial, which had become an international embarrassment, hundreds of local people protested forcing the police to use batons and pepper gas. The memorial ended up at the private Museum of the Fight for Estonian Freedom, established in Parnu by an apologist for the Estonian division called Leo Tammiksaar.[7] In 2004 300 veterans of the Estonian 20th SS Division paraded through Tallinn, and in 2007, representatives of the veterans demanded the removal of a new synagogue in Tallinn claiming its existence was an 'insult'. Until that year, Estonia had been the only country in Europe without a single synagogue. Since 2007, SS veterans have staged reunions at Siminäe, the site of clashes between Soviet and German armies in the summer of 1944.[8] In 2009, the Estonian publishing house Grenader Grupp published a calendar illustrated with German propaganda images of Estonian SS men. It sold out within three days.

According to publisher Aimur Kruuse: 'The members of the legion tried to bring freedom to Estonia, or to give their families time to escape to the west before the Red Army returned to kill them or send them to Siberia.'[9]

Freedom fighters or war criminals? Or both at the same time? In 2004, wealthy Estonian farmer Lembit Someril sponsored yet another memorial, this one dedicated to SS-Standartenführer Alfons Rebane. A bronze statue of Rebane was built on private land, but the unveiling was attended by Estonian MP and former Foreign Minister Trivimi Velliste. The ceremony was condemned by Jewish organisations but many Estonians regard SS Volunteer Rebane as a national hero. British intelligence agency MI6 once held him in high regard too: Rebane escaped to the west after the German defeat and was recruited by British intelligence. He played an important role as one of the co-ordinators of Operation Jungle, which backed anti-communist resistance in the Baltics. Rebane died in Germany in 1976, and after the fall of the Soviet Union, his body was taken back to Tallinn and interred with full military honours. Five hundred people attended the ceremony including the commander of Estonia's defence forces, Lt Gen. Johannes Kert.[10]

Who was Alfons Rebane? According to Soviet documents, Rebane's career as a collaborator began when he commanded the 658th Eastern Police Battalion that took part in attacks on villages near the town of Kingisepp and the village of Kerstovo in the Leningrad region (where many of the Baltic battalions were deployed).[11] German-controlled Schuma police battalions recruited in Eastern Europe took part in the mass murder of Jews and other civilians. Rebane was a practiced 'bandit hunter' before he joined the Waffen-SS. In the Baltic states and Ukraine, Himmler used his 'Eastern Legions' as vanguard troops in the German 'war on bandits', as well fighting against the advancing Soviet armies. This means that men like Rebane fought both as agents of genocide and as 'freedom fighters'. The debate provoked by the rash of Baltic memorials is a false one. The 'freedom fighter' and the killer of civilians may be one and the same.

Even Latvians or Estonians who are embarrassed by the old veterans who march through their streets every year vehemently deny that the Baltic SS legions had any involvement with German war crimes. The web site of the Latvian government devotes a great deal of bandwidth to refuting such allegations. Their case appears to be buttressed by one incontrovertible and terrible fact: in the Baltic, the Germans and their native collaborators like the Arājs Commando and scores of other Schuma battalions did their work with pitiless diligence. By the end of 1941, all but a handful of Estonian Jews had perished: only 50,000 remained alive in Latvia and Lithuania, most of them quarantined inside ghettoes. By mid-1942, when Himmler authorised the formation of the first eastern SS division, the majority of Baltic Jews had been murdered. How then, the apologists argue, could the SS legions, which were formed after this period, have any connection to the Holocaust? In *The Holocaust in Latvia*, Andrew Ezergailis concludes that 'The Latvian Legion is outside the scope of this study [of the Latvian Holocaust]'. He goes on: 'no single event has ever been adduced associating the ['Latvian Legion'] with atrocities against civilians.'

Ezergailis' account of the German occupation is in many respects exemplary. He provides a wealth of detail about Latvians, such as Viktors Ārajs, who collaborated with the German occupiers and refuses to pull punches. But his argument that the formation of the Latvian SS divisions in 1943 had no connection with the events of 1941–42 is simply wrong. The different strategies of SS recruitment in occupied territories reflected the changing needs of Himmler and the SS in the occupied Soviet Union. During the first so-called 'wild' genocide, or the 'Holocaust by bullets', the SS recruited mobile Schuma police battalions, such as the Arājs Commando, that carried out 'special actions' close to where their victims resided. In 1942, the Germans began systematically transporting Jews to specialised extermination centres: the Reinhardt camps in the General Government and Auschwitz-Birkenau. To facilitate this new strategy, the SS began recruiting guard units known as 'Trawniki men' – mainly Ukrainians like Ivan Mykolayovych Demyanyuk, now better known as John Demjanjuk, but also Latvians and Estonians. At the same time, from the summer of 1942, Himmler simultaneously authorised recruitment of Waffen-SS non-German combat divisions in occupied Eastern Europe. Many of these Waffen-SS recruits had previously served in the Schuma units and now took part in so-called anti-bandit operations, which in many cases served to liquidate any Jews who had somehow survived the 'Holocaust by bullets'. Since the recruitment of Schuma battalions and foreign Waffen-SS legions or divisions formed part of the same evolving genocidal strategy, it is quite wrong to argue, as Ezergailis and others have, that the combat divisions have no connection with the Holocaust.

Take the case of Latvian Juris Šumskis, cited by Ezergailis as a typical 'Latvian Legion' recruit. Šumskis was a young man (b. 1925) with a mediocre education who joined the infamous Arājs Commando. He said later that a friend had told him that pay and service conditions were good. Many of the first recruits who joined Arājs in the summer of 1941 were ideologically driven students and intellectuals. Šumskis, who volunteered on 29 April 1942, was typical of later batches. Few of these men had been to university and when questioned after the war offered quite banal reasons for their decision. As part of his training, which was conducted under German supervision, Šumskis was shown how to use light weapons and had 'political lessons' four or five times a week. That meant he was introduced to the doctrine of National Socialism and the evils of Bolshevism which had no doubt been merely instinctual before he joined up. Šumskis was not considered well educated enough to be sent to the SD school at Fürstenberg in Germany. But even before he had completed his training, Šumskis participated in special actions: in June, he took part in the slaughter of several hundred mentally ill patients at the Sarkandaugava Hospital in a neighbourhood of Riga. Some of the patients had difficulty walking on their own, and Šumskis was forced to carry one elderly woman on a stretcher to the execution site, where she was shot by a German officer. When all the patients had been liquidated, the SD commander Rudolf Lange made his Latvian auxiliaries swear an oath of secrecy. At the barracks, Šumskis and the other men who had taken part in the Sarkandaugava action received 500g of vodka.

For Šumskis, life as an SD auxiliary settled into a routine of tiresome guard duty – and routine murder. At the end of the year, he was assigned to a Latvian anti-partisan unit. Bandit warfare meant attacking villages suspected of harbouring partisans and setting them alight. If partisans had killed German troops, then a proportionate number of villagers would be shot. After this period, Šumskis' activities are poorly documented. We know that in October 1943 he was assigned to dig a mass grave in sand dunes at Liepaja, well known as an execution site. He escorted a party of political prisoners to the grave and helped execute them. In March the following year, he took part in another mass killing. In April, Šumskis was in Riga, where he joined a border guard battalion, which was absorbed by 15th Waffen-Grenadier-Division der SS in June. Not long afterwards, he was captured by the Russians. By the autumn of 1944, most of Arājs' men had been assigned to different units of the Latvian Waffen-SS divisions.[12] In short, there was an evolutionary relationship between the SD Schuma battalions and police auxiliaries and the combat SS divisions. With mass killing assigned to the extermination camps like Sobibór and Auschwitz-Birkenau, Himmler no longer needed field executioners, but 'bandit hunters'. The SS inaugurated this second stage of their foreign recruitment strategy in Estonia – the most privileged region of the occupied Ostland.

According to German racial theory, Estonians had a special status in German plans. This partly reflected the influence of the Eastern Minister, Alfred Rosenberg, who had been born in Tallinn (Reval). Fresh data accumulated by German race experts implied that racially desirable characteristics were more strongly represented in Estonia than in Latvia and Lithuania. This implied, naturally, that Estonians could be more readily 'Germanised' than other Baltic peoples and this had a decisive impact on occupation policy and Waffen-SS recruitment. In the occupied east, the many different Reich agencies and potentates appointed by Hitler waged internecine war with their rivals. But in the Ostland, which incorporated the Baltic nations and Belorussia, Rosenberg's Eastern Ministry was able to exercise a more powerful influence than in Ukraine, where Hitler consistently backed the despotic Commissar Erich Koch. Estonia is, of course, the most northerly Baltic state and in 1942 the German forces besieging Leningrad still straddled the old border with the Soviet Union. The presence of German Wehrmacht on Estonian soil meant that that the SD was forced to share jurisdiction with the army. In the confusing world of German occupation strategy this bolstered the power of Rosenberg's Estonian representative SA-Obergruppenführer Karl-Siegmund Litzmann, who frequently challenged the authority of his superior Hinrich Lohse. It helped that Hitler revered Litzmann's father – a general who had served in the German Imperial Army in the First World War. He was also on very good terms with Himmler and used his well-oiled connections to shore up his Estonian fiefdom. Both Rosenberg and Litzmann regarded Estonians as blood kin (literally in Rosenberg's case) and the 'light touch' manner of Litzmann's administration favoured ambitious Estonians who wanted to carve out their own little empires under German administration.[13]

Estonia had been the last Soviet domain in the Baltic to be conquered by the Wehrmacht. The gap between Soviet withdrawal and German occupation had therefore been somewhat longer than in Latvia and Lithuania. This gave many patriotic Estonians time to escape into the forests and organise militias called Waldbrüder (Forest Brothers). When the Wehrmacht crossed the Estonian border, the Waldbrüder sent resistance units called Omakaitse (home guards) to harass the retreating Russians. Just as in Lithuania and Latvia, an SD Special Task Force commander, in this case Dr Martin Sandberger, then took over the Estonian units and, once the Soviet forces had been pushed back across the River Narva, began deploying Estonian auxiliaries to carry out 'cleansing' operations against 'hostile elements'. Under German tutelage, these Omakaitse would be expanded to become a formidable pseudo-national militia that could muster up to 40,000 men. They were mainly recruited from farm workers who had become accustomed, over many centuries, to taking orders from Germans.

As the Wehrmacht pushed on towards Leningrad, Himmler appointed Dr Sandberger as 'Kommandeur der Sicherheitspolizei und des SD Estland'. The 30-year-old Sandberger (1911–2010) was an SD high flyer. Like many of his Special Task Force colleagues, he had a doctorate in jurisprudence. He had made his mark as an NSDAP student activist in Tübingen in southern Germany, risen fast through SD ranks and bagged a top legal job in Württenberg by tirelessly exploiting his party connections. After the destruction of Poland, Himmler appointed Sandberger to head the Central Immigration Office North-East (Einswandererzentralstelle Nord-Ost) to racially 'evaluate' ethnic German migrants. At his post-war trial in Nuremberg, Sandberger testified that before the invasion of the Soviet Union he had attended a meeting called by RSHA departmental head Bruno Streckenbach outlining Hitler's order to liquidate Jews, gypsies and Russian 'commissars'. At the beginning of July, the Special Task Force A commander Stahlecker dispatched Sandberger to carry out the 'Führer order' in Estonia. To do this, he would turn to the Estonian Omakaitse.[14]

According to Anton Weiss-Wendt, 'three million Polish Jews make one thousand Estonian Jews a drop in the sea of sorrow'.[15] Perhaps so – but this 'drop in the sea' was the majority of Estonian Jews, who perished between 1941 and the end of the war at the hands of the German SD and their Estonian collaborators. The Germans concluded that of out all the territories they occupied, Estonia had the highest levels of active collaboration and the lowest levels of resistance. No Estonian ever took up arms against the German occupiers. In contrast to Lithuania, actual pogroms were rare. Instead, the Estonian 'self-government' agencies set about 'self-cleansing' through standard police procedures, arrests, hearings and sentencing. Estonians put Jews to death as 'individuals subversive to the current regime'.[16] In fact, over half their victims were women, children and the elderly: the Estonian 'subversion rationale' was evidently a smoke screen. Overt anti-Semitism was broadly absent from newspapers and official pronouncements. Nevertheless, the German administration spoke openly about what 'had to be done'. The German commandant of the Narma concentration camp put it: 'I fought a duel with the Jews.' Estonians quietly backed German racial policy. And yet it is believed that some 11,000 European Jews died in Estonian camps. It has been called 'murder without hatred'. Why was Estonia a special case? And why did Himmler authorise an Estonian SS Legion in the summer of 1941?

Special Task Force commander Dr Sandberger believed firmly in '*Sympathiegewinnung in der Bevölkerung*': winning the sympathies of the Estonian people. He regularly invited Estonian officers to German dinner parties – an exceptional gesture in the occupied east. He later testified that 'from the beginning, great store was set by establishing close personal ties, in a comradely spirit, to promote mutual trust. These personal relationships facilitated smooth co-operation, and enable us to direct the large Estonian security apparatus with the help of only a few officers.' This was the 'British India model' that Hitler often claimed to admire. Estonian SD men wore the same uniforms as their German counterparts and Sandberger sent the most diligent to Germany for ideological training. He insisted that junior German officers treat Estonians of senior rank with respect and he forbade expressions of German racial arrogance.[17] Sandberger handed his Estonian security police an astonishing level of autonomy, as well as comradely friendship. Eager collaboration deserved reward. After the war, Sandberger tried to blame his former comrades in arms for the murder of Estonian Jews.

German immigrants had moulded Estonian society and culture for half a millennium. As a consequence, Jews had not settled here in such large numbers as they had in Lithuania and Latvia. In 1939, according to historian Eugenia Gurin-Loov, Estonia had just 4,500 Jewish citizens.[18] Approximately half had settled in the capital Tallinn, but there were significant communities in Tartu and Pärnu. Poorer Jewish families ended up in small towns and villages. In 1941, the Russians deported

400 Jews to Siberian camps. Most would survive the war. On 10 July, Wehrmacht forces arrived in Tartu, and the onslaught of Estonian Jews began. Accompanying the German troops was an Estonian Omakaitse unit, the Southern Estonian Forest Brothers, commanded by Friedrich Kurg. The Estonians rounded up and arrested Jews then locked them up in the local Kuperjanov barracks, which was rapidly turned into an improvised internment camp. Five days later, Sanderberger and his Special Commando 1a arrived in Tartu; they immediately transferred the Jewish families incarcerated at Kuperjanov to another barracks which was soon designated as the 'Death Barracks'. From here, the Estonian Forest Brothers, led by a few German officers, took their captives to the Tartu–Riga Road. Close to the road, the Russians had constructed an anti-tank ditch which now provided a convenient execution site. Other shootings, probably of the women and children, took place close to Tartu's Jewish cemetery.

Sandberger's Einsatzkommando and their Forest Brothers then moved on to Pärnu; here more executions took place close to the local station and at sites in the forest close to the town. Sandberger raced on towards Tallinn. Here in the capital city, Omakaitse officers had already prepared lists of 'Jewish communists' and other suspects. When Sandberger arrived, the Estonians arrested at least 200 men and a smaller number of women; all were murdered immediately. According to Gurin-Loov, the sequence of events after this first spasm of shootings is unclear. According to a prison guard Karl Tagasaar, who was interrogated after the war, large-scale executions by Omakaitse and Sandberger's SD men certainly took place inside Tallinn prison in September. The SD also built a camp at Harku where they held Jewish women and children until the end of 1941. On 5 June 1942 an Einsatzgruppe report concluded: 'Today, there are no more Jews in Estonia.'[19] Denunciations and executions continued even after that report had been submitted and in 1943 the SS began transporting Jews from other countries to Estonia, where they were incarcerated and then murdered by Estonian guards. When Himmler closed the Estonian camps, Sandberger ordered Estonians to carry out mass shootings of the prisoners who remained alive.[20]

Today Estonians are loath to accept that their nation had any involvement with the Holocaust at all. In national myth, the Forest Brothers and Omakaitse are celebrated as freedom fighters, not murder squads.

Estonians were the first Eastern Europeans permitted to serve not just as policemen but as soldiers in the Waffen-SS. Before the summer of 1942, when recruitment began in Estonia, Himmler had appeared reluctant to authorise the formation of combat divisions in the occupied east. Historians assume that he regarded the Eastern Europeans as Slavic *untermenschen* and that his allegedly reluctant decision to authorise recruitment in the east was a desperate response to the collapse of the German war effort.

This argument does not stand up to serious scrutiny. In the summer of 1942, when the first eastern legion was authorised, neither the Germans nor the subject peoples of occupied Europe had any expectation that the Reich might be defeated. The Wehrmacht had, to be sure, suffered its first serious set back when Operation Typhoon had failed in the winter of 1941–42, but Hitler's formidable war machine was bruised, not mortally wounded. Nor, as we have seen, did Himmler regard Eastern Europeans as a homogeneous sub-racial mass. His own fascination with the diversity of eastern ethnic groups had been reinforced by the 'scientific' findings of the Abel mission. What led Himmler to hesitate was not race, but nationalism. SS recruitment was an instrument of racial domination – of Germanisation. Himmler feared that authorising national militias would be interpreted as a precondition of political demands which would have worked against the onward flow of Germanisation. After June 1941 recruitment of Schuma battalions and other kinds of police auxiliaries channelled nationalist passions into the mass murder of shared racial enemies. By the summer of 1942, that gruesome process had run its course. The 'Final Solution of the Jewish problem' would be enacted mainly through different means. The crucial significance of the Estonian case is that it demonstrated to Himmler that as both police and Waffen-SS recruitment could be managed in a way that neutralised nationalist sentiments and permitted the realisation of SS racial ambitions.

After all, the idea of an Estonian SS legion had come from an Estonian. At the beginning of August 1941 Professor Edgar Kant, the acting rector of Tartu University, wrote to the German military administration. He claimed that many Estonian students backed Hitler's 'crusade against Bolshevism' and urged the German authorities to consider recruiting an Estonian legion or some other kind of military unit.[21] Shortly afterwards, General Otto-Heinrich Drechsler met members of the Latvian 'Land Self-Administration' who also urged him to consider recruiting Baltic legions. When Drechsler reported his discussions to SS headquarters in Berlin, he received very short shrift from Recruitment Chief Berger, who dismissed the proposal as a 'political trick' – meaning that it was a step too far towards genuine Latvian self-rule. Commissar Lohse too was hostile to any kind of native autonomy and joined in the attack on Drechsler's proposal. He sent Rosenberg a fifty-one-page memorandum arguing that all administrative power should forthwith be concentrated in the office of commissar, namely himself, and that the office should be made hereditary. It will be recalled that Lohse yearned to found his own dynasty and was nicknamed Herzog or duke. Thanks to Lohse, the Latvian proposal to form national legions was thwarted – for now.[22]

In Estonia, however, the proposal to form a national legion was welcomed. The German Commissar Litzmann replied warmly to Professor Kant's letter and, in defiance of Lohse, backed his proposal to recruit ardent young Estonians who wished to fight the Soviets. To begin with, Rosenberg refused to support Litzmann, claiming feebly that he was too busy struggling to fend off the aggressive commissar of Ukraine, Erich Koch. But Litzmann soon won the backing of Field Marshall Georg von Küchler, the commander of Army Group North, who 'shared' Estonia with Litzmann and then, more surprisingly, the German Foreign Minister Ribbentrop. Litzmann's refusal to submit to Lohse, his immediate boss, and his active encouragement of Professor Kant heralded a decisive step change in German occupation policy.

By mid-1942, the SS had already set up scores of Estonian police and Schuma battalions, and managed 40,000 Omakaitse men. But the service contracts signed by the Estonian recruits would run out on 1 September 1942. The Estonian police commander, Dr Heinz Jost, met Litzmann to discuss the impending crisis and the following day sent an urgent message directly to Himmler, who was on board his mobile headquarters, the *Zug Heinrich*. Jost revealed that many of the Estonians had made it clear that they would not sign another service contract. Many had come to regard their service as auxiliary Reich policeman as shameful. But Jost had a solution. The Estonian policemen would, he believed, almost certainly agree to join an SS legion, especially if pay and conditions were improved. Jost rounded off his appeal by stressing that at least 70 per cent of the Estonians were 'racially suitable'.[23]

A week later Himmler replied to Jost. He refused to make an immediate decision, but ordered Litzmann to proceed with preparations to form an Estonian SS legion. Himmler's involvement now set off alarms at German army headquarters. Major General Hans Kruth, who feared that Jost's plan was the thin end of an SS wedge, soon called on Litzmann and urged him to persuade the Estonians to renew their contracts, thus retaining the auxiliary police as part of the shaded SD/Wehrmacht administrative apparatus and fending off the SS. Alarmed by Kruth's ploy, Jost sent a second telegram to Himmler, urging him to make an immediate decision. The next day, Berger informed Rosenberg that following a meeting with Hitler, Himmler had agreed to form an Estonian SS legion.[24] On 28 August 1942, a rally was organised in Tallinn's central square. Litzmann called on the young men of Estonia to volunteer to join the new SS legion. He added a powerful enticement: anyone who accepted his challenge would be exempted from compulsory labour service in the Reich.

In January 1943 Himmler arrived at an SS training academy to meet the first Estonian Waffen-SS NCOs. He was pleased to discover that 'racially they could not be distinguished from Germans' – the Estonians, he reported, were one of the few races that can 'after the segregation of only a few elements be merged with us without any harm to our people'.[25] Himmler had crossed the Rubicon.

In Riga, Latvian army officers closely followed developments in Estonia. All were ardent nationalists but also fervently pro-German. They did not regard Latvians as any less deserving of Himmler's favours than their Estonian comrades. In 1940, many Latvian army officers had fled the Soviet terror to Germany. Among them were Generals Rūdolfs Bangerskis, Artūrs Mihails Silgailis and Oskars Dankers. In Berlin, Silgailis and a few other Latvian officers had been recruited by the Abwehr and received training at a camp in East Prussia. At the end of June 1941, Abwehr head Admiral Canaris promoted Silgailis to Sonderführer (special leader), outfitted his men in German uniforms and, as the Wehrmacht rolled into Lithuania, dispersed them among various German units to pursue clandestine missions behind enemy lines. One of the most devoted pro-Nazis in the Latvian administration was Lt Col Voldemsrs Veiss, who had formerly served as the Latvian military attaché in Finland. Veiss was a fanatical anti-communist. The 'Jewish' NKVD had, he claimed, murdered members of his family; he thirsted for revenge. In the volatile period immediately following the German invasion, SD commander Walther Stahlecker depended on both Veiss and Viktors Arājs to recruit Latvian auxiliaries to push forward anti-Jewish 'cleansing' operations. By mid-July, Veiss and his equally brutal henchman Lt Col Roberts Osis had seized control of all Latvian militia in the Riga area with the exception of the Arājs Commando. Armed with captured French and Czech weapons, these Schuma battalions, like the Arājs men, scoured the countryside, hunting down Jews and communist 'bandits'. At the end of October, the SS began deploying their Baltic Schuma men on the front line – first of all on the Leningrad front, near the Staraya Russa province, then in Ukraine and southern Russia to combat partisans. Veiss and the other Latvian commanders hatched up a plan to traffic volunteers for political favours. In December 1941, Rosenberg inadvertently handed the SS a useful recruitment tool. As Eastern Minister, he issued a decree making labour service in the Reich (RAD) mandatory for all Latvian men and women. Latvian students who wanted to continue university studies would have to rack up a year of labour service. When Latvians reported to RAD offices, the SS and SD pounced, offering young Latvian men a 'choice' between hard labour in Germany or service with a police unit or Schuma battalion.

It was increasingly evident that for many young Latvians the Schuma battalions had lost their allure. Since the majority of Latvian Jews had been killed, and the process of mass murder reallocated to the camps, security duties had become mundane. By the same token, the SD had less need of their services as man hunters. In these new circumstances, military service appeared much more attractive. In June 1942, HSSPF Friedrich Jeckeln, who had masterminded the Rumbula massacre, hinted to Latvian army officers that if they provided enough recruits for a Latvian SS legion they might be rewarded with some level of political autonomy. Himmler rapped him sharply on the knuckles for taking this initiative: 'political bargaining', he insisted, was 'fraught with danger'. But it was not long after slapping down Jeckeln that it became known in Latvian military circles that Himmler had ratified the formation of the Estonian SS Legion. It seemed that the stable door was, at the very least, ajar. In November, a delegation from the Latvian puppet administration led by Dankers, Veiss, and Silgailis met with Jeckeln's subordinate, the SS and Polizeiführer (SSPF) Walther Schröder in Riga. He was known to be sympathetic to their cause, and the Latvians again urged the Germans to consider authorising a Latvian SS legion. But Schröder's boss Jeckeln, now wary of antagonising Himmler, flatly turned them down and repeated his demand for more Latvian police recruits.

It was stalemate. But the changing fortunes of war would soon strengthen the Latvian's hand. In the summer of 1942, well-organised partisan armies in the occupied Soviet Union and the Balkans, acting on orders from Moscow, began to escalate their struggle against the German occupiers. Frequent and often highly effective attacks on German supply lines began to wreak havoc in army rear areas. In August, Hitler finally recognised the scale of the problem. He issued Führer Directive 46: 'Instructions for Intensified Action against Banditry in the East', which formally criminalised partisan attacks and handed control of 'bandit warfare' to Himmler and the SS.[26] On 9 September, Himmler called the architect of the Pripet Marshes mass slaughter, Erich von dem Bach-Zelewski, to his headquarters for lunch, and appointed him Chief Inspector of Bandit Warfare.[27] Both Himmler and his chief bandit hunter had used anti-partisan warfare as a means to camouflage racial mass murder. Where there is a partisan, there is a Jew, Bach-Zelewski often proclaimed, and where there is a Jew there you will find a partisan. In German thinking, the Soviet Union was a Jewish bandit state. The new partisan armies that now plagued the German war machine were, however, multinational militias. Many Polish and Ukrainian partisans hated Jews, and often murdered Jewish refugees. Many thousands of Jews, to be sure, fought the Germans as partisans, but after the summer of 1942 anti-bandit warfare was necessarily directed at genuine militias that had no explicit racial identity. Nevertheless, the equation between 'The

Jew' and the 'The partisan' continued to infest SS anti-bandit strategy. According to German army reports on anti-partisan actions, the number of Jews executed 'as partisans' frequently outweighed any other ethnicity.[28]

Just as the renewed partisan war erupted, and Himmler tightened his grip on security, the massive German offensive against Stalingrad which had begun that summer, began to bog down. In November, the Russians launched Operation Uranus with the intent of encircling the German 6th Army. As the brutal Russian winter began to grip, the entrapped and starving German troops buckled under relentless Soviet attack. As this catastrophe began unfolding in south-east Russia, Soviet partisans stepped up their campaign in the region near Minsk in Belorussia, on the border with Latvia.

It was at this critical juncture that Alfred Valdmanis, the former Latvian Minister of Finance and the most outspoken member of the Latvian puppet self-administration, came up with a fresh proposal which would open the way to the formation of an SS legion in Latvia. Even today Valdmanis evokes ambivalent responses among Latvians.[29] Was he hero or villain? A shrewd manipulator of mightier powers or an abject Quisling? Or all of these? Valdmanis (a superbly skilled chess player) was, to be sure, a survivor. He was born in Liepāja in 1908, then part of the Imperial Russian Empire, and from his student days he would pursue careers in three different nations and under no less than ten regimes: Tsarist, Imperial German, liberal democratic, Latvian authoritarian, Soviet and German National Socialist. He was both political chameleon and a charismatic manipulator of enemies and allies alike. In 1940, he cut his political cloth to fit the Soviet occupation – then in 1941, turned that coat inside out and joined the Latvian self-administration.

Following a series of frustrating meetings with SSPF Schröder, Valdmanis came up with a new initiative that he set out in informal document 'The Latvian Problem'.[30] He reminded the German occupiers 'With what love the German soldiers were received and guided on in Latvia'. But in return, he went on, Latvians had been insulted and humiliated: 'Have the Germans really come as liberators?' Valdmanis insisted that Latvians desperately wished to join the war on Bolshevism, but as citizens of a free, autonomous state. As a model, Valdmanis proposed Slovakia: a free nation, he asserted, but closely bonded with Nazi Germany. Since the Slovakian government was dominated by the anti-Semitic Hlinka Guard, Valdmanis was clearly signalling that Latvian would continue to play a part in the 'solution to the Jewish problem'. Valdmanis proposed that in return for 'Slovakian style autonomy', he and the SA would offer up at least 100,000 Latvian volunteers to serve in a new national legion. He warned Schröder: 'All of Latvia is in the grip of a sullen paralysis ... Many don't seem to care whether they are swallowed up by the

Bolsheviks or sucked up by the Germans ... From where shall we procure volunteers?' In other words, if the Germans wanted Latvian recruits they would have to make some serious concessions in return.

As it turned out, both Himmler and Rosenberg had already discussed granting limited autonomy to the Baltic states, and Himmler had brought the matter up at a meeting with Hitler. This meant that when Schröder met Valdmanis and the SA he could truthfully tell them that Latvian autonomy was being considered at the highest levels. In Berlin, Berger also flicked through Valdmanis' *The Latvian Problem*. He scorned the 'Slovakian option' and, in any case, was unconvinced that Valdmanis could come up with the promised 100,000 volunteers. But he urged Himmler to consider the propaganda benefits of a recruitment campaign apparently led by Latvians.[31] We cannot be sure, but Berger may have believed that if Latvians, say, were seen to join the German side, other Eastern Europeans, who might have thrown in their lot with the Soviet partisans, would follow their example. In January 1943, Himmler paid a rare visit to the front line near Leningrad and watched Latvian Schuma battalions in action with the 2nd SS Motorised Infantry. They fought well. Himmler made a snap decision to reorganise the three Schuma battalions as a single Latvian SS Volunteer Brigade. On 23 January, Himmler met Hitler in Posen and commended the Latvian units. Their discussion took place just as Soviet forces commenced their final decisive move against the German 6th Army trapped at Stalingrad. On 31 January, General Field Marshall Friedrich Paulus surrendered to Soviet forces, 'besmirching', Hitler spat, 'the heroism of so many others at the last moment'.

On 10 February, as 91,000 German and Romanian soldiers and twenty-two German generals, including the shattered Paulus, were marched into captivity, Hitler formally signed an agreement ordering the formation of a Latvian SS Volunteer Legion: '*Ich befehle die Aufstellung einer Lettischen SS Freiwilligen-Legion*.'[32] But in Riga, skirmishing between the Latvian SA and the Germans was far from over – and according to an SD report 'The people [of Liepsja, Latvia's second largest city] are all talking about a general mobilization of Latvian men. The common talk is that the Germans want to get Latvia's younger generation into the army, seemingly with good reason, just to use them as cannon fodder at the front and eliminate them.'[33] The Latvian negotiators showed signs of getting cold feet; but immediately after receiving Hitler's order, Berger formally announced the formation of the 15th SS Latvian Volunteer Division, and appointed a German SS-Brigadeführer, Peter Hansen, as commander. This naturally further dismayed the Latvians who had been led to believe that General Rudolfs Bangerskis would be appointed as a matter of course. As Dankers and Valdmanis bickered with Jeckeln and Commissar Lohse, the Germans cynically went ahead with plans to start inducting Latvians into the legion. To circumvent the Hague Convention of 1907, which proscribed drafting citizens of occupied nations, they used the old trick of decreeing obligatory labour service and then forcing the Latvian Department of Labour to act as a conscription agency. Once Latvians had registered with the department and been declared fit, they had to declare a choice between labour and military service. If they chose the latter, they had to sign a form declaring that they had made a voluntary selection.[34]

In March, after recruitment had begun, Hitler rejected the proposal for a 'Slovakian solution' – and Valdmanis, its most outspoken advocate, was sent off Berlin, where he enjoyed the comforts of the luxurious Adlon Hotel and was effectively neutralised. In Riga, the Germans continued to run rings round the confused and compromised Latvian leaders. Himmler grudgingly agreed to appoint General Bangerskis as 'Legion-Brigadeführer und General Major der Lettische SS Freiwilligen Legion'. So Hansen stood down, and it seemed as if the Latvian negotiators had won a round. But on 7 April, Jeckeln recalled Bangerskis and informed him that a mistake had been made. Since he was not a Reich citizen, he had no legal authority to command a German division. So Hansen was brought back and Bangerskis demoted to 'Inspector General of the Latvian legion'. To muddy the water still more, Berger promoted Bangerskis to SS-Gruppenführer. For the rest of the war, the outsmarted old man sat at a desk in an office in Riga, doing what he could to promote the cause of the Latvian SS legions.[35]

On 26 February the SS-Fuhrungshaupamt in Berlin, Wilmersdorf, issued a final memorandum: 'Formation of the 15th Latvian SS Volunteer Division'. Although it preserved a few of the concessions demanded by the Latvian SA, the oath vitiates any kind of 'autonomy': 'I swear by God this holy oath, that in the struggle against Bolshevism I will give the commander of the German Armed Forces, Adolf Hitler absolute obedience and as a fearless soldier, I will lay down my life for this oath.'[36]

What about numbers? Despite the cynical tactics of the Germans and 'negative propaganda' promoted by a few Latvian dissidents, 67,584 Latvians reported and registered. Of these some 27,000 were assigned to the RAD; the rest signed up for military service in the proposed legion. But the SS could still afford to be fastidious recruiters. The German race inspectors passed as fit less than 3,000 men. Notwithstanding the application of such rigorous admission criteria, Himmler made clear to General Dankers and Inspector General Bangerskis that the legion must be referred to as a 'Waffen-Grenadierdivision *der SS*'. This meant that it was not strictly an SS division ('das SS'), but 'belonged to' the SS. Latvian officers could not be designated as 'SS-Scharführer', say, but as 'Leg. Scharführer'. Himmler denied Latvian recruits any of the normal privileges enjoyed by his German officers: SS social clubs and brothels were off limits. In October and November, further mobilisations were demanded, and on each occasion Himmler refused to consider any concomitant 'political concessions'.

Visvaldis Mangulis, who observed some of the meetings between the SA and the Germans, tried to rationalise the dilemma of the Latvian SA: 'If they did not mobilize and the Germans won the war, then the Latvians would have a weak claim to independence ... If they did not mobilize and the Germans lost the war, then surely Latvia would be occupied by the Reds once more.'[37] Mangulis' analysis is hardly logical, but it is repeated every year by the organisers of the 'Latvian Legion' commemorations in Riga. If the Soviet army could crush the mighty German war machine, how could a single Latvian division resist its bulldozer-like advance into the Baltic? The fact of the matter is that the Latvian self-administration 'directors' had been 'hoisted by their own petard'. Collaboration made Latvians more, not less, vulnerable to any renewed Soviet terror. By agreeing to recruit young Latvians, the SA condemned an entire generation to certain death on the battlefield and decades of persecution as 'traitors' by the Soviets who would successfully reoccupy the Baltic after 1945. The dirty mirage of national autonomy, of 'Latvia for Latvians', led Valdmanis and the other SA stooges into a foolish trap. In return for an illusory share of power, they gambled away Latvian lives. The Germans, for their part, took a completely cynical view of the Latvian self-administration. The Mayor of Riga, Hugo Wittrock, writing to his close relative Alfred Rosenberg, disparaged the insatiable appetite of the Latvian 'clique': 'First "Volkshilfe", then directorate general, now protectorate, then á la Slovakia ... The end, I don't want to spell it out! The arrogance of these well known gentlemen [Valdmanis et al.] ... has now reached its peak.' Wittrock had no doubt about the appropriate solution: 'once kicked in the teeth, that gang quickly takes cover, which is what actually happened.'[38] Many ordinary Latvians, Abwehr intelligence revealed, regarded the German occupiers more realistically as incompetent, corrupt, selfish, narrow-minded, conceited and uncultured. Collaboration was a one-way, dead-end street.

In 1943, the Latvian SA had begun to fear that Germany could lose the war – but this did not necessarily mean that in early 1943 they had any reason to believe that they could negotiate from a position of strength. To be sure, the German front line was by then under intense pressure and the surrender of the German 6th Army at Stalingrad was a calamity. But Hitler and the Nazi elite did not believe the war was lost. Even after Stalingrad, Hitler's generals continued to fight an offensive war and laid plans to launch a massive new spring offensive against the Soviet armies. The Latvian misperception of German vulnerability led them to fall for an old trick: the promise of future autonomy in exchange for recruitment. So while Hitler was supposed to be considering Latvian autonomy, Berger cynically pushed ahead with drafting Latvians with the connivance of the Latvian authorities. It is customary for historians to pour scorn on 'chaotic' German administrations in occupied territories. In practice, the multiplication of different, competing authorities often wrong-footed indigenous nationalists with a kind of confusing hard cop/soft cop routine. That may have been inadvertent, but the effect was to encourage compliance with the promise of a better deal to come. The Latvian collaborators had fallen for this kind of practice from the day German SD men marched into Riga.

Today, Latvians who defend the annual commemoration of the 'Latvian Legion' stress that Latvians were conscripted: service was not voluntary. At the same time, defenders of the legion argue that Latvians joined to defend their nation against a second Soviet occupation. They want it both ways: the SS recruits as both heroes and victims. In reality, at least 25 per cent of the Latvian recruits did volunteer. But even if we accept that the Germans drafted the majority of recruits, they fought in any case for a Latvia 'cleansed' of fellow Jewish citizens.[39] In 1941–42, many of these conscripts, like the aforementioned Juris Šumskis, had taken part in the SD special actions against Jews and the mentally ill. In 1943, they took up arms against the Soviet Union – in defence of a Latvia founded on chauvinism.

The records show that in order to form the core units of a new SS division, Berger yoked together the Latvian Schuma and SD police battalions that had so impressed Himmler on the Leningrad front. Many of these men had not only murdered Jews in Riga and elsewhere, but also participated in 'bandit operations' in the Minsk region. As we saw in a previous chapter, the Arājs Commando had been militarised at the end of 1941 and deployed to fight partisans in Belorussia. The Schuma battalions had purged Latvia of unwanted Jews; it was logical to use their skills outside Latvian borders to continue their 'work'.

In November 1941, the 18th Latvian Police Battalion took part in the liquidation of the Jewish ghettoes at Barisov and Slonim. In January the following year, Latvian Schuma men joined in a renewed round of slaughter that, according to an SS report, left over 20,000 unarmed civilians dead. The Latvians proved to be such effective 'bandit hunters' that Himmler doubled the number of Schuma battalions. In February 1942, the Latvian puppet self-administration willingly took over the management of the Schuma and appointed a 'Committee of Latvian Volunteer Recruitment', headed by Pērkonkrusts fanatic Gustavs Celmiņš. It will be recalled that Celmiņš had fled to Berlin before the war, then returned to Latvia with the German armies in June 1941. It is significant that he and the other committee members presided over the redeployment of the Latvian Schuma battalions outside national borders as 'bandit hunters'.

In many areas behind the Eastern Front, as I have emphasised, SS anti-partisan actions frequently (but not consistently) provided opportunities to liquidate those Jews who had escaped from ghettoes or camps. If SS anti-bandit units captured Jews, it was customary to torture them in the vilest way before they were executed. Five Latvian battalions took part in one such 'bandit operation': Operation Swamp Fever under the notorious 2nd SS Infantry Brigade. Another operation led by HSSPF Jeckeln and known as 'Winter Magic' also deployed Latvian units that had been incorporated into the Kampfgruppe Jeckeln – and 'cleared out partisans' in a 55-mile-wide strip of territory along the Latvian border near Lake Osveya. A succession of search-and-destroy 'sweeps' led by these Latvian battalions invariably left burning villages and murdered civilians in their wake. In many cases, it is the officially reported 'kill numbers' that tell the real story. After one attack, the Germans reported that over 7,000 people had received 'special treatment' (i.e. immediate execution) and 3,300 Jews liquidated; the Germans lost two men.[40] It was during Operation Winter Magic that Walther Stahlecker, the former Special Task Force commander who had recruited Arājs and Veiss, attacked the village of Sanniki with a battle group comprising Germans, Latvians and Estonians. In the course of the attack, Stahlecker was shot dead and his men sought revenge. They burnt Sanniki, and killed every villager, both Christian and Jew.

To begin with, the new 15th Waffen-Grenadier-Division der SS undertook the same kind of tasks – as we find in a report filed by Lg.-Standartenführer Artūrs Apsītis on 17 November 1943. On 14 November his echelon, severely under-equipped, had been deployed to the south of Ostrov, where the Latvians received orders to 'clear out partisans and possible Red Army units ... from six villages', then to proceed further south to 'clear out a wider area occupied by partisans'.[41]

So let us be clear about the origins of the 'Latvian Legion' and their role in Hitler's 'war of annihilation'. In 1943, Himmler amalgamated a number of Schuma battalions serving on the Leningrad front into the 2nd Latvian Volunteer Brigade; this would become the core of the 15th Latvian Waffen-Granadier-Division. By mid-1944, many of the Schuma battalions had been transferred to the Waffen-SS divisions, at the same time as German Order Police battalions were being absorbed into the Waffen-SS. In 1944, the Arājs Commando was amalgamated with the 15th Latvian. So whatever the modern apologists for the legion claim, it is simply a matter of fact that men who had committed the most gruesome atrocities serving with the Arājs Commando and other Schuma battalions both in Latvia and later in Belorussia ended up serving in the two SS divisions known as the 'Latvian Legion'. As the Nazi programme of mass murder focused on the extermination camps rather than the 'rifle-and-ditch' method of earlier phases, Himmler had no further use for Schuma brigades and diverted their activities to 'bandit warfare' and the front line.

From the spring of 1943, a succession of military setbacks in the Mediterranean as well as on the Eastern Front had a powerful impact on German decision making. Mounting losses of German troops, the surrender of the 6th Army at Stalingrad and the security crisis engulfing the occupied territories led cumulatively to increased levels of non-German recruitment by the German armed forces and especially the Waffen-SS. But in Himmler's mind, the expansion of recruitment followed a completely logical path that was shaped by his racial creed. Foreign recruitment began with ethnic Germans in Romania, expanded to include Nordic Europeans in Scandinavia and Holland and then, in mid-1942, embraced Estonians who were considered a heavily Germanised Eastern European people. As the lessons of the Abel study of Russian POWs sank in, the potential pool of recruits broadened further still.

The mission's findings had a decisive impact on German anthropological ideas about race and subsequently on Waffen-SS recruitment. In 1942, Berger's recruitment office introduced a three-tier recruitment scheme based on Professor Abel's reports. It split foreign recruits into three hierarchical categories as follows:

> 'klassischen SS Divisionen'
> SS Freiwilligen-Divisionen (Germanische Freiwillige)
> SS Waffen-Divisionen (nicht-germanische Freiwillige)

Nicht-germanische here simply meant ethnic groups usually excluded from the Aryan family but that were, as Abel had showed, potential bearers of Germanic blood. In theory, there was no limit to the process of 'Germanisation', with the exception of Jews, Poles and Roma. In April 1944, Himmler delivered a lecture about how the selection process worked:

> The European peoples, the Latvians and Estonians, the Galicians [Ukrainians], the Bosniaks [Bosnian Muslims], Croats and Albanians are joining us, the senior peoples of Europe. The Latvians and Estonians will form divisions, the so called 'Waffen-Divisionen' of the SS. Their youths will attend our *Unterführenschulen*; if they are racially equal to us, our Germanic *Junkerschulen*, and without wanting to hurt or insult them, *Waffenjunkerschulen*, if they are racially different.[42]

What this statement proves is that as late as spring 1944, Himmler was still thinking in terms of racial hierarchies. He continued to view Waffen-SS recruitment as a means to 'gather Germanic blood'.

One last point is in order here. In late 1943, the Germans launched a campaign to recruit Lithuanians into the Waffen-SS. Dr Adrian von Lenteln, a Baltic German who had been appointed General Commissar for Lithuania, had every reason to expect success. Lithuanians had proved themselves eager executioners in the period following the German occupation in June 1941. As one American report put it: 'The Lithuanian military police, Litauische Schutzmannschaften is organised into SS units ... [and] used by the Germans to perform executions.' But in 1943, as Estonians and Latvians rushed to join the SS legions, the Lithuanians balked. Efforts to establish a Lithuanian SS legion ran into the ground. In March 1943, Himmler and von Renteln travelled to Kaunas to persuade the Lithuanians to start recruiting but got nowhere. Lithuanians, Himmler complained, were 'not worthy to wear SS uniform'. Today, Lithuanians celebrate this refusal as 'heroic resistance'. It was nothing of the kind. It was merely a shrewd recognition that Germany was losing the war and there was no point going down with the Reich. Many of the bureaucrats who successfully fended off Himmler and von Renteln had been responsible either directly or indirectly for murdering tens of thousands of Jews. The Lithuanian refusal teaches a very different lesson. The Estonian and Latvian administrations had a choice. Reinforcing Hitler's war and sacrificing the lives of tens of thousands of young men was not inevitable.[43]

As the tide of war turned relentlessly against Germany and the frontiers of the 'Greater German Reich' began to shrink, Himmler's SS empire bloated. As the Wehrmacht began its long retreat to the borders of the old Reich, SS propaganda proclaimed that the Waffen-SS would be the 'fire brigade of the Eastern Front'. The Nazi elite turned against what Goebbels called the 'fat, big paunched majors in the Bendlerblock [German army headquarters in Berlin]', but Himmler's star began to ascend to its zenith. At a meeting in September 1943 Hitler informed his loyal paladins that 'The best thing I leave to my successor is the SS'.[44] Himmler would become the emperor of defeat.

11

Nazi Jihad

> All Islam vibrated at the news of our victories. The Egyptians, the Iraqis and the whole of the Near East were all ready to rise in revolt. Just think what we could have done to help them, even to incite them.
>
> Hitler, April 1945

The town of Villefranche-de-Rouergue sprawls along the banks of the l'Aveyron River that winds through the Mid-Pyrenees in south-west France. Clinging to the wooded slopes that rise steeply from the river is the little chapel of Calvaire St Jean d'Airgrement, which commands a broad view of the wide, flat plain that rolls northwards. In the early morning, the l'Aveyron valley is quiet, tranquil, even bucolic. A few farm vehicles putter along narrow rural roads. In 1943, this was 'Maquis' country – for the German occupiers, a region to be feared and mastered. In most of the villages and towns of the Mid-Pyrenees stand memorials to French resistance fighters who died here fighting the Nazi terror machine. But in September 2006, the town council of Villefranche-de-Rouergue unveiled a very different memorial. It is dedicated not to wartime French heroes, but '*jeunes soldats de Croatie et de Bosnie-Herzégovinie tombés lors du soulévement du 17 Septembre 1943 à Villefranche-de-Rouergue contre leurs oppresseurs Nazis*'.

Why and how did young men from Croatia and Bosnia come to die so far from home in this pretty French village? The memorial tells us some of their story. They served in an SS division known as the 'Handschar', which had been recruited in Bosnia. Most of the recruits were Muslims. Here in Villefranche, a handful rose up against their officers. For a few hours they held the might of the SS at bay. Himmler, enraged, meted out violent retribution. The mutineers were hauled in front of a kangaroo court then shot dead by an SS firing squad. The drama that unfolded seventy years ago in Villefranche-de-Rouergue provides a surprising insight into Himmler's quest to harvest Germanic blood 'wherever it might be found'. Few Bosnian Muslims regarded their people as 'Germanic' at all – and might never have enlisted in Hitler's war had they not fallen under the malign influence of one of the most notorious of all wartime collaborators: the Grand Mufti of Jerusalem, Haj Amin el-Husseini. It was el-Husseini who convinced these Bosnians to become SS men and join the Nazi Jihad against the 'Jewish World Enemy'. Himmler's pact with this malevolent Arab cleric poisons the battleground of Middle Eastern nationalist ideologies to this day.

Writing in his diary at the end of the war Albert Speer recalled:

> I never saw Hitler so beside himself as when, as if in a delirium, he was picturing to himself and to us the downfall of New York in towers of flame. He described the skyscrapers turning into huge burning torches and falling hither and thither, and the reflection of the disintegrating city in the dark sky.

One of the longed-for miracle weapons imagined by Hitler at the end of the war was the 'Amerikabomber', a Daimler designed, four-engine giant that in theory could bring terror to faraway New York City. Hitler despised New York as the capital of world Jewry. His destructive fantasy was finally realised by Mohammed Atta and his fellow pilots when they flew their fuel laden aircraft into the World Trade Centre on 11 September 2001. Might the Jihad of our own time have been inspired by plans hatched in Third Reich?[1]

On 29 April 1945 Hitler, his mind and body ruined by drugs and disease, hidden away beneath the ruined Chancellery in Berlin, ordered Traudl Junge, his favourite secretary, to accompany him to the conference room. Hitler, his limbs shaking uncontrollably, leant on the abandoned map table and began to dictate his 'Last Testament'. Frau Junge struggled to keep pace with the torrent of poison. Hitler raged against the Jewish conspirators whom he claimed had brought down the Reich. He insisted time and again that he had wanted only to defend Germany against its sworn Jewish enemies in Moscow and Wall Street. Then, in an astonishing outburst, Hitler listed all the lost opportunities torn from his grasp by cowardly and treacherous subordinates:

> All Islam vibrated at the news of our victories. The Egyptians, the Iraqis and the whole of the Near East were all ready to rise in revolt. Just think what we could have done to help them, even to incite them … We had a great chance of pursuing a splendid policy with regard to Islam.[2]

It is one of modern history's most troubling counterfactuals: suppose, in 1941, Hitler had abandoned plans for the invasion of Russia and sent his forces to the Middle East instead. The German Wehrmacht could have easily have swatted aside the enfeebled, poorly equipped British forces based in Egypt and Palestine. Heydrich's Special Task Forces would have rampaged through Tel Aviv and Jerusalem. In Syria, the Vichy administration would have waved the German army through the border with Iraq while the Luftwaffe pounded British garrison forces there and swept on to the borders of the Raj. In Persia, Hitler's war machine could have feasted on oil reserves as rich and deep as any in Romania or the Caucasus. It was a prospect Churchill feared. In his darkest moments, he imagined 'Hitler's hand' stretching as far as the Indian border 'beckon[ing] to the Japanese'. Churchill's nightmare never became real, although Luftwaffe aircraft did bomb Baghdad. Hitler's mind was fixated by the Bolshevik enemy and a future empire in the east. His vision of empire was brutal, but profoundly parochial. He spurned most efforts to undermine the British Raj or empower to dark-skinned, 'inferior races'. But as Hitler's Reich flexed its imperial muscles, the British Empire had long been in decline. In the pink regions of the world map, new nationalist movements demanded freedom from British rule. At home, the moral authority of the empire was no longer taken for granted. In India and Palestine, where the imperial crisis was most acute, some of those waging war on the British Empire turned to Hitler's Reich to speed the collapse of foreign rule. For Subhas Chandra Bose, the Indian nationalist who spent much of the war lobbying the German Foreign Office in Berlin to back his cause, it was a case of 'my enemy's enemy is my friend'. Bose had few illusions about Hitler's racist world view and, when he met the Führer at the Wolf's Lair, had the guts to criticise his pejorative comments about Indians in *Mein Kampf*. Bose remained silent about Hitler's hatred of Jews.

Haj Amin el-Husseini, the Grand Mufti, was very different species of collaborator. He too spent the war in Berlin. Unlike Bose, he fully embraced the Nazi racial vision and backed the 'Final Solution of the Jewish problem'. After 1945, el-Husseini escaped back to the Middle East. In the 1960s, the old man became a mentor to a young Palestinian called Mohammed el-Husseini, no relation, who would soon be better known as Yasser Arafat. From the ruins of the Third Reich, the Grand Mufti brought back to his lost homeland the virus of European anti-Semitism. If you seek his monument, then watch an Arab satellite channel like Al-Manar (the Beacon). Here is an extract from *Diaspora*, an ambitious twenty-nine-part 'history' series: 'Listen!' says a Rabbi to a young Jew. 'We have received an order from above. We need the blood of a Christian child for the unleavened bread for the Passover feast.' A petrified boy is seized, and, in a gloating close-up, his throat is cut and his blood drained into a metal basin.[3] This wicked nonsense resurrects one of the most enduring and potent myths in western anti-Semitism: the 'Blood Libel'. As Anthony Julius points out, this medieval fantasy has become one of the most virulent anti-Semitic myths in circulation in modern Islamic discourse.[4] Al-Manar belongs to Hizbollah (Party of God). The *Diaspora* series, made with Syrian government backing, was shown for the first time during Ramadan in 2003. At least 10 million people a day tune in to Al-Manar's round-the-clock broadcasts recorded in Beirut.

The same message was once transmitted from a radio station built south-east of Berlin in Zeesen, a suburb of Königs Wusterhausen. From studios buried deep beneath a towering mast, the exiled Grand Mufti broadcast to his fellow Muslims in coffee houses, bazaars and public squares all over the Arab world. Radio Zeesen became the most popular radio station in the Middle East: el-Husseini and his colleagues used music and quotations from the Koran mixed together with Nazi propaganda that insisted that the Allies were lackeys of the Jews – and that Jews were dangerous enemies of Islam: 'The Jew is the enemy and it pleases Allah to kill him.'[5]

Then in 1943, Haj Amin el-Husseini took on another mighty task for the Reich. The Grand Mufti was already well known to the SS. He corresponded obsequiously with Himmler. Now recruitment chief, Gottlob Berger, wrote to the Mufti. He had an unusual request. Berger hoped the Grand Mufti would agree to travel to Sarejevo in Bosnia – then incorporated by the puppet state of Croatia (NDH) – and assist with a new SS recruitment drive. Hitler had recently authorised recruiting Bosnian Muslims as SS warriors and el-Husseini's task would be to make a series of public appearances designed to persuade young Bosnians to join up and 'cleanse the land'. The Grand Mufti's campaign was astonishingly successful. By 1944, over 20,000 young men had volunteered to join the new 13th Waffen Mountain Division of the SS 'Handschar' (1st Croatian). The SS issued their new recruits with a custom-designed uniform that included a fez. Himmler promised the Mufti that his Muslim warriors would be spared any exposure to 'pork, pork sausages and alcohol' and that their spiritual needs would be attended to by Bosnian Imams, specially trained at colleges in Dresden and Göttingen.

For his part, el-Husseini fervently hoped that the SS 'Handschar' and other Muslim divisions recruited by the Reich would spearhead the destruction of the British mandate in Palestine and the liquidation of all Jews who enjoyed its protection. From his Berlin radio studio, the Mufti demanded: 'Kill the Jews wherever you find them. This pleases God, history and religion. This saves your honour. Allah is with you.' Himmler and the Mufti both believed that the destruction of their Jewish enemies depended on sacrifice and martyrdom, the shedding of blood. The

Grand Mufti proclaimed in one of his radio talks: 'The spilled blood of martyrs is the water of life. It has revived Arab heroism, as water revives dry ground. The martyr's death is the protective tree in whose shadows marvellous plants again bloom.' Why did the Grand Mufti pledge allegiance to the Third Reich? How did the Bosnian Muslims, a Slavic Balkan people, come to figure so prominently in Himmler's master plan?[6]

Although Bosnian Muslims and Haj Amin el-Husseini had the faith of Islam in common, their political worlds did not overlap until the middle of the Second World War. If they looked outside their own homeland, the majority of Bosnian Muslims followed events in neutral Turkey but had little interest in the broader Islamic movement or noticed the protests of Palestinian Arabs against Jewish immigration. These matters naturally obsessed Haj Amin el-Husseini, however, and led him in due course to seek an alliance with the anti-Semitic German Reich.

In the period after the First World War, many influential Arabs admired Germany. They recalled the close bond between Imperial Germany and the Ottoman Empire and regarded the Weimar Republic as a potential ally against the British and French mandate governments in Syria and Palestine. After 1933, Hitler's widely publicised anti-Jewish proclamations had a seductive appeal for many in the Arab world and 'Hitler frenzy' spread across many parts of the Middle East and North Africa like a nasty rash. The most prominent and influential pro-German in the Arab movement was Haj Amin el-Husseini, who became the Grand Mufti of Jerusalem, the highest Islamic post in Palestine, in April 1921. A frail and diminutive man with a fluting, high-pitched voice, el-Husseini soon established himself as de facto leader of the pan-Arab movement by promoting a violent campaign against Jewish immigration to Palestine. He had discovered 'The Protocols of the Elders of Zion' just after the war, and lost few opportunities to denounce the wickedness of Jews and the misery they brought to the entire world. He bitterly opposed any Palestinian Arabs who dared refer to the benefits brought to Palestine by immigrant Jews.[7] The Mufti's campaign culminated with the eruption of the Arab Revolt in July 1937; Arab gangs targeted not only Palestinian Jews and British mandate officials but also moderate Arabs and the Mufti's political rivals. As violence engulfed Palestine, the British resolved to get rid of this troublesome cleric and the Mufti fled to Beirut in the French mandate of Syria on 14 October.[8] From a succession of opulent villas, the exiled Mufti organised a terror campaign against the British and the Palestinian Jews. Although the French and then Vichy authorities thwarted efforts to have him arrested, the Mufti eventually fled to Baghdad, probably with French connivance. Here he conspired with the pro-Nazi Iraqi nationalist Rashid Ali el Gaylani, who launched a coup against the pro British Hashemite regime. In May 1941, on the eve of Hitler's attack on the Soviet Union, Iraqi troops surrounded the British RAF base at Habbaniya, backed ineptly by special forces recruited by the Arabist, and former German ambassador in Iraq, Fritz Grobba and a handful of Luftwaffe bombers that ran out of fuel after flying a few ineffectual missions against the besieged British base. In London, the government somewhat reluctantly dispatched a relief force from Palestine to snuff out the uprising. Hitler's Gulf War ended less than a month after it had begun. El-Husseini and his followers then fled to pro-Axis Tehran. In the aftermath of the uprising, fanatical Iraqi nationalists turned on Baghdad's Jews as they celebrated the festival of Shavuot and in a two-day frenzy killed nearly 200 people. Jews refer to this forgotten pogrom as the Farhud.[9]

In October 1941, the Mufti, disguised as a woman, fled through Turkey to fascist Italy. Here he had a brief meeting with Mussolini then boarded a train to Berlin. He arrived in the capital of the Third Reich on the night of 6 November 1941, and was whisked off to a suite in the Adlon Hotel on Unter den Linden. Two weeks later, Foreign Office officials moved the Mufti and his entourage into a splendid villa on fashionable Klopstock Strasse that runs through the Tiergarten not far from the 'English Garden'.

As well as the inept architect of the Iraq uprising Fritz Grobba, el-Husseini's most enthusiastic backer in the German Foreign Office was Erwin Ettel – an ardent National Socialist. Ettel had worked as an aviation expert for Junkers in Turkey and then joined the NSDAP a year before Hitler seized power. He joined the Foreign Office and served in Rome between 1936 and 1939. At the end of October 1939 Ettel was dispatched to Tehran, where he first met the Mufti and his entourage, who had just fled Baghdad. A few German diplomats had some involvement with the chaotic and ineffectual resistance to Hitler. Ettel was not one of them. In 1937, he had joined the SS and by the time he began cultivating the Grand Mufti he had reached the rank of SS-Brigadeführer. He assured el-Husseini that German interests and his were 'completely overlapping' 'in this struggle against world Jewry'. Throughout the war, the Germans invested heavily in radio propaganda to the Arab world.[10]

Once the Grand Mufti and his entourage had been comfortably installed in their Berlin villa, his supporters in the Foreign Office lobbied for a conference with Hitler. The Grand Mufti did not have to wait long. In the early afternoon of 28 November, Ettel escorted the man British intelligence referred to as the 'Arab Quisling' in a large Mercedes across the Tiergarten to 6 Voss Strasse, Hitler's

Chancellery. The big car with its fluttering pendants pulled into the enormous Court of Honour. Two hundred black-clad duty guards of the SS 'Leibstandarte' Adolf Hitler' snapped to attention and a military band began playing. Ettel led the Mufti up a grand marble staircase into a small reception room where they met the ubiquitous Grobba. From here, 17ft high double doors opened into a mosaicked hall that led through a domed space into a gallery 480ft long and sheathed with mirrors. The Mufti's long walk through this succession of ever more overbearing spaces to Hitler's colossal study provided, as its architect Albert Speer intended, a wordless demonstration of the power of the Reich.[11] The Mufti was left in no doubt that he was in the presence of a great leader who could, if he chose, liberate the Arab world from its British and Jewish oppressors.

Haj Amin el-Husseini had courted Hitler's favour ever since the outbreak of the Arab Revolt. He had dispatched a stream of long-winded pleas to Berlin for help to throw off the British yoke and throw back the Jews. Now as the long-awaited meeting began, Hitler remained silent, sipping lemonade, as the Mufti lisped his way through a wordy account of the long Arab struggle. Finally, he urged Hitler to issue a joint statement with Mussolini that committed the Axis powers to the defeat of the British and their Zionist allies. Then it was Hitler's turn. He spoke, or rather lectured, without interruption, for over an hour. Hitler assured the Mufti that he fully supported the Arab cause. But Hitler refused point blank to issue any official Axis proclamation. He dismissed the Mufti's rather vague aspirations for a pan-Arab alliance. After all, if German armies pushed into the Middle East and Central Asia, he had no desire to confront a modern reincarnation of Saladin. Hitler's message to the Mufti was to be patient. He sketched his plan for a 'Final Solution of the Jewish problem': 'My struggle is with the Jews. The elimination of the Jewish people is part of my campaign. The Jews want to establish a state which will be the basis for the destruction of all nations of the world.' He promised that this 'Final Solution' would not neglect to deal with the Jews of Palestine.

At the end of the meeting, Hitler promised:

> At some not yet precisely known, but in any case not very distant point in time, the German armies will reach the southern edge of the Caucasus. As soon as this is the case, the Führer will himself give the Arab world his assurance that the hour of liberation has arrived. At this point, the sole German aim will be the destruction of the Jews living in the Arab space under the protection of British power.[12]

Hitler's promise was not merely diplomatic flannel: he meant what he said. But as they worked out what to do with him in the short term, the Germans would use the Mufti as a radio propagandist. Several times a week he was driven south from the Tiergarten across the Berlin Ring to Zeesen, where a giant short-wave radio mast transmitted Nazi poison to the world.

At Radio Zeesen, the Mufti shared his microphone with a squalid crew of collaborators like William Joyce (Lord Haw-Haw) and the American Mildred Gillars (Axis Sally). The Mufti rammed home to huge Arab audiences the perfidy of Jews: it was they who tried to poison the 'praiseworthy prophet' and 'plotted against him'.[13] German diplomat Wilhelm Keppler (who also promoted the cause of the Indian nationalist Bose) was impressed by the Mufti's impassioned rhetoric and encouraged him to think about reviving Berlin's Islamische Zentral-Institut, which had been founded in 1927 but was now moribund. It was a flattering offer that the Mufti could not resist. The institute would become both campaign headquarters and pulpit. At well-attended weekly sermons, el-Husseini demonstrated how well he had grasped Nazi racial rhetoric:

> In England and America, Jewish influence is dominant. It is the same Jewish influence that lurks behind godless communism ... That Jewish influence is what has incited the peoples, plunging them into this destructive war of attrition, whose tragic fate benefits the Jews and only them. The Jews are the incorrigible enemies of the Muslims.[14]

SS Chief Himmler lavished time and attention on the Mufti. He invited him to stay at his lavish East Prussian estate and introduced him to top SS dignitaries and generals like Gottlob Berger. Himmler and the Mufti reputedly spent hours discussing the nature of Jewish 'evil' and the insidious way they provoked conflict and war. According to the Mufti, Himmler disclosed that Germany was developing fearsome new atomic weapons. Final victory, he assured the Muslim cleric, was certain.

The Grand Mufti Haj Amin el-Husseini was not content merely to preach and broadcast. Like every other nationalist who came cap-in-hand to Hitler's court, he hoped to persuade the Germans to recruit armed militias to fight their shared cause. He turned first to General Hellmuth Felmy who had also taken part in the failed coup in Iraq. With Hitler's permission, Felmy set up the Deutsch-Arabische Lehrabteilung (German-Arab Training Department) to recruit and train an Arab legion. The Mufti rounded up a few eager Arab students in Berlin, and dispatched them to a training camp at Cape Sounion near Athens. These young Jihadists never saw action of any kind but, as they sunned themselves on the Aegean, back in Berlin a war of words erupted between the Mufti and his former ally Rashid Ali el Gaylani. Their venomous quarrel split the German diplomatic community into factions that supported one or the other Arab leader, with the exasperated Grobba trying to act as a mediator. This episode drove the Mufti and his party closer to the SS and its sister organisation the RSHA, whose 'Special Section Arabia' was headed by Sturmbannführer Wilhelm Beisner, a Middle East expert. In early 1942, Beisner assigned his deputy Obersturmführer Hans-Joachim Weise to liaise with the Mufti and to organise his personal security on trips to German-occupied territory.

Haj Amin el-Husseini may also have sought out the man who would become, in David Cesarani's words, 'the managing director of the greatest single genocide in history' – Adolf Eichmann. When the fugitive Eichmann was captured and brought for trial to Jerusalem, his former subordinate Dieter Wisliceny (who organised the deportation of the Jews of Slovakia) claimed that the Mufti visited the notorious Referat IV B4 at Kurfürstenstrasse 77 one day early in 1942. Here Eichmann spent a few hours explaining to his esteemed Arab guest, with maps and lists of figures, how Germany proposed solving 'the Jewish problem'. Eichmann himself vehemently denied Wisliceny's allegation and asserted that he had met the Mufti only once and at a reception, not at his Berlin headquarters. The Mufti's meeting with Eichmann may be apocryphal. It hardly matters. We know that the Mufti was directly involved with another murderous plan hatched up by the Germans to murder the Jews of North Africa.

In 1492, Spanish monarchs Ferdinand and Isabella expelled the Jews of Spain. They had fled to every corner of Europe and across the Straits of Gibraltar into North Africa. Now, five centuries later, this vast desert land had become a battleground between the Allies and the Axis. Once protected by the Ottoman caliphate, the Sephardic Jews who had settled here after 1492 now faced a terrible peril. In Morocco and Algeria, French officials of the anti-Semitic Vichy regime closed the gates of the old ghettoes and, under pressure from Berlin, passed a series of discriminatory laws. They began to evict middle-class Jews from their homes and forced them to work in labour camps. But it was further east in Libya and Tunisia that the blow would fall hardest. At their meeting in Berlin, Hitler had promised the Mufti that he would actively seek the destruction of 'the Jews living in the Arab space under the protection of British power. Historians Klaus-Michael Mallman and Martin Cüppers have now shown that Hitler's promise came perilously close to fulfilment.

On 21 June 1942 Field Marshall Erwin Rommel, the commander of the Panzer Army Africa (Afrika Korps) had swept through Libya and seized the deep-sea port of Tobruk, capturing 28,000 British troops. Soon after this spectacular triumph, at the end of the month, Rommel had penetrated on as far as El Alamein in Egypt, just 60 miles west of Alexandria. Here he would face the British 8th Army. Beyond, tantalisingly close, lay the Suez Canal – and the desert road to Palestine.[15] On

20 July, SS-Obersturmbannführer Walther Rauff flew to meet Rommel in Tobruk. Already well known as a technical expert, the 36-year-old Rauff had pioneered the use of 'gas vans' and mobile gas chambers to 'take the burden off' German execution squads and naturally to make economies. He was short and irascible. At Rommel's headquarters, Rauff received 'necessary instructions' regarding the deployment of a Special Commando (Einsatzkommando) in North Africa. The idea had in fact been suggested by the late RSHA Chief Reinhard Heydrich the year before. (The 'Blond Beast' had been assassinated in Prague in May. The new RSHA chief was the Austrian Ernst Kaltenbrunner.) Under Rauff's leadership, the commando would undertake the same kind of 'special tasks' as the SD Special Task Forces had in Poland and, after 22 June 1941, on the Eastern Front. As the Mufti discovered from his RSHA contacts, the Rauff Kommando would be deployed to begin with in Tunisia. But as soon as Rommel's army had captured the Suez Canal and begun to cross the Sinai desert, the Rauff Kommando would target the Jews of Palestine.

Rauff had very few men at his disposal: just twenty-four. But the Einsatzgruppen reports had suggested that, in the right circumstances, suitable local 'partisans' could be enrolled to help carry out the mass murder of Jews and communists. In Egypt and Palestine, the Mufti assured his German friends, Arabs would welcome Rommel's forces as liberators – and many would be eager to take on other auxiliary tasks. The Mufti's oft repeated claim was widely believed by the Germans. According to Walther Schellenberg, Himmler's spy chief: 'The exceptionally positive attitude among Arabs toward Germans is largely connected with the hope that "Hitler will come" to drive out the Jews ... Thus it is that Arabs today long for a German invasion, and repeatedly ask when the Germans will arrive.' A German liaison officer based in Syria reported that:

> The friendly mood to the Germans among the Muslim Arabs continues unabated. In general, they express the wish that the Germans might soon arrive and liberate the country from the occupying forces and from its misery. To speak about Hitler publicly, the Arabs use a number of pseudonyms. The newest code name for Hitler is '*Hajj Numur*,' the tiger. Wishes for Hitler's victory often serve as a form of greeting.[16]

On 1 July, Schellenberg briefed Himmler concerning the tasks of the Rauff Kommando. Later on the same day, Himmler travelled to Hitler's military headquarters at Rastenburg, where he brought up the matter of the 'Einsatzkommando Egypt'. Hitler authorised Rauff's mission soon afterwards: 'the SS Einsatzkommando will receive its operational orders from and will carry out its assigned tasks on its own responsibility. It is authorized, in the framework of its writ, and on its own responsibility, to undertake executive measures against the civilian population.'[17] Rauff now brought in other veterans of the Special Task Forces to join his 'Egypt Kommando' – including Hans-Joachim Weise, who was assigned to liaise with the Grand Mufti.

El-Husseini was, naturally, delighted to hear that Rauff was already in North Africa and preparing his campaign. He offered to travel to Egypt to prepare the ground and even transfer his own staff to Cairo once the British had been driven out. The Mufti urged Weise to:

> Set up bands of Arabs as a fighting force and equip them. They will march to Egypt and other Arab countries in order to disturb and harass the enemy by destroying roads, bridges and possibilities for contact more generally, and to promote uprisings inside the country ... Dispatch weapons and munitions to Egypt behind enemy lines, and then to Palestine, Syria and Iraq – in order to lay the groundwork for uprisings and to harass the enemy.[18]

The Mufti's message was in effect a Jihad – a call to arms. Everything hinged on Field Marshall Rommel and the Deutsches Afrikakorps.

Field Marshall Bernard Montgomery and the British troops he commanded in North Africa had no idea that the Rauff Kommando existed or that the Grand Mufti saw the German campaign as the opening salvo in a master plan that would engulf the entire Middle East and its Jewish citizens. But it was this ruthless, perhaps mentally unbalanced British egomaniac who scuppered the German Jihad. Field Marshall Rommel had expected merely to pause when he reached El Alamein, a few score kilometres west of Alexandria, before pushing through Egypt towards Suez. In November, at the second Battle of El Alamein, the British 8th Army smashed the Afrikakorps and their Italian allies and drove them back across Libya and into Tunisia. Fleeing alongside the Italians was Rauff's Kommando.

Once Rommel had crossed the Tunisian border, he dug in behind the old French 'Mareth Line'. Neither Hitler nor Mussolini would not permit him to abandon North Africa without a fight to the death. Rommel desperately hoped to avoid the shameful fate of his counterpart Field Marshall Paulus who had surrendered at Stalingrad. So the Germans dug in and prepared for a long fight. Rommel's last stand provided Rauff and his men with a last opportunity to strike at the Jews. On 6 December, Rauff summoned Moïse Borgel, the President of the Jewish community, and Chief Rabbi Haim Bellaïche to his headquarters and read out a decree obliging Jewish men to provide labour service to Axis forces 'defending' Tunis. Rauff insisted that Borgel begin making a list of the names of at least 2,000 workers. A few days later, the Germans demanded 3,000 names – and so it went on. At 8 a.m. on 9 December 1942, just 128 men gathered outside German headquarters on the Avenue de Paris. During the night, thick dark clouds had rolled in from the sea. When Rauff arrived, one witness recalled: 'he became apoplectic'; 'he foamed with rage'. His men instinctively raised their weapons. 'Pigs, dogs, deaf-mutes [sic] … you will be shot within the hour,' Rauff ranted. By then the skies had opened and a warm rain had begun to fall. But apparently fearing the reaction of local Arabs, Rauff ordered his men to hold fire. Instead he ordered a general round-up. Meanwhile, vengeful German soldiers broke into the Great Synagogue where Jewish families, made homeless by the fighting, had taken refuge. The Germans assaulted some of the women and dragged all the men outside into the drenching rain. Anyone judged too old or sick they locked up in the military prison. The bulk of the captives – some 1,500 men – they marched 40 miles across the desert to an improvised labour camp. One was a young man called Gilbert Mazuz, who was handicapped and wore an orthopaedic brace. When the German SS men ordered a halt for the night, Mazuz collapsed. A German officer strolled up and shot Gilbert Mazuz dead.

By the beginning of 1943, approximately 5,000 Jews had been transferred to forty German labour camps in the region around Tunis. They cleared rubble and built desert fortifications. Work was backbreaking and cruel. The German guards did not speak Italian and communicated with beatings and shootings.[19] The persecution and torture of the Tunisian Jews continued until the beginning of May, when Allied armies at last began a final assault. On 9 May, RSHA Chief Kaltenbrunner withdrew the Rauff Kommando to Naples. In North Africa, Axis troops surrendered four days later.

The 1943 Allied victory in North Africa saved the lives of half a million Palestinian Jews. Few histories of the Second World War commemorate this miracle – and no post-war prosecutor ever took up the case of the men of the Rauff Kommando. Haj Amin el-Husseini, who would have rejoiced had Rommel's army ever reached Tel Aviv and Jerusalem, was bitterly disappointed by the rout of Hitler's African armies. Rauff had the opportunity to pursue his murderous plans on a limited scale in Tunisia and he did so with minimal Arab involvement. The great Arab uprising never materialised.[20] But as the North African theatre of operations shut down, another one opened, on the other side of the Mediterranean. Soon the Mufti would pursue his Jihad in the German puppet state of Croatia.

By spring 1943, the Mufti had become very grand indeed. In March, the NSDAP paper, *Völkischer Beobachter*, devoted its front page to reporting a lecture given by the Mufti at the Islamische Zentral-Institut to celebrate the birthday of the prophet: 'Appeal of the Grand Mufti against the deadly enemies of Islam, Arabs will fight for their freedom on the side of the Axis.'[21] El-Husseini's speech had been read and vetted by Foreign Minister Joachim von Ribbentrop. The *Beobachter* reporter described the Grand Mufti as 'one of the great personalities of the Islamic world who had led the struggle of the Palestinian Arabs against onrushing Jewry'. The article makes it abundantly clear that in the two years since he fled Baghdad, el-Husseini had thoroughly absorbed the language of the Third Reich. A pan-Arab nationalist had been moulded into an anti-Semite. From his Berlin pulpit, the Grand Mufti declared: 'today world Jewry leads the allied enemies into the abyss of depravity and ruin, just as it did in the age of the Prophet ... The Muslim's bitter enemies are the Jews and their allies, the English, the Americans and the Bolsheviks.'[22] Sitting prominently in the congregation, Propaganda Minister Goebbels nodded and smiled. Himmler's head of recruitment, the 'almighty' Gottlob Berger, followed the exiled Mufti's rise to prominence with interest. He was convinced that the Mufti could do more for the Reich than spout racist homilies. Early in March 1943, Berger summoned the Grand Mufti to his headquarters. He had a special and perhaps dangerous mission in mind for the little Muslim cleric.

Unlike Himmler, Berger had no interest in understanding the wisdom of the Prophet. He was preoccupied instead with the deteriorating security in the vast swathes of occupied Europe. At the beginning of 1943, as Soviet troops crushed the German 6th Army, a new wave of partisans, galvanised by this stunning turn of events, rose up against Hitler's armies on the Eastern Front and in the Balkans. These new partisan armies were better trained, better equipped, better supported and, most important, better motivated. They scented victory in the cold winds blowing through the ruins of Stalingrad. But all over occupied Europe and the Soviet Union, these resurgent partisan armies were fractured by civil wars – and no more bitterly than in the former Kingdom of Yugoslavia. By the time Berger summoned the Mufti from his Tiergarten villa, the German south-east flank that guarded the Adriatic against Allied attack had become a running sore. In the aftermath of the German invasion in April 1941, Ante Pavelić's puppet Ustasha regime had unleashed a ferocious assault on 'foreign' Jews and Serbs. In occupied Serbia, ruled by another German quisling, the German army with minimal SS support proved equally as murderous. By early 1942, the majority of Yugoslavian Jews as well as tens of thousands of Serbs had been murdered. A Serbian revolt had exploded even as Hitler was proclaiming victory. But the partisan war against the German occupiers was fractured by an internal battle between royalist Chetniks led by Dragoljub 'Draža' Mihailović and Soviet backed partisan units commanded by Josip Broz Tito. The

Chetniks made war on fellow Serbs – and sometimes joined forces with the German occupiers. But notwithstanding this lethal bickering, and with Allied backing, by the spring of 1943, Serb partisans relentlessly hammered the German garrison troops and their collaborators. Since July 1941, Himmler and the SS had been charged with managing this Bandenbekämpfung (war on bandits). In the Balkans, Himmler and Berger had set up a large number of Schuma type battalions and formed the SS 'Prinz Eugen' division to quash the partisan revolt. For the Germans, the increasingly irksome Mediterranean flank had become a strategic nightmare apparently without a solution. Berger and Himmler would turn to another beleaguered community in the Balkans to defend the interests of the Reich.

The Bosnian Muslims are not a 'Semitic' people, like Jews and Arabs, and had hitherto taken little or no interest in pan-Arabism. They are South Slavs like their Serbian and Croatian neighbours and speak the same language. The Grand Mufti of Jerusalem meant little to their inward-looking community. At the end of the fifteenth century, Bosnia and Herzegovina became designated frontier provinces, or 'pashaliks', of the Ottoman Empire – and many South Slavs were 'Islamised'. In the sultan's sprawling domains, conversion to Islam was a no-brainer. In the Islamic world, state and faith were inextricably bonded and the Mosque opened the door to wealth and high status. The Ottoman rulers and their Slavic converts in the Balkans nevertheless tolerated other faiths, as far as the Koran permitted. In Sarajevo and other Ottoman strongholds in the Balkans, Jews, Christians and Muslims rubbed along well enough together for many centuries. Attitudes to Jews in Catholic Croatia were markedly less tolerant.

As the Ottoman Empire weakened in the eighteenth century, Bosnian Muslims retreated into what has been described as 'mental and cultural separateness' and reacted with dismay when, in 1878, the Congress of Berlin handed Bosnia and Herzegovina to the Hapsburgs. Bosnian Muslims naturally feared the changes that might be wrought by a Catholic state and many upped and fled to Turkey. But the Austrians, applying – as every imperial power will – a policy of divide and rule, had no desire to upset their new Muslim subjects. What they had gained from conversion centuries before, the Muslims by and large kept, angering the Catholic Croatians and Orthodox Serbs. As a faith community rather than an ethnic one, the Bosnian Muslims showed only moderate interest in the nationalist passions that gripped hearts and minds in Serbia and Croatia. In the Balkans, Muslims benefited from Austrian rule and feared the increasingly belligerent Serbs and frustrated Croatians. When war broke out in 1914, Bosniaks turned on Serbs and formed a militia, the Schutzkorps, to fight the Serbian army, allied with the Entente Powers, on behalf of the Austrians. They had, of course, chosen the losing side. As both the

Hapsburg and Ottoman Empires crumbled, the Bosniaks found themselves friendless. In the new Yugoslavia they felt insecure and marginalised. The older generation resented loss of wealth and privilege.

In the spring of 1941, the Axis powers Germany and Italy imposed the ultra-reactionary Croatian Ante Pavelić as puppet ruler of a new Croatian state. The Axis sanctioned Pavelić's demands to absorb Bosnia–Herzegovina and quashed feeble Muslim demands for independence. Many Muslims, however, backed the Croatian war on Serbs and some joined Ustasha murder squads. Pavelić went out of his way to woo Muslims, calling them the 'flower of Croatia'. The Croatian Ustasha squads made sure that their Muslim comrades were highly visible in any attack on a Serbian village and Serbs retaliated by attacking Muslim villages. The German invasion unleashed a bloody cycle of revenge upon revenge, atrocity upon atrocity. By the spring of 1942, Bosnian Muslims' religious and political leaders had had enough. They perceived themselves as hapless victims not of Nazi aggression but of a civil war. They felt trapped between the grindstones of Serb and Croat aggression. The German garrison, which had been severely depleted when troops were transferred to the Eastern Front, appeared to be bystanders rather than provocateurs. Since Bosniaks looked back fondly to the 'golden age' of Austrian rule, many were predisposed to seek 'German speaking' assistance. The German authorities sent out favourable signals – especially when these pro-German Muslims helped recruit a Croatian legion for the Russian front. In return, the Germans refused to allow the Italians to station troops in Sarajevo, then as now a Muslim stronghold, to call attention to their pro-Muslim sympathies.[23]

The strategy worked well. On 1 November 1942, a Muslim faction calling itself the 'People's [or National] Committee' sent a memorandum 'To His Excellency Adolf Hitler'. Although it was unsigned, its main author was probably Muhamed Pandza, a well-known theologian who had translated the Koran into Bosnian (or Bosniak, a Serbo-Croat dialect to be precise) and a leading figure in the Reis-ul-Ulema, the Muslim religious authority in Sarajevo. Pandza was on good terms with an SS officer called Karl von Krempler who seems to have encouraged him to make an appeal to the Reich.[24] The committee explained that after the German invasion, Bosnian Muslims had expected and indeed hoped to be granted autonomy under German 'supervision'. Instead, they complained, the 'insane' Pavelić regime had grabbed Bosnia–Herzegovina by force and then ignored every Muslim plea for even the most limited autonomy. Ustasha battalions wearing fezzes, they claimed, had carried out raids on Serb villages, deliberately provoking reprisals against Muslims who had no means of fighting back. The 1 November memorandum reflects the immaturity of Bosniak nationalism. The committee demanded not autonomy but a nostalgic return to the kind of status that Bosnia had enjoyed after 1878 in the Austro-Hungarian Empire. The committee proposed a new Bosniak administrative body, with its seat in Sarajevo and ruled by a chief appointed 'solely by you, our Führer'. To guarantee Muslim hegemony in Bosnia–Herzegovina, the committee obsequiously suggested that the 'Muslim Legion' formed by Major Muhamed Hadziefendic should be upgraded as a 'Bosnian Guard' that would then serve with the Wehrmacht, which would arm, supply and train Bosniak units.

In their final paragraphs, the committee made a direct appeal to German racial obsessions:

> We are of Gothic origin and that bonds us to the German people ... Islam has much in common with our old Gothic religion ... In 1463, we welcomed the Turks as saviours because the Serbs, Croats and Hungarians wanted to destroy us ... In the First World War, we were connected to Germany through our blood relation and with Turkey through Islamic religion and history. For our blood brethren, the Germans, we Muslims were to be a bridge from the West to the Islamic East.[25]

Hitler was completely indifferent to Bosnian Muslims and despised their history. He did not trouble to reply to this obsequious epistle. A few of the German consular staff in Zagreb, like General Edmund Glaise von Horstenau, had some sympathy for the Bosnian Muslim plight – but he believed that they were much too closely attached to the Pavelić regime, which Glaise von Horstenau abhorred. Himmler took a different view. Soon after the memorandum had been circulated in Berlin, SS officer Rudolf Treu had meetings with Muslim leaders and sent an urgent report to Himmler, warning that although the Muslims were, for now, warmly disposed to Germany, it could not be discounted that many would shift their allegiance to the Yugoslavian partisans or even the Allies. This was alarming. Himmler acted quickly, and according to SS records, made a decision to form a Bosnian Muslim SS just a few weeks after reading Treu's report. He assigned Artur Phleps, the commander of the SS 'Prinz Eugen', to establish a recruitment staff in Zagreb.

Himmler's swift response to the Bosnian Muslim entreaty sheds a great deal of light on his long-term thinking about a future SS state. According to historian George H. Stein, the 'Handschar' was the 'first Waffen-SS formation to be recruited without regard for racial and ethnic factors'.[26] A closer look at the evidence reveals that this is nonsense. Himmler did not view Bosniaks as a Slavic people. Just as he believed that Estonians, for example, 'could not be distinguished from Germans', he accepted the claim, first made by Croatian ethnologists as well as the authors of the memorandum, that Bosnian Muslims were a 'Gothic' people descended from ancient Persians. In other words, they were Aryans. In Himmler's mind, the racial origins of many Muslim peoples explained their traditions of martial bellicosity that he contrasted with Judeo-Christian 'softness'. He informed Goebbels that 'Islam ... promises heaven if they [Muslims] fight in action and are killed: a very practical and attractive faith for soldiers'.[27] The very same sentiment informs modern Islamic fundamentalism.

Armed with these handy fantasies, on 6 December 1942 Himmler approached Hitler to discuss the idea of a Bosnian Muslim division. The military rationale he offered was that the new division would assist the SS 'Prinz Eugen' Division to fight Tito's Soviet-backed partisan army that now threatened ethnic German communities in the all important Srem region, the 'Granary of Croatia'. Germany desperately needed to import grain and so the rich farms of Srem had to be defended. For Hitler, getting bread on to German tables was a pragmatic reason to rethink the strategy in the Balkans – rather different from Himmler's fantasies about recruiting 'Islamic warriors'.

At the end of the meeting, however, Hitler refused to make an immediate decision; he knew that the Croatian government would fight tooth and nail to resist any suggestion of Bosniak autonomy. Hitler had no interest in upsetting the hitherto pliant Poglavnik Ante Pavelić. But Himmler did not regard Hitler's prevarication as a serious setback – and by January 1943 active discussions were taking place between the SS and Siegfried Kasche, the pro-Pavelić German envoy in Zagreb. Himmler had little respect for Pavelić, who was after all supposed to be a client ruler. On 13 January, a second conference took place at Rustenburg – and this time Hitler agreed to permit the formation of the new Bosnian Muslim SS division. 'I hope to reach out to a people,' Himmler wrote to the German Plenipotentiary Glaise von Horstenau, 'who stand apart from the Croatian state and have a long tradition of attachment to the Reich, which we can utilize militarily.' By 'the Reich', Himmler meant in this case the defunct Hapsburg Empire. He believed erroneously that the Bosniaks had served in the Austro-Hungarian army. He would use the very same argument later that year when he began recruiting Ukrainians in the old Austrian province of Galicia.

The German SS representatives based in Zagreb now had to placate Pavelić, who was violently opposed to any 'autonomist' concession to Bosnian Muslims. Himmler should have understood the Croatian dilemma: if you arm any group of separatists, you risk turning them into a nationalist militia – and the Poglavnik desperately needed Bosnia–Herzegovina to remain part of the NDH. Glaise von Horstenau warned Himmler: '[the Croatians] saw this as a dangerous blow against their false principle of a national unified Croatian state.' To soothe Pavelić, Hitler dispatched von Ribbentrop, who it will be recalled had supported Croatian independence in 1941. The Foreign Minister ordered Kasche to go back to Pavelić and insist that 'the enemy has to be dealt with as forcefully as possible. It would be in the best interest of the common war effort that this German-led division be formed. I hope that that the Poglavnik will agree.' To Ribbentrop's great irritation, the Poglavnik certainly did not agree – and he may well have been encouraged to resist Ribbentrop's appeal by Kasche himself, who was an ardent admirer of the murderous Ustasha regime. Vjekoslav Vrančić, one of Pavelić's closest advisors, told a Muslim friend: 'We cannot give a No answer to the German request, but we can make it impossible for them to succeed. Allowing the Germans to establish a Bosnian division in Bosnia ... would be the same as losing Bosnia.'[28] Pavelić now made a counter proposal: a Croatian SS division that would recruit both Muslims and Catholic Croatians, but have Croatian officers and use Serbo-Croat as the language of command. The Germans refused; discussions again bogged down. Ribbentrop, sensing another humiliating impasse, beat a hasty retreat.

Only Hitler could break the deadlock. Croatia was in principle a sovereign nation state and an ally of the Reich. But Hitler had been unimpressed by Croatian soldiers both on the Eastern Front and as partisan fighters in their own backyard. So he now insisted that the SS proceed immediately to form a Muslim division that had no connection with the discredited 'Croatian' militias. In February, Phleps flew to Zagreb to hammer out details with the Croatian government, represented by Dr Mladen Lorković, the Foreign Minister. Phleps was astonished to discover that Lorković had evidently made a decision to fight Himmler every inch of the way. Now he would agree only to a 'Ustasha SS division' – the only concession Lorković was prepared to make was to adopt regional names for regiments, one of which would be called 'Bosna'. Phleps turned Lorković down flat and retreated to seek advice from Himmler, who refused to contemplate any compromise with the Ustasha government. So Phleps returned to Zagreb to meet Pavelić himself, who turned up accompanied by his tame deputy Dr Džafer-beg Kulenović, who declared himself a 'Croat of Muslim Faith'. Pavelić again refused to budge, and Phleps, who did not hide his astonishment that the Poglavnik still dared to resist the Reichsführer-SS, stormed out of the meeting, slamming the door behind him. In March, Himmler seemed to blink. After more discussions (this time without Phleps) Himmler agreed to permit a Croatian SS Volunteer Division, to be recruited by the Croatian government jointly with Waffen-SS, but with German as 'the language of command'. It appeared that Pavelić had won.

How then did Himmler finally end up with the Bosnian Muslim division that he had wanted all along? The answer is simple: he brushed aside the agreement with the Croatian government and ordered Phleps to begin a recruitment campaign in Bosnia that would exclusively entice Muslims. And it was at this juncture, in the spring of 1943, that Berger and Himmler turned for help to the Grand Mufti. They had chosen their moment well. The Grand Mufti was still locked in a battle with his rival in Berlin, Rashid Ali el-Gaylani, and had been bitterly disappointed by the German withdrawal from North Africa and the recall of the Rauff Kommando. He yearned to take action against the Jews and their purported allies, the British. He was well aware that the hated British had backed Tito's communist partisans in the Balkans – so the new SS division offered a means to strike at two mortal enemies on the same battleground. On 24 March, Berger and Phleps met the Mufti at his villa in Berlin. A week later, on 30 March, he was driven to Tempelhof Airport to board a special flight to Sarajevo. The Mufti's Bosnian crusade had begun. In a sermon delivered on the eve of his departure, he preached: 'The hearts of all Muslims must today go out to our Islamic brothers in Bosnia who are forced to endure a tragic fate. They are being persecuted by the Serbian and communist bandits, who receive support from England and the Soviet Union.'[29]

In Bosnia, the Grand Mufti was treated like a Sultan. When he swept into Sarajevo, his hosts installed him in the sumptuous palace of the former Austrian governor. It was inside the palace that on 28 June 1914, Austrian Archduke Franz Ferdinand had expired from wounds inflicted by Bosnian assassin Gavrilo Princip. 'The Mufti was an extremely impressive personality,' SS officer Balthasar Kirchner recalled. 'His reddish blond beard, steady motions [sic], expressive eyes and charismatic facial features gave him more the look of a philosopher than a revolutionary.'[30] The Mufti's bodyguard, a Bedouin, appeared never to sleep, eat or rest. According to the German Consul Dr Winkler: 'The faithful recognized [the Mufti] as a true Muslim; he was honoured as a descendant of the prophets. Friends from his theological studies in Cairo and pilgrimage to Mecca [the Haj] welcomed him.' Kirchner recalled that the Mufti was 'quite reserved' with respect to 'fighting Bolshevism' – 'His main enemies were the Jewish settlers in Palestine and the English'. He had not perhaps completely grasped that the Germans regarded the Bolshevik and the Jew as virtually coterminous. In Sarajevo's main mosque, the Mufti delivered a sermon and urged Muslims to support Germany and 'take weapons from them'. His grand tour was managed by the SS, but el-Husseini was determined to further his own cause. He was committed, Glaise von Horstenau later explained to Himmler, to the establishment of 'a United States of Islam extending all the way from Morocco to Bosnia' and the destruction of Zionism. His counsel to 'take their weapons' shows that he hoped that SS recruitment of Muslims would further his own Jihad just as much as Hitler's crusade.

On 12 May, following his triumphant grand tour, the Mufti met Himmler at SS headquarters in Berlin and made a number of astonishingly naïve proposals. He wanted agreement that the mission of the new SS division must be to protect the Muslim families 'of the volunteers'. The division must therefore never be deployed outside Bosnia–Herzegovina. The officers must be Muslims and the SS should not poach men from Hadziefendic's Muslim Legion, which should be left intact.[31] Himmler listened politely but refused to make a decision. However, on 19 May, he signed a formal agreement with the Mufti that guaranteed that Imams would be appointed and charged with the ideological training of recruits. This implied that the Bosniak recruits would not receive 'political' instruction from SS officers, but from the Mufti's own clergy.[32] As we will see, Himmler would exploit el-Husseini's Imams for his own purposes. As to the Mufti's specific proposals, he left it to Berger to officially reject all of them. The Mufti was no longer essential. His job had been to play the role of figurehead: he was the barker, not the ringmaster.

Although the Mufti had impressed the Bosnian Muslim community, the campaign to recruit Bosniaks got off to a poor start. By mid-April, just 8,000 men had come forward. Berger had been hoping for numbers well above 30,000. Appraised of this mortifying result, Himmler was at last forced to eat humble pie. He immediately flew to Zagreb to announce that Catholic Croatians could be accepted as recruits, provided that the numbers of Muslims exceeded that of Catholics by a proportion of ten to one. Even this proved hard to achieve: nearly 3,000 Catholics were eventually inducted, which made nonsense of the ratio Himmler had demanded. Although the SS chief had on balance won the battle, the frequent renaming of the SS division made clear that the recruitment had been a strategic fudge: the Kroatische SS-Freiwilligen-Division (Croatian SS-Volunteer Division) was amended to Kroatische SS-Freiwilligen-Gebirgs-Division (Croatian SS-Volunteer Mountain Division), then the SS-Freiwilligen-Bosnien-Herzegowina-Gebirgs-Division (Kroatien) or 13. SS-Freiwilligen-Bosnien-Herzegowina-Gebirgs-Division (Kroatien). It was only in May 1944 that the Germans settled on 13. Waffen-Gebirgs-Division der SS 'Handschar' (kroatische Nr. 1) – the 13th SS Mountain Division 'Handschar'. Recruits took an oath of loyalty to both Hitler and Ante Pavelić as head of the Croatian state. The Muslim SS division would never shed its titular link to Hitler's puppet state.

Smarting from his battle with Pavelić, Himmler was all the more determined to puff the Bosniak credentials of this new division. *Handschar* derives from *Handzar* (Turkish *hancar*) – a Bosnian fighting knife. The basic uniform would be field grey with special collar patch showing a scimitar (the *Handschar*) twinned with a swastika. The national arm shield, on the other hand, used the Croatian red and blue checkerboard. But as if to distract attention Himmler ordered recruits to wear, instead of the usual field caps, a most picturesque kind of headgear: a fez, made of crushed felt which bore both the *Hoheitszeichen* (German eagle and swastika) and the SS skull and crossbones, complete with a tassel. (In fact two kinds of fez were issued: one field grey, the other a dark red.)[33] These highly visible, mandatory fezzes loudly proclaimed the division's Muslim ethos and identity and at the same time its allegiance to the Reich. To please the Mufti, Himmler also guaranteed that Muslim recruits would enjoy a diet that conformed to Muslim dietary laws and, crucially, that they would have their own divisional Imams.

The recruits were not quite Islamic warriors. According to German officer Wilhelm Ebeling (in an unpublished memoir cited by historian George Lepre), 'most were dirt poor and illiterate ... It proved difficult to record their personal information for many didn't know how old they were, so we had to estimate. Some had several wives. In these cases, it had to be determined which wife was to receive the man's military benefits.' According to another SS officer, tuberculosis, epilepsy and other serious illnesses were endemic and so 'a large number of the candidates could not be accepted'. Erich Braun remembered that they 'arrived in clothing that was simply indescribable. When they received their new SS uniforms, they were overjoyed ... Some of the men took their newly-issued uniforms and sold them on the black market. They would then report in the next day as if they were new.'[34] On his return from Bosnia, the Mufti had assured Himmler that many Muslims had served in the Austrian army. But as recruitment got under way, it became all too clear to SS recruiters that most of those volunteers who claimed to have military experience were largely decrepit. This shortfall in officer class recruits had two consequences. The Germans had to promote a larger proportion of Catholic Croatians than Bosnian Muslims – and Berger was forced to transfer unusually high numbers of German and ethnic German officers borrowed from the SS 'Prinz Eugen' to form the officer corps of the new division.

Himmler had assured the Mufti and other Muslim leaders in Bosnia, as well as the Croatian government, that the 'Handschar' would be trained and deployed in Croatia; to be exact, at the Zemun camp on the Duna River, south-east of Novi Sad. But on 6 June 1943, Himmler reneged on his promise – a decision that would have fateful consequences. By now Phleps had returned to the depleted SS 'Prinz Eugen' and another Austrian, SS-Standartenführer Herbert von Obwurzer had been appointed to command the SS 'Handschar'; he would later take over command of the 15th Latvian SS Division. This tall, overbearing and choleric Austrian was an experienced 'bandit hunter' and had no doubt that his task was to turn the Bosnian Muslims into ruthless anti-partisan fighters. German occupation authorities had reported with mounting concern a steady drift of young Bosniaks to Yugoslavian insurgent forces. So to ring fence the Muslim recruits, von Obwurzer urged Berger to move the 'Handschar' out of harm's way. One might have expected the Waffen-SS leadership to transfer the division to a German training camp, such as the one at Wildflecken, near Frankfurt. Instead, Berger ordered von Obwurzer to move his SS division to south central France. In the rolling green hills and villages of the mid-Pyrenees, German SS officers like Gerhard Kretschmer and Anton Wolf would hammer these mountain boys of the Balkans into shape as SS men.

Most German officers assigned to the 'Handschar' despised the young Bosnian men who had enlisted in the elite Waffen-SS. Relations between the German commanders had also become fractious. On 23 July, von Obwurzer arrived in the French town of Mende and immediately began rowing with his fellow Austrian Erich Braun. Himmler disliked Obwurzer and news of the rift forced his hand. His first choice as a replacement was Hermann Fegelein, who had led the SS cavalry into the Pripet Marshes in July 1941. But he eventually settled on an obscure Wehrmacht colonel: Karl-Gustav Sauberzweig. It was a distinctly odd choice. Sauberzweig was a Prussian of the old school, who had lost an eye in the First World War and, by the time he assumed command of the SS 'Handschar', was a physical wreck. He spoke no Serbo-Croat, had never tried to learn any and had never served in the Balkans or with a 'Gebirgs' (mountain) division. But he had a reputation for efficiency (his nickname was *Schellchen* – speedy) and, according to reports, much liked by his officers and men. On 9 August, Sauberzweig arrived in Mende to take over command of the 'Handschar'. It was said that he called the Bosnians 'his children'. But Sauberzweig's new military family would never be a happy one.

In the meantime, Berger had flown to Zagreb in an effort to acquire more recruits for the 'Handschar', which remained under strength. At a meeting with the Foreign Minister Lorković, Berger insisted that all Muslims be released from the Croatian armed forces to serve in the 'Handschar'. The forceful SS recruitment chief got his way and 3,000 new recruits were soon boarding trains for training in France. Berger's bullying left Pavelić reeling. In Sarajevo, Muslim community leaders complained bitterly that the perfidious Waffen-SS had not just stripped farms and villages of their young men, but dispatched them to another faraway country. Now their homes and families were in grave danger – for by the summer of 1943, the unrelenting and vicious war between Germans and partisans and between Chetniks and communists had turned the former Yugoslavia into an abattoir.

The murderous activities of Phleps' SS 'Prinz Eugen' made matters a great deal worse – as even Himmler would belatedly acknowledge. That July, a 'Prinz Eugen' battalion entered the Muslim village of Kosutica where they discovered the remains of a dead SS man. Revenge followed swiftly and without mercy. The SS men, all ethnic Germans from the Banat, pushed and shoved the people of Kosutica into the village square then opened fire with machine guns, killing forty women, children and old men.

Some of the dead were, as it turned out, the fathers, wives, daughters and sons of the men now being trained in France.[35]

Rumours about the atrocities soon reached the men of the SS 'Handschar'. They had sworn oaths of loyalty to Hitler and the Croatian government, but SS commander Phleps' 'root-and-branch' tactics had in return robbed them of loved ones. Himmler shed a few crocodile tears and lectured Phleps on 'the old discipline and training'. But for the German occupying forces, reprisal was a military norm. Soon after the Kosutica atrocity, Phleps killed at least 3,000 unarmed civilians in villages along the Dalmatian coast – and this time he was careful to report them to his SS masters as 'enemy dead'.

In France, emotional turmoil among the SS recruits put enormous pressure on the new Imams. It will be recalled that the Mufti's agreement with Himmler obligated the Germans to appoint and train a divisional clergy. This had advantages for both parties. Most importantly, Himmler could rely on the Mufti to use the Imams as ideological educators, not just guardians of faith. Part of the job description was to ram home a simple message: 'kill all Jews'. Himmler regarded the 'Handschar' Imams as 'trustees of Islam' who would turn the raw Bosnian recruits into 'good soldiers and SS men'.[36] The Wehrmacht was served by Christian chaplains, but Himmler rejected any such pastoral care for Waffen-SS recruits. Himmler was attracted to a mishmash of pagan faiths and regarded Christianity as a 'Jewish' creed that would undermine 'hard' SS values. He professed himself a *Gottgläubiger* – a 'believer in God' – and many SS recruits followed his example. But he took a different position on Islam. In a long oration delivered to both German and Bosnian SS men in January 1944, Himmler elaborated on the close affinity between Islam and National Socialism:

> In the past two centuries, Germany, its government and leaders were friends of Islam on the basis of conviction, not opportunism or political expediency ... almighty God – you say Allah – ... sent the Führer to the tortured and suffering people of Europe ... It was the Führer who first freed Europe and later will free the whole world from the Jews, this enemy of our country ... They are also your enemies for the Jew has always been your enemy.[37]

Historians have tended to assume that these sentiments had equal significance for the Bosnian recruits as they had for their German officers; this is unlikely.

Anti-Semitism had limited appeal in the old Ottoman territories of the Balkans; it was, as we have seen, much more potent in Catholic Croatia. Pogroms had been unknown in Bosnia–Herzegovina – and it was a Muslim scholar Dervis Korkut who successfully hid the renowned 'Sarajevo Haggadah' from greedy German eyes.[38] This is why Himmler and el-Husseini invested so heavily in the divisional Imams and their training. The Muslim religious authorities in Sarajevo, the Ulema, had enthusiastically backed the formation of the 'Handschar', hoping it would protect their villages from attacks by Croatian and Chetnik murder squads – and it was the Ulema that had recruited the thirty or so Imams who would serve with the division. Both Himmler and the Mufti insisted that training would take place in Germany at a large villa in Berlin-Babelsberg, near Potsdam. Most of the new Imams had been schoolteachers; many had been educated in Cairo and Alexandria where some had been exposed to the radical teachings of the fundamentalist 'Muslim Brotherhood'. El-Husseini took a close interest in the young men. They were enthusiastic smokers and whenever he visited Babelsberg, he brought extra cigarette rations. The Imams also visited the Mufti at his splendid villa in Zehlendorf. 'What splendour and oriental beauty,' one recalled. He insisted that each Imam must be 'an example and ideal in his ways, actions and posture'; he must make his comrades 'despise death and achieve a full life'.

Lectures at the Imam school in Babelsberg included 'The Waffen-SS: its organisation and its ranks' and 'The History of Nationalism', as well as the rudiments of the German language. The Mufti lectured: 'Never in its history has Germany attacked a Muslim nation. Germany battles world Jewry, Islam's principal enemy. Germany also battles England and its allies, who have persecuted millions of Muslims, as well as Bolshevism, which subjugates forty million Muslims and threatens the Islamic faith in other lands.'[39] The course was a rush job: it took just three weeks to complete, but it had a powerful impact on some of the young Bosnians. Writing in *Handzar*, the division's newspaper, Husejin Dzozo, one of the Imams, celebrated the mission of the SS as follows. Aping his German teachers, he denounced the 'Versailles-Diktat' which allowed 'Jews and Freemasons' to corrupt European governments. 'Communism, capitalism and Judaism stand shoulder to shoulders against the European continent' – only the SS, the Imam concluded, can build a new Europe. Under Ottoman rule, Jews and Muslims had lived amicably in Bosnia–Herzegovina; now Himmler and the Grand Mufti rode roughshod over centuries' old traditions of tolerance.[40]

The Imams had undeniable impact. The divisional commander Sauberzweig claimed that the Bosniak recruits 'gladly accepted' Nazi doctrine, and that they had begun to regard Hitler as 'a second prophet' after Mohammed. But not every 'Handschar' recruit took the same view as Sauberzweig and Imam Dzozo. Quite the contrary – they joined the division to wage war against the Reich, rather than its 'Jewish-Bolshevik' foes. The allegiance of these men was to the Yugoslav partisan cause, not Hitler or the Grand Mufti. Berger warned Himmler that Tito had 'issued an order that everyone [that supported the partisans] should report for police duty in Croatia', which implied that partisans would try to infiltrate the new SS division.[41] It was the fear that the 'Handschar' would prove 'porous' to hostile elements that had persuaded Berger and Himmler to transfer recruits 'out of harm's way' to France. But as events would shortly prove, they had closed the stable door much too late.

Ferid Dzanic was a clever young man from a notable Muslim family who had joined Tito's Yugoslav National Liberation Army (JANL), but then had been captured and incarcerated in German camp near Sarajevo. On 1 August, Muslim recruiters came calling, looking for 'Handschar' recruits, and Dzanic applied to join. He was well educated and had been an officer cadet in the Royal Yugoslav Army, so he was eagerly accepted. German SS officers were just as impressed and awarded him a commission. He was dispatched to Dresden where the Mufti had set up another 'Mullah Training School', which was also used to indoctrinate promising 'officer-class' recruits. It was only later, after Dzanic had revealed his true allegiance, that the Germans characterised him (using an odd set of conflicting epithets) as 'power hungry, subservient, corrupt and vague … possessing a strong will and power of persuasion'.[42] In Dresden, Dzanic ran into Bozo Jelenek, a Catholic Croatian and Communist Party member, and Eduard Matutinovic. Nikola Vukelic was another Catholic who, although not yet 20, had also been highly praised by his German commanding officer. Along with another Muslim, Lutfija Dizdarevic, Dzanic formed a secretive cadre determined to derail German plans. It was impossible to do much more than talk and plot in Dresden. But all four conspirators served in the same battalion (SS Gebirgs-Pionier Bataillon 13) and ended up billeted together when they arrived in the French town of Villefranche-de-Rouergue. And it was here that the Bosniak conspirators hatched up a daring plan to wreck the SS division.

Dzanic left no record of his intentions, but it is pretty clear that the conspirators planned to arrest and execute their German officers in Villefranche, then march their battalion to Rodez and the other towns in the *département* where the rest of the SS division was billeted. Then the SS 'Handschar' would either join Allied forces in Italy or return to Croatia. The plan was both daring and naïve – for Dzanic had not taken into account the loyalty of the Imams. Just after midnight on 17 September, Dzanic and 'about ten armed men' burst into the barracks where some twenty-five German NCOs lay fast asleep. The mutineers roughly shook them awake and locked them in a storeroom. The revolt took the Germans completely by surprise. Once the first barracks had been secured, Dzanic and the mutineers rushed to the École Superieure, the girls' school that served as battalion headquarters. They arrested and disarmed the German officers and locked the commander Heinrich Kuntz in his room. Then they proceeded to the Hotel Moderne, where other officers had been quartered. One was the unit doctor Willfried Schweiger, who later wrote a detailed report. He heard the rebels demand: '"Are you with Germany or with us?" Seconds later, a shot ... a loud crash. Then it was the turn of SS-Hstuf. Kuntz. The same question ... another shot.'

Meanwhile, Dzanic returned to the Hotel Moderne, armed with a pistol, submachine gun and a knife, where he roused Imam Halim Malkoc. 'All the German officers are under arrest and will be shot,' he was informed. Dzanic demanded: 'Imam come with us, for if you do not you are our enemy.' The Imam was then left alone to dress. He said later: 'I was well aware of what the consequences of this action would be.' As soon as he had dressed, the Imam started talking with other Bosnians and tried to persuade them Dzanic and the other ringleaders had deceived them. Imam Malkok assembled a small party of Reich-loyal Bosnians and released some of the German officers and NCOs. According to Dr Schweiger's report, the Imam called out 'Heil Hitler! Long live the Poglavnik [Pavelić]!' In a series of confused skirmishes, the Germans and loyal Bosnians tracked down the mutineers and shot many of them dead. Thanks to the Imam, by morning the Germans had restored order. The mutiny was over, but the shock waves had begun rippling towards Berlin. It was time for recrimination – and vengeance.

All but one of the ringleaders, Nicola Vucelik, had already been shot or escaped, but Imam Malkoc helped Sauberzweig and the German prosecutor, Dr Franz von Kocevar, identify many others who had joined in the revolt. By midday 18 September, von Kocevar had handed down fourteen death sentences – and that afternoon, loyal Bosnians assembled in an open field opposite the town cemetery to receive a lesson in SS justice. As an SS squad loaded their standard-issue Mauser rifles, a Croatian interpreter called out the names of the condemned one by one when it was their 'turn at the stake'. Afterwards, SS-Rottenführer Hans-Wolf Renner stepped forward to administer a 'mercy shot' to finish off anyone who had not succumbed. The bodies were immediately buried in shallow graves dug in the rocky soil. A few weeks later, dogs unearthed the corpses. On 28 September, four Bosnians escapees were tracked down hiding out near Villefranche, and shot dead on the spot.[43]

From Berlin, recrimination followed swiftly. Himmler blamed the relocation to France: Villefranche was positively crawling with unreliable foreigners, including 'Jews from the Balkan lands' who had corrupted his noble 'Mujos'. The solution was obvious. Potential traitors must be weeded out of the division and the rest sent to benefit from proper Germanic instruction 'governed by the law of drill, the law of obedience' – in Germany.[44] The revolt had caught Sauberzweig on the back foot. But with Prussian zeal, he set about weeding out unreliable 'dark elements'. On 27 September, at the railway station in Mende, he watched with some satisfaction as loyal 'Handschar' men herded more than 800 alleged 'unreliables' into cattle trucks. Their destination: Dachau.

February 1944: the 'Handschar' men completed training in Germany and began the long journey back to their distant homeland. They came back to a traumatised land, scarred by Axis occupation and savage civil war. At their SS training camp in Neuhammer, the SS 'Handschar' had been reformed root and branch – in theory. Now they would be put to the test in combat.

In Germany, the Bosnians had been subjected to a bizarre cultural experiment designed to weld together Nazism and Islam. At the end of October 1943, the 'Handschar' celebrated Bairam, the Turkish equivalent Id al-Fitr, the joyous end of Ramadan. The Bosnians 'ate good food and halva'. Sauberzweig did what he could to inspire an esprit de corps: he declared that 'Your fate is Germany's fate'. Sauberzweig had little to say about ideology. That was the job of Imam Abdulah Muhasilovic, who had, like Imam Malkoc, become a fanatical SS propagandist. When he took the podium, wrapped in oak leaves and a giant swastika, he made sure the 'Handschar' understood who was to blame for the atrocities committed in their homeland: 'An entire army of our brothers, our refugees, wander about from city to village, wrapped in rags, barefooted, hungry and cold. Their Bairam feast will be spent in misery and distress … Chetniks and Partisans carry on their activities, murdering and plundering wherever they go.' Raising a gloved hand, the Imam went on:

> The world's Muslims are engaged in a life-or-death struggle … The entire world has divided itself into two camps. One stands under the leadership of the Jew, about whom God says in the Koran 'They are your enemy and God's enemy!' … On the other side stands Nationalist Socialist Germany, with its allies, under the leadership of Adolf Hitler, who fight for god, faith, morality.

At the beginning of 1944, Himmler visited the Bosniaks at Neuhammer. His lecture to the division can be found in the National Archives in Washington DC: 'Today the world knows what the SS is. We have more enemies than friends … The enemy knows that we are soldiers from the heart of Europe.' He called for his German officers to embrace their Bosnian and Croatian comrades: 'there is to be no difference between a German from the Reich, a Bosnian, Croatian, or a German from the south east … We have sworn the same oath to the same leader.'[45]

Another important visitor was, inevitably, the Grand Mufti, who arrived at the SS camp accompanied by Muslim notables from Sarajevo and from Albania, where the SS had recruited another Muslim division with the Mufti's backing. El-Husseini spent three days at Neuhammer and presented Muslim soldiers with packets of tobacco and pots of honey. According to a radio broadcast, the Mufti 'inspected troops in training and prayed with them'. He said he had been reminded of his own soldiering days during the First World War (he fought in the Turkish army). The Mufti, though, was evidently well informed about the military situation in the Balkans. As soon as he had returned to Berlin, he met with Berger to discuss conditions of service for the 'Handschar' when they returned to fight in Bosnia. He reminded Berger that in September (just as Dzanic was planning the mutiny in France) the entire Home Guard garrison of Tuzla had defected to Tito's partisans. This was shrewd, for the loss of Tuzla had been a wake-up call for the German occupation authorities.[46] Imam Hasan Bajraktarevic and two other clerics who had just returned from Bosnia made the same point. Many Muslims, they warned, had acquired a 'negative impression' of the Germans and feared that the 'Handschar' men would be sent to serve on the Eastern Front as cannon fodder. Chetnik attacks on Bosnian Muslim villages continued unabated and Tito had redoubled efforts to recruit disaffected Muslims. Himmler sent money and clothes to a Bosnian Muslim welfare association – on condition that his gifts would be distributed after the 'Handschar' had returned to Bosnia.[47] But his grudging generosity would prove too little, too late. As the 'Handschar' men completed training in Germany, thousands of Bosniaks rushed to join Tito's armies.

As the big, slow-moving German troop trains taking the SS 'Handschar' home rumbled across the Croatian border, the young men on board had few illusions about they would find.[48] The big trains steamed on ponderously, passing through station after station fortified to repel partisan attack. Alongside the line lay twisted train wrecks, some still smouldering. The German trains had to halt frequently to check the line. The 'Handschar' men knew much better than their German officers that their mountainous homeland favoured their foes. This is an unforgiving region of dense forests, high rocky ridges, plunging ravines and fast-moving torrents. Bosnia is a nightmare for even the best-equipped conventional military forces. For the agile partisan, however, forests can become natural fortresses; rock-strewn ridges (the *balkans*) rip vehicle tyres and the toughest boots to shreds. Flat, open land is rare – a gift to secretive, fast-moving bands who need to avoid aerial surveillance. This is a land of great beauty that repels mighty armies and favours lightning strikes and swift retreat. The Bosnian *balkans* can be hostile to intruders in many different ways. An endless chain of serrated ridges generates a tortuous mosaic of rain shadows. Shattering, unexpected downpours are frequent and demoralising. In the forests and ravines, nights are truly pitch-dark. Military historian Jonathan Trigg (who served in Bosnia in the 1990s) emphasises that 'mountain fighting, just like fighting in urban areas and woodland, soaks up men on a huge scale'.[49]

Although he was now in charge of a 'Gebirgs division', 'Speedy' Sauberzweig had no experience of real 'mountain combat'. Sauberzweig travelled to the mustering point in Mostar in much greater comfort than his men. In an open letter addressed to the 'Handschar', Sauberzweig tried to express empathy.[50] He described passing ruined fields and burnt out villages. In the blighted land, anyone who remained alive lived like troglodytes in cellars and shelters. Thousands more starved in refugee camps. Naturally, Sauberzweig wrote, this was all the fault of partisans not the German occupiers. He concluded:

> I also saw some of your fathers. Their eyes, when I told them that I was your division commander, shined as brightly as your own. Before long, each of you shall be standing in a place that you call home … standing firm as a defender of the idea of saving the culture of Europe – the idea of Adolf Hitler.

In fact, little love was lost between Sauberzweig's German officers and their Bosnian recruits. German officers had fallen out with both their ethnic German comrades and the Bosnians, whom they regarded as 'substandard': 'The complete inability of the Prussians to deal with soldiers of other nations is clear,' an informer told Glaise von Horstenau, who was no friend of the SS: 'No one makes an effort to learn [Serbo Croat]. [Germans] become angry when ethnic German officers speak [with the enlisted men] in their own tongue! Little can be expected from this division.'[51]

The men of the 'Handschar' would soon discover what the 'idea of Adolf Hitler' meant on the battlefield. At the beginning of March, the 'Handschar' celebrated Mevlud, the birthday of the Prophet – a Balkan speciality which orthodox Muslims (who call the day Mawlid) despise as 'too Christian'. Sauberzweig made sure his men enjoyed the day in high style. Then two days later, he ordered reconnaissance patrols to enter the Bosut Forest, a near impenetrable partisan stronghold north of the old Bosnian border which had to be cleared before the division could commence the main campaign to secure north-east Bosnia. To completely liquidate the partisan units, it was vital to block any escape routes; for Tito's men had perfected the art of the tactical retreat often by fleeing along the narrow river valleys. Sauberzweig commandeered an old Austrian gunboat, the *Bosna* (which had been sunk and raised at least twice), to guard the Sava River and sent a 'Handschar' regiment north to patrol the main road. Sauberzweig ordered the main attack on

10 March: five 'Handschar' spearheads battled their way into the Bosut Forest, forcing the partisans to retreat towards the Sava. Here the *Bosna*, its mainly Croatian crew commanded by Hermann Schifferdecker, surprised a partisan unit trying to cross the river and began firing on them. Heavy return fire erupted from the bank and the Croatian captain of the *Bosna* turned the boat around and fled at full speed. Dismayed, Schifferdecker forced the Croatian crew to turn the *Bosna* round too. By this time, the wounded crewman had died. As Schifferdecker cautiously chugged up river, urging on the terrified captain, they again came under heavy fire from the south side of the river. Panic erupted. Officers and crew threw themselves on to the deck. In the chaos, one of the German officers fell overboard and had to be retrieved under fire with a grappling hook. As bullets smashed into the *Bosna*'s superstructure or whined menacingly over head, Schifferdecker found himself under 'friendly fire' from a 'Handschar' unit on the other (north) side of the Sava. It was only with the greatest difficulty that Schifferdecker managed to retrieve his men and return to headquarters. That night, the *Bosna* vanished; its crew had had enough. Like the Americans in Vietnam, the SS 'Handschar' faced a determined and wily guerrilla force that could run rings around any armed force sent to attack.

The SS 'Handschar' eventually managed to cross the Sava and entered their own homeland at last. Here they had to fight what twenty-first century military strategists define as an 'asymmetric' war, where the enemy is not only in front of you but behind, to the sides and even amongst you. In the Balkans in 1944, asymmetries of many kinds multiplied. Muslim partisans fought Muslim SS men, and even victory proved very hard to define, for as the 'Handschar' knocked out one partisan brigade, scores of smaller units remained at large in the deep Balkan forests and in mountain hideaways. Muslim rage against Serbs escalated.

Sometime between 10 and 12 March, a large-scale massacre of Serbs took place in the town of Bela Crkva. Details about what happened are sparse. Jörg Deh, a German officer serving with the 'Handschar', claimed that he led the 'Handschar's' Spearhead F squad into Bela Crkva to scout for partisan forces. He reported that he then discovered 'the enemy gone, having murdered all the town's inhabitants'. But why would Serbian partisans murder fellow Serbians on such a scale? In fact, Deh's Spearhead F was not the first 'Handschar' unit to reach Bela Crkva. According to another German report, a 'Handschar' reconnaissance battalion and a company from the 27th Regiment had reached Bela Crkva two days earlier. Since it is unlikely that Serbian partisans would murder Serbs (or indeed have the time to do so on such a scale with the German SS troops pressing hard on their tail), the most likely explanation is that the first 'Handschar' Spearheads carried out the massacre, which was later discovered by Deh's second probing operation.[52] Rumours about

SS 'Handschar' 'excesses' reached Hitler's military headquarters. These lurid stories came from a surprising source: Hermann Fegelein, the SS cavalry officer who had once been mooted as 'Handschar' commander. Fegelein regaled Hitler and his fellow officers with tales of Muslim horror. He claimed that some 'Handschar' men had 'cut out the hearts of their enemies'. At this point Hitler 'reprimanded' Fegelein with an abrupt '*Das is mir Wurst*' (it doesn't matter to me).[53]

In the former Yugoslavia, Croatian militias and SS divisions like the 'Handschar' and its ragged Albanian counterparts waged a barbaric war. They fought simultaneously as both German proxy troops and as agents of a violent civil conflict that had been unleashed by the German invasion of Yugoslavia. Acting under German orders, the Muslim SS recruits and their Catholic Croatian comrades sought to completely exterminate their Serbian foes. Since Himmler regarded Serbs as Slavic *untermenschen* and his Muslim warriors as Aryan descendants of Persians, the Balkan war was just another means to liquidate the race enemies of the Reich. A few historians have claimed that the 'Handschar' men murdered tens of thousands of Jews when they returned to Croatia after 1943.[54] This is exaggerated. The majority of Jews had already been murdered by the Ustasha squads and by German Wehrmacht units and their Serbian collaborators. In the case of the 'Handschar', we have just two documented cases that led to the killing of Jews. In the summer of 1944, 'Handschar' men killed twenty-two Jews in Tuzia. Later that year, a 'Handschar' punishment detail was assigned to guard Hungarian Jewish forced labourers in the Austrian village of Jennersdorf. According to eyewitnesses they treated these men with great cruelty. Some who were considered unfit to work were taken away and shot.[55] These documented atrocities are shameful enough, but wild claims about much larger scale killings by 'Handschar' men of Jews serve no historical or indeed moral purpose.

Haj Amin el-Husseini's grand plan was to forge a pan-Islamic state, rivalling the defunct Ottoman Empire, that would be rendered *Judenfrei*. Had it succeeded, this German-backed Jihad would have unleashed a human catastrophe on an unimaginable scale as SS murder squads like the Rauff Kommando and Muslim militias reached out far beyond the Mediterranean to do the Grand Mufti's bidding. The SS 'Handschar' might possibly have played a part in the Mufti's plan, though the evidence for this is less than convincing. As it turned out, the Bosnian Muslims became pawns in a merciless German-instigated civil war that pitted Croatians and Bosniaks against Serbs. That was the Balkan tragedy.

12

The Road to Huta Pieniacka

> I know you will not disappoint the SS ... At the end of this war, the Führer will be able to say that the division set up by the brave people of Galicia [western Ukraine] has always done its duty ... Your homeland has become much more beautiful since you have lost – on our initiative, I must say – the residents who were so often a dirty blemish on Galicia's good name, namely the Jews.
>
> Heinrich Himmler, speech to Ukrainian SS recruits at Neuhammer, 16 May 1944

Shortly before dawn on 27 February 1944, the 2nd Battalion of the 4th SS 'Galizien' Police Regiment, clad in white winter uniforms, slipped quietly along a winding forest path that led the village of Huta Pieniacka in the Tarnapol district in east Galicia.[1] It was dark and bitterly cold and the SS men's breath hung thickly in the air. Boots crunched through ice. Leather belts creaked. The 200 or so dwellings of Huta Pieniacka lay silent under a thick blanket of snow. Chimneys, lit early, propelled ash high into the freezing air. A dark green veil of thick forest enclosed the village. German aerial reconnaissance had revealed a grid of narrow, twisting streets clustered around a crossroads and the Catholic church. The Germans called this borderland between the General Government and Ukraine 'Partisanen gebiet': bandit country. The province of East Galicia was crisscrossed by vital road and rail links, all vulnerable to attack from well-organised Soviet partisans who operated in small, fast-moving units. They could strike at will from their mobile strongholds hidden behind an impenetrable screen of birch and alder forest that stretched from the Prussian border to the eastern edge of the vast Pripet Marshes. No German commander relished engaging such an elusive enemy who could slip away like quicksilver through this labyrinth of forest and swamp.

A few very dim lights glowed as villagers began to get ready for the day. In 1944 Huta Pieniacka was considered to be a Polish village – and hence unreliable. Since November 1943 a few hundred Jews had also sheltered in Huta Pieniacka, as well as in a triangle of other tiny 'Polish' hamlets in the region.

When the SS men reached the shallow ditch at the edge of Huta Pieniacka, the German officer in charge called a halt. After a short wait, he signalled his men to begin encircling the village following the line of the ditch. The men whispered to each other in Ukrainian. For this SS regiment was attached to the 14th Grenadier Division of the Waffen-SS known as the 'Galizien'. They had marched to Huta Pieniacka with murder in mind. What happened here is still disputed by historians and Ukrainian officials and, in 2010, remains under investigation by the Polish Institute of National Remembrance. The institute's investigators have interviewed nearly a hundred witnesses since 2001, leading to a preliminary conclusion that 'there is no doubt that the 4th battalion "Galizien" of the 14th division of SS committed the crime'.[2]

According to eyewitness testimony:

> The SS surrounded Huta from three sides, shooting at a distance, set buildings on fire and entered the village. They plundered the belongings of inhabitants. People were gathered in the church or shot in the houses. Those gathered in the church, men women and children were taken outside in groups; children were killed in front of parents, their heads smashed against tree trunks or buildings, and then thrown into burning houses.

Another eyewitness testified: 'Then they took groups of people one by one to barns and houses, poured petrol over them and burned them. The screaming and crying was terrible.'[3]

According to villager Miecelslaw Bernacki: 'Finally, the village was set on fire. The only people who saved themselves were those who on finding out about the approaching Ukrainian SS managed to hide in the forests (only men).'[4] Bernacki told the investigators:

> They burned and killed 850 people and, you know, we could not recognize who was who as they were burned in the barns, houses and stables. You could only recognize somebody if they weren't burnt completely, and only then by their clothes. Because if one corpse stuck to another, the clothes stayed and you could recognize the colour. And otherwise only the bones remained.[5]

The massacre at Huta Pieniacka at the end of February 1944 was not an unusual occurrence. Three weeks later, on 23 March, another Polish village in the same district of East Galicia, Huta Werchobuska, suffered the same fate. All over the German-occupied east, as well as in the Balkans and occupied France, Waffen-SS anti-partisan squads like this Ukrainian regiment waged a deadly 'war on bandits'. In Ukraine, the road to Huta Pieniacka is a long and twisted one. It begins with the first Ukrainian battalions the 'Roland' and the 'Nachtigall', which were recruited by German military intelligence and murdered many hundreds of Jews in the city of L'viv in July 1941. After these battalions had been disbanded, the Germans continued to recruit Ukrainians to serve as auxiliary policemen in both the Reichs Commissariat Ukraine and the Galician district in occupied Poland, the General Government. Many Ukrainian Schuma battalions, like their counterparts in the Baltic, participated in special actions against Jews, most notoriously at the Babi Yar ravine in Kiev when the German Sonderkommando 4a and auxiliary police murdered 33,771 Jews between 29 and 30 September 1941.

In 1942, the Germans made a decision to liquidate all Jews who remained alive in the General Government. Operation Reinhardt may have commemorated the RSHA chief Reinhard Heydrich, who had been assassinated in Prague at the end of May.[6] To accomplish this monstrous plan, Himmler abandoned the so-called 'wild genocide' or 'Holocaust by bullets' that had been entrusted to the SD Special Task Forces and Schuma squads, and ordered the construction of specialised extermination camps in the Lublin District of the General Government. These camps would use new gassing technologies, developed by the Aktion T4 euthanasia experts such as Christian Wirth and SS-Obersturmbahnführer Walther Rauff of the RSHA Technical Department, to liquidate the Jews of the occupied eastern territories.

The trials of the Ukrainian Ivan (John) Demjanjuk first in Israel and then two decades later in Germany have made the names of these Reinhardt camps a litany of grotesque horror: Sobibór, Treblinka and Bełżec. The German managers of these mass-murder camps employed many thousands of Eastern Europeans as guards, mainly Ukrainians, but also Latvians and Estonians. The very few survivors of the Reinhardt camps have never forgotten the sadistic behaviour of these Eastern European camp guards. The Germans trained the majority of camp auxiliaries at a camp close to the Polish village of Trawniki. They were known as 'Trawniki-Männer'. As well as working as Reinhardt camp guards, these Eastern European recruits took part in other SS special actions. New research, using documents discovered in Russian archives, shows that 'Trawniki men' took part in the liquidation of the Warsaw ghetto in 1943, as well as other Jewish ghettoes, and many later joined the Waffen-SS 14th Grenadier Division, the 'Galizien'.[7] Likewise, Latvian men who had served in the murderous Arājs Commando later joined the Latvian SS divisions that Himmler began recruiting at the beginning of 1943. In the same way, Trawniki men recruited to serve at the Reinhardt camps ended up enlisting in Waffen-SS military divisions.

For many years, a Dr Swiatomyr Fostun served as the General Secretary of the Association of Ukrainian Former Combatants in Great Britain (the SS division's old comrades' club). Fostun lived comfortably in a London suburb. Dr Fostun was happy, indeed proud to talk about his service with SS 'Galizien', which he joined in 1944. According to Canadian researcher Michael Hanusiak, whose findings have been confirmed by British documentary producer Julian Hendy, 'Dr Fostun' was in fact Mychalio Fostun, who in 1943 had been trained at the Trawniki SS camp – and had taken part in many 'ghetto liquidations'. Dr Swiatomyr Mychailo Fostun and Mychailo Fostun shared the same birth date, 22 November 1924, for example. Other evidence is even more telling. In the 1970s, Dr Fostun had his photograph taken for inclusion in the *Almanac of the Association of Ukrainian Former Combatants*, which openly commemorates the veterans of the SS 'Galizien'. Three decades earlier, Mychailo Fostun had been photographed at the Trawniki camp. The resemblance is striking.[8] The unashamed 'Galizien' veteran, Dr Swiatomyr Mychailo Fostun died in London after a road accident; it was a violent end to a life that may once been devoted to murder.

The recruitment of the Trawniki men, as the Fostun case implies, provides a missing link connecting Operation Reinhardt with the formation of the 'Galizien' SS division. Understanding this evolutionary tie between the SS 'Reinhardt' camp guards and the Waffen-SS 'Galizien' is crucial. After Ukraine became an independent nation, many historians recast the SS division as a 'national liberation army', whose sole intent had been to resist the Soviets. This is a misrepresentation. Both the formation and the conduct of the 'Galizien' reflect its origins in the German plans for mass murder.

Although the precise timing is still debated, it appears very likely that in the winter of 1941, Hitler secretly authorised a massive escalation of the German 'war on Jews'. This 'unwritten order' was hammered into a practical plan at the Wannsee Conference convened on 20 January 1942 by RSHA chief Reinhard Heydrich to examine a 'Final Solution of the Jewish problem'. For the German bureaucrats who gathered at the villa overlooking Lake Wannsee, like the SD commander in Latvia Rudolf Lange, the numbers looked daunting. Military conquest had delivered 11 million Jews as well as other 'undesirables' into the hands of the German occupiers. The 'solution' had to be 'final' – but how? The mass shootings in the countryside would continue for some time, but streamlining the killing process now became an urgent priority. Heydrich had no doubt that the recruitment of native mass killers would need to be significantly ramped up.

The 'Final Solution' dreamt up in that lakeside villa presented a formidable task. Heydrich had precisely defined what 'Final Solution' meant in practice: 'Europe is to be combed through from west to east. The evacuated Jews will be brought, group by group, to the so called "transit ghettoes" to be transported from there further to the east.' Himmler had already decided to use the Lublin district 'further to the east' as a vast execution site. To manage what would soon be referred to as Operation Reinhardt, Himmler appointed one of his most repellent favourites, Odilo Globocnik (or Globotschigg) (b. 1904), who had already proved himself a highly proficient, as well as venal, *génocidaire*.

As well as Globocnik, Himmler recruited SS personnel who had worked on the T4 euthanasia programme. These men, responsible for the deaths of more than 70,000 men, women and children, 'lives not worthy of life', would be assigned to a cluster of new camps to be built in the Lublin district in the General Government. These would be specialised extermination camps, rather than mixed labour/murder camps like Auschwitz-Birkenau. The first to be constructed was close to the Polish town of Bełżec and situated at the end of a railway spur. Euthanasia veteran SS-Hauptsturmführer Christian Wirth, a 'gross and florid man' known as 'Savage Christian', was appointed camp commander. According to one of his staff: 'Wirth told us that in Bełżec "all the Jews will be struck down."' Soon after the Wannsee Conference, Wirth carried out the first 'experimental killings' at Bełżec using carbon monoxide. His engineers disguised the new gas chambers as 'Bath and inhalation rooms' – and the camp was ingeniously laid out so that large groups could be rushed from the railway head to their deaths in the most orderly way. Bełżec became fully operational in March 1942 and Wirth's system was replicated at the other Reinhardt camps at Sobibór and then Treblinka.

The successful completion of Operation Reinhardt depended on the smooth running of these slaughter houses. Under Globocnik and his second-in-command, SS-Hauptsturmführer Herman Höfle, the camp commanders (many of them Austrian) wielded absolute power in their obscene camp worlds. But the insatiable demands of Hitler's war machine meant that Himmler could allocate only a handful of German staff to the new extermination camps. He ordered Globocnik to find 'persons who seem to be especially trustworthy and therefore can be used to rebuild the occupied territories'.[9] These foreign assistant executioners would be trained at a former POW camp built not long after the Soviet invasion near the town of Trawniki. Its shabby wooden barracks would now become Himmler's college of genocide.

As Hitler's armies smashed through Soviet defences in the summer of 1941, millions of prisoners fell into German hands. Many would die of neglect, starvation or torture. Others, desperate to save their lives – and it would seem that Demjanjuk was one of them – offered to work for the Germans to do so. The majority of these 'Hiwis' (*Hilfswillige*, helpers) came from Ukraine, but Latvians, Estonian and Lithuanians also ended up at Trawniki. Vladas Zajanckauskas, for example, who is currently under investigation by the American OSI, had served in the Lithuanian army.[10] The Germans lumped them together, however, as 'Ukrainians', 'Trawniki-Männer' or 'Askaris' (a term first used during the First World War in German East Africa). Polish Jews called them 'blacks' (referring to the colour of their uniforms) or *Karaluch* (cockroach). Under the command of SS-Hauptsturmführer Karl Streibel, the 'Ukrainians' were trained to be tough and completely ruthless. The Germans armed them with whips and Russian carbines.

To procure recruits for the SS Bataillon Streibel, SS recruiters ransacked the hellish German POW camps still dotted across occupied Poland. Alerted by SS anthropologist Wolfgang Abel, they tracked down many Soviet prisoners who plainly had German origins. Some claimed descent from Germans who had settled in Russia in the latter part of the eighteenth century. Many more originated in the 'Volga German Autonomous Soviet Socialist Republic'. It seemed that German migrants had somehow ended up in just about every corner of the former Russian Empire. In the first wave of recruitment, these ethnic Germans enjoyed a privileged status. Ukrainian Feordor Fedorenko testified to a court in Florida in 1978: 'One day at Chelm [POW Camp], the Germans assembled the Soviet prisoners and walked down the line selecting 200 to 300 [ethnic Germans] who were sent to Trawniki ... These Volksdeutsche also wore black uniforms but theirs were well tailored and of better material.'[11] By 1943, as Operation Reinhardt began to wind up, Ukrainian recruits had come to dominate Trawniki and the camps. Regardless of their ethnic origin, the Trawniki men were, as Sobibór survivor Jules Schelvis wrote, 'overzealous'. Their task was to herd Jewish victims from the bogus railway station where they disembarked through the camp and finally into the gas chambers. At every stage of this journey to death, many of the Trawniki guards indulged in the grotesque cruelties. They routinely plundered money and jewellery, and traded it for alcohol in nearby villages. Many seized terrified Jewish girls from the crowd and raped them. At the Treblinka camp the most feared was a Ukrainian called Ivan Demaniuk: 'Ivan the Terrible'. According to survivor Eli Rosenberg, he 'took special pleasure in harming other people, especially women. He stabbed the women's naked thighs and genitals with a sword before they entered the gas chambers.'[12] As Operation Reinhardt picked up momentum in summer of 1942, some 1,000 Trawniki men, organised in two battalions of four companies, were stationed in Trawniki. These Trawniki men were not only deployed in the camps. According to a prosecutor in one of the post-war Trawniki trials: 'Trawniki men didn't just provide the great majority of camp personnel for the extermination camps ... but they took part ... in the liquidation of many ghettos. They were used in the most revolting and shocking operations and were known and feared for their cruelty.'[13]

At the beginning of 1943, German Trawniki recruiters arrived in East Galicia, a district of the General Government then ruled by the Austrian Baron Otto Gustav von Wächter. Galicia was also the stronghold of the Ukrainian ultranationalist movement, the OUN. It was at this time that Mychailo Fostun, who was almost certainly Dr Swiatomyr M. Fostun of Wimbledon, Surrey, joined up the Trawniki man.[14] Fostun, like many of new Trawniki recruits, came from the Tlumacz district. His identity documents include the required undertaking to serve the Germans for the duration of the war and confirm that he denied any Jewish ancestry and had never been a member of the Soviet Communist Party. From February until April 1943, after completing training, guard 3191 Fostun, served at the Jewish slave labour camp attached to the training camp at Trawniki. Fostun must have proved himself a diligent camp guard. On 17 April, the Germans selected Fostun and 350 other Trawniki men to take part in a special assignment. It would be commanded by one of Himmler's favourite SS generals.

When American troops broke into the empty villa in Wiesbaden, formerly occupied by SS-Gruppenführer and Generalleutnant der Waffen-SS und Polizei Jürgen Stroop in 1945, they stumbled on one of the most chilling of all accounts of mass murder. The Stroop Report is compilation of communiqués and photographs, bound together with an elegant cover that is emblazoned in gothic lettering: 'Es gibt keinen jüdischen Wohnbezirk in Warschau mehr' (The Warsaw Jewish ghetto is no more). Stroop's proud and meticulous report spells out in detail how Hitler's foreign executioners, trained at the Trawniki camp and including Mr Fostun and other Ukrainian volunteers who would later serve in the SS 'Galizien', liquidated the Warsaw ghetto in April 1943.

Since the 1990s, trials of some of the SS auxiliaries who took part in the destruction of the Warsaw ghetto have revealed many details about the role played by the Trawniki men; it is evident that they were kept busy at every stage of Stroop's attack, from the first encirclement to the final deportations. At assembly points, they guarded the prisoners on the trains, beat and humiliated them – and on arrival at Treblinka, herded them into the gas chambers. Witness statements by survivors refer frequently to the black-uniformed Trawniki men, who usually spoke Russian or Polish. As well as Ukrainians, Latvians and Lithuanians also took part in Stroop's 'Grand Action'. His report claims that his men 'destroyed' 56,065 Jews, 7,000 being killed immediately during fierce street and house to house battles. An unquantifiable number died in buildings razed by SS flamethrowers or destroyed by artillery shelling. The Trawniki men deported 6,929 ghetto prisoners to Treblinka, where all were gassed on arrival.

Stroop described his foreign executioners as 'nationalists and anti-Semites but the best soldiers. Young, mainly without education, wild at heart and with a tendency towards base things. But nevertheless obedient.'[15]

The following year, 1944, Swiatomyr Fostun and many other Trawniki men enlisted in the 14th SS Division 'Galizien'.

Many historians of the Second World War regard Himmler's decision to recruit a 'Ukrainian' SS division as a last resort – an act of desperation. It is assumed that he abandoned any pretence that the Waffen-SS was an elite Aryan corps. This view is quite wrong and demonstrates a misunderstanding of German racial ideology and the way the ideas of race scientists changed as the war unfolded. I have shown that following Professor Abel's studies of Soviet POWs, Gottlob Berger's SS recruitment office was persuaded to broaden the catchment area where Germanic blood might be, as Himmler put it, 'harvested'. As the SS empire expanded in mid-1942, Berger turned first to 'Germanic' Estonians and then 'suitable' Latvians to bolster his military divisions. Himmler next turned to the Bosnian Muslims, having been convinced that Bosniaks were a 'Gothic' people descended from Persians. Did the SS recruitment of Slavic Ukrainians fit the same evolving pattern? And if so how?

The most important clue is the name of the new SS division: the 'Galizien'. In just about every document concerning the SS 'Galizien', Himmler insisted that the division was not Ukrainian but Galician. This was not semantic window dressing. It will be recalled that the former Kingdom of Galicia was annexed by the Austrian Hapsburgs in 1772 and become the most easterly province of their sprawling empire. Today, the Galician region occupies part of eastern Poland and the western edge of independent Ukraine. This is a liminal territory that stretches, in memory at least, from the regional capital L'viv, formerly Lemberg, south towards Chernivtsi in Bukovina; the Carpathian Mountains form a western border while its eastern margin flows along the Zbruch River. For modern Ukrainians, as historian Omer Bartov discovered, poor, muddy, backward Galicia remains 'somewhat foreign and suspect'. Indeed Galicia is not very Ukrainian, especially its western half. All over Galicia, cities and villages still show traces of a history that has drawn together Poles, Jews, ethnic Germans as well as ethnic 'Ruthenians'. This was the birthplace of a rich Jewish culture that flourished for centuries alongside chauvinist Ukrainian nationalism. For the Hapsburg emperors in Vienna, this most remote territorial possession was an exotic backwater huddled on the western edge of the Russian Empire. Galicia has always been a volatile borderland, often prey to the slings and arrows of territorial upheaval. At the end of the First World War, as the Austrian Empire collapsed and Russia was engulfed by revolution and civil war, Galicia became a battleground. At the Paris Peace Conference, the victorious powers had frustrated Ukrainian demands for independence and the Soviets held on to much of their disputed homeland, while leaving Galicia up for grabs. Poles had, in any case, always dominated the western part of the old Austrian province – and in 1923, a resurgent Polish nation annexed the east as well, renaming it 'Eastern Little Poland'.

In the period before the First World War, the light-touch rule of the Hapsburgs had encouraged Ukrainian nationalism to flourish in Galicia. Conversely, in eastern Ukraine, any expression of national separatism had been ruthlessly stamped out by the tsars through a policy that would be maintained by Lenin and then Stalin. As a consequence, Ukrainian nationalists tended to be both pro-German and, especially after 1917, aggressively anti-Semitic, since they identified Bolshevism with Jews. Ironically, Polish-ruled Galicia and its cultural hub, the city of Lwów (formerly Austrian Lemberg, now L'viv), became the crucible of a radical Ukrainian nationalism that was both anti-Semitic and dedicated to the destruction of Poland.

In 1939, the Germans and Soviets once again split Galicia according to the secret protocols of the Nazi-Soviet Non-Aggression Pact. Then after the German invasion of the Soviet Union, Hitler assigned Galicia to the General Government as the 'Distrikt Galizien'. The eastern region (Ostgalizien) would be ruled by an Austrian SS professional, Otto von Wächter, who in 1943 would become the main architect of the SS 'Galizien'. Wächter's fiefdom bordered the vast Reichskommissariat Ukraine that was ruled by former Gauleiter Erich Koch. Koch was a gluttonous despot who waged an unending turf war with his nominal boss, the despised Eastern Minister, Alfred Rosenberg.

For both Himmler and the Austrian-born Hitler, Galicia had a special historical significance as a reservoir of Germanic blood. This was reinforced by Himmler's RuSHA race experts, who claimed that about 25 per cent of the 'Ruthenian' population (i.e. Ukrainians) possessed a significant quantity of Germanic blood. Galicia was thus ripe for 'Germanisation', which provided the underlying logic of Waffen-SS recruitment. We have evidence that Galicians fitted the SS recruitment plan in the records of an extraordinary meeting that took place at Hitler's 'Werewolf' headquarters near Vinnitsa in Ukraine. The purpose of the meeting was to resolve the unending battle between Reich Commissar Erich Koch and his superior, the Eastern Minister Rosenberg. Rosenberg had complained bitterly and often that Koch ruled his fiefdom with excessive cruelty, thus damaging his efforts to exploit Ukrainian anti-Bolshevik and nationalist aspirations. Most pointedly, he accused Koch of inciting attacks by Ukrainian partisans which was for the Germans a very sore point in the spring of 1943. Hitler turned to Rosenberg. He reminded him that Wehrmacht attempts to recruit eastern troops (Osttruppen) in the occupied Soviet Union had usually ended in calamity. He would not permit recruitment of Ukrainians in the Reich Commissariat. But the people of Galicia, he pointed out, had lived for more than a century under Austrian rule. They had close connections with the old Hapsburg Empire. 'It is therefore possible,' he concluded, 'for the SS to set up a Ukrainian division in Galicia.'[16] Historians who have interpreted Himmler's decision to recruit a Ukrainian division in 1943 as an expedient response to German losses on the Eastern Front have failed to take account of this crucial distinction he made between the broad mass of the Ukrainians and those born and bred in 'Austrian' Galicia.

For the Germans, there was just one stumbling block. As we have seen, the main Ukrainian nationalist faction the OUN had made its autonomist aspirations all too plain at the end of June 1941, when OUN leaders had rashly declared independence after occupying the Galician capital L'viv. It was this impertinent gesture that had led Himmler to reject the idea of a 'Ukrainian' SS militia when it was proposed by Berger in 1941. Even after the Galician region was absorbed into the General Government, on Hitler's orders, Himmler refused to consider recruiting an SS division – although, as we have seen, thousands of Ukrainians served in the Schuma battalions. In Himmler's view, the nationalist OUN, albeit devoutly anti-Semitic, had a stranglehold on Galician political culture. Then in the early spring of 1943, the Governor of East Galicia, SS-Gruppenführer Otto von Wächter, decided that the time had come to try a new approach. Wächter would become the main driving force behind SS recruitment.

At the beginning of March 1943 Wächter flew from his headquarters in L'viv to Hochwald in East Prussia to meet Himmler, who was headquartered at a railway siding in his official train, *Heinrich*. They had much to discuss. The minutes taken at the meeting refer to progress with ghetto clearances and similar matters to do with the 'Final Solution'. For the Nazi governors of the occupied east, this was business as usual. Much more pressing was the increasingly precarious state of the German front line. For Wehrmacht commanders and the Nazi elite in Berlin, the destruction of the 6th Army and the loss of half a million men at Stalingrad was 'the most catastrophic hitherto experienced in German history'. 'Imagine it,' one Russian soldier wrote to his wife, 'the Fritzes are running away from us.'[17] Closeted inside his military headquarters at Rastenburg, Hitler raved about the cowardice of his generals, while Goebbels tried to spin the bad news, telling the German people that the 6th Army had been 'annihilated' so that 'Germany might live'.[18]

Stalingrad had shown that the 'invincible' Germans could lose a battle; it did not, as some historians claim, 'decide the war'. Shortly before Wächter met Himmler at Hochwald, Field Marshall Erich von Manstein had launched a successful strike against Soviet forces at Kharkov and captured the city by 11 March. Manstein's formidable panzer armies restored the German front to more or less the same line reached in December 1941. Hitler's war machine had by no means lost its offensive capabilities and that summer his commanders would muster enormous forces at Kursk to unleash the last great offensive of the war in the east, Operation Citadel. As it turned out it was the rout of German armies at the cataclysmic Battle of Kursk that was fought months later in July 1943 that truly signalled the beginning of the end for the German campaign.

Early in March, what focused the minds of Himmler and Governor Wächter was a different kind of crisis. By the spring of 1943, Soviet backed partisan units had become a serious threat to the German rear areas in the Balkans and on the Eastern Front. In the early months of the German attack, Hitler had used the 'partisan threat' to rationalise his radical conception of war in the east. He told a meeting of senior aides: 'The struggle we are waging against the partisans resembles very much the struggle in North America against the Red Indians.' For both pragmatic and ideological reasons, Hitler made Himmler's SS responsible for pacifying German army rear areas and waging war on 'bandits'. To begin with, as we have seen, Himmler used Bandenbekämpfung (bandit warfare) as a cover for the liquidation of Jewish civilians and communist officials. This equivalence of Jews and bandits would continue to shape Himmler's 'bandit war'. Field reports submitted by the German army, as well as by Himmler's chief bandit hunter HSSPF Erich von dem Bach-Zelewski, frequently included a count of any Jews who have been killed whether or not they were considered to be actual partisans.[19] Now in the spring of 1943, the partisan war had become a lot more menacing. According to Goebbels: 'The activity of partisans has increased noticeably ... The partisans are in command of large areas ... and are conducting a regime of terror.'[20] Later that year Hitler would reiterate Himmler's responsibility for the Bandenkampf und Sicherheitslage (Bandit Fight and Security Situation). For Himmler, the reaffirmation of SS security management was another step towards complete domination of German occupation strategy. But it was also an overwhelming responsibility, which carried a tremendous risk of failure. This gave Governor Wächter a distinct advantage. He could claim with facts and figures to back him up that his own fiefdom, the 'Distrikt Galizien', was, so far, relatively free from 'bandit activity'. But, he warned Himmler, he had recently noted a troubling rise in the number of 'bandit attacks' – and he blamed the insurgent Ukrayins'ka Povstans'ka Armiya, the UPA. Wächter had a radical solution to propose. He wanted Himmler to authorise the formation of a new SS division to be recruited in East Galicia. This would, he argued, siphon off support for the UPA and reinforce security in this strategically vital borderland region.

It was remarked that Wächter 'understood Ukrainians'.[21] This is generous. As an Austrian, Wächter retained a national memory of Galicia as a former Hapsburg province. Like the administrators of the Austrian Empire, he toyed with Ukrainian aspirations but according to very strict rules. It was a family tradition. Gustav Otto von Wächter was born in July 1901 in Vienna. His father General Josef Freiherr was a Sudetendeutscher (an ethnic German from the Sudetenland) who had fought in the western Ukraine during the First World War. Described in his SS file as 'highly intelligent', Wächter like so many of Himmler's top SS managers had a doctorate in jurisprudence. A tall, trim man with curiously Slavic eyes, he soon became a big player in the Austrian Nazi movement and fled to Germany in 1934 in the chaotic aftermath of the abortive putsch. Following a short period of detention, Wächter began cultivating the SS elite with his usual energy and cunning. His subsequent ascent was, as his personal file records, rapid: March 1935 Untersturmführer; 1 June Obersturmführer; 9 November Hauptsturmführer; 20 April 1936 Sturmbannführer; 30 January 1937 Obersturmbannführer; 30 January 1938 Standartenführer.[22] In March 1938, after German troops had marched into Austria, the new Governor Artur Seyß-Inquart appointed Wächter head of the police in Vienna. The 'Wächter Commission' took charge of expropriating Jewish property and 'cleansing' the Austrian bureaucracy by removing all Jews from public office.

Wächter was a zealous bureaucrat. In 1939, Hitler appointed him Gauleiter of the Kraków district in the new General Government, a region that would become one of the epicentres of the Holocaust. Governor General Hans Frank, who, it will be recalled, had been Hitler's personal legal advisor, welcomed Wächter's appointment hoping that he could use him to work against Himmler. Kraków was the administrative and communications hub of occupied Poland and, after 1939, was thoroughly 'Germanised'. Wächter set about removing all traces of Polish culture and began deporting Jews. He believed that Jews had 'no native place' and often discussed ways of achieving 'total Jewish extermination'. According to one of his subordinates, 'I have to say that my impression is that [Wächter] represented the point of view of the Master Race, that is the SS point of view towards the so-called Fremdvölkischen (foreigners). He was a high SS leader, constantly running around in his SS uniform.'[23] The Italian journalist Curzio Malaparte, who witnessed the German-inspired pogroms in Romania, provides us with a vivid portrait of Wächter in his book *Kaputt.*[24] In one chapter Malaparte describes an evening at the venal court of the 'King of Poland', Governor Hans Frank. In the course of a lavish dinner, Wächter and his wife refer to the 'filthy state' of Polish Jews. Later, Malaparte listens to Frank's critique of the massacres in Iaşi: 'the Romanians are not a civilised people,' he grumbles. 'We must be surgeons, not butchers.' Wächter concurs: 'Germany is called upon to carry out a great civilising mission in the East.'

After his appointment as Governor of the 'Distrikt Galizien', Wächter was able to pursue his 'mission' on a much bigger scale. From his new headquarters in L'viv, he masterminded the liquidation of the last Jewish communities in his fiefdom. At the beginning of July, 1941, German troops and the Ukrainian 'Nachtigall' battalion had massacred thousands of Galician Jews. In December, the Germans had incarcerated the survivors in the L'viv ghetto, but in the summer of 1942, soon after Wächter had been appointed governor, he accompanied Globocnik to a meeting with Himmler to discuss Operation Reinhardt. When Wächter returned to L'viv, he authorised mass deportations to the Reinhardt camps. As the last train left L'viv for Treblinka, Wächter dispatched Ukrainian police battalions to flush out any survivors still hiding in the ghetto. Like all the SS top brass in the east, he had a great deal of blood on his hands.[25]

Himmler valued Wächter highly. But the General Government was, like every other German administration, a political battleground. Hans Frank hated Himmler and the SS as much as he despised his Polish subjects, and he fought every SS incursion into his fiefdom. He accused the HSSPF Friedrich Wilhelm Krüger of trying to 'build an SS state within a state'. By 1943, a succession of brutal skirmishes between Frank and the SS had begun to tear the General Government apart. The wily Gauleiter Wächter, who was also an SS-Brigadeführer, adroitly cultivated allegiances in both camps – and by spring 1943 he had a powerful card to play. His 'Distrikt Galizien' was the eastern 'wall' of the German General Government. If the Soviet armies ever managed to overrun the Commissariat Ukraine, then East Galicia would become the next line of defence. If that happened, the internal security of Wächter's domain was a matter of overriding importance. If your enemies are already behind you, there is no point building a wall and shutting the fortress door. So far, Wachter had maintained friendly relations with the pro-German

Ukrainian elite and successfully kept the lid on UPA activities in East Galicia. This was the Austrian way; according to his personal assistant Dr Heinz Georg Neumann, 'we favoured the Ukrainians because of political expediency. Wächter therefore tried to carry out as much as possible the old tradition of Austrian policy in Galicia.' Himmler commended him warmly:

> One thing I would like to point out ... Galicia has remained quiet and in order. This is to your credit and can be attributed not least to your harmonious work with the brave [Friedrich] Katzmann [SSPF Lemburg] and ... to the real cooperation of your administration with SS and police.[26]

In short, Wächter had cleverly pushed all the right SS buttons. But as Stalin's armies pushed westward, the security of the 'Galician wall' could never be taken for granted.

When he met Himmler in March, Wächter's main objective was to raise the delicate matter of recruiting Ukrainians and forming a new SS division. This, he argued, would bind the Ukrainians in East Galicia to the Reich and provide a means to attack and neutralise the partisans. As Wächter had anticipated, Himmler, who was the Reich's Chef der Bandenbekämpfung (Chief Bandit Hunter), was obsessed with security. But he had to work hard to convince Hitler and the Wehrmacht high command to commit recourses to this shadow war.[27] At the same time, Himmler exploited the Bandenbekämpfung to continuously expand the SS and its military and police forces – including non-German auxiliaries. It was this SS 'war on terror' that drove the dramatic expansion of non-German recruitment after the summer of 1942, and why in March the following year he listened attentively to Wächter's proposal to begin recruiting in Galicia. He had long been fascinated by the racial ancestry of the Galician peoples and would soon come to believe that this fresh wave of SS recruitment provided another opportunity for 'harvesting Germanic blood'.

Nevertheless, Himmler still had to square the nationalist circle. In other words, although Galicians possessed some quantum of Germanic blood, many were ardent, indeed fanatical nationalists. Wächter knew very well that the main Ukrainian nationalist faction, the OUN, wielded a good deal of influence and power. Fortunately, from his point of view, the OUN was split between rival clans. Its radical wing, the OUN-B, led by Stepan Bandera, resolutely opposed collaboration with the Germans and fed recruits into the insurgent movement, the UPA, which threatened to become the most dangerous insurgent force in the General Government. But Andreas Melnyk's rival OUN-M took a more conciliatory line. This more conservative wing of the nationalist movement traditionally leant towards Germany – and as historian Taras Hunczak admits, Wächter's plan proved 'easier and more successful than anyone could have possibly anticipated'.

Why? One answer is that Himmler reaped what General Governor Frank and Wächter had already sown. Wächter was convinced, like Rosenberg, that in the Reichskommissariat Ukraine, Erich Koch had foolishly squandered the good will that Ukrainians had shown in 1941, when they greeted German troops as a liberating army. After visiting the Commissariat and observing at first hand what Koch's rule meant in practice, Wächter wrote angrily to Martin Bormann. He did not pull any punches:

> The Bolsheviks have done useful preparatory work for us ... They are hated and regarded as oppressors ... Even if we only provide some relief for the population [in Ukraine], in contrast to the terror of the Soviet regime, then we will win them to our side. Unfortunately, we are not doing this.[28]

Bormann is unlikely to have passed on Wächter's letter to Hitler, who consistently backed Koch's draconian methods. Certainly, Wächter never received a reply. But in any case, German administrators in the General Government had deliberately raised the status of Ukrainians and demoted Poles as well as Jews. Frank established a Ukrainian Central Committee (UCC) chaired by the Ukrainian geographer, Professor Dr Volodymyr Kubijovych, and maintained good relations with its members. The UCC was analogous to the puppet 'self-administrations' in the Baltic states and its power was limited to welfare and cultural promotions. In his Galician fiefdom, Wächter cleverly exploited the UCC; he often reminded Kubijovych about the good old days under Austrian rule – and of the treachery of the Poles when they snatched statehood from the Ukrainians in 1919. He diligently cultivated the Melnyk wing of the OUN and the hyper reactionary Ukrainian 'Front of National Unity'. He made sure the most compliant pro-German Ukrainians got the best jobs and excluded anyone suspected to be linked with the rival OUN-B. Wächter and Kubijovych had already begun discussions about an SS legion a few weeks before the meeting with Himmler. He did not, of course, reveal to his Ukrainian friends the real purpose of the proposed new SS division: 'to be utilised as much as possible in the spirit of an introduction of the Ukrainian population of Galicia *to the strategic concept of being Germanic*.'[29]

On 28 March Himmler wrote to Wächter informing him that Hitler had agreed to the formation of the proposed SS Freiwilligen-Division 'in principle'.[30] He went on: 'we should proceed in stages as follows.' First would come a kind of bribe: those Ukrainian farmers who had delivered quotas to his satisfaction in the spring would be permitted to buy their own land from the old Soviet style collectives that the Germans had retained. Those 'not up to mark', of course, would remain Reich property. After that, 'your summons will be issued to the able-bodied youth of Galicia'. Wächter now had to get the Ukrainians completely on board. The biggest stumbling block was semantic. Ukrainians did not, of course, regard themselves as Galicians. For them Galicia had vanished along with the Hapsburgs. But Kubijovych readily agreed that his fellow Ukrainians should be persuaded to fight the Bolsheviks as German allies. He poured scorn on the idea of a Galician division and proposed instead a 'Ukrainian legion' modelled on the Latvian and Estonian ones. Wächter knew that Himmler would never tolerate any use of the term 'Ukrainian' so he proposed a compromise. If the proposed SS 'Galizien' division performed well on the battlefield, then he might consider offering some kind of Ukrainian self-rule in the Galician district of the General Government. It was the usual Faustian offer: non-specific autonomy in exchange for shed blood. Wächter sugared the pill by promising that the UCC would have a primary role in the recruitment and formation of the division through a new military board, to be headed by Kubijovych himself. He promised to consider appointing Ukrainian priests to serve with the division. When Kubijovych continued to prevaricate, Wächter removed the velvet glove. He warned Kubijovych that if he could not secure an agreement, the UCC would be the only losers; as the occupying power, Germany would draft Ukrainians with or without an agreement. Kubijovych had few options and caved in.

Now it was Himmler's turn to prevaricate. Three days after the meeting with Kubijovych, Wächter met SS recruitment chief Gottlob Berger in Berlin to discuss the new SS 'Galizien'. At this point, an unexpected difficulty emerged. Because the Waffen-SS had a chronic shortage of training staff, Berger proposed that the German Order Police (ORPO) take charge of the formation of the new Galician division and their training. Wächter knew that Berger's plan would be anathema to the UCC, who vehemently opposed allowing Ukrainians to join another SS police battalion. Nevertheless, Berger informed Himmler that he had agreed with Wächter to 'approach the Ukrainian population with a large scale call to arms' and referred to 'this new *Ukrainian* SS-Polizei-Schützen-Division'.[31] Berger's no doubt inadvertent reference to 'Ukrainians' immediately set off alarm bells. A few days after the Berlin meeting with Berger, Himmler's adjutant Karl Brandt sent a telex to Wächter ordering him to 'move slowly … do not yet issue the call to arms'. Why the cold feet? Wächter discovered that shrill and powerful voices had been raised against his idea of a Galician SS division. Kurt Daluege reminded Berger that, in 1918, Ukrainian nationalists had assassinated Field Marshall Erich von Eichorn, who had recruited and trained Ukrainians to fight the Red Army just as Himmler now proposed to do again. Surely it was certain that any Ukrainians recruited by the SS would once again turn their weapons on their new German benefactors? Commissar Erich Koch, fearing that Wächter's scheme threatened his own Ukrainian kingdom, the Commissariat Ukraine, soon weighed in and scolded Berger so abusively that he complained to Himmler. Himmler was forced on the defensive. He knew that although Hitler had authorised the new division, he remained sceptical about the reliability of any kind of 'Eastern' recruits. Himmler defensively fell back on semantics: he reiterated that new division was Galician not Ukrainian; he insisted that terms such as 'Ukrainian division or Ukrainian nation' must never again be used in any discussion or memorandum.[32]

Back in L'viv, Wächter ignored the rumpus in Berlin and refused to apply the brakes. On 12 April, he chaired a meeting with SS and General Government officials to set a timetable for the recruitment of the SS Freiwilligen Division 'Galizien'.[33] Not a single Ukrainian delegate was invited. One of the keynote speakers was SS-Brigadeführer and Generalmajor der Polizei Jürgen Stroop, who had recently masterminded the destruction of the Warsaw ghetto. His attendance emphasises that Wächter planned to use the new SS division as an 'anti-bandit' militia that would take over the work of auxiliary police battalions. In any case, Himmler had agreed to provide funds from the coffers of Kurt Daluege's German Order Police, not the Waffen-SS, so the ORPO would be paying for the new division. This bond with the hated German Polizei would be concealed from Ukrainians 'for psychological reasons'. Wächter and his SS henchmen agreed that for propaganda reasons, the public face of the new division would be the Wehrausschuss Galizien (Military Board), to be headed by Professor Kubijovych and staffed mainly by respected Ukrainian elders who had served with the Austrian army.

Wächter was setting the same kind of trap that had already ensnared the Latvian collaborators. Also present at the meeting was Colonel Alfred Bisanz, who would become Wächter's Trojan horse in the Ukrainian camp. He had excellent credentials: born in Przemyśl to German and Ukrainian parents, he had served in the austrian Army during the First World War. In 1918, he had led the Ukrainian Lemberger-brigade against the Soviet Red Army. After the German invasion of Poland, he joined the Bevölkerungswesen und Fürsorge department in the General Government to look after the welfare of ethnic Germans resettled in the east. Bisanz also 'managed' the 'evacuation' of 30,000 Jews from the Galician district in 1941 and the following year participated in the 'March Action' when the remnant Jewish population was dispatched to the Reinhardt camps.[34] Bisanz had one disadvantage: he did not get on well with Himmler, who called him 'a big pig'. But Bisanz was fervently anti-Polish and anti-Semitic and could strategically play up his Ukrainian credentials whenever needed. As the foolish Professor Kubijovych got to work on Wächter's behalf, Bisanz quietly awaited his chance.

Wächter set 28 April 1943 for a public proclamation (*Festakt*) that would officially inaugurate SS recruitment in the 'Distrikt Galizien'. He knew he had to keep the pressure on Himmler, which he did through Berger. On 16 April, Berger telexed Himmler: 'In line with orders, I have put myself in contact with [Wächter] and made the following discoveries ...' He emphasised that with increasing turbulence in the General Government, any 'cessation of canvassing for the division would bolster resistance activities and give support to enemy propaganda'. This greatly reassured Himmler and matters now moved quickly. On 18 April, Kubijovych convened a conference of prominent Ukrainians in L'viv. He urged that they take advantage of Himmler's offer and exploit a fresh opportunity to recruit a Ukrainian national army that would be trained by German officers. He pointed out that the German occupiers had already recruited tens of thousands of Ukrainians for their own ends; now they, the Ukrainians, had an opportunity to turn the tables. If Germany was defeated, and this was no longer out of the question, Ukrainians would need an army to combat the resurgent Bolsheviks and their Jewish lackeys.

After listening to the impassioned Kubijovych, the other delegates agreed in principle to co-operating with Wächter. But Kubijovych drew up a list of ten demands to present to the Germans as follows:

1. The division would be used exclusively against the Bolsheviks
2. Its name and markings be Ukrainian
3. Its officers be Ukrainians (attached German officers to act as liaisons with German high command staff)
4. The division be provided with religious ministration by Ukrainian priests
5. The division be attached to the Wehrmacht (that is the *Heer* or regular army)
6. The division be considered as the first unit in the creation of the Ukrainian national army, into which it would be eventually incorporated
7. That all Ukrainian political prisoners in German prisons and concentration camps be released under a general amnesty, including former officers of 'Nachtigall'
8. That all other Ukrainian military units be dissolved
9. The division to be a fully motorised unit possessing a full range of weapons, including tanks
10. Forced labourers who fled from Germany should be granted amnesty

Wächter listened politely. But the Ukrainians failed to recognise that the Germans had no intention of genuinely negotiating. So Wächter agreed to just two of Kubijovych's conditions, 1 and 4, neither of which would trouble Himmler. Wächter readily agreed to taking on a few priests.[35] It was better to have them on side and on message rather than preaching against the Germans from the pulpit. He threw out every other proposed condition. But to let that useful idiot Kubijovych save a little face, he proposed appointing a balanced cadre of Ukrainian and German officers. Distracted by Wächter's token concessions, the Ukrainians caved in. They even agreed to Bisanz's proposal that the division use the old Austrian symbol for Galicia, the golden lion, rather than the Ukrainian national emblem of a trident.

Wächter evidently possessed unusual powers of persuasion. Secretly he emphasised to his subordinates that recruitment must appear to be a Ukrainian led affair, not 'simply as an organ of the German authorities'. But he stressed that 'the formation of the division is to be utilised as much as possible in the spirit of *an introduction of the Ukrainian population to the concept of being Germans* [my italics]'.[36] Wächter's statement is powerful evidence that Himmler's recruitment of Ukrainians in the Galicia district fitted his master plan to Germanise the east.

Wächter cunningly appealed to Ukrainian chauvinism. This is evident in the recruitment posters issued by the Military Board and now held by the Bundesarchiv in Berlin. These graphic images often show Ukrainian SS recruits bayoneting stereotypical 'Jewish-Bolshevik' soldiers. In *Novi Visti*, a Ukrainian bulletin board newspaper published in late 1943, the headline proclaims 'To Arms' next to a Galician golden lion. Against a background of long lines of eager Ukrainian recruits, a knight holding a shield emblazoned with SS Sig runes slays a Bolshevik dragon. The accompanying text exhorts recruits to 'annihilate the Jewish-Bolshevik monster'; consecrate your freedom with the enemy's 'evil blood'. German ideology and Ukrainian nationalist aspirations coalesced around the figure of the 'Jewish-Bolshevik'. As in the Baltic states, nationalist collaborators wished not merely to defend their nations against the Soviet aggression but to rid them of hated outsiders, above all Jews. It is this that connects the motivations of nationalist collaborators who joined forces with the SS during the first wave of genocide between 1941 and the end of 1942 and those who volunteered to join Waffen-SS divisions after the summer of 1942.

On 28 April, at the Festakt in L'viv to launch recruitment, Wächter stood alongside German and Ukrainian dignitaries to make an emotional appeal for volunteers. For Goebbels, the beleaguered Minister of Propaganda, the new Galician division was very welcome news: he sent camera teams from Berlin to record events. In his speech, Wächter praised the determination of the Galician people to join the struggle against Bolshevism. The Führer had rewarded them by agreeing to form the new SS 'Galizien' division. All volunteers, he promised, would be on 'equal footing' with German soldiers and would be provided with priests of their confession and nation. 'For centuries you stood on the side of Europe against the marauding Orient, prove yourselves now in this hour of destiny.'[37] Professor Kubijovych then affirmed the wish of the Ukrainian people 'to fight against Bolshevism with weapons in our hands'. After the main event, Wächter led the German delegation to the Cathedral of St George, where he was greeted by Rev. Dr Joseph Slipyj and Rev. Dr Wasyl Laba, who gave their blessing to the battle against Bolshevism. Himmler feared the influence of the Ukrainian church and its priests. So he arranged for Laba's sermon to be tape-recorded and transcribed in Berlin. They noted that he had spoken of the resurrection of the 'Ukrainian people's army' and alluded to the famous 'Sich riflemen' recruited 'twenty years ago' (during the First World War). Alarmed, Berger ordered Wächter to 'remove this sore tooth'. But Wächter was already one step ahead. Now that recruitment was officially under way he could begin sidelining the Ukrainians. On the same day as the Festakt, he sacked Professor Kubijovych as head of the Military Board and replaced him, as planned, with Bisanz. The wretched professor had just sent a grovelling telegram to Governor General Frank thanking him 'on behalf of the Ukrainian people of Galicia' and passing on his 'gratitude to the Führer'.[38] Recruitment began immediately: 'Everyone to the Division!'

Soon after the Festakt, Wächter began reporting very impressive figures to Berger and gloatingly passed them on to his old rival Krüger. By 14 May, it appeared that 38,569 Ukrainians had registered to join the new SS division; by the end of the month this figure had risen to 67,210. Finally on 2 June, a delighted Wächter recorded a final figure of 81,999 registrations. On a cold, wet Sunday, Wächter staged a spectacular showpiece rally at the Kolomyia Stadium on the banks of the River Prut, which a German cameraman filmed in colour. The Waffen-SS had, it appeared, sufficient human 'capital' to form more than one Galician division. Governor Wächter must be a miracle worker.

Then Himmler's medical examiners got to work, and they applied the same physical criteria as they did with other Waffen SS volunteers. The SS experts assessed the volunteers using a broad range of anthropometric tests, including skeletal proportions, hair and eye colour, and height. At the beginning of June, Berger reported to Himmler that:

> The number of voluntary registrations for the Legion [sic] 'Galizien' has risen to about 80,000. Of these approximately 50,000 have been provisionally accepted. Of those 50,000, 13,000 have been examined, half of these are [suitable for combat use] and may

> be enlisted. Consequently, one may expect after final completion of the examinations, a total of approximately 25,000 suitable individuals measuring at least 165cm.[39]

This latter measurement referred to the famous SS minimum height requirement. A closer look at the figures shows that a significant proportion of the Ukrainian volunteers 'failed' that crucial test and the criteria was later lowered. Berger was not especially dismayed.

What conclusions can we draw from this confusion of numbers? First, even if we revise down Wächter's claim of 80,000 registrations to a more sober figure of about 50,000, it is plain that many Ukrainians in the Galician district rejected anti-German propaganda that was disseminated by both Polish and Ukrainian insurgents. Kubijovych and the UCC propagandists had done a fine job. A young volunteer from L'viv wrote to his parents: 'Dear Mother and father, I have joined the [Galizien] division. I trust you will not object to it, but I have followed my conscience. I want to fulfil my duty to my people ... the moment has arrived and if we lose it we shouldn't call ourselves a nation.'[40]

Second, Himmler had not abandoned strict recruitment criteria. Many Ukrainians may have been a few centimetres shorter than the average German SS recruit, but Himmler still insisted on a rigorous weeding out of unsuitable candidates. In October, Wächter finally abolished the Military Board and handed responsibility for recruitment to the Ergänzungsamt der Waffen-SS Ergänzungsstelle Warthe (XXI) Nebenstelle, headed by SS-Hauptsturmführer Dr Karl Schulze. The new body applied even more stringent standards. On 30 October, Schulze reported that out of 80,000 volunteers, only 27,000 had been passed by SS medical boards. Of these 19,047 had been called up and 13,245 had actually reported for service. By the end of October, just 11,578 Ukrainians were in training – numbers barely sufficient to muster a division. These figures provide no support for the frequently made claim that 'after Stalingrad' the Waffen-SS recruitment bodies tore up the rule book and accepted 'all comers'.

On 18 July 1943, 740 new SS recruits assembled in front of the Grand Opera House in L'viv to begin service with the 14. Waffen-Grenadier-Division der SS (Galizische Nr. 1). It was a warm, still day; Nazi swastika banners hung limply alongside the Galician lion. On the podium, Governor Wächter and other German military and civil dignitaries stood, their arms stiffly raised in the *Hitlergruss*. Over 50,000 Ukrainians had travelled to L'viv; they sang, cheered and threw flowers. After the usual speeches, the new recruits marched off to Chernovetsky station, where they would board trains taking them to a former Polish cavalry training camp in the General Government called Heidelager. A few scratched 'Free Ukraine' slogans on the sides of wagons. The young men had provisions for three days, as well as one packet of undergarments, one sewing kit, one knife, a watch and writing equipment for letters home. As night fell, the German trains began steaming slowly westwards.

There was no disagreement about the purpose of the new SS division. It would maintain the security of the General Government – and this would mean fighting a dirty war. As we have seen, Himmler's Bandenbekämpfung was genocide disguised as counter-insurgency. His foreign recruits frequently targeted civilians and Jews who had somehow survived the ghettoes and camps. To ensure that the SS 'Galizien' performed, Himmler packed its German officer corps with experienced 'bandit' fighters well versed in mass murder of unarmed civilians. Wächter had turned to Jürgen Stroop to advise on the formation of the new division. The first commander of the division was SS-Brigadeführer and Generalmajor der Waffen-SS Walther Schimana. As HSSPF Central Russia after 1942, Schimana had fought Himmler's war in Belorussia, where anti-partisan and anti-Jewish actions were closely intertwined. But Schimana was not 'hard' enough for Himmler. In October he replaced him with SS-Oberführer Fritz Julius Gottfried Freitag. Freitag had joined the Waffen-SS in 1940 and been assigned to Himmler's personal staff then taken part in the mass killing of Jews in the Pripet Marshes in 1941. Later he was appointed commander of the 2nd SS Brigade. By the time he took over the SS 'Galizien', Freitag was an experienced *génocidaire*. He made sure that his subordinates had equivalent experience: Franz Lechthaler, for example, formerly commander of the 11th Battalion of the German reserve police, had led Lithuanian auxiliaries in Kaunus and oversaw the murder of 5,900 Jews in Belorussia; Siegfried Binz, another anti-partisan 'expert', had reportedly liquidated more than 10,000 alleged bandits and Jews; Friedrich Dern had also served with notorious SS brigades.[41] These German commanders would be charged with moulding raw Ukrainian recruits into dedicated 'bandit hunters'.

SS training was harsh, brutal and basically pitiless. SS 'Heidelager' was a primitive affair, much of it still under construction, with the crudest facilities. Built by slave labour on a flat, sandy triangle of land at the confluence of the San and Vistula rivers, the camp embodied the SS empire in miniature. Here, Waffen-SS officers trained their foreign recruits as the new German rockets were tested at a firing range nearby. Flemish Belgians drilled alongside Bosnian Muslims, Estonians and Latvians, all clad in SS uniforms. For the Germans assigned to the 'Galizien', fine distinctions between Galicians and Ukrainians counted for nothing. Neither the Germans nor the Ukrainians received any kind of language instruction; only those Ukrainians who had served with the Austro-Hungarian army spoke German with any kind of proficiency. For most recruits, SS discipline was a shock. One recruit recalled that his platoon commander 'was always ill tempered, excelled in yelling, abusing and insulting, and was always trying hard to persecute and abuse us'. The abusive drill sergeant has always been a cliché of the military life, but this particular martinet, one Scharführer Brandscheit, did not trouble to disguise his hatred for Ukrainians. On Sundays, the recruits chose between Mass or a bath, inspiring Brandscheit to laugh heartily: 'I don't know which is dirtier, your souls or your arses.' Wolf-Dietrich Heike was one of a small number of German army (as opposed to SS) officers who served with the SS 'Galizien'. In his memoir, Heike talks about the clash between Germanic order and Ukrainian spontaneity: 'Among Ukrainians emotion overshadows reason. Not reason, but emotions that well up from the depth of the soul constitute the leitmotif of the Ukrainian's life. This characteristic seems to apply to all Slavs.' Heike's Ukrainian driver had a temper 'like a powder keg' and drove accordingly. Himmler's Galicians, Heike recalled, never forgot for a moment that they were Ukrainians – certainly not Galicians. In their barrack huts and vegetable gardens, the trident, the Ukrainian national symbol, popped up everywhere, much to the German officers' displeasure.

The commander, Freitag, enjoying pouring salt on Ukrainian wounds. He was ill tempered and volatile, possessing, it seemed, a talent for immiserating the daily lives of his subordinates and tormenting non-German recruits. He viewed his new assignment as punishment for some obscure misdemeanour. He had already served briefly with the Latvian SS 15th Division and had made no secret of his contempt for 'foreigners'. Freitag demanded more German officers and removed allegedly unreliable Ukrainians. One SS veteran he recruited was SS-Obersturmbannführer Franz Magill, another veteran of the Pripet Marches special action. Since those glory days with the SS cavalry, Magill had fallen on hard times. The SS 'Galizien' was not a prestige posting.

Many of the Ukrainian officers, like SS-Hauptsturmführer Michael Brygidyr, had previously served in SS Schuma battalions, routinely used to kill partisans, burn down villages and, when the opportunity arose, murder Jews. Although Freitag, Dern and the other German officers distrusted their Ukrainian colleagues, they shared this bond as SS executioners. At the hub of this uneasy alliance was the sinister figure of SS-Hauptsturmführer Dmytro Paliiv (b. 1896) – an unnervingly hollow-cheeked, almost skeletal former journalist who liaised, semi-officially, between the Military Board and the choleric Freitag. In the late 1920s, Paliiv had split with the OUN and formed the Front of National Unity. He was, even in Ukrainian terms, an extremist 'integral nationalist'. In the early 1930s, Paliiv had published a series of articles whose coarse and brutal anti-Semitic content has been compared to Julius Streicher's ravings in the German propaganda rag *Der Stürmer.*[42]

By the winter of 1943/44, Hitler's eastern empire was crumbling. At the beginning of the new year, General Governor Hans Frank warned delegates at a security conference that the front was just 50km away. (A year before, the distance had been 1,700km.) On the borders of the General Government in the 'Distrikt Galizien', tens of thousands of refugees fled the Soviet army into Reich territory, carrying with them the menace of typhus and cholera. In January, the Soviets launched a major new offensive in Ukraine, putting tremendous pressure on Wächter's fiefdom.[43] As Stalin's armies pushed relentlessly westwards, Soviet-backed partisans intensified their campaign. Now the Wächter's SS 'Galizien' would receive its baptism of fire and blood.

In February 1944, the 2nd Battalion of the 4th Galizien SS-Freiwilligen-Regiment was assigned to anti-partisan duty in the Cholm district, near Tarnopol, a region governed by Poland until 1939. After 1941, German rule had sparked a vicious civil war between Polish insurgents and the Ukrayins'ka Povstans'ka Armiya, the UPA. The Germans and their Ukrainian auxiliaries regarded many Polish villages in the region as bandit strongholds and as a matter of course punished them accordingly. According to the *Chronicle of the Halychnya Division* (Galizien division), held by the State Archive of Kiev, on 23 February the 'Galizien' regiment attacked the village of Huta Pieniacka. A Polish brigade killed two Ukrainians, Roman Andrichuk and Oleksa Bobak, and allegedly mutilated their bodies. When the news reached Wächter's headquarters in L'viv he was enraged and organised ceremonial funerals for the two men. Five days later the Germans and the Ukrainian 'Galizien' men returned to Huta Pieniacka thirsting for revenge. They encircled the village and subjected it to a brief artillery bombardment. Then they closed in. They herded villagers into barns and set them on fire. Eyewitnesses watched horrified as the Ukrainian men dashed children against walls and cut open the stomachs of pregnant women. The entire village was torched. The Germans and their Ukrainian executioners tormented villagers until nightfall. By then most were drunk and returned to camp bellowing nationalist songs.

The massacre was first reported a few days later by Polish sources. Ever since, Poles and Ukrainians have argued bitterly about what happened and who killed between 800 and 1,000 mostly unarmed people in a single day. The attack on Huta Pieniacka precisely fitted Himmler's Bandenbebekämpfung doctrine. Huta Pieniacka was targeted for two reasons: it was Polish and the SS had discovered that the villagers were sheltering Jews. We can establish this as certain fact because one of the Jewish survivors of the attack left a memoir, which has been published by Yad Vashem.[44]

Its author, Zvi Weigler, had been born in the East Galician town of Sasov. In 1943, he and a number of other Jews had escaped from a labour camp in Zlochov. Hardened though they were by ghetto and labour camp life, Weigler and his companions now faced a battle for survival, especially in the winter. The Jews came to depend on Polish farmers and on the generosity of two villages in particular. One was Huta Werchobuska and the other: Huta Pieniacka. In his memoir, Weigler tells us that the Polish farmers and villagers provided shelter and food even though Governor Wächter warned village heads (*soltisses*) that any village found to be sheltering Jews would be collectively punished. Weigler and the other Jewish fugitives always shunned Ukrainian villages, fearing that they would be betrayed to the SS.

That February, the Polish resistance reported that the Germans had begun planning a punitive action against Huta Pieniacka and other villages for providing aid to partisans and Jews. As soon as he heard the news, Weigler and other Jews retreated back into the forest. Then Weigler states: 'The warning ... was not an empty threat. The punitive action came in February, 1944.' Weigler and the other Jews fled after the first attack. Then on 28 February, Weigler watched from his forest shelter as German and Ukrainian SS men returned to Huta Pieniacka. As other eyewitness accounts have stated, the village was surrounded then raked by machine-gun fire from all sides. The SS men hurled hand grenades through the doors of houses then herded the farmers and their families into barns. They slammed shut the doors and set them on fire. The Ukrainians stood by to make sure that no one escaped. When night fell, Weigler and the Jews who had hidden in the forest returned to the still burning village and began searching for anyone who remained alive. The following morning, Polish farmers from nearby villages arrived with wagons and took the handful of survivors to the hospital in Brody. Weigler later discovered that few had survived. Three weeks later, on 23 March, Huta Werchobuska 'suffered the same fate'.

That a massacre happened in Huta Pieniacka has never been disputed. But Ukrainian SS veterans and some historians continue to raise doubts about the role of the SS 'Galizien', and some have attributed the atrocity to a German Schutzpolizei battalion. This phantom has never been identified.[45] Since 1989, memorials erected to the victims at the site of the village of the attack have been vandalised, rebuilt then vandalised again. Huta Pieniacka remains a painful running sore. The hard evidence that the SS 'Galizien' did take part can be found in a Ukrainian document. In 1944 the Ukrainian Military Board met in L'viv and heard testimony from a Captain Khronoviat who had encountered the 'Galizien' men shortly after Huta Pieniacka was attacked. His report has been published by the Canadian Ukrainian Veterans' Association, presumably because it shamelessly exonerates the massacre as an act of revenge:

> The attack began at 6 a.m. ... The soldiers fought well. Inhabitants of Huta got it's 'fame' [sic]: they were maltreating the Ukrainians, murdering our peasants, they tore out the jaw of one of our priests. The village was set on fire. Every house was hoarding ammunition, cracking was heard, grenades were bursting. By the way, the Jews were hiding in the village.[46]

Khronoviat's report removes any doubt that Ukrainian SS men attacked Huta Pieniacka and that one reason that they did so was that Jews had been reported in the village and neighbourhood.[47] This was not simply a *Vergeltungsmassnahme* (revenge action) provoked by the 'massacre' of the two soldiers. For Himmler and his commanders 'bandit warfare' was still a means of 'ethnic purification'. It will be recalled that at meeting convened by General Max von Schenkendorff to discuss anti-partisan tactics with Himmler's 'bandit expert' HSSPF Erich von dem Bach-Zelewski, SS doctrine was crystallised in a slogan: 'Where there's a Jew, there's a partisan and where there's a partisan there's a Jew.'[48] To the very end of the war, Himmler often stressed this bond between 'bandits' and 'Jewish-Bolshevism': 'In the concept of the partisan, Bolshevism tries to promote banditry to a national status. We have challenged this newly coined status by the Jewish-Bolshevik Untermensch [sub-humans] and have fought to remove the bandits from the population.'[49] As Hitler's armies retreated west, Himmler's war on bandits 'mopped up' survivors of the camps and ghettoes. Waffen-SS foreign recruits, whether they had volunteered in L'viv, Paris or Riga, needed little persuading that, as Himmler claimed, the Jew and the partisan were one and the same enemy – to be exterminated as vermin.

On 16 May 1944 Himmler visited the SS 'Galizien' at a camp in Neuhammer. After Wächter had shown off the prowess of his recruits, Himmler spoke to the officer corps in German, with Paliiv translating for the Ukrainians. Himmler offered congratulations to the German officers and men of the SS 'Galizien': 'the designation "Galician" has been chosen according to the name of your beautiful homeland ... [which] has become even more beautiful since it lost, through our intervention, those inhabitants who often sullied the name of Galicia, namely Jews.'[50]

Officers and NCOs, German and Ukrainian, applauded loudly.

Part Three:
March 1944–April 1945

13

'We Shall Finish Them Off'

> For the past five weeks, we have been fighting for Warsaw ... We'll get through and then Warsaw, the capital city, the brain, the intelligence of this ... Polish nation will have been obliterated.
>
> Himmler, 21 September 1944[1]

In the spring of 1944, a succession of catastrophic hammer blows overwhelmed Hitler's armies along the Eastern Front from the Baltic in the north all the way to the Black Sea in the south. The armoured might of Stalin's armies remorselessly drove the Germans back along every sector of the Eastern Front. In the Crimea, the Russians cut off 120,000 German and Romanian troops and crushed them without mercy. In May, Stalin and his generals turned their attention to the central Belorussian sector of the German line that had been caught in a pincer movement to north and south leaving a giant protrusion eastwards. A massive new strategic push, Operation Bagration, named after a Georgian prince, was set in motion to hurl the Germans back across the Polish border and into Romania. As Marshall Georgi Zhukov mustered prodigious numbers of troops, tanks and artillery, Soviet partisans unleashed a wave of deadly attacks in the German rear, targeting railway lines and roads to cut off supplies and reinforcements. As the Wehrmacht battled to restore order, a million Soviet troops closed in on the German bulge, backed by deafening barrages of Katyusha rockets, known as 'Stalin Organs' by panic-stricken German troops. Zhukov's forces gobbled up territory, driving ever closer to the strongholds of the General Government. On 17 July, a supremely confident Stalin staged a victory parade in Moscow to show off 57,000 shamed and wretched German prisoners of war.[2]

The spectacular success of the first phase of Operation Bagration shattered the German front line. By the autumn, Soviet forces had pushed into the Baltic overwhelming German defences in Latvia and Estonia. On 23 July, Soviet troops crossed the border of the General Government and encircled the old Galician capital of L'viv. Three days later, the local governor Otto von Wächter, who had recruited the SS 'Galizien' division, telexed Hans Frank in Kraków to announce that he had lost control of the 'Distrikt Galizien'.

Hitler's Axis began to shed allies. In Romania, Marshall Ion Antonescu, who had masterminded the slaughter of Jews in the summer of 1941, was ejected from power by Carol's son King Michael who signed an armistice in Moscow on 12 September. German forces began pulling out of the Balkans, and Bulgaria belatedly declared war on Germany. In Hungary, the regent Admiral Miklós Hórthy, who had been reluctant to commit more than a few light divisions to the German war effort, began secret peace negotiations with the Allies, provoking a full-scale German occupation in March. The German ambassador and Plenipotentiary Edmund Veesenmayer appointed a new government headed by the compliant Dominik Sztoja to keep Horthy in line. For the Germans, Hungary was unfinished business. Its Jewish community was still largely intact. Goebbels had always believed that Hórthy was 'reliable' with respect to what he called the 'rhythm of the Jewish question'. The Hungarian was, he said, 'murderously angry with the Jews'. But Hitler was convinced that his Hungarian ally had dragged his feet; with a more compliant regime in Budapest the matter could be settled at last. That summer, Adolf Eichmann arrived in Budapest with a Sonderkommando to begin organising deportation of Hungarian Jews to Auschwitz. Hórthy prevaricated. In October, with the Soviet army fast approaching his borders, Hórthy sent General Béla Miklós de Dálnok to negotiate an armistice directly with the Russians. Enraged, Hitler dispatched Otto Skorzeny to Budapest with orders to depose Hórthy. Skorzeny kidnapped the astonished regent and flew him to Germany. In the meantime, Arrow Cross fanatic Ferenc Szálasi, acting on Hitler's orders, seized power and took control of the Hungarian army. Arrow Cross death squads began murdering Jews and Eichmann returned to Budapest to begin his thwarted programme to deport the surviving Hungarian Jews to Auschwitz. In December, the Germans incarcerated thousands of Jews in the Budapest ghetto, where they were subjected to unrelenting deadly assault by SS units and their Arrow Cross allies.

As Hitler's Reich shrank, Heinrich Himmler's power as SS chief bloated. In July, a colonel called Claus Schenk Graf von Stauffenberg botched a plan to assassinate Hitler at his headquarters near Rastenburg. In the aftermath of the bomb plot, Hitler lashed out at the Wehrmacht and Himmler, and Gottlob Berger seized the chance to 'gather up' all the 'foreign' units and divisions into the Waffen-SS. Their catch included the ill-fated Indian Legion, recruited by nationalist Subhas Chandra Bose, as well as hundreds of thousands of Osttruppen, including Russians, Turkmen, Azerbaijanis, Armenians, Mongolians and Georgians. In September, Himmler turned the most ruthless of these new SS warriors against the detested Polish capital of Warsaw.

During that last summer of Hitler's Thousand-Year Reich, chronic stomach pain compelled Himmler to repeatedly summon his Baltic German masseur Felix Kersten, who often found his patient bedridden and tormented by gastric cramps. Himmler screamed: 'I can't bear this pain any longer.' At one visit, Kersten noted that Himmler had a copy of the Koran open on his bedside table.[3] But Himmler's public rhetoric still replayed the fatuous old myths. On 26 July, he addressed a new division of SS 'Volksgrenadiers' (Infantry Division 545) and gloried in 'a belief shaken by nothing, the belief in the Führer, the belief in the future of this greater Germanic Reich, the belief in our own worth, in ourselves'.[4] In Himmler's imagination, the Greater Germanic Reich clung tenaciously to life. So too did his impassioned conviction that providence would look after the bearers of Germanic blood. Himmler's twisted optimism was nourished by the fantasy that new super weapons, *Wunderwaffen*, could soon be unleashed on Stalin's 'Asiatic' hordes. Like many in Hitler's court, Himmler hoped that the alliance between Bolshevik Russia and the British and Americans would sooner or later break down; then he, rather than the plainly ailing Führer, would be called upon to lead a new Germany.

As Kersten waged war on Himmler's stomach cramps, Soviet armoured divisions ground relentlessly on towards the River Vistula, the last major natural barrier before they reached the borders of the Old Reich. Less than a quarter of the General Government now remained in German hands. By August, the Soviet vanguards had begun to approach the east bank of the Vistula. On the other side lay the city of Warsaw. The Polish capital had been Hitler's first prize in the 'war of annihilation' that had begun on 1 September 1939. 'The fact that we are governing,' Himmler informed General Keitel, 'should enable us to purify the Reich territory of Jews and Polacks.' In the course of the next five years, German rule killed 6 million Poles, more than half of them Polish Jews.[5] The Germans had turned occupied Poland into a slaughterhouse. Auschwitz and the Reinhardt death camps were all built on Polish soil. By the summer of 1944, Soviet General Rokossovsky and the hard-fighting First Byelorussian Front had reached Brest-Litovsk, where Hitler's troops had massed on 22 June 1941. That month, Soviet troops entered the Lublin district which had been the epicentre of Operation Reinhardt, the systematic mass murder of the Jews in the occupied east. In 1943, Himmler had ordered

'Action 1005' to remove all traces of the mass murder camps, like Treblinka, or turn them into forced labour concentration camps. Despite the best 'cleansing' efforts of the Germans, enough damning evidence remained to shock the Soviet soldiers and journalists who discovered this ruined archipelago of death. Then the Soviet forces and a small pro-Soviet Polish army led by General Zygmunt Berling pushed on towards the former Polish capital Warsaw, situated at the confluence of the Vistula and Narev rivers.

As the Soviets came ever closer, a grotesque procession began pouring back across the Vistula bridges and then through the old centre of Warsaw. Clad in frayed and tattered bloodstained uniforms, German SS men fled west alongside Ukrainians, Hungarians and Cossacks. Few had any transport; most walked or hobbled painfully. Extreme fatigue and hunger scarred their mud-caked faces. Few still carried arms. In the wake of these shattered relics of Hitler's war machine came a much bigger wave of civilian refugees. Inside Warsaw, the Germans began dismantling factories and sending industrial plants back to the Reich. Although Hitler had refused to arm any Poles, the German SS governor Ludwig Fischer harangued the Warsovians to 'demonstrate their anti-Bolshevik sentiments' as they had when they sent Lenin's army packing in 1920. Warsaw, he ranted, had become the 'breakwater for the Red flood'. He ordered Poles to begin building defences. 'One hundred thousand volunteers immediately!' But not a single Pole was prepared to lift a finger to defend the hated German garrison.

As this human tide streamed through and past Warsaw, SS staff began incinerating the accumulated paperwork that documented five years of occupation. On 29 July, Poles watched with astonishment as immaculately turned out (and well-fed) soldiers of the Heeresgruppe Weichsel that comprised the Hermann Göring Panzer Division and the SS 'Wiking', its ranks crowded with Swedes, Estonians, Danes and Norwegians, marched east across the Vistula and through the satellite town of Praga on the east bank. Hitler's multinational armies dug in to defend Fortress Warsaw (Festung Warschau) and repel the 'Red tide' that bore down towards the city – unaware that another hostile army was being mustered in their midst.

From his secret command centre in an old tobacco factory, Polish general Tadeusz Komorowski (codename *Bór*, the forest) had mustered a secret 'home army', comprising some 40,000 fighters. For months, Komorowski had been impatiently observing the retreat of demoralised German soldiers and refugees. Now, surely, it was time to strike hard and fast against the crumbling Reich. Born in L'viv, Komorowski (who had served in the Austrian army and spoke perfect German) was convinced that he and the Polish government in exile could not afford to dither. The Polish resistance was divided. Stalin, who had joined forces with Hitler in 1939 to dismember the Polish state, now backed the communist Polish Committee for National Liberation to ensure that, after the destruction of the Reich, Poland would be securely locked up inside a new Soviet empire. As the war raced to a climax in the east, General Bór and his advisors concluded, after impassioned debate, that seizing the former Polish capital provided the only way to resist the Soviet tide. So they began to hatch up a plan codenamed 'Tempest' to seize the initiative and destroy the German garrison. Tempest was planned in secret and no attempt was made to warn the ordinary people of Warsaw of the fire storm that would consume their city. In his underground fortress, General Bór fretted and argued.[6]

It had been a blistering summer. Now as the leaves in Warsaw's battered public parks began to turn, Marshall Rassovetsky's artillery could be heard growling on the eastern horizon. Warsovians apprehensively gazed upwards as Soviet aircraft roared low over the city with increasing frequency. The pressure on the home army leaders was almost unbearable. A decision had to be made – and soon. At 6 p.m. on 31 July, General Bór, after last-minute agonising, sent home army runners out across the city with orders to launch the revolt the following day at 5 p.m. – codenamed W-hour (*wybuch*, outbreak). At W-hour, Warsovians would be leaving work to get home or dashing to cafes and bars, and the city streets would be packed. General Bór hoped that the timing would catch German garrison troops off guard. As the home army messengers, most of them women, fanned out through every district of Warsaw, it was as if a tremendous electrical current hummed and then flashed from point to point. As the clock ticked down to W-hour, streets and trams began brimming with people hurrying homeward in the sultry heat, oblivious to the troglodyte home army units dashing to their positions beneath their rushing feet. A few shop and cafe owners had been forewarned and they slammed shut their doors and shutters.

The Warsaw Uprising began precisely on schedule at 5 p.m. Home army fighters poured out of their hideouts and on to the streets and some 180 strategic German positions came under attack within minutes: bridges, aircraft runways, the railways stations and military and police headquarters. The fighting was intense and bloody. Less than half of General Bór's troops had working firearms. But few hesitated to hurl themselves at the hated occupiers. They threw up street barricades, made from anything that lay to hand: bricks, furniture, street-carts, even typewriters and picture frames. In the first few hours, the Germans reeled. By 8 p.m., three hours after W-hour, a number of landmarks had been captured, including the trophy prudential high rise, the main post office and some city power plants. Many German strongholds held out, but as darkness fell, the Polish national flag could be seem fluttering fitfully from the prudential tower.

Telex reports from the besieged German garrisons in Warsaw flooded into General Governor Frank's headquarters in Kraków. On the fragile German front line, Warsaw occupied a pivotal position – and the home army had already severed vital supply lines to the German troops who must somehow resist the Soviet onslaught on the eastern side of the Vistula in Praga. German dismay soon metamorphosed into craving for vengeance. The last entry in Frank's diary, made on 5 August, reads: 'The city of Warsaw is in flames for the most part. The burning of houses is also the surest method of getting insurgents out of bolt holes. After this insurrection and its crushing, Warsaw will be completely destroyed as it deserves.'[7]

The uprising had alarming implications. The Soviet army might join forces with the Polish rebels and turn a blaze into a conflagration. General Heinz Guderian, the new Wehrmacht Chief of Staff, urged Hitler to immediately remove Warsaw from Frank's jurisdiction and make the city a militarised zone, under Wehrmacht control. Hitler, who was still in a state of shock following Colonel von Stauffenberg's explosive visit to the Wolf's Lair, brushed Guderian's suggestion aside.[8] Instead, he summoned his SS chief. The destruction of the Polish people that had begun in 1939 could now be finished once and for all. News of the uprising had enraged Himmler. At a meeting of some very worried Gauleiters in Poznan, he had insisted that the uprising changed nothing: '[Racial reconstruction] is irreversible ... It is irreversible that we create a garden of Germanic blood in the East.' He then travelled to Rastenburg for a crucial meeting with Hitler, and reported to high ranking SS officers: 'I should like to tell you this as an example of how one should take news of this kind quite calmly.' 'The moment is a difficult one,' he had somewhat pointlessly informed Hitler. But, he had continued, the uprising was also 'a blessing'; an opportunity, above all, to punish the Poles who have 'blocked us in the East for seven hundred years and stood in our way since the first Battle of Tannenburg'. He vowed to end the 'Polish problem' forever: 'In five or six weeks it will all be behind us. Then Warsaw will have been extinguished, the capital, the head, the intelligence of 16 to 17 million Poles.' The only possible remedy was total destruction: 'Every block of houses is to be burnt down and blown up.' Impressed by Himmler's resolve, Hitler ordered him to crush the uprising with maximum force. Himmler blithely remarked: 'You may well think that I am a frightful barbarian. I am, if you like, when I have to be.'

Back in Poznan, Himmler summoned his most battle hardened and dedicated SS generals. Heinz Reinefarth had been born in Germany's eastern borderlands and studied law at Jena University, where he had also acquired an impressive collection of duelling scars. Reinefarth had enjoyed a glittering SS career, serving in Bohemia-Moravia and at the Order Police Main Office in Berlin. At the beginning of 1944, Himmler had appointed him SS and Police Leader (SSPF) in the Reichsgau 'Wartheland', where he had energetically pursued the task of murdering any recalcitrant Poles. Himmler ordered Reinefarth to form a battle group (Kampfgruppe) and proceed immediately to Warsaw where he would join forces with SS General Erich von dem Bach (formerly Bach-Zelewski). It will be recalled that it was Bach-Zelewski who, in July 1941, had masterminded the destruction of Jewish villages in the Pripet Marshes and who had later been appointed Chief of Bandit Warfare (Chef der Bandenkampfverbände, Ch.BKV) in June 1943. Bach's codename was 'Arminius', after the Germanic warrior who defeated the Roman legions at the Battle of the Teutoberg Forest. His chubby appearance and professorial manner belied his expertise as a ruthless killer of Jews and 'bandits'.

In 1959, when German prosecutors finally caught up with Bach-Zelewski, who had been reduced to working as a parking garage guard, he provided a detailed account of the orders he had received from Himmler: 'captured insurgents must be killed' whether or not they are fighting 'in accordance with the Hague convention'. Even those 'not fighting' – the women and children – 'should likewise be killed'. All Warsaw, Himmler instructed, 'must be levelled to the ground'. Nothing could be left standing. The razing of Warsaw would provide a lesson for the rest of Europe: 'then,' Himmler concluded, 'the Polish problem will no longer be a large problem historically for our children, who come after us, nor indeed for us.' Bach-Zelewski's defence strategy was the most banal. He was just 'following orders' and, as the court soon discovered, he had taken Himmler's writ as gospel.

To accomplish this monstrous task, Himmler chose his forces with care. He wanted 'hard men' who would relish the duty. At his headquarters, Himmler set about building an SS army of vicious German criminals who would be set to work alongside vengeful 'Eastern troops' among them Cossacks, Azerbaijanis and a few Ukrainians. These renegades would join forces with other European SS volunteers, serving in the 'Wiking' division already stationed on the east bank of the Vistula. Bach's Korpsgruppe (that incorporated Reinefarth's Kampfgruppe) comprised some 8,000 men backed by diverse SS and police battalions.[9] Himmler and Bach assigned a vanguard role to the notorious SS 'Dirlewanger' brigade, commanded by a perverted fanatic called Dr Oskar Dirlewanger. The Sonderkommando Dirlewanger recruited hardcore criminal types dredged from German prisons and army punishment cells: poachers, petty criminals, SS men on punishment duty and a few hundred foreign SS recruits who had ended up on probation. The lugubriously featured Dirlewanger had once been convicted of raping a minor. His good friend SS recruitment chief Gottlob Berger had got the charges dropped. Dirlewanger and his vicious crew had proved their worth to Himmler's 'bandit war' on countless occasions, most recently in Slovakia where they had fought alongside the 'Galizien' SS division. At his trial, Bach-Zelewski defended Dirlewanger's cutthroats: 'Although their moral qualities left much to be desired, their fighting ability was extremely high. They had nothing to lose and everything to win. They gave no mercy in battle and did not expect any.' He insisted that: 'To remove [the Dirlewanger Brigade] from the battle would have been nothing less than to give up any idea of an offensive.'[10] He demonstrated, perhaps inadvertently, that the military ethics of the Dirlewanger Sonderkommando defined the ethical horror of the SS assault on Warsaw: it was by intent a criminal act.

On 3 August, SS General Bach settled into an abandoned Polish mansion at Ożarów, 8 miles west of Warsaw's Old Town. From here, he would direct operations under the blank gaze of long-dead landowners, whose portraits still hung on the dirt-streaked walls. Bach reinforced Dirlewanger's two battalions with an Azerbaijani regiment (the 111th) and added to this poisonous mix three Cossack regiments, two more Azerbaijani battalions, the 22nd SS Cavalry Brigade 'Maria Theresa', made up of Hungarian ethnic Germans, and the 29th Waffen Grenadier Division RONA, led by a Soviet deserter SS-Oberführer Bronislaw Kaminski. In Warsaw, the RONA men showed no mercy or restraint. How had these men and their brutish commander ended up serving in Himmler's SS?

In a speech made to Waffen-SS troops in Stettin in July 1941, Himmler elaborated on the meaning of 'struggle of races': 'beautiful, decent and egalitarian', Germany faced 'a mixture of races, whose very names are unpronounceable, and whose physique is such that we can shoot them without mercy or compassion'. He then spelt out his core argument: these races had been 'welded into one religion, one ideology, that is called Bolshevism' by Jews.[11] In 1941, SS murder squads targeted Turkmen who had survived the Soviet liquidation of independent Turkestan. The Einsatzgruppe reports also refer to the execution of 'Asiatics' and in Kiev, German physicians used lethal injections to murder 'Turkmen' and 'low grade Caucasians'.[12]

In Germany, powerful voices rose to protest against this plainly short-sighted and murderous policy. Not all 'Asiatics' were Bolsheviks and Jews, they insisted; many Central Asian peoples were devout Muslims who hated Stalin as much as any decent German. Gerhard von Mende was the most prominent member of this cabal and he would revolutionise German relations with the peoples of Central Asia and the Caucasus. Like Alfred Rosenberg, von Mende was a Baltic German and an ardent anti-Bolshevik.[13] When I interviewed Gerhard's son Erling in 2007, he remembered a frequently absent father who dressed with dapper good taste and possessed a mordant sense of humour. Born in 1904 in Riga, Gerhard had seen his father shot dead by Bolshevik soldiers when he had just turned 14. The family escaped to Germany, where Gerhard struggled to get an education; he finally entered Berlin University at a mature 24. His fellow alumni remember a blonde, blue-eyed and strikingly scrawny young man with a wandering left eye. In Berlin, then the holy grail of 'Oriental Studies', Gerhard's star rose fast. He was an able and hard-working scholar who spoke Latvian, Russian, Swedish, Turkish and different Turkic dialects. Germans had long been fascinated by the rich Muslim cultures threaded along the 'Silk Road' – a term that was invented by a German scholar. German ideologues like Rosenberg and scholars like von Mende despised the so-called 'Muscovite' centre, which had become, in their view, the headquarters of 'Jewish-Bolshevism' and looked instead to the more exotic and turbulent Muslim peripheries. For centuries, Imperial Russia had struggled to master their troublesome minority peoples – and the 'nationalities question' haunted Stalin's nightmares. The tsars had bequeathed to their Soviet heirs two sprawling regions where Russians formed a minority. Central Asia, which we refer to today as 'the 'Stans', comprised Kazakhstan, Kyrgyzstan, Tajikistan, Turkmenistan and Uzbekistan. In the mid-twentieth century, geographers simply referred to the entire region as 'Turkestan' and its Muslim peoples as 'Turkmen'. The Caucasus (known to linguistic historians as the 'Mountain of Tongues') had nurtured a profusion of ethnicities and distinct languages: from Christian Georgia and Armenia in the south to the Muslim enclaves of the north, inhabited by fractious and tribal-minded Dagestanis, Kalmyks, Chechens and Ossetians.

In 1936, von Mende published his magnum opus: *Der Nationale Kampf der Rußlandtürken: Ein Beitrag zur nationalen Frage in der Sovetunion.* He argued that the Turkic peoples (by which he meant Uzbeks, Kazakhs, Kyrgyz and Tatars) might, with suitable guidance, provide a kind of reserve army of anti-Bolshevik forces. He argued that these fissiparous peoples were unlikely to turn into genuine nation builders. Furthermore, the Soviet Union would need to suffer a 'severe shock' before any Turkic peoples dared rise up against their 'Muscovite' oppressors. In 1933, von Mende had joined the Sturm Abteilung (SA) (simply because NSDAP membership lists had been temporarily closed as opportunist 'March Violets' rushed in their application forms) and began using his SA contacts to make friends with influential party members in the revamped German ministries. Although his son Erling von Mende vehemently denies the fact, his father was an unashamed anti-Semite, denouncing the 'exceptional Jewification of ... the Soviet Union' in a pamphlet and passing on damaging information about a Jewish colleague who worked at the Reich Education Ministry.[14] This contempt for Jews infected many of von Mende's academic publications, and no doubt he shared these views with his many Turkic colleagues and friends. Von Mende formed a close bond with a community of Turkic exiles known as the Prometheus Movement – originally founded to rid the Russian Empire of Russians. Based in Paris and Warsaw, the Prometheans had reformed as an anti-Bolshevik faction. They referred to Mende as 'Lord Protector' – and even during the period of Nazi-Soviet rapprochement he helped his friends set up 'national committees' with ambitions to become governments in exile.

Immediately after the German invasion of the Soviet Union, Gerhard von Mende was recruited by Alfred Rosenberg's Eastern Ministry, where he took charge of the Caucasus division, working under Georg Leibbrandt, who later represented the Ministry at the Wannsee Conference. At the Eastern Ministry headquarters, located not far from his new home in prosperous Charlottenberg, von Mende recruited staff from among his Promethean circle of friends. One of these men was an Uzbek called Prince Veli Kayum Khan, who would become von Mende's most important protégé.

At the end of 1941, von Mende received reports that described the treatment of Soviet POWs in the vast German camps constructed in the General Government. With Rosenberg's backing, he sent Kayum Khan to investigate. Accompanied by Mustafa Chokai, Khan arrived in the Kraków district of the General Government at the end of the year. Inside vast, sprawling camps the Germans had caged more than 3 million Russian POWs.[15] It was the dead of winter, and the German gulag was a perilous realm even for visitors. The unfortunate Chokai was struck down by typhus soon after he arrived. Kayum Khan battled on: he reported that he had discovered thousands of fellow Muslim Turkmen slowly starving to death in appalling conditions. Most of the prisoners had been forced to dig holes in the ground to use as pitiful shelters. Every day, he was informed, SS execution squads roamed the camps searching for prisoners who had been circumcised or had 'slit eyes'. The SS squads removed these 'subhumans' and shot them in nearby woods. With von Mende's backing, Kayum Khan successfully petitioned the German authorities to improve the conditions for 'Turkic' prisoners and had the better educated transferred to a training camp near Berlin. Here Kayum Khan began moulding these former Soviet citizens into an anti-Bolshevik army.[16] He had another backer as well as Rosenberg: Germany army intelligence, the Abwehr, which set up Operation Tiger B to manage the training of the Turkmen.

Gerhard von Mende and Prince Veli Kayum Khan tried to re-educate the German Wehrmacht commanders, who regarded every Central Asian as a dangerous 'Asiatic'. They had great success. In December 1941, the OKW authorised the formation of two Muslim units: the Turkestan Legion and the Caucasian

Muslim Legion. Recruits exhibited striking diversity: Uzbeks, Kazakhs, Kyrgyz, Karakalpaks, Tajiks, Azeris, Dagestanis, Chechens, Ingusges and Lezgins all joined up.[17] In early 1942, as Hitler's forces thrust deep into the Crimea, Rosenberg and von Mende set up liaison offices to co-ordinate recruitment of other Osttruppen units. In the Crimea, the new campaign proved another triumph for von Mende: at least 20,000 Tatars volunteered, accounting for the entire male population aged 18–35 not already conscripted by the Soviets.

In early 1942, Hitler himself made a remarkable pronouncement at his military headquarters: 'I consider,' he said, 'only the Mohammedans to be reliable. All the others [Caucasians] I consider unsafe. It can happen [to us] anywhere – one has to be incredibly careful. I consider setting up units of purely Caucasian peoples to be very unwise.'[18] The new recruits offered another advantage: many were fiercely anti-Semitic. According to German diplomat Otto von Bräutigam, the Turkic recruits took a keen interest in attacking any Jews they encountered.[19] Assisted by von Mende and Kayam Khan, Wehrmacht propaganda units enlisted the brightest and best Turkmen to staff liaison offices to churn out newspapers and pamphlets, many saturated by anti-Jewish sentiment.[20] On 20 January 1942, Georg Leibbrandt, von Mende's boss, took part in the Wannsee Conference to plan the 'Final Solution of the Jewish problem'. A week later, Leibbrandt called follow-up meetings at the Eastern Ministry – all attended by von Mende. The purpose was to discuss 'the definition of the term "Jew"' in the Eastern Territories'.[21]

In the second half of 1942, the German summer offensive Operation Blau brought the North Caucasus under German occupation, with Army Group A under Field Marshall Wilhelm List achieving deepest penetration in mid-November. German plans to recruit Soviet minorities were stepped up, alongside the destruction of entire Ashkenazim Jewish communities by Einsatzgruppe D and German army units.[22] After the debacle of Stalingrad in early 1943, it became increasingly difficult for von Mende to convince his Turkic protégés like Kayum Khan that the Reich could guarantee them political independence. This dilemma worsened, naturally, when German forces began their long retreat and the Soviet Army advanced into formerly occupied regions of Central Asia and the Caucasus. This meant that von Mende increasingly emphasised the pan-Islamic identity of his eastern recruits. In 1943, he turned for help to the increasingly influential Haj Amin el-Husseini, the Grand Mufti, who was by then on excellent terms with SS Chief Himmler and his recruitment head, Berger. As we have seen, the Mufti regarded German anti-Jewish ideology as the vital common ground between Muslims and the Reich. Now he would help groom von Mende's eastern warriors as Islamic crusaders. By the spring of 1943, no fewer than twenty-one Osttruppen battalions, commanded by thirty or more German officers, had been thrown against Stalin's armies in the Caucasus. It was the evident eagerness of the Muslim recruits to shed blood for the anti-Bolshevik cause that especially intrigued Himmler. Now SS recruitment head Gottlob Berger began to covet the Wehrmacht's Osttruppen.

A vile Austrian major called Andreas Mayer-Mader would play a leading part transferring the eastern battalions to the SS. Mayer-Mader had lived in Asia for many years before the war, and served in Chiang Kai-Shek's Chinese National Army. His service record was not a distinguished one. Mayer-Mader was fascinated by the exotic peoples of Central Asia – and especially by Gerhard von Mende's beloved Turkmen. The German Wehrmacht commander in the Caucasus assumed that Mayer-Mader's 'Asian expertise' made him the perfect commander for the 'Turkestani battalions' that had been raised in the German POW camps in the General Government. Mayer-Mader's military competence was minimal, but he soon forged a close relationship with Kayum Khan and the 'Turkestan National Committee'. According to Wehrmacht reports, Mayer-Mader secretly backed Khan's plan to turn the battalion into a 'Turkestani National Army', staffed mainly by Turkmen officers and organised in different 'tribal' units. Under Mayer-Mader's command, the thuggish Turkmen troops had been let loose against Soviet partisans. They performed their duties with unremitting savagery, murdering unknown numbers of non-combatant civilians in the most sadistic manner. The lazy and venal Mayer-Mader appeared to have simply let his men run amok – and, for this, the Wehrmacht sacked him.

After his departure, Battalion 405 went to pieces, and in August 1942 the unit was disbanded and officers and men sent back to the Kalmyk steppe. Then at the end of 1943, the wily Mayer-Mader offered his services to Himmler. Mayer-Mader boasted that he could recruit at least 30,000 Turkmen to form a 'Neu-Turkestan' SS division, poaching most of them from Wehrmacht legions. In Berlin, Mayer-Mader, now an SS-Obersturmbannführer, conferred with the Mufti, who happily blessed his plan to form a Turkic-Muslim battalion and promised to promote recruitment – just as he had in Bosnia–Herzegovina. At the beginning of January 1944, Himmler amalgamated a batch of Wehrmacht 'Turkestanische' and 'Azerbaijanische' battalions with new Muslim volunteers as the 'Ostmuselmanische SS Regiment', numbering about 3,000 Turkmen, Azeri, Kyrgyzi, Uzbeki and Tadjiki volunteers.

A year earlier, in March 1943, Himmler had signed a far-reaching agreement with Rosenberg that assigned to the Waffen-SS responsibility for the ideological training of the eastern troops. The agreement referred to the 'strong instinctive anti-Semitism of the eastern nations'.[23] As he had with the SS 'Handschar', recruited in Bosnia, Himmler agreed to attach Imams to each SS Muslim battalion. The

Mufti handpicked most of these trainee Imams and sent them to the German Imam schools in Dresden and Göttingen. Inspired by the Mufti and Mayer-Mader, Himmler spoke excitedly to his SS officers about creating an entire SS division from the Turkic tribes – and he began sending a few Muslim units to Trawniki for special training.

As the war turned against the Germans and it became evident that Stalin, not Hitler, would win the war, these eastern recruits began deserting in large numbers, either to the Soviet army or to one of the thousands of partisan units now causing havoc behind German lines. By March 1944, Mayer-Mader commanded a relic eastern army, which was demoted to become the 1st 'Ostmuselmanische' SS Regiment, or the 94th SS Regiment. In reports sent to Himmler by RSHA agents, Mayer-Mader was denounced as a poor commander who promoted equally lazy and ineffectual officers. But Himmler remained convinced that this new Muslim SS regiment would be ideal for the 'bandit war' he waged in the east: burning villages, rounding up civilians for labour service, mopping up any surviving Jews and destroying crops. So he ignored the RSHA warnings and dispatched Mayer-Mader to Belorussia, which was overrun with Soviet partisans who played havoc with German supply lines. Himmler assigned the 94th SS Regiment to HSSPF for Belorussia, Curt von Gottberg. Under the overall command of Erich von dem Bach-Zelewski, Gottberg formed a new anti-partisan Kampfgruppe with the Turkic regiment, the Dirlewanger Brigade and the Kaminsky or RONA brigade. Gottberg had proved himself a ruthless anti-partisan commander, who in December 1942 had insisted to his officers that 'Each bandit, Jew, Gypsy is to be regarded as an enemy'. In the course of Operation Nürnberg, Gottberg claimed a kill of 799 bandits, and at least 1,800 Jews.[24] Now in the 'Bandengebiet' (partisan country) in the region near Minsk, Gottberg unleashed Mayer-Mader's rabble, Dirlewanger's gangsters and Kaminsky's brutes against partisan forces. Once Mayer-Mader's men had done their job a predictable chain of events took place. German officers attached to rival counter-insurgency forces reached Berlin, accusing Mayer-Mader of losing control of his forces. SS officers from Dirlewanger's units took over command, and on 2 May Mayer-Mader was shot dead near the town of Hornowo-Wiercinski. He was the single reported German casualty that day and it is probable that Dirlewanger, acting on orders from Himmler, had ordered his assassination. Order had been restored – for the time being.

The relics of von Mende's 'Turkic Muslim division' detrained on the outskirts of Warsaw at the end of July 1944, to wage war on the Polish home army at the side of the Russian Kaminsky Brigade (RONA) and the criminal troops of Dr Dirlewanger.

At the beginning of August 1944 Warsaw baked under cloudless, blue skies. The Vistula ran low and sluggish, a thin film of cracked mud forming on its banks. In Warsaw's parks, the parched lawns yellowed and tree leaves rattled like parchment. Inside their oven hot bunkers or hidden behind paving stone barricades some 600 Polish companies now awaited the German assault. General Bór had, of course, no artillery or air force – and his fighters would confront Himmler's SS forces with revolvers, antique rifles, a few hand grenades and homemade petrol bombs.

At the rail heads in the western suburbs of Warsaw, Himmler's fearsome war machine rapidly put on muscle. As SS generals Reinefarth and Bach-Zelewski completed preparations, Soviet guns positioned on the eastern side of the Vistula abruptly fell silent and Russian fighter aircraft vanished from the skies above Warsaw. They would not be seen again for many weeks. Stalin's mighty armies would not come to the assistance of the Polish home army.

On 5 August , the German assault began with a massive artillery barrage. Bach-Zelewski's strategy depended on seizing home army strongholds in the centre of Warsaw, located deep beneath the winding streets of the Old Town. Since Polish units still controlled the main road arteries that led from the western suburbs east towards the heart of the city, the SS 'Attack Group' would first need to smash enemy strongholds in the Wola district. Warsaw resembled a medieval vision of hell. A thick column of black smoke rose high above the Old Town. German garrison troops, still besieged in the Old Town, began shooting Polish civilians, usually on the pretext that home army insurgents had been spotted firing from their houses; a flood of terrified women, children and the elderly, streamed west – straight into the path of the SS assault. No mercy would be shown to any Pole armed or unarmed. Heading the attack, the Dirlewanger Brigade and their Azerbaijani accomplices battered their way through Wola with unrelenting savagery. One eyewitness saw German troops pushing screaming women and children back into a burning apartment block. Many Polish families fled inside Wola's old macaroni factory. It was a deathtrap. SS troops set fire to the factory then machine-gunned anyone who tried to escape or let loose their hunt dogs. At the macaroni factory 3,000 unarmed civilians are thought to have been killed. At another factory nearby, SS soldiers murdered some 5,000 people in the same way.

Hospitals too offered Himmler's SS warriors special opportunities to wreak havoc. At Wolski Hospital, German troops fanned out through the wards, shooting some of the sick and injured where they lay. Anyone who could walk or run they drove onto a nearby railway viaduct where a machine gun had been set up. The

Wolksi was burnt to the ground. Many other hospitals suffered the same dreadful fate. Just one escaped destruction. This was where Dr Dirlewanger set up a temporary headquarters.

The assault on Warsaw, in other words, was the climax of Himmler's Bandenbekämpfung – fought this time not in the forests and swamps of Belorussia and Ukraine, but in city streets by Chief SS Bandit Hunter, Erich von dem Bach-Zelewski. A German army observer recalled that Dr Dirlewanger seemed to be 'everywhere': 'a fellow with a pronounced, remarkable look of a hanging bird.' As they fought their way towards the Old Town, building by building, street by street, the Dirlewanger's men hurled both the living and the dead from apartment windows on to the streets where they 'lay still'. On another occasion, the same witness watched as Dirlewanger rounded up Polish women and children and drove them forward as a human shield.[25]

In the neighbouring western suburb of Ochota, Bach-Zelewski unleashed the RONA or Kaminsky Brigade. Commanded by SS-Brigadeführer Bronislav Kaminsky, these wild-eyed, habitually intoxicated young Russian men attacked city hospitals with as much relish as Dirlewanger's thugs. Inside the famous Madame Curie Institute, they robbed and looted patients, raped and murdered nurses. When the RONA men tired of this sport, they corralled any surviving patients and staff in the hospital garden. As the RONA men raced through the hospital setting it ablaze, SS men began firing wildly into the defenceless crowd of hospital workers and patients.

The RONA was the largest single group of non-Germans serving in Himmler's army. Like the Azerbaijanis, they began as German army recruits, not Waffen-SS men. In the summer of 1942, Wehrmacht commanders set up a *Selbstverwaltungsbezirk* (self-administration area) in the region, centred on the small town of Lokot, south of Bryansk.[26] When partisans stepped up raids in Lokot the following year, Kaminsky, a local strongman who was leader of the National Socialist Russian Workers Party, approached the Germans with a suggestion to form a local defence militia. Kaminsky was a swaggering loudmouth who impressed the Germans as someone who could get things done. Born in Vitebsk, Russia, to a Polish father and a German mother, Kaminsky studied chemistry in Leningrad and after graduating found work as an engineer. But in July 1935 he was arrested by the NKVD, accused of espionage and sentenced to ten years' penal servitude. Although he served half his sentence, and even joined the Soviet army when he was released, Kaminsky was fiercely anti-Bolshevik and welcomed the German occupation.

By the end of 1942, Kaminsky's RONA (Russkaia Osvoboditel'naia Narodnaia Armiia), or Russian Army of Liberation, had swollen to division strength, fitted out with all the trappings of a modern army at the cost to the German occupiers of 3–4 million roubles a month.[27] The RONA men rampaged through Kaminsky's fiefdom, burning villages and murdering anyone suspected of sheltering partisans. The German occupiers occasionally expressed mild disapproval. One disgruntled army officer complained that Kaminsky behaved 'like an African chieftain': he pilfered German supplies and, when challenged, was 'insolent'. He used German staff cars to drive his 'female companions' to the theatre and cinema while displaying a sign that read 'In the service of the Wehrmacht'. His generous fuel allowance was used taking *Hurenfahrten* (whoring trips). The German commanders, however, would hear nothing wrong. They spoilt Kaminsky with fine wines, spirits, cigars and, for his many female admirers, perfume.[28] Hitler sent a personal message of congratulations following an especially 'successful' engagement with Soviet 'bandits'.

In September 1943 the Germans closed down the Lokot experiment. Soon afterwards, as the Soviet army pushed ever closer, German troops began to pull back from Bryansk – and Kaminsky and his RONA troops followed, along with tens of thousands of their camp followers. Association with any of the Wehrmacht 'self-administrations' was a ticket to a gulag camp or a firing squad. With the Soviet army baying at their heels, the desperate RONA caravanserai fled to the Vitebsk region of Belorussia, and it was here that Kaminsky began courting the SS. By early 1944, the region was convulsed by Soviet partisan attacks. Although RONA was by now plagued by desertions, Kaminsky led a number of attacks on Belorussian villages and managed to impress Himmler, who awarded him an Iron Cross 2nd Class. In June, Himmler agreed to absorb the Kaminsky Brigade into the Waffen-SS as the Waffen-Sturm-Brigade RONA – later renamed (on 1 August) the 29th Waffen-Grenadier-Division der SS (russische no 1). Kaminsky himself was appointed Waffen-Brigadeführer and general-major.[29] At the end of July, Himmler dispatched the RONA to Warsaw.

As the SS attack group battered their way into Warsaw's western suburbs, the RONA SS men rampaged from street to street, building to building, dragging anyone they discovered alive from their hiding places for summary execution. Like malicious magpies, they ransacked the dead and dying, gloating over their hoard of gold watches, rings and dental fillings ... For elite German SS officers, Kaminsky's men were barbaric *untermenschen*. But Himmler had demanded the systematic plunder of the Polish capital and Kaminsky's troops, in their fashion, obeyed. As the RONA men murdered and robbed, at Warsaw's rail depots German auxiliary troops loaded Reichsbahn wagons with their booty and dispatched it to warehouses in Germany.

Even Bach-Zelewski was forced to report that 'The Poles are fighting like heroes'. He was painfully aware that jealous Wehrmacht commanders like Guderian had begun to grumble that the SS 'Attack Group' was taking much too long to crush the home army. The toll on home army companies rose every hour, but in the Old Town the famous statue of King Sigismund III still stood proudly on his marble plinth: a precarious symbol of resistance that taunted the German SS commanders. The strategic dilemma was that Himmler's doctrine of Bandenbekämpfung obligated the SS troops to carry out mass murder of Polish civilians – not simply to rout the home army. 'Bandit warfare' which conflated civilians, above all Jews and combatants, had provided the logic behind the deployment of the RONA and Dirlewanger brigades – but to fight Himmler's 'bandit war' in a modern city took time. Bach-Zelewski faced complex strategic obstacles. Home army tactics forced the Germans to engage them on two closely linked fronts: on the surface in streets, apartments and public buildings, and underground in cellars, sewers and the maze of tunnels that the Polish commanders used with great cunning to both conceal and deploy their highly mobile forces. Above ground, the Germans used heavy ordnance and flamethrowers to pulverise buildings and incinerate their defenders. Below, they resorted to pumping an odourless, colourless gas called 'A-Stoff' into sewer and tunnel entrances. This was then ignited – with gruesome consequences for anyone caught in the explosion. As John Erickson aptly put it: 'The German command fused ingenuity with bestiality to fight one of the ghastliest battles of the war.'[30]

On 9 August, after more than a week of brutal combat, General Bach's licensed killers finally reached the Vistula and swung round to encircle the Old Town. As Allied negotiators implored Stalin to intervene, the German ring tightened remorselessly, and air and artillery bombardment reduced Warsaw with terrible speed to what General Bór admitted was a 'city of ruins', where there were only 'ruins left to burn'. As the final German assault began, a fog of cloying dust thickened the sultry air. A new kind of 'terror weapon' now rumbled on to the streets, pushing aside smashed barricades and piles of foetid corpses. A Polish journalist nicknamed this new horror the 'bellowing cow' – it emitted a tremendous roar 'like some kind of monster before the flood'; its 'hellish breath' incinerated buildings, turning their occupants into living torches.[31]

In the subterranean strongholds, home army commanders could see that their position was hopeless. On 19 August, General Bór ordered a 'fighting withdrawal' from the Old Town and 2,500 Polish fighters began to slip through sewers and other escape routes, carrying or dragging their 'moveable wounded'. When the Germans finally broke their way into General Bór's empty stronghold at the end of the month, they discovered a mound of corpses and a few wounded men. Bach-Zelewski's licensed killers showed them neither mercy nor respect. They emptied cans of petrol over the living and the dead and set them on fire.

The terrible events that consumed the city of Warsaw in the autumn of 1944 have a troubling resonance in Polish and European history and memory. Most historians believe that the anti-communist Polish home army was betrayed by Stalin and his generals and that Great Britain and the United States offered not military aid but mere hand-wringing. According to another view, the uprising was a tragic mistake that offered Hitler and the SS the opportunity to 'finish the job' of extinguishing the Polish nation state – and Stalin the means to destroy the anti-Communist Polish home army. There can be no doubt that the destruction of Warsaw, its defenders and tens of thousands of Polish civilians was the final act of Himmler's 'war on bandits', waged by his most fearsome bandit hunter, Erich von dem Bach-Zelewski. Himmler admitted this to his SS officers in September: 'I ... gave orders for Warsaw to be totally destroyed ... As a result, one of the biggest abscesses on the Eastern Front has been removed.'[32]

Resistance formally came to an end on 30 September. Historian Andrew Boroweic in his book *Destroy Warsaw* has published reliable figures, based in part on Bach-Zelewski's war diary, that unequivocally demonstrate the full horror of one of the last and most terrible battles of the Second World War. In 1939, the population of Warsaw stood at 1.2 million, including 352,000 Jews. After the 1943 ghetto uprising, that figure had been reduced to 974,745, including 24,222 German occupation officials and hangers on. The SS killed 1,559 identified Poles, but estimated actual Polish losses at over 100,000, with 15,000 wounded and taken prisoner. According to Bach-Zelewski, German SS forces lost 73 officers and NCOs, and 1,453 men, with just over 8,000 wounded.

Hitler awarded the Knight's Cross to Bach-Zelewski and to Dr Dirlewanger. Reinefarth, who had already had the award, received the Oak Leaves cluster. At Bach-Zelewski's headquarters, the German conquerors held a riotous 'medal party'.[33] In his diary, Bach-Zelewski commented: 'In these days I have become part of history and I am so proud for my sons.'

Hitler's demonic lust for destruction was not yet satisfied. He dispatched teams of engineers called Brandkommandos (fire commandos) to the smouldering ruins of Warsaw. They fell on the smouldering ruins with flamethrowers and explosives, obliterating anything left standing. For weeks, these engineers of annihilation worked their way across a landscape of desolation, ripping the old Polish capital apart, brick by blackened brick. Scrap metal, however, was loaded on to carts and sent to Germany.

But not all of Himmler's warriors shared in the triumph. SS-Brigadeführer Bronislav Kaminsky was not awarded a German medal. Kaminsky and his RONA fighters had learnt their trade as partisan hunters in the occupied Soviet Union. In

Warsaw, they had applied those lessons with drunken dedication. Kaminsky clearly understood what Bandenbekämpfung meant, whether it was waged in a forest or a city. But at some point during the Warsaw campaign, Bach-Zelewski had tuned into the BBC and discovered with dismay that he had been added to the Allied list of war criminals. To begin with he expressed astonishment – after all, he had, as he put it himself, 'always shown extreme humanity before God'. Then Bach-Zelewski began to wonder who might dare testify against him. Dr Dirlewanger was, of course, a German SS officer bound by Prussian military codes of conduct. If the Allies chose to investigate Bach-Zelewski's wartime conduct, Dirlewanger could be relied on not to betray his commanding officer. But Kaminsky was a different matter. He would surely betray even his own Russian grandmother. So Bach-Zelewski ordered Kaminsky's immediate arrest along with three of his staff officers and had them court-martialled. The SS court found Kaminsky guilty of 'failing to obey orders' and sentenced him to death.[34] The sentence was duly carried out but some kind of subterfuge was needed to prevent a mutiny by the surviving RONA men. SS men upended Kaminsky's Mercedes in a ditch and riddled it with bullets. Photographs were taken of the scene and shown to a few trustworthy RONA officers. Their beloved leader had been assassinated – by 'bandits'.

Former SS General Erich von dem Bach-Zelewski was not brought to justice for many years after the war ended. But at the end of the 1950s, he was successfully tried and convicted of war crimes by a German court. He died, a shabby relic of Himmler's war, in a prison cell. Heinz Reinefarth, his fellow commander, had much better luck: he enjoyed a comfortable and lucrative post-war career as mayor of Westerland on the island of Sylt – and in 1962 was elected a representative of Schleswig-Holstein. Reinefarth died peacefully at his luxurious Sylt manor in 1979.

14

Bonfire of the Collaborators

I do not regret at all my past ... Every morning I can look into the mirror without shame or pangs of conscience.

French SS volunteer[1]

On 23 March 1945 Adolf Hitler summoned a group of German army generals to his private apartments in the battered Chancellery in Berlin. He was not happy: 'We don't know what all is strolling around out there [sic]. Now I hear for the first time, to my surprise, that a Ukrainian SS Division has suddenly appeared. I didn't know anything about this Ukrainian SS Division.' Hitler was just getting started: 'The Indian Legion are a joke!' He was referring to a German army infantry regiment, the Indische Freiwilligen-Legion Regiment 950, which had been recruited by the Indian nationalist Subhas Chandra Bose in German and Italian POW camps. The legion had been taken over by the Waffen-SS in August 1944. 'There are Indians who couldn't kill a louse, who'd rather be eaten themselves. What are they still supposed to be fighting for, anyway?' Hitler had a good idea who was to blame for this dire state of affairs: 'Tomorrow I'd like to speak with the Reichsführer [Himmler] right away. He's in Berlin anyway ... We can't afford the luxury of having units like that.'[2] The next day, Himmler was forced to endure a ferocious dressing down. A month later, on 28 April, Hitler was informed that the 'loyal Heinrich' had tried to open 'peace negotiations' with the Allies through a Swedish intermediary, Count Folke Bernadotte.[3] Reuters had put out the story and it had been picked up by the BBC. The news provoked one of the last and most violent of Hitler's infamous temper tantrums. He sent orders to have Himmler banished from the movement and executed. Two days later Hitler was dead. The order was never carried out.

The excommunication of Heinrich Himmler seemed to be a precipitate fall from grace and power set in motion by his futile attempt to negotiate with the Allies. But Himmler had known for more than a year that Hitler's game was up. After the failure of the bomb plot in July 1944, the SS and its armies had dominated the Reich. Germany had at last become an SS state. In August, Hitler had chosen the Waffen-SS to destroy Warsaw and snubbed the German army. In Hitler's court, grandees like Hermann Göring had fallen from grace; Martin Bormann and Propaganda Minister Goebbels were Himmler's last rivals. In public, Himmler continued to profess absolute loyalty to the Führer. In private he knew that radical changes would need to be made to ensure that his SS empire somehow survived the certain defeat of Germany. Hitler was a broken man – but his hold on power would need to be loosened if the tide of collapse was to be stemmed. In the bitter winter of 1944/45, the old bonds between Hitler and Himmler began to crumble. In Hitler's mind, the SS chief had let him down once too often. Given command of Army Group Vistula, he had failed and ended up feigning illness in a sanatorium. Above all, he had contaminated the once mighty armies of the Reich with the impure blood of 'Ukrainians' and other foreign recruits – and for this he could never be forgiven. To unfold the downfall of Heinrich Himmler we need to return to the summer of 1944, as Waffen-SS recruitment strategy entered its final phase.

The Nazi Reich had been shattered. Germany was under assault from east, west and south. Armadas of Allied bombers battered German cities day and night virtually unopposed by the Luftwaffe. More than 3.5 million German soldiers were missing or dead. On the home front, morale continued to plummet. Hitler had not been seen in public for many months and his voice was rarely heard on German radio. He refused to visit ruined German cities or wounded troops in military hospitals. In every corner of Europe, there was chaos and misery. From the east, great refugee armies fled the relentless advance of Stalin's armies. In a futile attempt to cover up traces of the worst genocide in human history, SS men drove tens of thousands of camp survivors on death marches back towards the old German borders. Chronic disarray plagued the Nazi elite and the high command of the Wehrmacht. Hitler, drugged on amphetamines and other medications, deluded into fantasies of victory by the promise of new 'wonder weapons', spent his waking hours raging at his generals. In the hell of the collapsing Reich two men still prospered. One was Albert Speer, Hitler's architect and now Minister of Armaments, whose ruthless application of 'total war' management strategy and ruthless exploitation of slave labourers kept German factories at work. The other was SS chief and Minister of the Interior, Heinrich Himmler, the second most powerful member of the Nazi elite. As he scrapped with his Wehrmacht generals, Hitler enjoyed singing the praises of their hated rivals. The Waffen-SS was, he told lunch companions at the Chancellery, 'an extraordinary body of men, devoted to an idea, loyal unto death'. But a barely acknowledged tension crackled between the two Nazi leaders. The seeds of Himmler's downfall had been sown.

By the summer of 1944, just over half the men serving in the Waffen SS were not 'Aryan Germans'. The bloating of the Waffen-SS had begun in the summer of 1942 and had depended on non-German recruitment. The 'Nordland' division alone consisted of Dutchmen, Danes, Norwegians, Flemings, Swedes, Swiss and ethnic Germans. By the summer of 1943, Bosnians, Ukrainians, Latvians, Estonians fought in Waffen-SS uniforms. In the summer of 1944, Himmler acquired yet more SS warriors by gobbling up the Wehrmacht Osttruppen divisions recruited in the occupied territories of the Soviet Union, as well as the Indian Legion and a shabby bunch of British fascists known as the Britisches Freikorps.

Even as the German armies fell back towards the old borders of the Reich, the lure of Himmler's elite, black-uniformed *ubermenschen* continued to entice many young men. One late convert was a French idealist who, in June 1944, made his way to the recruiting office of the SS 'Charlemagne' division in Paris. His name was Christian de la Mazière.[4]

In August 1944, Paris was paralysed in torpid ennui. The city baked in a brutal heat wave. Metro stations stood silent and empty; electricity came on only between half past ten and midnight. In theatres and music halls, actors and entertainers performed by candlelight. Noxious odours bubbled up from the slow, murky waters of the Seine. In the shadow of the Arc de Triomphe on the Champs-Elysées, dazed Parisians and a few German officers sipped weak coffee under limp cafe awnings or trotted aimlessly between appointments, waiting for something to happen. From the roof of the old Napoleonic Naval Ministry, now the German Admiralty, a red and black swastika flag still hung. According to one reporter, 'the city was decomposing'.

In Normandy, the Allied armies had at last 'forced the lock of the door' and their supreme commander General Dwight D. Eisenhower had some hard decisions to make. For once, American and British strategists unanimously agreed that the main push should be towards the 'Siegfried Line' that defended the old German border. A single American army (the 12th) would be diverted to encircle but not attack Paris. Eisenhower had no desire to get bogged down in a protracted and bloody battle in the 'City of Light'. But General Charles de Gaulle, the leader of the French Free Forces, bitterly opposed this Allied plan to downgrade the French capital. He had always imagined marching in triumph down the Avenue des Champs-Élysées to the Place de l'Étoile to reclaim France and salvage her reputation. De Gaulle's relationship with the other Allied powers had never been harmonious. Now his spies had discovered that the French Communist Party planned to strike against the Germans and liberate Paris. This would mean that this proud and reactionary figurehead would be shoved to one side. Whoever liberated Paris would rule France. De Gaulle had to stop the communists at all costs.

In the meantime, Hitler had summoned Major General Dietrich von Choltitz to his Rastenburg headquarters. In the aftermath of the July bomb plot, Hitler was a wreck. After delivering a hysterical attack on the perfidious Prussian officer corps, he appointed the astonished von Choltitz 'Fortress Commander' for Paris with orders to punish any civil disobedience. Behind Hitler's decision to appoint Choltitz was an even more draconian plan. The German general had distinguished himself on the Eastern Front as a practitioner of 'scorched earth' tactics. As he retreated, his forces left in their wake a swathe of burned villages, mangled factories and ruined crops. 'Why should we care if Paris is destroyed?' asked Hitler – and that, in short, would be Choltitz's future task.

Paris had once been the most coveted posting for German officers and diplomats. German ambassador Otto Abetz was a noted bon vivant, who had spent many blissful hours at Maxim's and the Hôtel Bristol consorting with celebrity hangers-on like the actress Arletty and the opium-addicted poet Jean Cocteau. Now in August, with unseemly haste, their German friends had begun to prepare for the end. The 813th Pionierkompanie began setting explosives at key locations: at electrical and water facilities, and beneath the beautiful old bridges that spanned the Seine. German engineering units mined the Palais du Luxembourg, the French Chamber of Deputies, the Foreign Office, telephone exchanges, railway stations and factories. U-boat torpedoes were brought to Paris and positioned in tunnels beneath the city. On 16 August, Hitler ordered Gestapo and civilian administrators to leave Paris and three days later de Gaulle ordered free French forces to begin attacking German positions. The Paris uprising had begun.

Both the British and the Americans blamed Stalin for the tragedy of Warsaw. They had no desire to be held accountable in the eyes of the world if Paris now suffered the same grisly fate. Eisenhower was also aware that the French revolt was fast turning into a civil war waged between the Communist Forces Françaises de l'Intérieur (FFI), led by Colonel Henri Rol-Tanguy ('General Rol'), and the Free French Gaullists. The Bolshevik spectre stalked the grand boulevards. Allied anxieties worked in de Gaulle's favour. In any case, General Leclerc (the nom de guerre of Comte Philippe de Hautecloque), the commander of the Allied 2nd Armoured

French Division, had ignored American orders and begun advancing towards Paris. Inside the city, Gaullist Yvon Morandat outflanked General Rol and formed a Cabinet. Faced with a Gaullist coup, Eisenhower had few options left – and at last approved an attack on Paris, unaware that a French one was already underway. German SS men began executing their prisoners; in one Gestapo prison at Mont Valerien they shot 4,500 men. At Rastenburg, Hitler was heard to ask 'Is Paris burning?' At his headquarters in the 'City of Light', von Choltitz vacillated.

Robert Brasillach, a pro-German fellow traveller, wrote later: 'You could feel that everything was at an end.' After five years of docile occupation, astute Parisians began polishing their credentials as résistants. A few well-known 'collabos' were summarily shot and some pitiful women who had enjoyed horizontal liaisons with German officers were publicly shaved and shamed. But even as the German occupiers prepared to abandon the city, the French far right was not a spent force. On 21 April, an Allied air raid had badly damaged Sacré Coeur, the church built to atone for the sins of the 1871 Paris Commune, and the Parti Populaire Française (PPF), led by a veteran of the Eastern Front Jacques Doriot, staged a noisy, well-attended rally to protest against Allied barbarism. A star speaker was Léon Degrelle, recently promoted SS-Sturmbannführer and resplendent in black Waffen-SS uniform, displaying an Iron Cross. On 3 May, French fascists gathered en masse in Pére Lachaise cemetery to commemorate the centenary of their prophet, the Catholic anti-Semite Éduard Drumont.

As the flame of revolt sputtered into life all over France, the Germans responded by applying the techniques of Bandenbekämpfung to the Western Front. In western central France, the multinational SS 'Das Reich' (Alsatian French nationals served in its ranks) descended on the village of Oradour-sur-Glane and executed male villagers and burned women and children to death in the village church. Even as Himmler's elite troops turned their weapons on unarmed French civilians, Himmler launched a new recruiting campaign. In Paris, posters declared '*La SS t'appelle!*' and, in the last summer of Hitler's war, thousands of young Europeans responded to the final SS call to arms.

Despite its grand sounding name, the Waffen-Grenadier-Brigade der SS 'Charlemagne' (französische Nr. 1) had been stitched together from an older French SS unit, the 7th Storm Brigade, which had suffered very high losses on the Eastern Front, and the notorious French police, the Milice Française. Commanded by Joseph Darnand, this paramilitary force, set up in 1943 by Vichy Prime Minister Pierre Laval, had worked closely with the German authorities to root out and deport tens of thousands of French Jews. In July, Himmler had summoned Darnand to Berlin, promoted him to Obersturmführer and put him in charge of forming a new SS division. It was one of the paradoxes of wartime collaboration that Darnand was instinctually anti-German; but he hated and feared 'Jewish-Bolsheviks' a great deal more than the despised 'Fritzes'. Many of the Frenchmen who answered Darnand's summons would end their lives defending the Chancellery in Berlin, just a few hundred metres from Hitler's smouldering corpse.

Christian de la Mazière was a young Gallic volunteer who answered Himmler's call. Many of his comrades died on battlefields in Pomerania and Berlin – but de la Mazière survived and lived long enough to write a uniquely cogent account of why he volunteered to join the SS 'Charlemagne' in the dying days of the Reich. Like so many lives on the far right, Mazière's opened with an authoritarian and bigoted father. Claude-Nicolas de la Mazière descended from decayed aristocratic stock and, when Christian was born, managed an elite cavalry school in Saumer. Like many born into this embittered and faded class, Claude-Nicolas was Catholic, a patriot and, following French tradition, a dyed-in-the-wool anti-Semite. In the century since the Enlightenment, French Jews had been legally emancipated by Napoleon; some had prospered. This had hardened ancient hatreds especially among relics of the Ancien Régime in rural areas, where Catholic priests still promulgated their ancient libel that Christ had been murdered by the Jews. French anti-Semitism was energised in an even more menacing guise after the Franco-Prussian War in 1871 and the humiliation of the Second French Empire. Many well-known Jewish families had roots, if not relatives, in Germany or Eastern Europe and some used Yiddish at home. To men like Claude-Nicolas de la Mazière, French Jews would never be French enough.[5] According to the founder of L'Action Française, Charles Maurras, the public expression of anti-Semitism was a public duty, a matter of 'patriotic will'. For opportunist politicians, Jews provided useful scapegoats. They could be vilified as conspiratorial agents of international capital or on the other hand as Marxists, red in tooth and claw. In 1894, French anti-Semites found their cause in the trumped up Dreyfus Affair, and a leader in the shape of Charles Maurras. When Dreyfus was finally exonerated, the Jew-hating French far right dreamt of revenge.

Christian de la Mazière would grow up to reject his father but not his poisonous hatreds. Like Louis-Ferdinand Destouches, the writer who called himself Céline, he saw France as 'rotten to the core with Jews': its citizens had been enslaved by Jewish bankers. In the mid-1930s, de la Mazière joined L'Action Française and fell under the spell of its silver-tongued propagandist Léon Daudet, a literary jack of all trades who drummed into his disciples that 'One must be anti-Semitic ... there, in effect, is the psychological root of all the ideas and the feelings that have brought nationalists together'.[6] In 1937, de la Mazière and a party of French enthusiasts travelled to the annual Nazi 'Party Day' in Nuremberg. He wrote later that this first experience of Hitler's Germany 'through the banners and floodlights of Nuremberg' was a 'revelation'. This political equivalent of a High Mass, conducted by Adolf Hitler standing before more than a million Germans and international fellow travellers, the massed black and red banners, the severe black uniforms of the SS and the superbly drilled crowds of euphoric Germans, thrilled de la Mazière and his fellow supplicants. Nuremberg seemed to a celebration of faith not merely power. 'I began to dream,' de la Mazière confessed, 'of a new world in which Europe would become a beacon of National Socialism … I felt the sincerest need to sacrifice myself for an ideal.' This new world would be 'Jew-free' at last; in de la Mazière's words, Jews stood for 'a general force of evil'.

Nearly a decade later, de la Mazière had lost none of his crusading zeal. On the contrary, the German conquest and occupation had convinced him that this was a superior power and culture. He welcomed the deportation of Jews, organised by the SS and their French collaborators. Now, from the windows of his apartment in the Rue Chevert, de la Mazière watched half-naked German gunners dismantling anti-aircraft guns in front of Les Invalides. For two years he made a living writing for a German sponsored anti-Semitic journal. He liked to boast that his work had inspired a friend to enlist in the French volunteer SS division and sacrifice his life for the Reich. Now de la Mazière heard the news that SS Chief Himmler had called for fresh blood to join the SS 'Charlemagne' Division to 'save Europe from Bolshevism'. Was this, he reflected, his last chance for glory? Should he abandon his pen for a rifle?

One sweltering day, at two o'clock in the afternoon, de la Mazière made up his mind. He took lunch then strolled across the Pont d'Alma to the Hotel Majestic – the SS headquarters in Paris. Behind the Majestic's opulent facade on the rue Dumont d'Urville, de la Mazière discovered a spectacle of chaos and undisguised panic. SS men scurried from office to office, gathering up files and hurling them on to a bonfire that blazed in the hotel courtyard. But somehow a young SS man found time to procure the right documents. An SS-Hauptsturmführer appeared and ordered de la Mazière to find his way to the camp at Wildflecken, near Sennheim in Alsace, where the French volunteers would be being trained. Then the SS men bowed, made the *Hitlergruss*, clicked their heels and vanished. In his memoir, de la Mazière recalled: 'This raised arm. I felt I had crossed a threshold … these SS fascinated me and I wanted to be assimilated into their ranks. I saw them as a race apart … strong courageous and ruthless beings without weakness, who would never become corrupt.'

By the time de la Mazière made that symbolic *Hitlergruss* in the summer of 1944, many millions of Jews and other 'enemies of the Reich' had been murdered by men in SS uniforms. Himmler's empire resembled a sprawling multinational corporation whose business was pillage and mass murder. The SS empire was corrupt on an unimaginable scale. And yet as de la Mazière's account demonstrates, the SS was still able, even as the Thousand-Year Reich collapsed, to conjure up the alluring promise of ideological partnership in an ideal future world. Himmler possessed a refined gift for exploiting dreams, including his own. He was, after all, a fantasist himself: a former chicken farmer who fervently believed that he was the reincarnation of a medieval emperor. As de la Mazière would soon discover, the SS Junkerschulen like the one at Wildflecken were not just pitiless machines designed to turn out ruthless killers. They resembled monasteries, 'downright religious establishments', that reshaped minds and bodies.

It was there on the Wildflecken parade ground on 12 November 1944 that Christian de la Mazière, trembling with anticipation, stood alongside other French volunteers and took the Waffen-SS oath in German and French: 'I swear to you, Adolf Hitler, Germanic Führer and Re-maker of Europe, to be true and brave. I swear to obey you and the leaders you have placed over me until my death. May God come to my aid!' De la Mazière was marched off to the infirmary where a medical officer took a blood sample. An assistant heated up a tattooing instrument and branded the SS Sig runes, with blood group, just below his left armpit with a sharp hiss of singed skin. Many SS recruits would, very soon, have reason to regret taking this particular rite of passage. Christian de la Mazière had joined Himmler's elect.

The apparent triumph of the SS 'State within a State' nourished many delusions. After serving with the Wehrmacht Légion Wallonie, the Belgian demagogue Léon Degrelle had finally persuaded German race experts that the Belgian Walloons had as much right as Flemish Belgians to claim Germanic descent. In April 1943, he had been summoned to Himmler's headquarters a few kilometres from Hitler's Wolf's Lair near Rastenburg. This was the first time Degrelle had met the SS chief. He provides very few details of his impressions; he was much too fixated with Hitler. The purpose of the meeting was to agree details about the transfer of the légion to the Waffen-SS as the SS 'Sturmbrigade Wallonien'. Their conversation then turned to the political destiny of the Belgian Walloons. Degrelle assured Himmler that his political movement 'Rex' and the Walloonian people he claimed to represent would happily join the 'Greater Germanic Reich'. He reassured Himmler that any resistance would be dealt with by the new SS Sturmbrigade. Degrelle broadcast the same surreal message to his new recruits: 'Soldiers of the Führer, you will also be, after the war, political soldiers who will raise [in Wallonia] the banner of victorious revolution.'[7] Degrelle at last, it would seem, was within striking distance of the top levels of the Reich. But his progress had, as ever, been propelled by the over-heated fuel of whimsy. In any case, Himmler had the last laugh. For despite all Degrelle's posturing as a 'Germanic' hero, the légion had been accepted into the Waffen-SS as '*würdig Nicht-Germanen*' (worthy non-Germans).[8]

And so Léon Degrelle marched off to war again. On 2 November 1943, SS-Obersturmbannführer Degrelle, officially listed as an aide-de-camp, along with just over 1,000 Walloonian SS volunteers boarded military transport trains at the Wildflecken/Gersfeld station and began the long journey to the east.[9] SS-Sturmbannführer Lucien Lippert commanded. The German front line might have seemed the safer option: on the home front in occupied Belgium, the Belgian resistance, the Armée Secrète, had begun picking off Degrelle's supporters and collaborators in a well-organised campaign of assassination.[10]

By the time the Walloons reached the front, the Soviet army had penetrated to a position a hundred miles west of Kiev, biting deep into the Reichskommissariat Ukraine. As the Russians prepared to cross the Dnieper, German forces clung to a chain of strong points on the opposite bank. One was located in the industrial region of Cherkassy and defended by the SS 'Wiking' division, reinforced by the Estonian SS 'Narva' and the Flemish SS Sturmbrigade Langemarck. The 'Wallonien' reached the Cherkassy salient at the end of November. The supremely pompous Degrelle was immediately at loggerheads with his commanding officer Lippert, but the front line offered few opportunities for political skulduggery. Through December and into early January 1944 the 'Wallonien', fighting against overwhelming forces in sub-zero temperatures, suffered 'fearful losses'. Lippert's natural caution was frequently undermined by the reckless Degrelle, who was desperate to bolster his standing with Himmler.[11] By the beginning of February, some 250 SS 'Wallonien' men who had survived this relentless attrition had become trapped inside the 'Korsun-Cherkassy pocket' by the relentless Soviet advance. On 13 February, Lippert was killed by an explosive shell, a moment Degrelle recalled in grisly detail. Lippert, he wrote, uttered 'the superhuman scream of a man whose life is suddenly torn from him', but he still possessed 'the extraordinary lucidity to pick up his kepi from the ground and put it back on his head so as to die fittingly'.[12]

By then, armadas of Soviet T-34 tanks, backed by roaring batteries of 'Stalin Organs' relentlessly tightened the trap around the German forces. The new 'Wallonien' commander Jules̹ Mathieu ordered Degrelle and the surviving Belgians to 'break out' towards the south-east, where they could rendezvous with German reinforcements. At dawn on the morning of 17 February, Degrelle joined a 'fantastic jumble' of tanks, mechanised and horse-drawn vehicles, Ukrainian refugees and even Soviet POWs all trying to reach a narrow corridor of escape before the Russians slammed it shut. 'It was no lark,' Degrelle recalled. Ahead, German panzers, driven by their 'marvellous warriors' forced open a breach just a few hundred metres wide. As the Belgians, hauling their dead commander on a sled, raced after the tanks, thick snow began to fall making it almost impossible to see further than a few metres, but shielding the retreating troops from Soviet aircraft. Degrelle led his battalion deep into a ravine, where he halted, uncertain what to do next. At any moment, he feared, the Soviet 'Mongols' would discover them and start hurling down grenades. But somehow Degrelle extricated his men and they trudged painfully west, past twisted tank wrecks and the 'hot intestines of horses spilled on the bloodied snow'.

A day later, the fleeing German forces reached the east bank of the Gniloi-Tikitsch River. Huge blocks of ice crusted the rushing torrent, swollen by the spring melt. Every bridge had been destroyed. Close behind, Degrelle recalled, Soviet tanks had begun spilling over the ridge he had just descended – and he and the retreating Belgians had no choice but to take their chances in the icy river. Many vanished forever in the swirling torrent. Soldiers who made it to the other bank, some 'naked and as red as lobsters', huddled together by the frozen bank, but there was no time to spare. As the Russian tanks began firing from the opposite bank, Degrelle and his comrades made a dash for the forest. Looking back, he could see unceasing 'human streams' of German soldiers gushing from the woods, desperately dodging Soviet fire and throwing themselves into the freezing water.

For Degrelle, the 'Cherkassy breakout' was the zenith of his career as a collaborator. At the beginning of 1944, the Propaganda Ministry in Berlin battled with a relentless inundation of catastrophic news from the east. Goebbels seized on the *Tcherkassykämpfer* and turned a shambolic retreat into a spectacular triumph. Its most prominent hero was, of course, Léon Degrelle, who was summoned to the ailing Hitler's military headquarters and, in front of Germany's delighted press corps, awarded a Knight's Cross. Buoyed up on a wave of fatuous euphoria, Degrelle was flown back to Belgium in triumph on 22 February. In Brussels, Degrelle's most devoted supporters staged a rally at the Palais des Sports. Diehard Rexists filled the hall and applauded as the Chef reaffirmed his faith in Hitler and the Reich.

Degrelle squeezed every last drop of acclaim from his triumphal progress. On 5 March, he arrived in Paris and harangued a pro-German gathering at the Palais de Chaillot in SS uniform. On the platform with Degrelle stood the crème de la crème of the French collaborationist movement, among them Marcel Déat, Jacques Doriot and Joseph Darnand, the commander of the French SS 'Charlemagne' division. After the Rally, German ambassador Otto Abetz, who had long promoted

Degrelle's cause, threw a party at the German Embassy. At the beginning of April, Degrelle returned to Brussels and, riding in an armoured half-track with his children, and led the 'Wallonien' men to yet another rally. The Bourse had been decorated with SS Sig runes and swastika flags which proclaimed '*Honneur a la Légion!*' As the round of celebrations and parties continued, and expensive champagnes and cognac flowed at Degrelle's lavish home in the Drève de Lorraine, the Chef de Rex waited impatiently for the call from Hitler promoting him from war hero to Belgian Staatsführer.

The German military governor, Eggert Reeder, observed the triumphant spectacle with alarm. Reeder had long had the measure of Degrelle and despised him as a vainglorious fantasist. 'It is the case,' he reported:

> every time, that Degrelle can be judged to be erratic, easily influenced, clumsy in his actions and occasionally unreliable when it comes to political matters … Due to his temper and a number of other weaknesses, Degrelle often tends to drift into the realms of political fantasy and a self aggrandisement which has nothing to do anymore with proper optimism or realistic political judgement.[13]

The call from Hitler never came. For the Nazi elite, Degrelle was a merely a short-breathed propaganda windfall.

Even Degrelle could not compete with Himmler's phantasmagorical optimism. A remarkable snapshot of the SS chief in early 1944 can be found in the memoirs of Arturs Silgailis, a colonel in the Latvian 15th Waffen-Grenadier-Division der SS, which had been dispatched to the front in November 1943. Relations between the German officers and their Latvian recruits, many of them conscripts, had come perilously close to collapse and Silgailis had confronted the division's German commander SS-Brigadeführer Carl Graf von Pückler-Burghaus at his billet on a Latvian farm. Pückler-Burghaus claimed to sympathise with the Latvians and introduced Silgailis to his good friend, Hermann Fegelein, the priapic husband of Eva Braun's sister and a veteran of the SS Cavalry Brigade 'Florian Geyer' that had led the attack on Jewish villages in the Pripet Marshes in 1941. Fegelein had been appointed SS liaison officer at Hitler's military headquarters at Rastenburg, and he offered to arrange a meeting between Silgailis and Himmler. So it was that at the beginning of February 1944, Latvian collaborator Silgailis found himself at Himmler's headquarters, sited less than 10 miles from the Wolf's Lair.

Himmler entered, flanked by two aides and another SS officer, and immediately launched into a tirade about Pückler-Burghaus who had, he believed, 'mishandled' the Latvians. Himmler invited Silgailis to lunch and they sat down to consume a lavish meal, while Himmler (Silgailis claims) listened attentively to his complaints about the treatment of the Latvian troops. He proposed that the SS set up a military academy to train Latvian recruits. Himmler rejected that idea: a 'Latvian' academy would blinker its students. To rub in his point, he placed his elbows on the dinner table and covered his eyes. He went on: 'every SS officer, regardless of nationality, should politically look far beyond the boundaries of his country: *he must envision the whole living space of the family of German nations*' (my italics). By these, Himmler explained, he meant the Germans, the Dutch, the Flemish, the Anglo-Saxons, the Scandinavians and the Baltic peoples. The most important task of the present time was to combine all these nations 'into one big family': a union founded on equality, the separate identity of each nation and its economic independence 'adjusting the latter to the interest of the whole German living space'. Naturally, since it was the 'largest and strongest' nation Germany would take the leading role. Later, when the war was over, 'all German nations would be leading' – a point Himmler clarified by pointing to the current 'Grossdeutschland' comprising 'many Germanies' in which each one had an equal place. He went on: 'Germanic' nations produced hardier peoples because they had to contend with harsh northern climates. The softer 'Roman' peoples would have to acquire the same qualities through unification with Germany. After that had been accomplished, the Slavic nations would have to be tackled. Even they must be persuaded to join 'the family of white races'. All 'white people' faced a terrible menace: the increasing numbers of 'yellow people' whose 'born fanaticism – the disdain for death and belief in Nirvana' encouraged them to reproduce with fecund abandon. By promoting patriotism and large families, the white races would defend Europe against the 'Asiatic' menace. The Waffen-SS led the way, Himmler went on, by recruiting German, Roman and Slavic peoples – even Islamic units: 'every unit has maintained its national identity while fighting in close togetherness'. Returning finally to Silgailis' suggestion to form a Latvian military college, Himmler pointed out that in order to embrace this racial battle against the 'yellow peoples' all SS officers must be trained at the same college – in Germany. By receiving the same education, every SS officer would embrace the cause of equality among the white races.

This meeting is one of the very few documented encounters between Himmler and a representative of the SS foreign legions. Himmler speaks as if the rapid advance of the Soviet army presented merely a short-term setback: his faith in the basic principles of the 'General Plan East' remained intact. It is curious that

Himmler refers to the threat posed by 'yellow peoples', but Hitler too often lashed out at 'yellow Asiatics' and in anti-Semitic tracts Jews are often characterised as an 'Asian' race. Himmler was, of course, attempting to bind Silgailis to the Reich by promising the Latvians a place at the high table. He appears to have succeeded. Silgailis and his fellow collaborators wrung a few concessions from their SS masters and fought on for the Reich. But that lunchtime lecture smacks of Himmler's perverted idealism. His master plan could somehow still be made to work. Closeted inside his eastern headquarters hundreds of kilometres from the blood and ice of the front, Himmler drew on his last mental resources and the magic fingers of Dr Kersten to blunt the catastrophe that was engulfing the Reich.

By now every telex brought bad news. In Riga, Commissar Hinrich Lohse had suffered a breakdown; in Minsk, a chambermaid had inserted a land mine into the bed of HSSPF Wilhelm Kube – with fatal consequences. The Russians had finally overrun the Reichskommissariat Ukraine, and its despotic ruler Erich Koch had replaced the unfortunate Lohse. Koch had no intention of spending much time in Riga. Instead, he ruled the fast-shrinking Ostland from a lavish suite in Berlin's Adlon Hotel, a short distance from Rosenberg's Eastern Ministry. Koch liquidated the Baltic 'self-administrations' and began deporting tens of thousands of Latvians from northern Latvia (Vidzeme) to East Prussia as forced labour. Hitler, however, ordered Koch to hold the Kurland (Courland, also known as Kurzeme) in western Latvia. The Kurland protrudes deep into the Baltic, west of the Gulf of Riga. For the Latvian commanders like Silgailis and Bangerskis, the 'defence of the Kurzeme' offered a way to convince their wavering recruits that would be 'defending Latvia' against the Soviets. But the Kurland had a very different significance for the Germans; Latvian blood would be shed to serve only German interests. As the Soviet armies crossed into the Baltic, Admiral Karl Dönitz had met with Hitler and convinced him that new generation U-boats, which had the capacity to stay submerged longer, would soon become available and could be deployed to cut off the Allied forces in Europe. To launch this second 'Battle of the Atlantic', the Kriegsmarine would require the ice-free Baltic ports of Liepaja and Ventspils, both located on the Kurzeme coast. For Hitler, invariably entranced by the phoney promise of 'Wunderwaffen', Kurland was merely a bridgehead to launch another illusory front.[14] The Latvian SS men would be mere cannon fodder. So Hitler renamed the old Army Group North as 'Army Group Kurland' and ordered the 'Baltic fortress' must be defended at any cost. There would be no surrender.

As the Germans turned the Kurland into a fortress, 200 Soviet divisions swung south towards Riga, pushing the rest of the German forces back through Lithuania towards the East Prussian border. As they abandoned the Ostland, German civilian administrators seized what plunder lay to hand. Convoys steamed from Baltic ports, stuffed full of livestock, machinery and Latvian slave workers. Inland, the retreating Germans troops torched grain fields and hacked down forests. They dismantled factories and even 'Germanic' villas and churches which they regarded as property of the Reich. East of Riga, they blew up the Kegums hydroelectric power station on the Daugava, flooding farmland and drowning villages. The German barons had always regarded their Latvian peasants as expendable.

In the meantime, soldiers of the Latvian 19th Waffen Granadier Division of the SS dug in along an 8-mile segment of the 150-mile western Kurzeme front – destined to be the last German outpost in the east.[15] On the other side of the fragile fortress walls were up to 200 Soviet divisions of the three Baltic fronts and the Leningrad front. In the Baltic, which formed the western bulwark of 'Festung Kurland', torpedo boats of the Soviet Red Banner Fleet preyed on German convoys that steamed between the Kurzeme ports and what was left of Hitler's Reich. As well as the Latvians, Festung Kurland would be defended by half a million German troops: thirty divisions of the 16th and 18th Armies, commanded by the foul-tempered Generaloberst Ferdinand 'the bloody' Schörner. Schörner and his officers had little time for their Latvian allies and most did little to hide their contempt. Any soldier caught 'shirking' or trying to desert would be swiftly and brutally punished. Inside Festung Kurland punitive executions became a frequent occurrence. One German commander tried to make amends by proclaiming that the Latvian recruits would be celebrated as 'Latvia's next historic men', but desertions mounted daily.

Seventy years later, Latvian apologists claim that the two Latvian Waffen Granadier Divisions of the SS defended Latvia against Soviet assault. Their argument makes no sense at all from a strategic point of view since the Latvian legionaries allied themselves with the army of a power doomed to defeat. 'Defending the Latvian nation' in Waffen-SS uniform in any case conceals a hidden agenda. Many soldiers who served in the Latvian Waffen-SS divisions had served in the SS Schuma battalions that had liquidated the majority of Jews in the Baltic nations. One of these mass killers was Standartenführer Voldemārs Veiss, who had offered his service to the Reich when Special Task Force A, led by Franz Walther Stahlecker, had arrived in Riga in July 1941. Like Viktors Arājs, Veiss (who had once served as Latvian military attaché in Finland) proved himself a dedicated killer. He had been rewarded with a leading role as Director General of the Interior in the puppet Latvian 'self-administration'. The 'self-administration' was riddled with radical nationalists who had long wanted a 'Latvia for Latvians' cleansed of any 'foreign' interlopers. This ethnic cleansing had been successfully achieved by Latvians who voluntarily participated in German managed operations against Latvian Jews. When this murderous onslaught had been completed, with the destruction of the vast majority of Baltic Jews, it was the compliant Latvian 'self-administration' that took a leading role recruiting many thousands of young Latvian men to fight in what would become a Latvian civil war, since many other Latvians also served in the Soviet army and were equally determined to liberate Latvia from its German occupiers. In every German-occupied region, in the Balkans, Ukraine, France and Belgium, collaboration provoked deadly civil conflict.

On 15 October 1944 the Soviet onslaught began with a massive artillery barrage. The German commanders made sure the Latvians bore the brunt; some Latvian battalions fled en masse, and Soviet armoured forces broke through at several points. As the Germans struggled to stabilise the line, they pulled out the worst performing Latvian units and deported them to Danzig and Gotenhafen (Gdyna). At a training camp in Sophienwalde, SS officers refitted and retrained the shell-shocked legionaries, bulking up numbers with Latvian dissidents who had been imprisoned in Stutthof concentration camp, and other Baltic citizens on labour service. As the Soviet army shattered German divisions on the East Prussian border and poured into Pomerania, the Germans attached the Latvians to the Third Panzer Army to defend the Oder-Vistula canal. Here too the Russian assault was crushing. The German commander of the 'Latvian Legion' was killed and the Latvians again fled the battlefield. The 15th Waffen Latvian Division had also suffered terrible losses and the 'Latvian Legion' had ceased to be an effective combat unit in any meaningful sense: in the spring of 1945, battered German divisions cannibalised any survivors to defend the last citadels of the Reich; so much for defending Latvia.

In every battered sector of the disintegrating German line, the Soviet meat grinder began consuming Himmler's foreign legions. In the spring of 1944, the Soviets had trapped the relics of the SS 'Galizien' and a German division inside a snail-shaped pocket near the city of Brody that was under continuous artillery and air attack. Marshall Ivan Konev refused to delay his advance across Ukraine and assigned a Special Task Force to mop up any remnant enemy troops. These Soviet soldiers hated the SS and loathed the traitorous foreign legions 'der SS'; so tremendous firepower was focused on the SS 'Galizien' soldiers known to be hunkered down in the south-east tail of the Brody pocket. Some of the German troops occupied their time shooting Russian prisoners, including some elderly Ukrainian men who had sons fighting with the 'Galizien'. One Ukrainian officer remembered hearing screaming: '*Kamerad, kamerad nicht schiessen!*' (don't shoot, comrade!) As the German soldiers finished off the prisoners with the customary shot to the back of the head, a Ukrainian officer said (in German), 'That's the reason why the Germans have to leave Galicia ... That's why you're losing the war.'[16] For their part,

Wehrmacht officers turned their frustration on their Ukrainian comrades. General Lange reported to Berlin: 'As was anticipated, the Galicians in no way showed themselves to be the fanatical defenders of their homeland against Bolshevism.' He grumbled that German weapons had been wasted on a 'totally undisciplined, disordered mob'.[17] A few German troops and the Ukrainians eventually managed to break out of the Brody pocket, but with appalling losses. Among those who never returned was the vile anti-Semite Dmytro Paliiv.

The Soviet army had effectively destroyed the SS 'Galizien'. Out of 11,000 men trapped in the Brody pocket less than 3,000 escaped alive; an additional 7,400 were recorded as missing in action. Yet more defected to the Ukrainian insurgents, the UPA, or hid in Ukrainian villages. Himmler summoned the SS 'Galizien' commander, Fritz Freitag, to Berlin. To Freitag's astonishment, Himmler wanted the 'Galizien' division reformed as soon as possible. He airily dismissed Freitag's complaints about the Ukrainians' mediocre performance in battle, pointing out that German divisions too had failed to repel the Soviet assault. On 19 October, a reluctant Freitag announced: 'We all wish to renew our vow to the Führer, that we will pursue unto victory, our joint battle against the Bolshevist hordes and their Jewish-Plutocratic helpers.'

The Germans assigned the reformed SS 'Galizien' to Slovakia where a revolt had broken out against the pro-German Hlinka regime. The Ukrainians would serve alongside Dr Dirlewanger's criminal brigade that had recently been transferred from Warsaw. 'You can be absolutely certain,' Himmler made clear, 'that I will never reprove anyone for excess.'[18] Independent research carried out by Julian Hendy sheds light on the role played by the SS 'Galizien' in Slovakia.[19] Hendy discovered that, on 15 October, the German commanders of the 'Galizien' raised a Kampfgruppe to liquidate partisans who taken part in the uprising. The unit would be led by Friedrich Wittenmayer, a career police officer who had extensive experience of 'anti-bandit' operations. According to reports made by German intelligence, and held in the Czech State Archives in Prague, the battle group cleared railway tracks and roads and captured enemy weapons. But Himmler's 'bandit war' consistently punished civilians and this campaign by SS 'Galizien' men in Slovakia was no different from other German anti-partisan operations. On 26 October, a German commander reported that 'Battle Group Wittenmeyer in process of occupying Nizna Boca (10km south of Krl. Lehota). Road between Rosenberg and Poprad therefore now free of the enemy. The Ukrainian volunteers of the 14th Waffen Grenadier Division of the SS used in the operation fought excellently.'[20] It is alleged that in the village of Nizna Boca the Ukrainian SS men carried out summary executions of male villagers. Hendy interviewed a Slovakian historian who stated:

> They took part in terrorist operations and reprisals against Uprising fighters … I will be very specific: In the Smerycany area and in Nizna Boca/Maluzina it specifically attacked the civilian population. In Smerycany, the Wittenmayer unit from the Division burnt the village down using artillery and mortar fire. The civilian population was driven out of the village and 80% of the 120 houses were burnt down. During the raid on Nizna Boca, five people died. These are just the most telling examples of when this unit struck.[21]

The Nizna Boca killings were not on the same scale as the massacre in Huta Pieniacka. This was a relatively small-scale action at the tail end of a bigger operation directed at Slovakian 'rebels'. But we must bear in mind that the attack was just one more instance of 'bandit warfare' and such tactics would have been mimicked in other villages in Slovakia where Himmler's SS executioners were deployed.

Meanwhile, the Soviet war machine penetrated ever deeper into the lands Hitler imagined would become a German 'Garden of Eden'. On 1 August 1944, Otto von Wächter, the former Gauleiter of the Galicia district who had first proposed recruiting a Ukrainian SS division back in March 1943, sent a secret telex to Himmler. He explained that the Soviets had overrun the General Government, and his position as district governor had thus been abolished. He requested a transfer to the Waffen-SS. Himmler readily agreed; he appointed Wächter head of the Military Administration in Italy under SS-Obergruppenführer Karl Wolff. Wächter would, Himmler wrote, be of 'immense use' in this 'interesting and difficult field'.[22] The SS dominated the German occupation of northern Italy and, after 1943, Himmler dispatched there some of his most experienced *génocidaires*, including the odious Odilo Globocnik.

In the Balkans, Josip Broz Tito's partisan armies had seized the initiative in the summer, and desertions from the SS 'Handschar' had begun to escalate. Berger was forced to plead with Haj Amin el-Husseini, the Grand Mufti of Jerusalem, to intercede with the loyal German trained Imams who served with the division. El-Husseini did as he was told but his cadre of Imams had only a modest impact on plummeting morale. The turning point came in September 1944. As German garrison armies started withdrawing from Greece through the Balkans, the RAF's multinational 'Balkan Air Force' launched Operation Rat Week to disrupt road and rail links along the line of retreat. Tito's partisans backed the operation on the ground, successfully cutting supply lines to the 'Handschar'. As the division's mutinous 'penal battalion' struggled to get the railways running again, they came under repeated attack from the partisans, who could wreck 15 miles of track in a night. The Germans used their foreign SS units as buffers between retreating German armies and attacking Allied forces. As the chaos generated by Operation Rat Week got a grip, Tito ordered a final assault on the battered and demoralised SS battalions. Then on 6 September, the Chetnik leadership that had fought with Tito's insurgents so bitterly finally turned against the Germans and the 'Handschar'. As military conditions deteriorated sharply that month, Bosnian Muslims deserted in droves: in the first three weeks of September at least 2,000 men vanished into the mountains. Most took their weapons and equipment with them. To hasten the disintegration of the hated SS division, Tito announced an amnesty for all deserters from German forces.

On 18 September, as the crisis deepened, the 'Handschar's commander Karl-Gustav Sauberzweig flew to Berlin to meet Himmler. According to Sauberzweig's report, the SS chief reprimanded him 'in the sharpest and most disgraceful manner' and accused him of 'defeatism and incompetence'. But there would be no escape: Sauberzweig was ordered to return to Bosnia and somehow plug the leaks. In Berlin, he called on the Grand Mufti at his Tiergarten villa. El-Husseini was no fool: he had no illusions left about the likely fate of his Muslim armies; neither his nor Hitler's war could be won. But he promised the distraught Sauberzweig that he would fly to Sarajevo as soon as it could be arranged and would rally the shattered 'Handschar' men. The Mufti never returned to Sarajevo.

The German commanders of the 'Handschar' had few options left. On 21 October one of the Imams, Adulah Muhasilovic, deserted, taking with him more than a hundred other 'Handschar' men. The rebels fled in German vehicles equipped with anti-aircraft guns and drove north and surrendered to Tito's XVIII Croatian Brigade. When news of this latest calamity reached Berlin, an enraged Himmler ordered Sauberzweig to immediately disarm 'unreliable elements' and hand them over for labour service in Germany. Sauberzweig was once again summoned to Berlin and sacked. Sauberzweig was suffering from chronic exhaustion and Himmler had him admitted to the Charité Hospital in Berlin and then, when that proved ineffective, dispatched him to the Hohenlychen Clinic, where Albert Speer had been treated after suffering a nervous collapse.

By the autumn of 1944, Hitler's armies had been severely damaged. The Soviet army had overrun Romania and part of Hungary. After Romania's defection, Stalin swallowed sixteen Romanian divisions and in October fourteen Bulgarian divisions joined the Allies. The Soviets then linked up with Tito's partisan armies to drive German Army Group Serbia from eastern and north-eastern Yugoslavia, the region that would become modern Serbia. On 2 August, when Muslim Turkey broke off diplomatic relations with Germany, the Bosnians who had enlisted in Hitler's war shed every last illusion. An exultant Marshall Tito seized Belgrade on 20 October.[23] By then, the relics of the 'Handschar', which was now mainly

'German', had been transferred to Hungary in time to join the pell-mell flight of German forces. Although the new German commander of the SS 'Handschar' did not formally surrender until May, 1945, Himmler's Persian warriors had long before deserted the Reich. After the German surrender the following year, the new Yugoslav government requested Sauberzweig's extradition to Belgrade to stand trial with other captured German SS 'Handschar' officers. On the eve of his departure, 'Speedy' Sauberzweig swallowed a cyanide capsule.

The majority of historians have interpreted the recruitment of non-Germans by the SS as a symptom of desperation. In other words, Himmler and his recruitment chief, Gottlob Berger, simply wanted as many bodies in SS uniforms as possible to hurl at the Soviet army. This book has demonstrated that foreign recruitment served a crucial ideological purpose: the 'Germanisation' of specific European ethnic groups and the dissolution of national identities through blood sacrifice. The collapse of Himmler's master plan can be dated not to the main period of non-German recruitment between the summer of 1942 and the spring of 1943, but to 12 November 1944. For it was on this date that the SS 'Galizien' ceased to exist: Himmler approved its renaming as the 14. Waffen-Grenadier-Division der SS (Ukrainische no 1). He accepted that the 'Galizien' was no longer a valid term because 'the division became a melting pot for Ukrainian men from all regions of the Ukraine'. German statements naturally tried to spin the demise of the 'Galician project'. The 'Ukrainische no 1' was 'intended as the future military home for all Ukrainians' and its formation represented the 'escalation of the battle against Bolshevism by the entire Ukrainian people'. Himmler authorised a new oath: 'in the battle against Bolshevism, and for the liberation of my Ukrainian people and Ukrainian homeland, I will give absolute obedience to the Commander in Chief of the German armed forces and all fighters of the young European nations against Bolshevism, Adolf Hitler.'[24] The new Ukrainian SS division was permitted to display the blue and gold national flag and its orchestra was allowed to play the Ukrainian national anthem. At the end of 1944, Himmler finally abandoned any pretence that Ukrainians recruited in the old Austrian province of Galicia represented a distinct semi-Germanised elite. As Otto von Wächter, the SS architect of the 'Galizien', had belatedly realised, 'Galicia' was a redundant territorial concept, which in any case incorporated as much 'Polish blood' as German. Himmler's Galicians had finally become Ukrainians.

In the spring of 1945, as the Ukrainians fled west to escape Stalin's NKVD agents, Wächter made his final exit. As a leading German administrator, in short a mass murderer, there was a price on his head. Now he fled across the Alps into Italy and followed the Vatican 'rat line' to Rome. In the Eternal City, Wächter was hidden away by the pro-Nazi Austrian Bishop Alois 'Luigi' Hudal (who had once preached that Hitler represented the 'new Siegfried of Germany's greatness') and, with new identification papers, became 'Alfredo Reinhardt' – a final homage to the assassinated SD leader, Reinhard Heydrich. He died a few years later in Hudal's arms.

On 12 January 1945, Soviet marshalls Koniev and Zhukov launched the last winter offensive of the war. They concentrated their main attack on the central sector of the Eastern Front just to the east of Kraków and Łódź. In Hungary, the Germans clung desperately to Budapest. Hitler had spurned every proposal to withdraw and Soviet forces had encircled the city as Arrow Cross fanatics rampaged through the Jewish quarter. Hitler had stripped other sectors of the front in a futile effort to break the siege. Now the weakened German forces holding the residue territory of the General Government faced a titanic assault. Before dawn, expendable Soviet punishment battalions moved forward to ferret out the German positions – and then at 10 a.m., a stunning bombardment from Russian heavy mortars (350 for every kilometre of front) and the deadly Katyushas and Ivan Grozny rocket launchers began pounding the German lines, making the ground shake and blasting tanks, men, concrete, earth and rock high into the freezing air. As the blazing deluge tore open gaping wounds all along the German front, tank armies clanked forward into the mangled breaches. On 17 January, the First Polish Army captured Warsaw, and on the same day, Hitler fled his Rastenburg headquarters for the last time and took refuge in the Reich's Chancellery in Berlin.

'Shrivelling up like an old man', Speer recalled, Hitler never appeared in public again. On 30 January, he made a last radio broadcast to mark the twelfth anniversary of his appointment as Reich Chancellor. 'His voice sounded shrill with despair,' recalled one German listener. 'How heavy must be the burden he bears!' wrote another.[25] Hitler's state of mind, according to a small brigade of loyal memoirists, veered from suicidal despair to agitated optimism. As Hitler exercised his dog, Blondi, in the ruined gardens of the Chancellery, Russian armoured divisions raced towards the old German borders on the River Oder, often eating up over 300 miles every day. The Soviet armies smashed, split, encircled or simply swept past the shattered relics of Hitler's war machine. On the Western Front, the last German offensive in the Ardennes had finally fizzled out. A massive steel trap began to close around Hitler's Berlin fortress.

Now the Reich would be defended by old men, boys of the Hitler Youth and the last relics of Himmler's foreign legions. Many SS veterans have had nearly seventy years to rehearse their story and prepare their defence. But one French volunteer who survived the last days of the Reich, 'Pierre', tells his story with unvarnished rawness. For many decades after the liberation of France, Pierre naturally chose to remain silent about his service in the French Waffen-SS. But since his retirement, he has sought out other veterans; now, he says, they have little to lose by openly discussing their experiences. They do not discuss ideology and share a robust contempt for the patrician Christian de la Mazière, the SS 'Charlemagne' volunteer who made a highly publicised confession in the celebrated French documentary about wartime collaboration, *The Sorrow and the Pity*. Born into a lower middle-class Parisian family in 1920, Pierre attended mainly Catholic schools. The Catholic Church had always maintained close bonds with the anti-Semitic right. In the bitter aftermath of the Dreyfus Affair, the Catholic founder of the Ligue antisémitique de France, writer and journalist Édouard Drumont (1844–1917) accused Jewish bankers and speculators of destroying the traditions of French Christianity. Voltaire had denounced Jews as wicked relics of antiquity, but for Drumont they were heralds of a corrosive modernity. His vitriol spilled into the pages of the Catholic newspaper *La Croix*, which persisted smearing the name of Alfred Dreyfus long after he had been exonerated. When Pierre turned 14, he joined the French Blueshirts (Mouvement Françiste), founded by another poison-tongued hack, Marcel Bucard. He says that he was inspired to join by one of his school teachers, who had taught him a 'profound anti Bolshevism: a real threat to western civilisation'. Bucard banned Jews and Freemasons from joining the Blueshirts and his party was generously supported by the reactionary perfume magnate François Coty, as well as the Abwehr. In common with many young radicals in the 1930s, Pierre revered Adolf Hitler and was thrilled by his rise to power. Then in 1939, came the shattering news that the Germans had done a deal with the hated Bolsheviks. Pierre explains: 'The German-Soviet friendship treaty which led to the occupation of Poland by the Germans and the Bolsheviks was a big disappointment to me: at least, it led me to join the [French] army without any ulterior motive and with a free mind.'

Pierre was 18 when he joined up. In May 1940, as the German advance pushed his unit back across Belgium, he was captured along with six other French soldiers by an infantry convoy of the German 7th Panzer Division. Pierre points out that 'the six of us represented three different regiments – so you will have an idea of the chaotic circumstances following our retreat from Belgium'. The Germans herded their prisoners into trucks:

> my Germanic 'cousin' asked me whether I was hungry. I answered in the affirmative and he offered me a piece of bread and three pieces of sugar, apologizing and

> explaining at the same time that they had made such a fast progress that their supply of food had not been able to follow them! That was my first encounter![26]

Pierre was taken to the POW camp Stalag VIIIC, close to Sagan in Upper Silesia. It was here, at evening roll call on 22 June 1940, that he heard the stunning news that the French had signed an armistice at the railway siding in the Compiègne Forest, where the Germans had formally capitulated in 1918:

> Like myself, the people of my shack were filled with consternation and were absolutely quiet; that was not so in the [German] barracks where the news was received with cries of joy! I was equally happy and unhappy: happy because the horrors of war had ended for all the French civilians who were fleeing on all French roads and unhappy about the defeat of my country and its army.

As Pierre's comrades struggled to adjust to life behind barbed wire, many had difficulty coming to terms with the shame of the French surrender. Pierre shared their humiliation. Like other young Frenchmen, he had a ready explanation for the catastrophe. France had been brought to her knees by 'certain dark forces'. The French Prime Minister Léon Blum had led France to defeat. And no wonder: Blum was a Jew! He called himself a 'socialist' but the truth was he was a 'Bolshevik' and in league with the Russians. Since the Nazi-Soviet Pact remained in force when Hitler invaded Scandinavia and Western Europe, Pierre reconciled taking up arms against Germany with his fierce hatred of Bolshevism. But, he went on: 'The German attack on Russia in 1941 made me once again change my frame of mind. Without participating, I had become an "interested spectator".'

Pierre was released from Stalag VIIIC a year later. Back in German-occupied Paris, he was dismayed to find out that he was now liable for compulsory labour service (RAD) in a French factory, manufacturing goods for Germany. He hated the idea of working in a German factory. So he volunteered to join the Todt Organisation, the paramilitary engineering corps, as a Fremdarbeiter in OT-Einsatzgruppe West, becoming, in effect, 'a military worker in uniform'. At the Todt 'Information School', Pierre trained as a telephone operator and, not for the first time, swore an oath of allegiance to Adolf Hitler. The oath marked a 'definite break', he says, 'between yesterday and tomorrow'. Pierre won't say much about his time in the Todt Organisation ('boring'), but at the beginning of 1944, he made a decision to 'take up arms for the Reich' and in May began training with the Waffen-SS at a camp in Duisberg. Did he not realise that Germany was headed for defeat? 'We had hope,' he says, 'we had heard about the new wonder weapons. And besides, we hated the Bolsheviks.'

Pierre admits that SS training included 'lessons in political theory' but 'Infantry School comprised combat, marches, 'aiming at targets', singing led by marvellous instructor sergeants, all bearers of the insignia of combat of their speciality and the Iron Cross Second Class. The comparison with the French army, he says, was not to the latter's advantage! Once a week the officers took their meals at the same table as the men of their troop who were encouraged to talk freely and ask questions. 'Unthinkable in France!' Pierre had hoped to be assigned to the Kriegsmarine, the German navy, but in the last summer of the war the Waffen-SS needed not sailors but soldiers on the Eastern Front. Pierre ended up on a troop train to Greifenberg in Pomerania, where he joined the Légion des Volontaires Français contre le Bolshevisme (LVF), formed in 1941 by French rightists like Bucard, Marcel Déat, Jacques Doriot and others. He was not happy: 'Many of us were bitter and disappointed to find ourselves again under French orders. To calm us down we were sent into the neighbouring villages to dig potatoes!' The LVF was in any case a spent force. Sent into action near Moscow at the end of 1941, the legion had been virtually annihilated. The French volunteers would never completely recover. In September, as Pierre was digging up Pomeranian potatoes, Himmler transferred all French nationals serving the Reich in the Waffen-SS, the Wehrmacht and the Todt Organisation to a new SS division: the Waffen-Grenadier-Brigade der SS 'Charlemagne' (französische No 1).[27]

By now Pierre spoke excellent German, and at the beginning of 1945 received promotion to Rottenführer (sergeant) and 'translator to the High Command'. Pierre sums up the mood of the French SS men laconically: 'We did not think about whether the war was going to be won or lost, nor did we make any prognosis; our role, the role of every soldier, was to obey ... Without getting our hopes too high, we knew that Germany was always good for a surprise.' In other words, he had every expectation that one of the powerful new 'Wunderwaffen' would soon be deployed.

The SS 'Charlemagne', commanded by Brigadeführer Gustav Krukenberg, was, as Pierre soon discovered, undermanned and poorly equipped. With just over 7,000 recruits, the 'Charlemagne' would never reach divisional strength. In September, the French SS men joined Himmler's defence of Pomerania, and in the savage battles as the German armies fell back through East Prussia and Pomerania, the French suffered catastrophic losses. A 'Charlemagne' veteran, who transferred from the French 'Milice', remembered that 'We marched all night. During the day we hid in the woods. We tried to eat in abandoned farms. Russian and Polish cavalry hunted us down ... We went through Stettin, in flames ... still on foot still pursued.'[28]

According to the memoirs of Felix Kersten, Himmler exalted in bloodshed: he proudly informed Kersten that of the 6,000 Danes, 10,000 Norwegians,

75,000 Dutch, 25,000 Flemings, 15,000 Walloons and 22,000 French serving in the Waffen-SS, 1 in 3 had been killed in action. Losses among eastern volunteers from the Baltic and Ukraine had been even more impressive.[29] His sanguineous joy was driven by a perverted logic. For him, the catastrophe that engulfed the foreign volunteers had a different kind of significance. This was the blood sacrifice that would bind together the future 'Germanised' citizens of 'SS Europa'. It was a kind of Spartan folly that in death lay rebirth. Himmler had often spoken of 'harvesting Germanic blood' wherever it might be found. Now that harvest would be winnowed in the savage winds of war.

A long way from the slaughter, Himmler took refuge inside the Hohenlychen clinic, which was protected from aerial attack by two big red crosses painted on its roof. He was already in disgrace. At the end of the previous year, on 26 November, Hitler had rewarded Himmler, whom we should recall was both SS chief and the Minister of the Interior, by making him commander-in-chief of the Army Group Upper Rhine (Heeresgruppe Oberrhein).[30] According to the memoirs of Himmler's masseur, Felix Kersten, Himmler vowed to drive the Allies into the sea. But the new Supreme Commander had seen little real military action – and Hitler's appointment provoked a predictable response from Wehrmacht commanders like Guderian: 'Then whom do we get? Hitler appointed Himmler! Of all people – Himmler!'[31] When the Heeresgruppe Oberrhein was humiliatingly swatted aside by advancing American forces, Hitler recalled Himmler from his 'watch on the Rhine' and ordered him to rebuild the shattered 2nd and 9th Armies in North Prussia as a new Army Group Vistula (Heeresgruppe Weichsel). This second and, as it turned out, last promotion was, it would seem, a calculated slap in the faces of allegedly defeatist and treacherous Wehrmacht commanders.

Hitler's appointment led not only to military disaster, but to a humanitarian catastrophe as well. As the Soviet forces raced pell-mell towards the Oder, tens of thousands of German civilians fled west before them, desperately hoping to reach safe havens inside Germany. At the end of January, Himmler halted his headquarters train at Deutsche Krone in western Pomerania (now Wałcz in the west of Poland) and summoned the local Gauleiter Franz Schwede-Coburg. Shortly afterwards, Himmler issued orders that no German citizens would be permitted to leave East Pomerania. As loyal citizens of the Reich they must stand firm. On 4 March, Soviet tanks reached Kolberg and cut off close to a million and a half refugees, compelling them to flee north to the Baltic ports where they sought to escape by sea. The tragic consequences are, of course, well known: Soviet submarines torpedoed grossly crowded refugee ships, including the *Wilhelm Gustloff*, with dreadful loss of life. The role played by SS Chief Himmler and the grotesque mismanagement of the German administrators has been conveniently forgotten.

It was said by Waffen-SS General Felix Steiner that Himmler's 'notions of military affairs were devoid of any real solidity ... the most junior lieutenant could put him right'. His evaluation chimes with other eyewitness accounts – most notably, the diaries of Colonel Hans-Georg Eismann, who left a damning portrait of Himmler as military commander.[32] His handshake, Eismann recalled, was soft and feminine; his eyes had a distinctly 'Mongolian look'. Himmler was much preoccupied with health matters and petty obsessions dominated his daily routine, which had little to do with strategic necessity. He worked for an hour before lunch then retired for a siesta until mid-afternoon, then resumed work until 6.30 p.m. Himmler appeared to be exhausted, and unfocused. As a military commander, Eismann concluded, Himmler was like a 'blind man talking about colour'. But it did not take a military genius to see just how dire was the plight of Himmler's forces. Like the Wehrmacht commanders to whom he reported, it suited Eismann to point the finger at Himmler, but the harsh reality was that the Soviet tide could no longer be turned back. On 15 March, Zhukov launched a fresh offensive that took Himmler, and indeed the rest of the despised General Staff, completely by surprise, and smashed the surviving German forces in Pomerania. On the 19th, Soviet forces broke through to the Oder, south of Altdamm, and a despairing General Hasso von Manteuffel presented Hitler with an ultimatum: 'Either withdraw everyone to safety on the west bank of the Oder overnight or lose the whole lot tomorrow.'[33] For once, Hitler took a general's advice – and the German forces pulled back across the Oder, and destroyed the main bridges.

Himmler had by then been forced to relocate his headquarters to a villa, 80 miles north of Berlin and conveniently close to the SS clinic at Hohenlychen. It was here that Himmler had, probably through the intercession of Kersten, begun to plan secret meetings with the head of the Swedish Red Cross, Count Folke Bernadotte, to hammer out a deal with the Allies, though he continued to insist that he could not contemplate 'betraying Hitler'. Count Bernadotte had a single objective: to secure prisoner releases; he had no illusions that the Allies would ever contemplate a deal with the 'hangman of the Reich'.[34] Hitler had no idea that Himmler was contemplating such treachery. But the SS chief's days of military glory were coming swiftly to an end. On 7 March, Himmler suffered a severe attack of angina. Dr Karl Gebhardt (who was responsible for some of the most gruesome medical experiments conducted in the SS camps) had him admitted to Hohenlychen as a patient. In Berlin, a whispering campaign accused Himmler of seeking personal military glory and leading his army group to disaster. Locked away in his subterranean fortress, Hitler finally turned against the man he had long believed to be his most loyal paladin. According to Goebbels' diary entry for 11 March, Hitler raged that Himmler was guilty of 'flat disobedience'. On 15 March, a shamed Himmler was driven to the Chancellery to confront the apoplectic Hitler and received, head bowed, a 'severe dressing down'.[35] Humiliated, Himmler slunk back to his sick bed at Hohenlychen, and, after a visit from Guderian, surrendered his command of the Army Group Vistula. Hitler replaced him with the rather more competent General Gotthard Heinrici.

As Stalin's armies prepared to crush the last strongholds of the Reich, Gottlob Berger dispatched the last of the SS foreign legions to defend Hitler's capital – Fortress Berlin. The city would now become the bonfire of the collaborators. As the act of the Reich unfolded, Hitler called a military conference in Berlin and denounced Himmler's SS foreign legions. He demanded a meeting with the SS chief to explain why a 'Ukrainian legion' was fighting alongside German soldiers: 'There's still a Galician division wandering [about] out there. Is that the same as the Ukrainians? If it consists of Austrian Ruthenians [Ukrainians] we can't do anything other than take their weapons away immediately. The Austrian Ruthenians were pacifists. They were lambs, not wolves ... It's all just self deception.' The next day, an abject Himmler was forced to return to the Chancellery for another demented brow beating.

His master plan had at last been torn to shreds.

A week later, Soviet commander-in-chief, Georgi Zhukov, and his staff occupied a command bunker dug into the Reitwin spur, a fish hook of land on the eastern bank of the Oder near the town of Küstrin. From this vantage point, Zhukov contemplated the main axis of attack that led directly across the Oder then ascended, through a labyrinth of marshes, drainage ditches and streams, to the Seelow Heights. His plan was not especially subtle. The Soviet juggernaut, 11 armies and 8,000 artillery pieces, would simply batter the Oder strongholds, occupied by the 9th Army, to dust. Beyond lay the road to Berlin. More than a hundred miles to the south, in the Cottbus region, Zhukov's rival Ivan Koniev would lead his tank army across the River Neisse, under cover of 'night and fog', to strike at the 4th Panzer Army. In Zhukov's command bunker a scale model of Berlin had been specially constructed, and as he waited impatiently for Stalin's orders to strike, Zhukov obsessively scrutinised every detail of Hitler's fortress. Even now, the conquest of the Nazi citadel looked like a formidable challenge. On the other side of the Oder, Himmler's successor General Gotthard Heinrici had no doubt that the Russians would open proceedings with a massive artillery barrage and he prepared his defences accordingly. On 14 April, as agreed at the Yalta Conference, American forces halted on the Elbe, leaving Berlin to Stalin. Two days later, on 16 April at 5 a.m., a radiant explosion of light signalled the start of the Soviet bombardment. The barrage commenced soon afterwards. But the millions of Russian projectiles crashed and detonated in empty trenches. Heinrici had cunningly pulled back his front-line troops. As lethal fire rained down from the German positions on the Heights, tearing through Zhukov's men struggling to cross the Oder marshes, Koniev's engineers rushed to throw bridges and pontoons across the Neisse at 150 carefully chosen points, and his tank armies began rattling across, heading for the autobahn that led straight as an arrow towards the southern suburbs of Berlin.[36]

The German commander of the French SS division, Dr Gustav Krukenberg, had at his disposal two battalions that between them could muster just 700 men. On 21 March, the French SS volunteers mustered at Anklam railway station to wait for transportation to new billets in Mecklenburg. When no trains appeared, they set off on foot, singing. Three days later, the Frenchmen marched into Neustrelitz, about 50 miles north of Berlin, and found billets in surrounding villages. Krukenberg told the men: 'You may abandon the armed fight … I only want to have combatants with me now.' The majority of the French SS volunteers, including Pierre, now agreed to fight on. A hundred miles to the south-east, Zhukov's armies finally smashed through the Seelow Heights fortresses on the 19 April. Two days later, Koniev's forces captured the massive and abandoned concrete bunkers at Zossen that had been the headquarters of the German high command and the nerve centre of the German 'war of annihilation'. At dawn three days later, the two Russian armies joyously met close to Schönefeld Airport.

One evening, as the Soviet armies rumbled ever closer to Fortress Berlin, SS General Felix Steiner escorted the Belgian collaborator and celebrated hero of Cherkassy, Léon Degrelle, on a tour of Berlin's ringbahn. The light was fading, and Berlin's concrete ramparts were intermittently illuminated by the unremitting Soviet artillery barrage. Steiner pointed out Soviet tanks already crawling through the eastern suburbs. For the histrionic Walloon, the vast panorama evoked the last days of the Roman Empire. At 9 p.m. Degrelle drove back into the centre of Berlin in a battered Volkswagen. In the Hotel Adlon's still brightly lit restaurant, waiters in spotless tuxedoes bustled about, serving purple slices of *kohlrabi* on silver platters. Soon, Degrelle imagined, the grand old hotel would be set aflame by some 'large pawed barbarian'. Not far from the Adlon, deep underneath Voss Strasse, Hitler still reigned over his stifling underground empire. Degrelle and a small party of Belgian SS men drove on south towards Potsdam where the little party rested for a few hours.

For Degrelle, Germany still stood for civilisation, a bulwark against those Soviet barbarians he had watched gobble red worms and corpses on the Eastern Front. Here in Berlin, the capital of the Reich, noble, pale-faced young men of the Hitler Youth waited quietly for the enemy, clutching their Panzerfausts 'as solemn as the Great Teutonic knights'. To the east, along the rapidly disintegrating German front line, Degrelle's SS 'Walloonien' comrades now fought alongside Flemish SS volunteers and a battalion of Latvians. But the former Chef de Rex who inspired them to fight Hitler's war had had enough of heroics. On 30 April, at 8 a.m., he ordered one of his lackeys to pack some 'very heavy suitcases' and set off towards Lübeck in northern Germany, where he hoped to track Himmler down, leaving the Belgian volunteers to fend for themselves.[37]

The majority of Berliners, with the exception of a few diehard fanatics, had no doubt that to continue to defend the Reich was insanity. They waited stoically for the final cataclysm in stinking cellars and fragile shelters. For the foreign volunteers of the Waffen-SS like Pierre and his comrades realism was not an option. On 25 April, a French SS Sturmbataillon, armed with a few machine guns and Panzerfausts set off in the direction of the centre of Berlin, the citadel, in a ragged convoy of hastily commandeered private cars and trucks. Pierre recalls that 'there were 450 enthusiastic singing men leaving for the battle'. According to French veteran Robert Soulat, their morale was high: 'a strange flame burned in our eyes.' As the French SS battalion assembled in the Marketplatz in Alt-Strelitz, Krukenberg caught sight of a gleaming black Mercedes approaching at high speed, Nazi hub pendants fluttering wildly. Krukenberg realised that the bespectacled gentleman gripping the wheel with manic determination was none other than his commander-in-chief, Heinrich Himmler. The French SS men leapt to attention, right arms erect; but, staring rigidly ahead, Himmler swept past, apparently oblivious to the little group of loyal SS recruits. Later, Krukenberg realised that Himmler was returning from a meeting in Lübeck with Count Bernadotte. Their discussions had ended in stalemate and Himmler had begun building a bolthole for himself at Schloss Ziethen on Wustrow Island in the Bay of Lübeck. Soon after this troubling encounter, the SS battalion, numbering about 500 men, began marching south in Himmler's wake.[38]

As SS-Brigadeführer Krukenberg led the French SS Sturmbataillon into the outer suburbs of Berlin, they collided with a stream of refugees, flowing south through streets jammed with abandoned and burnt out trucks and carts. Soviet fighters roared back and forth a few hundred metres above roof tops, abruptly diving to strafe the terrified refugees. In the midst of this miserable human flood, men in black SS or green Order Police uniforms shuffled disconsolately, eyes cast down. As the French convoy reached the crossing on the Falkenrehde canal, the bridge exploded, wounding some of Krukenberg's men and crippling his car. They still had 15 very perilous miles to march before they reached the citadel. After fording the canal on foot, the battalion edged their way under the ringbahn and arrived at the Olympic Stadium about 6 miles west of the Chancellery. They broke into a Luftwaffe supply depot and were delighted to find that it was still crammed with hundreds of bars of Swiss chocolate. The next morning, Krukenberg requisitioned another motor vehicle and drove ahead of the battalion through Charlottenburg towards the Brandenburg Gate, where Speer's East–West Axis, leading east from the Victory Monument, had been turned into an airstrip. The citadel was eerily quiet. Just after midnight, Krukenberg drove across Pariser Platz then turned right along Wilhelm Strasse towards the Reich Chancellery. The sky over Berlin glowed a lurid red, and the SS men could hear the steady rumble of heavy artillery. Krukenberg was astonished: the citadel appeared to be undefended.

Krukenberg and a few officers entered the Chancellery garden and were admitted to the Führerbunker; descending a long concrete staircase they entered a strange, malodorous realm that Goebbels compared to a 'maze of trenches'. Krukenberg requested a meeting with Army Chief of Staff, General Hans Krebs, the last senior military figure left in Hitler's bunker community. The recycled air was warm and foetid: a stinging cocktail of oil, urine and sweat. A diesel engine spluttered and hummed. Hitler's SS 'Leibstandarte' guards sat drinking morosely in a dingy little room not far from his study. As the Russian guns ceaselessly pounded Berlin's shattered buildings, the lights inside the bunker flickered. Hitler had been ruined by this subterranean life. His face had turned pale, puffy and sallow; he was barely able to make his way from his private quarters to the military conference room unaided. He had become slovenly, his clothes splattered with food stains. Hitler still relished wolfing down his favourite Viennese cream cakes.

Krebs, as Krukenberg discovered, had not given up hope. Both he and Hitler had become convinced that German units were being mustered somewhere to the south by General Walther Wenck, referred to reverentially as 'Wenck's Army', that would soon march north, brushing aside allegedly weak Soviet spearheads, and 'rescue Berlin'. Before Krukenberg left that night to return to his men, he requested a meeting with Fegelein, who was, in theory, his commanding officer. A search was ordered – but Fegelein was nowhere to be found. A puzzled Krukenberg drove back towards the Olympic Stadium, which they reached just before dawn.

A few hours later, Krukenberg met with General Helmuth Weidling at his headquarters in Hohenzollerndamm in Wilmersdorf. Now, he told Krukenberg, he faced 2.5 million Soviet troops with just 2,700 Hitler Youth, 42,500 geriatric Volkssturm units, armed with a few Czech and Polish rifles – and dilapidated foreign Waffen-SS units.[39]

Berlin was now completely encircled. For the desperate inhabitants of Hitler's subterranean fortress, the single link to the outside was the East-West Axis airstrip. As Russian shells descended on the centre of Berlin; as American and Russian soldiers exchanged cigarettes at Torgau on the Elbe; and as RAF bombers pounded the abandoned Berghof in the Bavarian Alps, Hitler and Goebbels talked long into the night about how they might still win 'this decisive battle' and not leave the stage of history in disgrace. Goebbels conjured up a rosy posterity: if Europe was 'Bolshevised', then National Socialism would swiftly attain *ein Mythos* (a mythic status). History would look kindly on their hard-fought crusade against Bolshevism.

Shortly after dawn on 28 April, Soviet T34 tanks clattered across pontoon bridges thrown across the Landwehr canal close to Hallesches Tor, the gateway to the broad avenues that led straight to the citadel government quarter and the Reich Chancellery. The canal was the last line of defence – the moat around Hitler's citadel. The French SS Sturmbataillon, reduced to just sixty men but reinforced by Hungarian and Romanian SS troops, threw up a fragile defensive line across the broad avenues of the old Friedrichstadt. Ammunition was in desperately short supply and the Russian tanks just kept on rolling across the pontoon bridges. No word came from the longed for 'relief army' of General Wenck.

On 29 April, Hitler convened his final military conference. He ordered that all defenders of the 'capital city of the Reich' must break out of the Soviet encirclement, join up with other units still fighting and take to the forests. Not a whisper had been heard from Wenck. As it turned out, Wenck had never got any further than Potsdam and was withdrawing to the Elbe. The 'relief of Berlin' had been a chimera. In his 'Testament', dictated to his faithful secretary Traudl Junge, Hitler blamed the downfall of Germany on 'international Jewry'. He insisted that 'I do not wish to fall into the hands of enemies who ... will need a spectacle arranged by Jews'. At about 3.30 a.m. on 30 April 1945, Hitler and Eva Braun, who he had married the same day, committed suicide. Hitler's valet Heinz Linge hauled the corpses up to the Chancellery garden. The loyal paladins of Hitler's court, led by Goebbels and Bormann, watched solemnly as Linge emptied jerry cans of fuel over the remains of the Führer and his wife. As acrid black smoke rose into the air, Soviet shells began falling into the garden.[40]

As this macabre rite unfolded in the Chancellery garden, some twenty-five survivors of the French SS Sturmbataillon had taken refuge in the basement of a library on Friedrichstrasse. One of more fanatical French SS officers, Hauptsturmführer Henri Josef Fenet, lamented that its beautiful collections would soon be ripped to shreds by a drunken 'Mongols'.[41] It was a rather beautiful spring day. In the Tiergarten, sunlight dappled the shredded trees. Twisted ruins stood out against the blue, cloud-flecked sky. Smoke drifted across rubble filled streets. On Tuesday 1 May, SS-Brigadeführer Mohnke appeared at Krukenberg's headquarters at Stadtmitte U-bahn station. A party of German officers led by Colonel General Krebs had crossed the front lines to begin surrender negotiations with the Soviet commanders. He admitted that 'Army Wenck', the promised relief force, had been beaten back. But he chose not to reveal that Hitler was dead. It was a gross deception. Mohnke now ordered Krukenberg to take his men to Potsdamerplatz station and halt the Soviet advance. This was the last line of defence. The obedient Krukenberg fought his back way as far as Hermann Göring's Air Ministry – a massive granite edifice within sight of the Chancellery. Terrified Luftwaffe men cowered in the bombproof basement. Before Krukenberg could get any further along Leipzigerstrasse, he was called back to the Air Ministry building. Krebs had rescinded his orders. The battle for Berlin was over.

As night fell, the French SS men took refuge in the ruined vaults of Reinhard Heydrich's former bastion – number 8 Prinz Albrecht Strasse. Himmler's SS headquarters lay in ruins next door. Its cellars, once used to torture the enemies of the Reich, now provided sturdy shelters. At 7 p.m. Krukenberg furtively crossed Leipziger Strasse and descended the long concrete stairway into the Führerbunker for the last time. Inside he met Krebs who confessed that Hitler was dead, and by his own hand. Goebbels too had killed himself, his wife and his children.

In his memoir Krukenberg wrote: 'All the sacrifices of the troops had been in vain. The idealism of the volunteers had been abused in the worst way.'[42]

The last survivors of Himmler's 'Germanic army' now fled Berlin. A day later, Pierre arrived, alone, in the village of Gadebusch, halfway between Rostock and Lübeck. In the middle of the road, blocking his way, stood a solitary British parachutist 'who invited me, very politely, to enter a wonderful meadow which served as Assembly Camp where the number of one hundred German prisoners of war grew to several thousand within two or three days!'

A few badly shaken French SS men took refuge inside Postdamerplatz station, sheltering with elderly German veterans of the kaiser's war behind a pile of wicker baskets. Russian soldiers found them there the next day. A Soviet soldier brandished a gun in the face of Unterscharführer Roger Albert-Brunet shouting 'SS! SS!' then shot him through the head. He was the last foreign legionary to die in Berlin.

I had a question for Pierre. Does he regret pledging allegiance to Hitler – and did he feel, like Krukenberg, that he had been abused in the 'worst possible way'? When he replied after a few moments thought, Pierre had defiant look. 'I don't regret anything. Were we wrong about Bolshevism?'[43]

15

The Failure of Retribution

Certainly we had been beaten. We had been dispersed and pursued to the four corners of the world.

Léon Degrelle, *Campaign in Russia*

At the end of April 1945, Léon Degrelle reached Schwerin in northern Germany. Here, beneath the grand castle of the Dukes of Mecklenburg, a vast human torrent flowed away from the advancing Soviet juggernaut towards the Baltic ports; broken carts and abandoned clothes blocked the roads. Degrelle was in search of Heinrich Himmler. He knew nothing of Himmler's sudden downfall and would not learn that Hitler was dead for another two days. On 1 May, Degrelle and his small entourage, led by faithful bearer of luggage Robert du Welz ('promoted' to SS-Hauptsturmführer), moved on to Kalkhorst. His quarry was indeed very close, hiding inside the Schloss Ziethen some 35 miles to the north. Officially, he was still Reichsführer-SS, Chief of Police and Chief of the German Reserve Army. He was accompanied by a tiny company of faithful attendants that included Dr Rudolf Brandt, his doctor, Prof. Dr Karl Gebhardt, Werner Grothmann (Chief Adjutant of the Waffen-SS) and Otto Ohlendorf, who had once commanded Special Task Force D.[1]

On 2 May, Degrelle was at last able to corner his prey. He noticed a line of official Mercedes emerging from a side road and set off in hot pursuit. It was getting dark when he caught up. Pale skinned and wearing an odd leather cap, Himmler, Degrelle recalled, took both his hands and proclaimed: 'You have been among the faithful, you and your Walloons … Gain yourself six months. You must live.' Degrelle desperately affirmed his 'absolute loyalty unto death'; how should he continue the struggle? Himmler asked how many men he had with him: just three, Degrelle admitted. A smile hovered on Himmler's thin lips. Then he embraced Degrelle for the last time and drove away. As night fell, Degrelle doggedly pursued Himmler's party, but his sputtering VW, burdened with bulging suitcases, could not keep up with the powerful Mercedes as it vanished into the darkness. In a last conversation with Dr Kersten, Himmler had speculated: 'what will history say of me? Petty minds, bent on revenge, will hand down a false and perverted account of the great and good things I have accomplished for Germany.'[2] On 22 May 1945 the former Reichsführer-SS bit down on a potassium cyanide capsule concealed in a back tooth and was dead moments later.

On 4 May, Degrelle and Hauptsturmführer du Welz reached Copenhagen. Both Denmark and Norway remained officially under German control. The following day, Degrelle met Dr Werner Best, the German Plenipotentiary, and demanded access to an aircraft. Best sent him on to Oslo where he had meetings with Reichskommissar Terboven and Vidkun Quisling. Neither showed any interest in Degrelle's offer to prolong the struggle against Bolshevism. Instead Terboven offered Degrelle the use of a Heinkel aircraft 'belonging to Minister Speer' that had been abandoned at Oslo Airport, complete with pilot. That evening Degrelle and his loyal adjutant, still wearing their SS uniforms, drove to the airport. The aircraft took off at midnight and ascended over the North Sea before turning south-west. After a hair-raising flight, the Heinkel crash landed in the Bay of Concha not far from San Sebastian in northern Spain. A journalist from *Le Soir Illustré* photographed the wreckage. As the former Chef de Rex lay in a hospital bed recovering from his injuries, the government of liberated Belgium applied for his extradition. At the last moment, Degrelle vanished again, spirited into hiding by Spanish Falangist admirers. Mme. Degrelle had chosen not to join her disgraced husband and obtained a divorce. Later he remarried; his new spouse was a niece of the French Milice chief, Joseph Darnard. As 'José León Ramírez Reina', Degrelle embarked on a new career as a construction magnate and, thanks to some lucrative deals with Franco's government, made a fortune. By the end of the 1950s, Degrelle emerged from hiding and was busy forging bonds with the new European far right who revered him as the last link to Hitler and the Reich. He allied himself with the Spanish neo-Nazi faction, the Círculo Español de Amigos de Europa (CEDADE). The Madrid office published his rambling biography of Hitler, as well as his solipsistic recollections of the Eastern Front. Emboldened by new circle of admirers, Degrelle made a public appeal to Pope John Paul II to snub the 'fraud of Auschwitz' and entertained Jihadist anti-Semites at his opulent villa in the hills above Malaga. After Franco's death, the Belgian government renewed its application to have the corpulent and fast-fading

Degrelle brought back to Brussels to stand trial for treason. But Léon Degrelle, the former Chef de Rex, the 'hero of Cherkassy' and one of the most contemptible and self-deceiving foreign servants of Hitler's Reich was lucky to the very end. He died in Malaga, unrepentant, in 1994.

Degrelle fascinated and inspired resurgent neo-Nazis and Holocaust deniers. But he had little real impact on post-war European history. He was a relic, a ghost. Haj Amin el-Husseini, the Grand Mufti of Jerusalem, left behind a more menacing and malign legacy. German-produced Arabic radio broadcasts had saturated the Muslim world with hysterical warnings about Jews and global conspiracies, and at the end of the war, the Mufti's renown in Palestine and Egypt was undimmed. He would have a profound and malicious impact on the development of Islamic radicalism following the foundation of the state of Israel in 1948.

By the winter of 1944, Berlin was no longer a safe haven for men like the Grand Mufti. He had never been a brave man and was often found cowering under tables as the great armadas of Allied bombers pounded the capital of the Reich. His allies in the Foreign Office, like Erwin Ettel, did what they could to protect their esteemed Muslim guest and tried to coax him to escape Germany to whatever safe haven he chose by U-boat. The Mufti was simply too timid to contemplate such a journey and held on in Berlin to the very end. At the beginning of May 1945, the Grand Mufti and his entourage at last packed up and fled. He knew that once the British reached Berlin they would waste little time tracking him down. After many tribulations, they managed to reach Constance in the French zone of occupation. Recalling how well he had been treated after his flight from Palestine, when he escaped to French Beirut from British Palestine, the Grand Mufti surrendered to the French authorities. He was soon relaxing in an opulent villa near Paris.

The British urgently petitioned the French authorities to hand over the fugitive Muslim cleric who had slipped out of their hands so many times. But General de Gaulle was in no mood to oblige his ally and personally issued instructions that el-Husseini should be permitted to remain in France and resume, without interference, his political activities on behalf of the Palestinian Arabs. For the French, who blamed the British for the catastrophe of 1940, the Mufti offered a delicious opportunity to spite perfidious Albion. Since the French had reasserted their presence in North Africa, they had good reason not to wound Arab public opinion. The Mufti had little time to enjoy French hospitality. His protectors discovered that an 'Irgun' assassination squad had arrived in France. On 28 May 1945, el-Husseini bolted to Italy, then secretly boarded a British ship, the SS *Devonshire*, bound for the Egyptian port of Alexandria.[3]

The return of the Grand Mufti electrified the Arab world. At a rally at Heliopolis in Cairo exultant crowds swamped his convoy – and King Farouk offered him appropriately sumptuous accommodation at his 'Inshas Palace'. The leader of the Muslim Brotherhood, Hassan al-Banna breathlessly declared: 'The hearts of the Arabs palpitated with joy at hearing that the Mufti had succeeded in reaching an Arab country … The lion is free at last and will roam the Arabian jungle to clear it of wolves. The great leader is back.'[4]

In 1936, Haj Amin el-Husseini had embarked on an epic journey that had led him from Iran and Iraq to the hub of Hitler's Reich. Like the Indian nationalist Subhas Chandra Bose, he was obsessed with the liberation of his disputed native land from a colonial power. But Bose was a radical socialist who rejected Nazi ideology and even dared criticise Hitler's *Mein Kampf*. The Mufti's hatred of Jews provided a poisonous bond with Hitler and his fanatical elite, above all Heinrich Himmler. In Berlin, the Mufti's loathing had been deepened. His German masters taught him to believe that only systematic mass slaughter could solve the 'Jewish problem'.

Faced with chronic unrest in Palestine, the British decided to leave the Grand Mufti in peace. Arrest would only enhance his reputation and they had nothing to gain from his martyrdom. An American agent stationed in Cairo reported that 'it was unlikely that any strong action will now be taken … against former Axis collaborators'.[5] A British observer provided a vivid picture of a very relaxed Grand Mufti:

> His Eminence was in excellent mood, charming, joking. 'Put yourself in the Arabs' place. Remember yourselves in 1940. Did you ever think of offering the Germans part of Britain on condition they let you alone in the rest. Of course not, and you never would. To start with, you would have preferred to die defending it. In the second you know that they would never have kept their word to remain in the one part.'[6]

Requests by the new Yugoslavian government for the repatriation of the Mufti to stand trial for his participation in the SS recruitment campaign in Bosnia also proved futile. As the battle for Palestine intensified, pan-Arab organisations like the Arab League had every reason to protect their own. The discovery of German crimes in Europe after the opening of the camps had no resonance in the Middle East. The enemy remained the old colonial powers, especially Great Britain. Most Arab nationalists like Ahmad Husayn, leader of the semi-fascist Young Egypt Party, and many others allied with King Farouk sharpened their propaganda war on Britain and her Jewish partners. In November 1945 the reinvigorated Akhwan el Muslemin, the Muslim Brotherhood, incited attacks on Jewish homes and property in Alexandria. In the post-war period, Arab fantasies about the 'Jewish enemy' would be reinforced as the West German government reappointed many of the Mufti's Foreign Office supporters and stationed them at embassies in Baghdad and

Cairo. Muslim radicals revered Haj Amin el-Husseini. The late Fatah leader known as Yasser Arafat eulogised him as his political mentor.[7] El-Husseini brought back to the Muslim middle the vile poison of Reich anti-Semitism. The 'Final Solution' had been the Mufti's passport to power and influence in Nazi Germany. Ensconced in an opulent Berlin villa, he relished his new status as the Reich's pet Muslim eminence and became a zealous student of racial pseudoscience promoted by German universities and other Reich agencies. As Gottlob Berger's SS recruiter in the Balkans and the Caucasus, the Mufti frequently referred to 'killing Jews' as the main task of Muslim divisions like the SS 'Handschar' – though this wicked plan meant little to his Bosniak recruits. It was the divisional Imams who ranted about murdering Jews; and they had been trained at the Mufti's schools in Germany. Haj Amin el-Husseini brought this malevolent cargo back to Alexandria at the end of the war, like Bram Stoker's Dracula.

National Socialism moulded the thinking of Arab nationalists and profoundly influenced the despotic pseudo-moralism of Islamic extremism. In the early 1950s, the movement's martyr Sayyid Qutb made the link between Axis and Islam explicit in his book, *Our Struggle with the Jews*, which was republished in Saudi Arabia in 1970. Qutb adopts the conspiratorial analysis of anti-Semites like Rosenberg; his language is riddled with references to 'Jewish machinations and double dealing'; Jews are the 'worst enemies of Islam', 'sowing doubt and confusion and hatching plots'. Only the Koran and the Muslim community stood in the way of this Jewish offensive: a Manichean eschatology that mimics Hitler's rants about the Soviet Union.[8]

The Grand Mufti's bond with Hitler's Reich scars and mutilates Middle Eastern politics to this day. In the first decade of the twenty-first century, anyone who seeks the Grand Mufti's monument needs look no further than the conspiracy-mongering and 'Jew-hating' doctrines of Hamas and Hezbollah.

16 March 2010

At the Freedom Monument in Riga, the SS veterans vanish and the protestors begin to disperse. A few curious onlookers have stayed behind to gaze at the mass of wreaths and flowers propped against the plinth. The elderly veterans have vanished along with their shaven headed protectors. Young Latvians, who on the surface appear to have no connection with the far right and neo-Nazi supporters of the legion, take pictures of each other in front of the monument. I can hear one smartly attired young man telling a BBC reporter that he believes the legionaries were 'heroes'. It would seem that a new generation of young Latvians has begun to adopt the memory of the Latvian SS legion as an icon of national identity seven decades after the end of the Second World War. It is not hard to understand why. Latvian historians and educators have deliberately chosen to commemorate the 'three occupations' and to erect a prominent Museum of Occupation. There is no national museum dedicated to the Latvian Jews who were murdered all over Latvia after June 1941. Until memory of the Holocaust and the role of the Latvian collaborators are given appropriate moral clout, young Latvians will persist in miscommemorating the SS divisions as a 'national army'. The destruction of the so many tens of thousands of Jewish citizens of the Baltic nations must be understood as the focal catastrophe of the region's history, not a sideshow to foreign occupations.

The shadows of Himmler's foreign legions still fall over Europe. The terrible role played by SS executioners like the Latvian Viktors Arājs is in danger of being forgotten – or worse, rationalised as a necessary response to Bolshevism. Conservative intellectuals in Eastern Europe and Germany have launched a campaign, spearheaded by the 2008 Prague Declaration, to elevate the Soviet 'terror' to equal terms with the German Holocaust. They demand that Nazi and Soviet crimes should be jointly memorialised and invested with the same moral status. This is both bad history and bad ethics. By representing the premeditated plan to do away with an entire pejoratively defined race as somehow ethically one and the same as the cruelties of Stalin's regime, the signatories of the Prague Declaration knowingly erode the specificity of the Holocaust, which has been proven by scores of studies. Equivalence implies that alleged victims of Soviet occupation who collaborated with the Third Reich should be exonerated on the grounds that by doing so they defended their nation against Soviet terror. This pervasive rationalisation occludes the fact that many collaborators participated in ethnic cleansing in the occupied Soviet Union and were driven by the same exclusivist ideology. They backed the occupational powers not to defend their nations but to purge them of unwanted citizens.

Latvia is not alone in failing to exorcise the demons of wartime collaboration. It would be tempting to blame this on 'endemic anti-Semitism' – a natural tendency to blame 'the Jews' for any misfortune. But this would be wrong. The architects of this dangerous modern myopia were the Soviet liberators of Eastern Europe who rewrote the history of the German occupation for their own ends. It is well documented that in 1945, the Soviets tried and punished tens of thousands of 'collaborators' who had served in the Wehrmacht and Waffen-SS. This wave of retribution took place against the background of a resurgence of Russian anti-Semitism, masterminded by Stalin himself, that culminated with the 'Doctors' Trial'. This meant that the Soviets punished collaborators not for taking part in the German 'war on the Jews' and other lives 'not worthy of life', but for aiding 'Hitlerites' and killing non-specific 'innocent civilians'. Although the Soviet army had liberated death camps, including Auschwitz-Birkenau, only a very few Russian investigators, most notably Vasily Grossman, openly acknowledged the ethnic identity of the majority of victims. In official Soviet histories of the 'Great Patriotic War', the murder of millions of Jews in the occupied Soviet Union was pushed into the shadows, and any public discussion of a 'Jewish Holocaust' denounced as collusion with an American 'Jewish lobby'.[9] This fusion of open aggression against Soviet Jews, combined with a gross misrepresentation of the actual intent of the Nazi genocide, promoted congenital blindness to the historical experience of Russian and European Jews. This occultation of victims has had profound consequences for the way the war was remembered in the Soviet Union and its Eastern European satellite states – and is bearing very dangerous fruit in post-Soviet Eastern Europe.

In modern Romania, aggressive anti-Semitism 'continues to be a problem', according to a 2003 report by MCA Romania.[10] A Romanian television poll listed Cornelius Codreanu, the founder of the Legion of St Michael, as one of the hundred 'greatest Romanians'. Gigi Becali, the President of the New Generation Party and a wealthy businessman has revitalised old legionary slogans in his campaigns, and called for Codreanu to be canonised. In Hungary, the 'neo-Fascist' Hungarian Guard (Magya Garda), the paramilitary wing of the Jobbik Movement for a Better Hungary, explicitly acknowledges its descent from the Arrow Cross movement – and its black-uniformed supporters can stage large-scale, eye-catching parades. The emergence of these new extremist movements in Eastern Europe has the same roots as the Prague Declaration. Once someone is persuaded that 'Stalin was just as bad or worse than Hitler' then it is a short step to sanitising the ultranationalist far right and its ancestral militias like the Iron Guard and Arrow Cross which opposed Bolshevism. Mix in a slug of anti-Zionism and a shiny new kind of anti-Semitism no longer seems as shaming as it should. The rise of these aggressively xenophobic new movements following the collapse of the Soviet Union disconcertingly mimics the period after the First World War, when anti-Semitic demagogues emerged from the ruins of the German, Austrian and Russian empires.

Himmler and the German managers of the Holocaust dispatched their murder squads eastwards – into Poland, Ukraine and the Crimea. They built the apparatus of mass murder in the east – at Auschwitz, Treblinka, Sobibór. This geographical bias meant that Eastern European collaborators had a hands-on involvement in the killing, as Schuma riflemen or as camp guards. In the west, puppet administrations had actively collaborated with the management of the 'Final Solution' and tens of thousands of young men had volunteered to serve in the Waffen-SS, as well as police forces like the French Milice, which organised arrests and deportations to the east. In 1945, the new governments of France, Holland and Belgium confronted daunting problems as they struggled to come to terms with liberation. Asserting legitimacy depended on renouncing wartime collaborationist regimes and criminalising figureheads like Marshall Philippe Pétain or Vidkun Quisling. This punitive frenzy reflected the painful experience of the majority in occupied Europe between 1940 and 1945. 'Most,' the late historian Tony Judt noted, 'experienced the war passively – defeated and occupied by one set of foreigners then liberated by another.' Occupation was, above all, humiliating and degrading. Resistance, it must be said, was a noble but perilous course to take – and the German occupiers consistently responded with savage reprisals that decimated entire communities. But when the Germans fled, leaving short-lived power vacuums in their wake, formerly passive citizens donned the mantle of resistance by taking summary revenge on known or suspected collaborators. In France, 'extra judicial proceedings', meaning executions, claimed at least 10,000 lives; in two small regions of northern Italy, 15,000 alleged collaborators were killed in the final two months of the war.[11]

As this bloody inchoate wave of vengeful slaughter receded, new judicial apparatuses emerged to formalise the punishment of collaborators. Their task, with hindsight, was riddled with contradictions. It proved very difficult to come up with a legal definition of 'collaboration' that covered its many different forms. Should a jobsworth bureaucrat who pen pushed for the Vichy regime necessarily be judged less culpable than a French Milice volunteer who rounded up French Jews? Treason was an especially tricky judicial concept. Many accused collaborators argued that they had continued to faithfully serve the Motherland – albeit under the aegis of a foreign power. The case of Waffen-SS volunteers was just as twisted. Although recruits swore an oath of loyalty to Adolf Hitler, Himmler had cunningly allowed the foreign Waffen-SS legions and divisions to retain some scraps of national identity and many former foreign SS volunteers insisted that they had joined Hitler's war not as collaborators but as loyal Frenchmen or Netherlanders who fervently wished to defend their fellow countryman against Bolshevism. Treason was often a blunt and discriminatory legal weapon that tended to punish rank-and-file grunts.

As Churchill's Iron Curtain began descending, and rehearsals began for a new Cold War, the fractious Allies struggled to find common ground to bring the agents of Hitler's war to justice. The apparatus of justice and retribution was the first battleground of the Cold War.[12] The International Military Tribunal convened in Nuremberg successfully prosecuted the main actors of the German catastrophe, such as Rudolf Hess and Hermann Göring. But the process got tougher as the Americans worked their way down through the hierarchy through the so-called 'Desk Killers' in the German ministries to the doctors and a few of the Einsatzgruppe leaders. Judicial energies began to falter. A more radical denazification process was effectively resisted by the new West German government. Although Churchill himself had, in October 1942, proclaimed that retribution should be considered one of the 'major purposes of the war', by 1948 he was insisting that it was vital to 'draw the sponge across the crimes and horrors of the past'.[13] The consequence of this rapid diminution of righteousness was to allow the lesser players to quietly disappear, out of sight and mind. It did not take long for some energetic spongers to get to work. Just as US Army Intelligence recruited one of the Reich's most notorious 'Eastern experts', Reinhard Gehlen, who subsequently embarked on a career West Germany's spy chief, British intelligence offered lucrative jobs to former Latvian and Estonian SS officers like SS colonel Alfons Rebane, who had served with the 20th Estonian SS Division. He had been captured in the British zone, with 1,000 of his men, and had ended up working in a Bradford textile mill. In 1947, the British Secret Intelligence Service, the SIS, came calling. Rebane moved to London and, despite acquiring a severe alcohol habit, enjoyed a prosperous career with British intelligence; he died in West Germany. The SIS also recruited a number of Latvians who had served with Viktors Arājs. Plate 32 is a mug shot of Arājs himself taken by a British intelligence agency.

One of the more troubling consequences of Cold War expediency involved the former SS 'Galizien' men. More than 11,000 Ukrainians who served in the division had fled west and surrendered to the British in Italy. The Ukrainians had, of course, surrendered in German SS uniform, and as former 'Soviet citizens' faced immediate repatriation to the Soviet zone of occupation. The British temporarily lodged the Ukrainians at a DP (Displaced Person) camp at Rimini to await judgement. In the event, only about 3,000 would suffer that fate: they became British or Canadian citizens.[14]

At the Yalta Conference in early 1945, the Soviets had insisted that all Eastern European refugees should be repatriated once the war was over. Stalin feared that this vast and desperate mass of people might somehow coalesce into an anti-Soviet émigré block, so it was essential that they were retrieved before they could do any harm. As a consequence of the Yalta agreements, at least 50 million Eastern Europeans and former Soviet citizens were repatriated by the Allies to the east. By the winter of 1945, just 2 million still remained in the west. As the repatriation net edged ever closer to the DP camp at Rimini, the former commander of the SS division, General Pavlo Shandruk, looked for means of rescue. He got in touch with Archbishop Ivan Buchko, who specialised in Ukrainian religious matters in the Vatican. Shandruk described the plight of his men, emphasising that they were good Catholics and proven anti-Bolsheviks: the Soviets did not wish them well. Buchko went immediately to Pope Pius XII, who in turn approached the British military authorities. The Vatican delegation persuaded the British to change the status of the Ukrainians from 'displaced persons' to 'surrendered enemy personnel' – a sleight of hand that meant that the British would not be obliged to agree to repatriation. The former SS men stayed put.

Two years later, at the beginning of 1947, the Ukrainians faced a new problem. Italy was about to sign a peace settlement with the Allies, and once negotiations had been completed the British and Americans occupation forces would withdraw, leaving the DP camps in Italian hands. Shandruk feared that the Italian government would then honour its obligation to repatriate 'Soviet citizens'. Once again, Shandruk raised the alarm. In the British Foreign Office, opinion was split. Some officials, knowing something of the division's reputation, argued that they were not 'innocent dissidents'. But an anti-Soviet Whitehall faction, citing the disastrous repatriation of the Cossacks, took a more conciliatory line. While the British mandarins dithered, Ukrainian pressure groups in Britain, Canada and the United States, brought under one organisational roof by Canadian MP Anthony Hlynka, began to loudly insist that the Ukrainians be offered sanctuary. The British agreed to look again at the case of the 8,000 Ukrainians at the Rimini camp. A rationale was not hard to find. Both during and immediately after the war, the British had employed German POWs for labour service. Now many Germans were returning home and needed to be replaced. Britain had been bankrupted by the war and faced severe labour shortages. The Minister of Labour George Isaacs had already drawn up an emergency plan code-named 'Westward Ho!' to recruit at least 100,000 DPs and set them to work in British mines, farms and factories. Isaacs favoured hard-working Balts – but not Ukrainians, Poles, ethnic Germans and, shockingly, Jews. 'Jews of any nationality' would not be favoured because of 'opposition from public opinion at home'. Such shameful prejudices infected post-war British policy. The ministry repatriated black and Asian soldiers and workers, who had contributed so much to the war effort, preferring to solve Britain's labour shortage with white Eastern and Central Europeans.[15]

The British government promised to screen the Ukrainians to head off any protests by left wingers and appointed Brigadier Fitzroy Maclean, whose Special Refugee Commission was already in Italy vetting Yugoslavian refugees, to handle the task. Maclean had fought with Tito's partisans and already knew quite a lot about the 'Galizien'. He suspected that many of the Rimini Ukrainians might not be 'repatriable' and that 'there were war criminals amongst them'.[16] But London set the SRC a tight deadline of mid-March, giving Maclean just a few weeks to process 8,000 plus men. He protested vigorously that 'These men have had ample time in which to disguise their identity and this commission has no machinery in whatever form for criminal investigative work'.[17] Since he was also still obliged to screen Yugoslav POWs, Maclean delegated Ukrainian screening to his assistant D. Haldane Porter. In the time available, Porter made a decision to interview a representative selection. The SRC had problems recruiting Ukrainian speakers and so had to rely on the word of a fanatical former 'Galizien' officer who had no difficulty pulling some very thick wool over British eyes. The Ukrainians had destroyed their SS service cards and Porter had no time to access German records. Many of the POWs insisted that they had never been Soviet citizens and were from the Polish region of Ukraine, occupied by the Soviets in 1939. When it came to investigating the service records of 'Galizien' units, Porter was hamstrung: he had no doubt that every Ukrainian he interviewed 'may be all or in part lying'. He made some effort to cross check divisional rolls with Soviet lists of war criminals, but was forced to rely on a Ukrainian history of the SS division which claimed that it had been formed in September 1944, after the Battle of Brody, thus excluding any investigation of well-attested atrocities committed by the SS recruits before that date.

As Italian independence loomed ever closer, the pro-Ukrainian faction in Whitehall finally prevailed. This change of mind was enforced on the ground by one of Maclean's officers, the rabidly anti-Soviet Major Denis Hills. Hills accepted that the Soviets had a legal case, but as one of a new breed of Cold War warrior was determined to thwart their demands. He told journalist Tom Bower: 'I found myself shielding the Ukrainian Division [sic] of 8000 men from forcible repatriation ... but legally they should have been returned.' Hills exploited the vexed nationality issue by insisting that the interned Ukrainians could not be considered 'Soviet citizens'. Maclean's screening mission was in effect abandoned. On 23 March, the British Prime Minister Clement Attlee agreed to permit the Ukrainians to enter Great Britain as an 'innocent people'.[18]

This was not the end of the matter. The arrival of 8,000 former SS men could not be kept secret – and a journalist called Felix Wirth got in touch with the maverick Labour MP Tom Driberg. Wirth insisted that the Ukrainians had played a 'terrible role as Germany's faithful and active henchmen in the slaughter of the Jews of [L'viv] and other towns'. The 'notorious Ukrainian SS division' had perpetrated 'monstrous outrages'. In the House of Commons, Driberg tabled a question as to whether the Ukrainians had been properly 'checked' and somewhat tamely accepted Foreign Office assurances that they had. Soon afterwards, the first shiploads of Ukrainian SS men began arriving in British ports. 'Westward Ho!' officials in the Home Office began to prepare to 'civilianise' the DPs and began negotiations with Ukrainian lobby groups to allow some to move on to Canada.

As 'Westward Ho!' gathered momentum, more Eastern European DPs flooded into Great Britain. Many had originated in the Baltic nations and, as they passed through various screening processes, doctors noted that some had distinctive tattoos (i.e. 'blood group markings') under their arms. This was unique to Waffen-SS men; it had been customary for Allied officers to get POWs to lift up their left arms. At a DP hostel in Hans Crescent, London, a Polish doctor, who was all too familiar with the tattoo's significance, began asking unwelcome questions and provoked a minor riot. When the 'SS tattoo affair' threatened to escalate, an influential Latvian based in London called Karlis Zarins, aka Charles Zarine, sprang to his fellow citizens' defence. He launched a campaign to defend the Latvian Legion and bombarded the Foreign Office with memoranda and letters. He insisted that the 'Latvian Legion' had been 'taken over' by the Germans and that recruits had wanted only to fight the Russians. Zarins' defence sowed the seeds of the obfuscation that still blights discussion about the occupation of Latvia. Like its Ukrainian counterpart, the Balt lobby was well organised and supported by some influential anti-Soviet right wingers, like the Duchess of Atholl. When, in November 1945, the Russians tried to arrest former SS-Standartenführer Arvids Kripens, who was held in the British POW camp at Zedelghenin, Belgium, they were sent away empty handed. Echoing Zarins' relentless stream of propaganda, the British asserted that 'the fact that he belonged to an SS formation' did not justify handing him over to the Soviet authorities. The Russians had not troubled to assemble much evidence since in Soviet-ccupied Latvia; the fact that Kripens had been an officer in the 'Latvian Legion' would automatically have been enough to have him executed or deported. In one of his diatribes, Zarins alluded to one 'Arājs': 'a great national patriot … I should feel very relieved if His Majesty's Government would allow them to come to the safety of this country.' Our first Labour government did not choose to offer refuge to that self-proclaimed 'killer of Jews', Viktors Arājs.[19] That honour would go to West Germany.

In this way, Britain and Canada became a refuge for many thousands of individuals who joined a 'war of annihilation' that targeted a bogus 'Jewish-Bolshevik' foe.

That naturally brings us back to the question posed at the beginning of this book. Was Daniel Goldhagen's proposal in *Hitler's Willing Executioners* right? Was the Holocaust a consequence of German 'exterminatory anti-Semitism'? Can his argument account for the tens of thousands of non-German executioners who willingly took part in mass murder operations against Jews? Was the Holocaust a European rather than a German crime?

The answer to that question, on the basis of available evidence, must be yes. Following the invasion of Poland in September 1939, many non-German Europeans actively sought the destruction of Jews in German-occupied regions of the Soviet Union. After 1933, German agencies like the Abwehr and the SS promoted the cause of ultranationalist factions in most European nations. Every one of these factions broadly accepted that Soviet Communism was a manifestation of Jewish power, and must be eliminated. The mythical notion of a 'Jewish head' on a 'Slavic body' that Hitler and his followers had adopted from Russian anti-Semites became the shared ideological language of the European far right and was sanctioned by a powerful nation state: Germany. But Hitler had no interest in promoting the cause of any nationalist movements and remained suspicious of 'arming foreigners'.

SS Chief Heinrich Himmler went much further. He imagined a future 'SS Europa' which had dispensed with the NSDAP and its leader Adolf Hitler and was chopped up into SS-ruled provinces. Jews and other undesirables would be liquidated and a massive programme of 'Germanisation' would redraw the ethnic map of Europe. This master plan depended on SS recruitment of non-German ethnic groups: in Himmler's words, 'harvesting German blood'. The unfolding of this master plan commenced in 1940 with the Nordic peoples of Scandinavia and the Netherlands, but as German race science adapted to new data gathered in POW camps, Himmler's scheme would draw in other peoples, beginning with Estonians. Himmler believed that loyal service in SS police battalions and the Waffen-SS military divisions could fast track the process of 'Germanising' 'suitable elements' in occupied Europe, raising non-Germans up to the level of 'Germanic' peoples over time. The service he demanded as the price of a future place at the Aryan high table was mass murder. This was congruent with the political ambition of radical nationalists like the Lithuanian LAF and the Romanian Legion of St Michael, who had long sought the destruction of their fellow Jewish citizens.

When the Reich was defeated in April 1945, just a single stage of Himmler's master plan had been completed, at least in part. That is not to belittle the worst genocide in recorded history. Himmler's foreign executioners played a murderous part in the destruction of European Jewry between 1941 and 1945. The German SS learnt how to manage their auxiliary murderers. They wanted to recruit, as the governor of occupied Poland put it, surgeons not butchers. SS top brass conceived and built Trawniki and the Reinhardt camps for a single wicked purpose: to murder every Polish Jew. They recruited men like John (Ivan) Demjanjuk and many thousands of other Eastern Europeans to help realise this master plan. These men had been brought up to hate Jews. But the lethal application of this hatred was managed by Hitler's willing German executioners.

Appendix 1

Maps

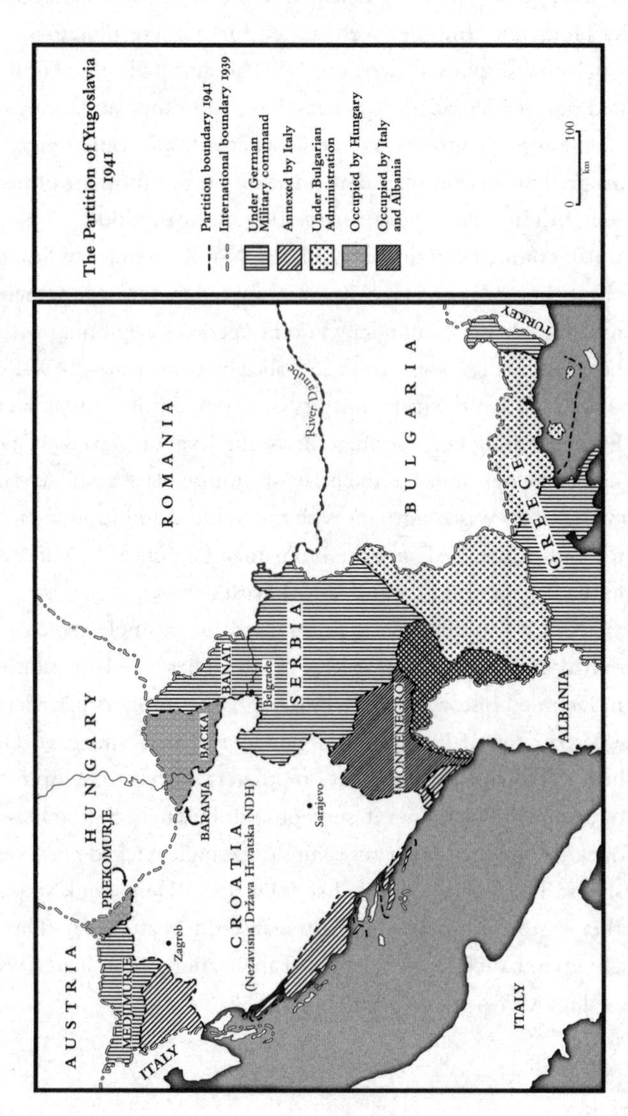

1 The political division of the Balkans following the German invasion, April 1941.

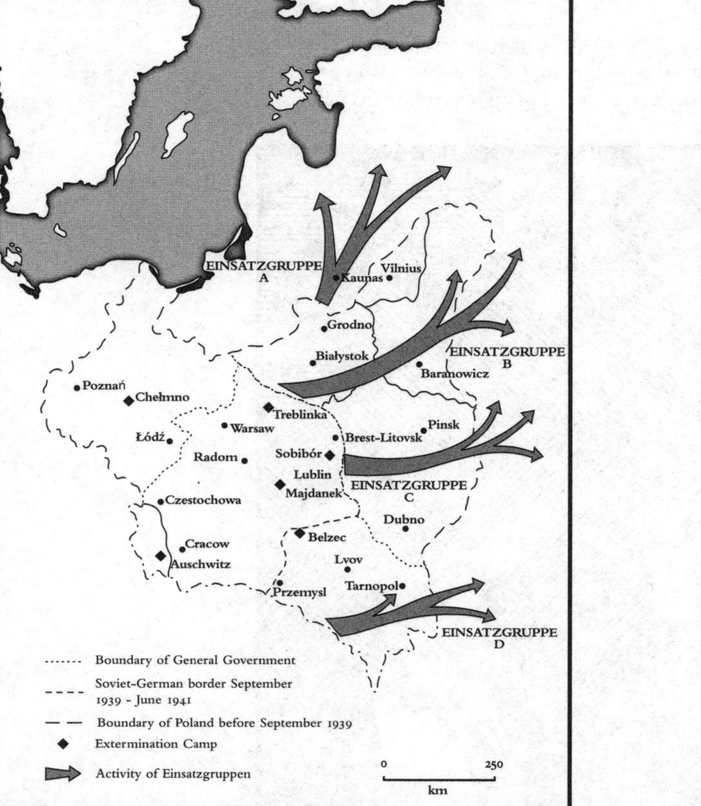

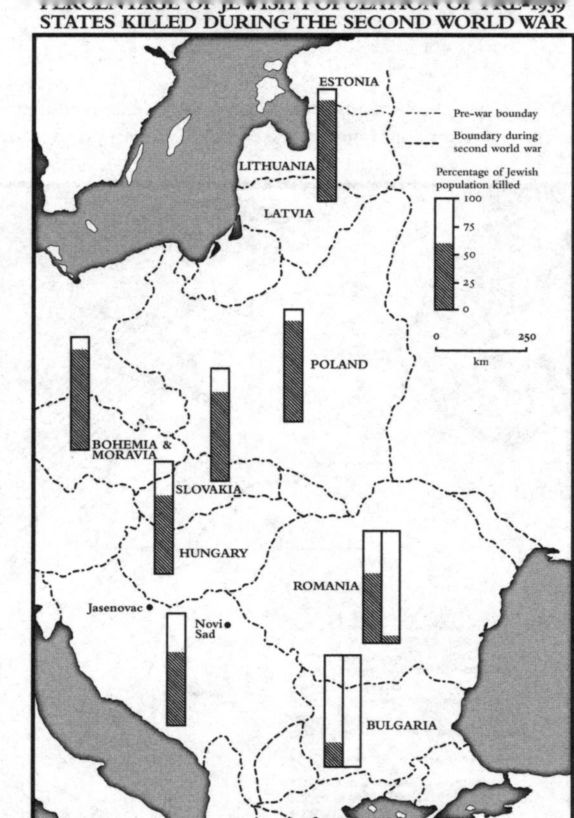

2 SD Einsatzgruppen followed German army groups across Eastern Europe and the Soviet Union (left). Along each route, the Germans recruited local collaborators to form auxiliary police squads that facilitated mass murder, as shown by the recorded percentages of Jewish communities who fell victim to the genocide (right).

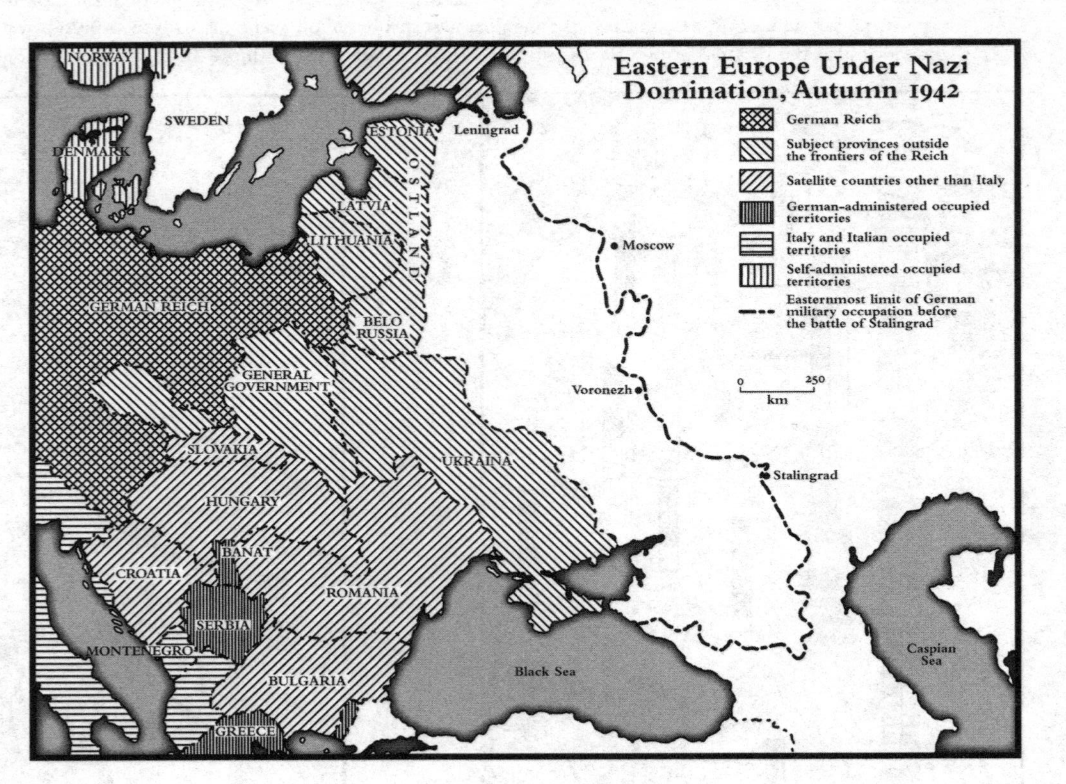

3 The German political division of Eastern Europe and the occupied Soviet Union following the 1941 invasion. Puppet administrations in each new administrative region assisted with the recruitment of non-German police and Waffen-SS units.

Appendix 2

Foreign Divisions Recruited by the Third Reich

Arab Nations

Deutsche-Arabische Bataillon Nr 845
Deutsche-Arabische Lehr Abteilung

Albania

21. Waffen-Gebirgs-Division der SS Skanderbeg (albanische Nr. 1)

Belgium

27. SS-Freiwilligen-Grenadier-Division Langemarck (flämische Nr. 1)
28. SS-Freiwilligen-Grenadier-Division Wallonien
SS-Freiwilligen Legion Flandern
SS-Freiwilligen-Standarte Nordwest
SS-Freiwilligen-Sturmbrigade Langemarck
6. SS-Freiwilligen-Sturmbrigade Langemarck
5. SS-Freiwilligen-Sturmbrigade Wallonien
SS-Freiwilligen-Verband Flandern
SS-Sturmbrigade Wallonien
Wallonisches-Infanterie Bataillon 373

Bulgaria

Waffen-Grenadier Regiment der SS (bulgarisches Nr 1)

Croatia

17. Air Force Company
369. (Kroatische) Infanterie-Division
373. (Kroatische) Infanterie-Division
392. (Kroatische) Infanterie-Division
Croatian Air Force Legion
Croatian Air Force Training Wing
Croatian Anti-Aircraft Legions
Croatian Legion

Croatian Naval Legion
Polizei-Freiwilligen-Regiment 1 Kroatien
Polizei-Freiwilligen-Regiment 2 Kroatien
Polizei-Freiwilligen-Regiment 3 Kroatien
Polizei-Freiwilligen-Regiment 4 Kroatien
Polizei-Freiwilligen-Regiment 5 Kroatien
Polizei-Freiwilligen-Regiment Kroatien - See Polizei-Freiwilligen-Regiment 1 Kroatien
13. Waffen-Gebirgs-Division der SS Handschar (kroatische Nr. 1)
23. Waffen-Gebirgs-Division der SS Kama (kroatische Nr. 2)

Denmark

Danish volunteers in Waffen-SS
Frikorps Danmark

Estonia

Estnische Grenzschutz Ersatz Regiment
Estnische Grenzschutz Regiment 1 (Polizei)
Estnische Grenzschutz Regiment 2 (Polizei)
Estnische Grenzschutz Regiment 3 (Polizei)
Estnische Grenzschutz Regiment 4 (Polizei)
Estnische Grenzschutz Regiment 5 (Polizei)
Estnische Grenzschutz Regiment 6 (Polizei)
Estnische SS-Freiwilligen-Brigade
3. Estnische SS-Freiwilligen-Brigade
Estnische SS-Legion
20. Waffen-Grenadier-Division der SS (estnische Nr. 1)

Finland

Finnisches Freiwilligen-Bataillon der Waffen-SS

France

French forces during WW2
French Volunteers and Collaborationist Forces
Französische SS-Freiwilligen-Grenadier-Regiment
Französische SS-Freiwilligen-Sturmbrigade
Légion des Volontaires Français (LVF)
Légion Tricolore – *see* Légion des Volontaires Français
Waffen-Grenadier-Brigade der SS 'Charlemagne' (französische Nr. 1)
33. Waffen-Grenadier-Division der SS 'Charlemagne' (französische Nr. 1)

Hungary

22. SS-Freiwilligen-Kavallerie-Division Maria Theresa
1. Ungarische-SS-Schi-Bataillon
1. Ungarische SS-Sturmjäger Regiment
25. Waffen-Grenadier-Division der SS Hunyadi (ungarische Nr. 1)
26. Waffen-Grenadier-Division der SS Hungaria (ungarische Nr. 2)
33. Waffen-Kavallerie-Division der SS (ungarnische Nr. 3)
Waffen-Schi Bataillon der SS 25
Waffen-Schi Bataillon der SS 26

India

Indische Freiwilligen Legion der Waffen-SS
Infanterie-Regiment 950 (indische) (Legion Freies Indien)

Ireland

Irish volunteers in the Waffen-SS

Italy

Italienische-Freiwilligen-Legion
Karstwehr-Bataillon
Karstwehr-Kompanie
1. Sturm-Brigade Italienische Freiwilligen-Legion
Waffen-Gebirgs-(Karstjäger) Brigade der SS
24. Waffen-Gebirgs-(Karstjäger-) Division der SS
Waffen-Grenadier-Brigade der SS (italienische Nr. 1)
29. Waffen-Grenadier-Division der SS (italienische Nr. 1)

Latvia

Lettische Freiwilligen Polizei Regiment - See Lettische Freiwilligen Polizei Regiment 1 Riga
Lettische Freiwilligen Polizei Regiment 1 Riga
Lettische Freiwilligen Polizei Regiment 2
Lettische Freiwilligen Polizei Regiment 3
Lettische Grenzschutz Regiment 1 (Polizei)
Lettische Grenzschutz Regiment 2 (Polizei)
Lettische Grenzschutz Regiment 3 (Polizei)
Lettische Grenzschutz Regiment 4 (Polizei)
Lettische Grenzschutz Regiment 5 (Polizei)
Lettische Grenzschutz Regiment 6 (Polizei)
Lettische SS-Freiwilligen-Brigade
2. Lettische SS-Freiwilligen Brigade
Lettische SS-Freiwilligen Legion
15. Waffen-Grenadier-Division der SS (lettische Nr. 1)
19. Waffen-Grenadier-Division der SS (lettisches Nr. 2)

Lithuania

Litauische Polizei Regiment 1

Netherlands

Landstorm Nederland – *See* SS-Grenadier-Regiment 1 Landstorm Nederland
SS-Freiwilligen-Grenadier-Brigade Landstorm Nederland
SS-Freiwilligen-Legion Niederlande
SS-Freiwilligen-Panzergrenadier-Brigade Nederland
4. SS-Freiwilligen-Panzergrenadier-Brigade Nederland
34. SS-Freiwilligen-Grenadier-Division Landstorm Nederland
23. SS-Freiwilligen-Panzergrenadier-Division Nederland (niederlandische Nr. 1)
SS-Freiwilligen-Standarte Nordwest
SS-Freiwilligen-Verband Niederlande
SS-Grenadier-Regiment 1 Landstorm Nederland

Norway

Freiwilligen Legion Norwegen (Den Norske Legion)
SS-Schijäger Bataillon Norwegen (Skijegerbataljon Norge)

Romania

Romanian volunteers in the Waffen-SS
Waffen-Grenadier-Regiment der SS (rumänisches Nr 1)
Waffen-Grenadier-Regiment der SS (rumänisches Nr 2)

Serbia and Montenegro

Polizei Freiwilligen-Regiment Montenegro
Polizei Freiwilligen Regiment 1 Serbien
Polizei Freiwilligen Regiment 2 Serbien
Polizei Freiwilligen Regiment 3 Serbien
Polizei-Selbstschutz-Regiment Sandschak
Serbische Freiwilligenkorps – *See* Srpski Dobrovoljački Korpus
Srpski Dobrovoljački Korpus

Spain

Esquadron Azul
250. Infanterie-Division (División Azul)
Spanische-Freiwilligen-Kompanie der SS 101
Spanische-Freiwilligen-Kompanie der SS 102

Soviet Union

Armenische Legion
Azerbajdzansche Legion
Böhler-Brigade
Freiwilligen-Stamm-Division
Galizische SS Freiwilligen Regiment 4 (Polizei)
Galizische SS Freiwilligen Regiment 5 (Polizei)
Galizische SS Freiwilligen Regiment 6 (Polizei)
Galizische SS Freiwilligen Regiment 7 (Polizei)
Galizische SS Freiwilligen Regiment 8 (Polizei)
Georgische Legion
162. (Turkistan) Infanterie-Division
600. (Russische) Infanterie-Division
650. (Russische) Infanterie-Division
Kalmücken-Kavallerie-Korps - See Kalmüken Verband Dr. Doll
Kalmücken-Legion – *See* Kalmüken Verband Dr. Doll
Kalmüken Verband Dr. Doll
Kaukasischer Waffen-Verband der SS
Nordkaukasische Legion
Osttürkischen Waffen-Verbände der SS
Russkaya Ovsoboditelnaya Narodnaya Armija (RONA):
Waffen-Sturm-Brigade Kaminski
Waffen-Sturm-Brigade RONA
Sonderverband Bergmann
Tataren-Gebirgsjäger-Regiment der SS

Turkestanische Legion
Waffen-Gebirgs-Brigade der SS (tatarische Nr. 1)
Waffen-Grenadier-Brigade der SS (weißruthenische Nr. 1)
14. Waffen-Grenadier-Division der SS (ukrainische Nr. 1)
29. Waffen-Grenadier-Division der SS (russische Nr. 1)
30. Waffen-Grenadier-Division der SS (weissruthenische Nr. 1)
Wolgatatarische Legion

United Kingdom

Britisches Freikorps (British Free Corps)

Appendix 3

Officer Rank Conversion Chart

Army	Waffen-SS	English Rank
	Reichsführer-SS	
Generalfeldmarschall		Field Marshal
Generaloberst	Oberstgruppenführer	General
General	Obergruppenführer	Lieutenant General
Generalleutnant	Gruppenführer	Major General
Generalmajor	Brigadeführer	Brigadier
Oberst	Oberführer	
Standartenführer		
Colonel		
Oberstleutnant	Obersturmbannführer	Lieutenant Colonel
Major	Sturmbannführer	Major
Hauptmann	Hauptsturmführer	Captain
Oberleutnant	Obersturmführer	Lieutenant
Leutnant	Untersturmführer	Second Lieutenant

Appendix 4

Terms & Abbreviations

Abteilung	department, battery, battalion
Abwehr	department of army intelligence
Allgemeine-SS	the main or general SS
Auslandsorganisation	NSDAP agency responsible for Germans in foreign countries
Auswärtiges Amt	Foreign Office, Reich Minisitry for Foreign Affairs
Einsatzgruppe	Special Task Force or special squad commanded by the SD or SIPO
Einsatzkommando	detachment of Einsatzgruppe
Feldgendarmerie	military police
Freiwillige	volunteers
Gau	NSDAP territorial entity in Reich and occupied or annexed territories, each with a Gauleiter
Gebirgsdivision	mountain division in the army or Waffen-SS
Geheime Staatspolizei	Gestapo, secret state police, department of the **Sicherheitspolizei (SiPo)**
Generalgouverment	General Government, German-occupied Poland
Germanische Leitstelle	attached to SS Main Office, responsible for Germanic SS and propaganda
Hilfspolizei	auxiliary police
Höhere SS und Polizeiführer (HSSPF)	Higher SS and Police Leader
Kreis	district
Nationalsozialistische Deutsche Arbeiter Partei (NSDAP)	Nationalist Socialist German Workers Party, Nazi Party
Oberkommando das Heeres (OKH)	High Command of the Army

Oberkommando Der Wehrmacht (OKW)	High Command of the Armed Forces, including army, navy and airforce
Ordungspolizei (ORPO)	Order Police. Regular uniformed Reich police, including **Schutzpolizei** and Gendermerie
Rasse und Siedlungshauptamt (RuSHA)	Race and Settlement Main Office, responsible for establishing and monitoring racial norms in SS and Waffen-SS, and among ethnic German settlers
Reichsarbeitsdienst (RAD)	Reich Labour Service: agency exploiting foreign nationals in occupied countries
Reichsführer-SS and Chef der Deutschen Polizei	Reich SS Leader and Chief of the German Police: from 1936, Heinrich Himmler
Reichskommissar für die Festigung Deutschen Volkstums	Reich Commissar for the Strengthening of Germanism. Himmler's title after 1939, making him responsible for resettlement of ethnic Germans
Reichskommissariat für das Ostland	Civil administration including Baltic states and White Russia (Belarus)
Reichskommissariat Ukraine	Civil administration of Ukraine excluding East Galicia
Reichsministerium für die besetzten Ostgebiete (Omi)	Reichs Ministry for the Occupied Eastern Territories, under Reichsleiter Alfred Rosenberg
Reichsleiter	highest rank in NSDAP
Reichssicherheitshauptamt (RSHA)	Reich Security Main Office
Reichswehr	German armed forces after the First World War
Schutzmannschaften (Schuma)	native auxiliary police units
Schutzpolizei	Protection Police, part of **Ordungspolizei**
Schutz-Staffel	the SS
Sicherheitsdienst des RfSS (SD)	SS Security Service, Reich intelligence organisation headed by Reinhard Heydrich
Sonderkommando	special detachment
SS und Polizeiführer (SSPf)	SS and Police Commander subordinate to HSSPF
Sturmabteilung (SA)	storm troops, or Brownshirts, the NSDAP militia
Totenkopfverbände	Death's Head units
Totenkopfdivision	Waffen-SS division formed from the **Totenkopfverbände** in 1939
Volksdeutsche Mittelstelle (VoMI)	SS agency responsible for Volksdeutsche (ethnic Germans) living outside the Reich

Notes

Preface: Riga, 2010

1 This is a counterfactual thought experiment. No such commemoration has ever been organised by a far-right British political party nor, as far as I know, been planned.
2 UK National Archives, HO 45/25773, 45/25781, 45/25801, 45/25817, 45/25819, 45/25820, 45/25822, 45/25834, 45/25835, 45/25836, KV 2/2828. 2/915, 2/254, WO 311/42; Weale, A. (2010, 2002).
3 R. West, *The Meaning of Treason* (1949), p. 190.
4 Weale (2010), p. 288.
5 http://praguedeclaration.org/
6 http://www.guardian.co.uk/commentisfree/2009/jun/10/europe-far-right-cameron
7 http://www.am.gov.lv/en/latvia/history/legion-kalnins/
8 http://www.am.gov.lv/data/file/e/HC-Progress-Report2001.pdf
9 G. Swain, 'The Disillusioning of the Revolution's Praetorian Guard: the Latvian Riflemen, Summer–Autumn 1918', in *Europe-Asia Studies*, Vol. 51 No 4 (1999).

Introduction

1 Quoted in Ezergailis (1996), p. 194.
2 NARA, T-175, 104/2626381, Gruppenführerbesprechung am 8 Oktober, 1938.
3 Between 1941 and the end of the war, SS divisional nomenclature underwent a succession of often confusing changes. I have used the most common terms.
4 See K. Farrokh, *Shadows in the Desert: Ancient Persia at War* (2007).
5 A. Zamoyski, *1812: Napoleon's Fatal March on Moscow* (2004), pp. 84–7.
6 Domarus (ed.), *Hitler* (1990), p. 1146.
7 Weale (2010), p. 307.
8 Quoted in 'Viking and Wehrbauer: SS Ideology and the Recruitment of Norwegians into the Waffen-SS', unpublished paper by Terje Emberland.
9 Karel C. Berkhoff, 'The Mass Murder of Soviet Prisoners of War and the Holocaust: How Were They Related?', *Kritika: Explorations in Russian and Eurasian History*, Vol. 6 No 4 (Fall 2005; New Series), pp. 789–96.

10 Siedlung Rasse, *Deutsches Blut: das Rasse- und Siedlungshauptamt der SS und die rassenpolitische Neuordnung Europas* (Heinemann, 2003), pp. 533–5.
11 Daluege directive, USHMMA, RG 06.025.63, KGB Archive, Box 59.
12 See list of Waffen-SS divisions on p. 387.
13 Goldhagen (1996), p. 162. There is a substantial literature in English and German about 'Hitler's Foreign Executioners': see bibliography.
14 Ibid., p. 389.
15 Quoted in Boog et al. (2000), v4, p. 445. Hoepner would be executed in 1944 as a conspirator in the plot against Hitler.
16 See Timothy Snyder's important essay, 'Holocaust: the Ignored Reality', *New York Review of Books*, Vol. 56 No 12. This book was completed before the publication of Snyder's monumental *Bloodlands* (2010).
17 Foreword to *Tapping Hitler's Generals* (Frontline Books, 2007).
18 Bartov (1991), p. 102.

1 The Polish Crucible

1 For accounts of the Berghof meeting: Nuremberg Documents 798-PS and 1014-PS, included in TMWC, Vol. 26, pp. 338–44, 523–4; shorthand notes were taken by General Franz Halder: see Documents on German Foreign Policy, 1918–1945, Series D Vol. 7, pp. 557–9. See also Longerich (2010), pp. 143ff.
2 Quoted in Rossino (2003), p. 7.
3 IMT, xxxix: 425ff, 172-USSR, 2 October 1939.
4 Kersten, *Memoirs* (1957), p. 54.
5 For an up-to-date account of Himmler's role in the SA purge, see Longerich (2008), pp. 180ff.
6 See Browder (1990), Chapter 14.
7 *Gesichtspunke fuer einer Uebernahme der Leitung der deutschen Polizei durch den RFSS im Rahmen des RuPrMdl*, BA/Schu 464.
8 A caveat is appropriate. Frick insisted that Himmler's office was 'in' the Interior Ministry – which was in any case countered by Himmler's inclusion of a 'concurrently' (zugleich) which implied that the relationship with Frick was an equal one.
9 See Browder (1990), pp. 225ff.
10 Quoted in Longerich (2010), p. 144.
11 ZStL, 1 Js 12/65 RSHA; see also Steinbacher (2000), p. 54.
12 ZStL, 1 Js 12/65 RSHA.
13 Burleigh (2010), pp. 124ff.
14 ZStL, VI 415 AR 1310/65 E16: cited by Rossino, p. 15, testimony of Lothar Beutel, July 1965.
15 For an exhaustive account of the part played by the ethnic German Self-Defence Corps, see C. Jansen & A. Weckbecker, *Der 'Volksdeutsche Selbstschutz in Polen 1939/40* (Munich, 1992).
16 There are many accounts of the attack on Poland. One of the best is the most recent Burleigh (2010), pp. 119ff. I have also used Weinberg (1994), Mazower (2008) and the *Penguin History of the Second World War* (1972).
17 Despite an aggressive wooing, both Lithuania and Hungary refused to join the German attack on Poland.

18 See Kosyk (1993), pp. 52–61.
19 See Blood (2006), passim.
20 Ibid., p. 46.
21 ZStL, 1 Js 12/65 RSHA ASA 179, p. 360;see also NARA RG 242, T-580, BDC File of Kurt Daluege. Cited by Rossino, p. 259.
22 See A. Rossino (2003), pp. 153ff.
23 The Ostrów incident was the subject of a post-war criminal investigation and reports and eyewitness accounts are now held by the Bundesachiv Ludwigsburg (Zentrale Stelle für Landesjustizverwaltungen) in 8 AR-Z 52/60, 201 AR-Z 76/59, 205 AR-Z 302/67, 206 Ar-z 28/60, 211 AR-Z 13/63.
24 Quoted in Browder (1996), p. 163.
25 Browning (2004), p. 16.
26 Rubenstein (2008), p. 5.
27 The best single account of the Waffen-SS is B. Wegner, *Hitlers Politische Soldaten: die Waffen-SS 1933–1945* (München, new edn, 2008).
28 IWM, AL 2704 E313.
29 Wegner (2008), p. 252 (BDC: PA Eicke).
30 This account of Eicke's campaign is based on Sydnor (1977) and Rossino (2003).
31 *Tätigkeitsbericht während des Einsatzes*, vom 13. Bis 26.9.1939, BA Koblenz, Einsatzgruppen in Polen. See also *The Black Book of Polish Jewry: an Account of the Martyrdom of Polish Jewry under Nazi Occupation* (New York, 1943).
32 Accounts held by USHMM, cited by Rossino (2008), pp. 157ff.
33 See USHMM Photo Collection WS 50414: a series of photographs depicts the murder of Jews in Konskie, an atrocity provoked by the actions of Reserve Lieutenant Bruno Kleinmischel. Hitler had appointed Leni Reifenstahl to make an official documentary about the Polish campaign. That day, she was working in Konskie and witnessed the murders as they took place. Reifenstahl can be seen in the background of one photograph, apparently distraught. The documentary was never completed.
34 Wette (2006), pp. 100–2.
35 See Domarus, 'Hitler: Speeches and Proclamations 1932–1945 …' Vol. 3, p. 1836.
36 Noakes & Pridham, Vol. 3 (2001), p. 319.
37 Ibid., p. 323.
38 BA NS 19/1791: SS Befehl. See also Ackermann (1970), appendix 4.
39 NARA, T175, Roll 37.
40 Hitler's speech is quoted in Friedrich Heiss, *Der Sieg im Western* (Prague, 1943).
41 Goebbels, *Tagebucher*, p. 273.
42 For an account of this meeting, see Halder, *Tagesbuch*, p. 49.
43 NARA T-175, 104/2626163ff.: *Ansprache des RFSS an das Offizierskorps der Leibstandarte SS 'Adolf Hitler' am Abend des Tages von Metz.*
44 Capitalised, the Diaspora refers to the historic exile of the Jewish people. Without a capital *diaspora* (from the Greek 'scattering of seeds') can be attached to the large scale movement of any people, and was first used to refer to the consequence of the nineteenth century Irish famines.
45 See Lumans, passim; and Ferguson (2006), pp. 36ff.
46 See Lumans (1993), pp. 23–5.
47 See Doris L. Bergen, 'The Volksdeutsche of Eastern Europe and the Collapse of the Nazi Empire, 1944-1945', in *The Impact of Nazism: New Perspectives on the Third Reich and its Legacy*, ed. Alan E. Steinweis & Daniel E. Rogers (Lincoln and London, 2003).

48 Mazower (1993), pp. 223–32.
49 F. Umbrich & A. Wittmann, *Alptraum Balkan* (Böhlau, 2003).
50 See Nuremberg documents NO-1605, NG-1112, NO-3362; USMT IV Case 11, PDB 66-G and PDB 43.
51 See Hague II: Article 44: Any compulsion of the population of occupied territory to take part in military operations against its own country is prohibited; and Article 45: Any pressure on the population of occupied territory to take the oath to the hostile Power is prohibited. http://avalon.law.yale.edu/19th_century/hague02.asp
52 See A. Estes, *European Anabasis* (Gutenberg, 2002).
53 Boog et al. (1983, 1990), pp. 1052ff.
54 These were small towns with large Jewish populations, mainly found in the Russian 'Pale of Settlement' – the region where the Russian tsars allowed Jews to settle permanently in their empire. Shtetl culture became a distinctive expression of Jewish culture and learning.

2 Balkan Rehearsal

1 Nuremberg Documents, VI, 1746-PS, Part II, pp. 275–8.
2 Zuroff (2010), pp. 146ff.
3 'Crimes in the Jasenovac Camp', the State Commission of Croatia for the Investigation of the Crimes of the Occupation Forces and their Collaborators, Zagreb, 1946.
4 According to Albert Speer: 'For reasons of safety, the train was drawn by two heavy locomotives and had a special armoured car with light antiaircraft guns following behind the locomotives. Soldiers all muffled up stood on this car … Then came Hitler's car … The walls were panelled in rosewood. The concealed illumination, a ring running around the entire ceiling, threw a bluish light that gave a corpse like look to faces; for that reason the women did not like staying in that room … Then came the guest cars, a press car and a baggage car. A second special car with antiaircraft guns brought up the rear of the train.' The train was destroyed in November 1944.
5 Ruth Mitchell, *The Serbs Chose War* (Holt, Rinehart & Winston, 1943).
6 Kershaw, p. 362.
7 Quoted in Mazower (1993), p. 16.
8 This brief account of the German occupation of Greece borrows details from Mark Mazower's *Inside Hitler's Greece*.
9 See D. Czech, 'Deportation and Vernichtung der Griechischen Juden im KL Auschwitz', *Hefte von Auschwitz*, 11 (1970), pp. 5–37, cited by Mazower.
10 See Tomasevich, op.cit., p. 345.
11 See Hoare, *The History of Bosnia* … pp. 190ff.
12 E. Serotta, *Survival in Sarejevo: Jews, Bosnia and the Lessons of the Past* (Vienna, 1994).
13 NARA, CIC Report APO 512.
14 Quoted in Tomasevich, p. 44.
15 See Ciano, Diaries, p. 200.
16 DGFP, 12: 425.
17 DGFP, 12: 513–7.
18 It is noteworthy that in the early 1990s, a recently reunified and aggressive Germany under Chancellor Helmut Kohl took a similar position as Ribbentrop – again with

bloody results. The German Foreign Minister Hans Dietrich Genscher, responding to intense anti-Serbian sentiment in Germany, and disdaining the views of better informed peace negotiators, pushed hard for recognition of Slovenian and Croatian independence. In January 1992, Genscher bulldozed through an agreement. Kohl proclaimed 'a great triumph for German foreign policy'. Croats sang a new song *Danke Deutschland.* The consequences, as many had feared, proved catastrophic.

19 YA Mil. Hist. Reg. No 51/6–2, box 155.
20 Quoted in Tomasevich (2001), p. 344.
21 Yeomans, *Of 'Yugoslav Barbarians'* … and Burleigh (2006), pp. 262–70.
22 To bolster his case, he argued for example that the word *Hrvat* (meaning Croat) derived from the Persian word for friend, *Hu-urvatha.* His other evidence was equally as flimsy. Iranians and Croats, he noted, both had a passion for equestrian skills.
23 Breitman et al., *US Intelligence and the Nazis* (2005), p. 206.
24 DGFP, 12: 605–6.
25 Hoare (2006), p. 21.
26 YA Mil. Hist. Reg. No 29/15–3, box 85.
27 Tomasevich, p. 593.
28 Burleigh (2006), pp. 263–4.
29 Tomasevich, pp. 212–3, 593. My italics.
30 DGFP, 12: 979 NARA T-120, Roll 5797.
31 Browning, 'The Wehrmacht in Serbia Revisited', in Bartov (2002), p. 36.
32 Hilgruber, *Staatsmanner und Diplomaten bei Hitler* (Frankfurt a.m., 1967), p. 611.
33 Quoted in Steinberg (1990), p. 30.
34 C. Falconi, *The Silence of Pius XII* (1970).
35 BA-MA Wehrmachtsbefehlshaber (AOK 12), RH20 12/153.
36 See Vladimier Dedijer, *Jasenovac: das jugoslawische Auschwitz und der Vatikan* (Freiburg, 1989).
37 Vladko Malek, *In the Struggle for Freedom*, (1957), quoted by Steinberg.
38 http://public.carnet.hr/sakic/hinanews/arhiva/9904/hina-15-g.html
39 http://www.writing.upenn.edu/~afilreis/Holocaust/wansee-transcript.html
40 Der Chef der Sicherheitspolizei und des SD to Reichsführer-SS und Chef der Deutschen Polizei (Berlin, 17 February 1942). Many other German reports are referenced in Hory, L. & Broszat, M., 'Der kroatische Ustascha-Staat 1941–1945', *Vierteljahrshefte für Zeitgeschichte*, No 8; Deutsche Verlags-Anstalt, Stuttgart, 1964.
41 29 June 1941, BAK R70 SU32.

3 Night of the Vampires

1 K. Reddemann (ed.), *Zwischen Front und Heimat: Der Briefwechsel des münsterischen Ehepaares Agnes und Albert Neuhas, 1940–1944* (Münster, 1996), quoted in Evans (2008), p. 179.
2 Report of the International Commission on the Holocaust (RICHR) submitted to President Ion Iliescu in Bucharest on 11 November 2004.
3 Randolph L. Braham, *Romanian Nationalists and the Holocaust: the Political Exploitation of Unfounded Rescue Accounts* (New York, 1998).
4 Ioanid (2000), p. 17.
5 Quoted in Ioanid, *Sword of the Archangel* (1990), p. 83.
6 Butnaru (1992), p. 45.
7 Ioanid, *The Sword of the Archangel: Fascist Ideology in Romania* (London, 1998).

8 Quoted in Payne (1995), p. 282.
9 Reproduced in 'Charisma, Religion and Ideology: Romania's Inter war Legion of the Archangel Michael' by Constantin Iordachi, in *Ideologies and National Identities* (Budapest, New York, 2003), p. 33.
10 Quoted in Butnaru (1992), pp. 49–50.
11 Sebastian, p. 59.
12 Quoted in Deletant (2006), p. 35.
13 Deletant (2006), p. 48ff.
14 DGFP, Vol. XI, No 205, p. 5058.
15 Quoted in Butnaru (1992), p. 74.
16 For an insider's view see Wilhelm Höttl, *The Secret Front: The Story of Nazi Political Espionage* (London, 1953).
17 Matatias Carp, 'Cartea neagră: Suferintele Evreilor din Romania, 1940–1944', Vol. 1, *Legionarii si Rebeliunea* (Bucharest: Editura Diogene, 1996), pp. 56–7.
18 Quoted in Ioanid (2001), p. 53.
19 Butnaru (1992), pp. 82–5.
20 Sebastian (2001), p. 310.
21 Ancel, *Documents*, Vol. 2, No. 72, pp. 195–7; *Jurnalul de dimineata*, No 57, 21 January 1945.
22 Ibid., p. 197.
23 Ibid., p. 308.
24 Ibid., p. 312.
25 See Schellenberg, *Hitler's Secret Service* (New York, 1971), p. 320. Schellenberg also claimed that Reinhard Heydrich had directly instigated the Iron Guard revolt.
26 Goebbels, *Tagebücher*, Vol. 4, (Munich, 1992), pp. 1524–5.
27 ADAP, Series D 1937–41, Band XIII, dok. 207, p. 264: Ambassador von Killinger to the German Foreign Ministry, 16 August 1941.
28 See *Rumänien und der Holocaust: zu den Massenverbrechen in Transnistrien 1941–1944* (Berlin, 2001), pp. 123ff.
29 Cable from Mihai Antonescu to the Romanian legation in Ankara, 1 March 1941. Romanian Foreign Ministry Archives, Ankara File T1, p. 108. Transcript from Cabinet meeting of 5 August 1941 (excerpt), Interior Ministry Archives, file 40010, Vol. 9, p. 40. Quoted in Ancel (HGS, 19, 2, 2005).
30 Ancel (2005), p. 253.
31 Ibid., p. 256.
32 Ancel, *Documents*, Vol. 6, No 1, p. 1.
33 Matatias Carp, *Cartea neagră*, Vol. 2 (Bucharest: Socec, 1948), p. 43. (Testimony of Eugen Cristescu, former head of SSI.)
34 Quoted in Ancel (2005), p. 252.
35 Figures are disputed – earlier accounts refer to half that number. The higher number cited and endorsed in Ioanid is based on data gathered from Iaşi synagogue lists gathered by the Romanian SSI in 1943. See Ioanid, p. 86.
36 http://www.hungarianhistory.com/lib/carp/carp.pdf
37 Malaparte, *Kaputt* (London, 1946), pp. 104ff. I have used a number of details from the chapter 'The Rats of Jassy'.
38 USHM/RSA, RG 25.002M, roll 18. Quoted in Ioanid, p. 272.
39 Note that according to Carp: 'the preparations for the pogrom in Iasi can be reconstructed only on the basis of evidence and individual testimonies collected

by judicial bodies. However, these are also in complete, since the testimonies of Germans and deceased Romanians are missing. Also missing is the testimony of General von Schobert – who died in an aviation accident near Kiev; and the testimonies of Generals von Hauffe and Gerstenberg, who headed the German military mission in Romania; no evidence was given by General von Salmuth, Commander of the 30th German Military Corps; nor by General von Roetig, Commander of the 198th Army Division; nor by Colonel Rodler, the Romanian head of the Abwehr; nor by his right-hand man, Hermann von Stransky; nor from Captain Hoffman, commander of the German garrison in Iasi; and absent above all others is the testimony of Baron Manfred von Killinger, German Ambassador to Bucharest. Similarly missing are the testimonies of certain Rumanian personalities, the most important of whom are: Becescu-Georgescu, the Director of the SSI – who died a few years ago; Major Emil Tulbure representative of the SSI in Iasi, who died of a heart attack a few days after the pogrom; his assistant, Major Gheorghe Balotescu, who disappeared in Germany after August 23, 1944.'

40 Ioanid, p. 66.
41 USHMM, RG 25.004M, roll 48.
42 Malaparte, pp. 46ff.
43 Ibid., p. 70.
44 http://www.jewishgen.org/Yizkor/pinkas_romania/rom1_00142.html
45 Interview by Francisca Solomon, November 2006. Quoted with permission.
46 http://nuremberg.law.harvard.edu/NurTranscript/TranscriptSearches/tran_about.php. NMT 04. Pohl Case – USA v. Oswald Pohl, et al., English Transcript (10 July 1947), p. 4128.
47 Figures from Carp (2000).

4 Horror Upon Horror

1 www.holocaustinthebaltics.com.
2 Zuroff (2010), p. 106.
3 Klee et al. (1991), pp. 28ff.
4 http://www.xxiamzius.lt/archyvas/xxiamzius/20021206/mums_04.html.
5 Klee et al., p. 31.
6 Baranauskus (1970), pp. 194–5.
7 Eidintas (1998), p. 17.
8 Rein, 'Local Collaboration …', JHGS (winter 2006).
9 Nathan Cohen, 'The destruction of the Jews of Butrimonys as described in a farewell letter from a local Jew', in *Holocaust and Genocide Studies*, Vol. 4, No 3 (Hebrew University of Jerusalem), pp. 357–75.
10 A. Tooze (2006), pp. 461ff.
11 Quoted in Burleigh (2002), p. 5.
12 See A. Tooze (2006); and Burleigh, M., *The German-Soviet war and other tragedies* (Burleigh, 1997), pp. 92ff.
13 See Snyder, 'Holocaust: The Ignored Reality', in *New York Review of Books*, Vol. 56, No 12 (16 July 2009).
14 Cited by Burleigh (2002), p. 194. English translations in Noakes & Pridham (2001), pp. 324–6; and Gigliotti & Lang (2005), pp. 167–9. See also Helmut Heiber, *Der Generalplan Ost* (1957), *Denkshrift Himmlers über die Behandlung der Fremdvolksischen im Osten.*

15 M. Burleigh, *Ethics and Extermination: Reflections on Nazi Genocide* (Cambridge, 1997). pp. 9–11.
16 Quoted in Burleigh (2010), p. 247.
17 Headland (1992), p. 75.
18 Stahlecker, 'Consolidated Report', 15 October, 1941: Nuremberg Document L-180. NARA: records of Case 9, USA versus Otto Ohlendorf et al., roll 8.
19 American Military Tribunal, No II, United States of America versus Otto Ohlendorf, Vol. 6, p. 2158.
20 The Jäger Report was discovered after the Einsatzgruppen Trial in Nuremberg had ended and is now in the Central Lithuanian Archives. Jäger himself was not a defendant in Nuremberg; he was arrested many years after the war but committed suicide in 1959 while awaiting trial.
21 Headland (1992), p. 155.
22 BA, R70 Sowjetunion/32, folios 4–10; Krausnick, *Die Einsatzgruppen*, pp. 163ff. for Heydrich's orders.
23 Stahlecker, Riga, 7 July 1941.
24 See MacQueen, 'The Context of Mass Destruction ...', in *HGS*, 12(1) (1998), pp. 27–48.
25 Quoted in Ezergailis (1996), p. 81.
26 Lumans, pp. 26ff.
27 Bellamy (2007), pp. 88ff.
28 For a detailed account see Dov Levin, *The Lesser of two Evils: Eastern European Jewry under Soviet Rule, 1939–1941* (Jewish Publication Society, 1995), pp. 272–3.
29 Freidlander (2007), p. 220.
30 See Kwiet, *Rehearsal for Murder* ..., pp. 12ff.; Bauer (1980), p. 36.
31 Levinson (2009), pp. 166ff.
32 Quoted in Winkler, H.A., *Germany, Volume 2: The Long Road West, 1933–1990* (Oxford, 2007), p. 100.
33 Quoted in Burleigh (1997), p. 41.
34 Extracts quoted in Boog et al. (1996), pp. 491ff.
35 See Kwiet, p. 9.
36 See R. Hilberg, *Documents of Destruction: Germany and Jewry 1933–1945* (Chicago, 1971).
37 Quoted in Kwiet, p. 17.
38 See Arad, *Ghetto in Flames*, pp. 35ff.
39 Ibid., p. 24.
40 See Stang (1996), pp. 114ff.
41 Burleigh (1997), pp. 92–3, citing Schumann & Nestler (eds), *Europa unterm Hakenkreuz*.
42 Arad et al. (1989), p. 4.
43 Quotations from *The Last Days of the Jerusalem of Lithuania: Chronicles from the Vilnius Ghetto and Camps, 1939–1944* ed. Benjamim Harshav (Newhaven and London, 2002).
44 See Arad, op. cit., p. 19.
45 Bauer, op. cit., p. 67.
46 Details from Bauer, op. cit., pp. 75ff.
47 Klee et al. (1988), pp. 38–45.
48 Michael MacQueen, *Lithuanian Collaboration in the 'Final Solution': Motivations and Case Studies* (Center for Advanced Holocaust Studies USHMM, 2004).
49 Quoted in Sutton (2008), pp. 126–7.

5 Massacre in L'viv

1 See Brandon & Lower (eds) (2008), p. 114.
2 In historical terms, distinguishing Galicia from its northern neighbour Volhynia is close to impossible. Strictly, one should refer to Galicia-Volhynia, but to simplify I have used 'Galicia' throughout.
3 Macmillan (2001), Chapters 6 and 17 provide an indispensable analysis of the Russian problem in 1919.
4 Littman, *Pure Soldiers*, p. 9.
5 See H. Torrè, *Le process des pogroms* (Paris, 1928).
6 Quoted in Golczewski, p. 122.
7 See Bruder, pp. 38ff. for a discussion of OUN ideology.
8 Ibid., p. 46.
9 For a full account see V. Shandor, *Carpatho-Ukraine in the Twentieth Century: A Political and Legal History* (Harvard, 1997); and Kosyk (1993), pp. 40ff.
10 Lahousen documents, IfZ, Fd 47 and IMT 3047-PS.
11 Golczewski, pp. 126–7.
12 BA NS 43/43.
13 Quoted in Headland (1992), p. 114.
14 See Rosenberg documents, IMT 1017-PS.
15 Another Abwehr officer who liaised with the Ukrainians was Helmut Groscurth. Like Canaris, Groscurth was a conservative nationalist who somewhat half-heartedly turned against the Nazi regime. He called Heydrich a 'criminal'. On 19–22 August at Bjelaja Zerkow, Groscurth resisted SS demands to shoot Jewish children who had been captured. He held out for a day then capitulated to SS demands under some pressure to 'avoid acrimony'. See Friedlander, in Bartov et al. (2002), pp. 18ff.
16 Kosyk, pp. 82ff.
17 Quoted in Meinl, p. 314.
18 BA-MA, RH 20-11/485, 12 June 1941.
19 NARA, T175, roll 10, Himmler to Berger, 31 April 1941.
20 Gigliotti & Lang (2005), pp. 167–9.
21 IfZ No 2860 & IMT XXII 32.
22 Friedman (1980), p. 180.
23 Dallin (1981), p. 118.
24 Boog et al. (1996), pp. 547ff.
25 Quotations from Boll, in Bartov et al. (2002), op. cit., p. 67.
26 Ibid., p. 64.
27 De Zayas, *The Wehrmacht War Crimes Bureau: 1939–1945* (Rockport, Maine, 2000), Chapter 20. According to de Zayas, the Bureau collected data from various sources and published a report 'War Crimes of the Soviet Forces' in November 1941. The Bureau's evidence was also used by the German Foreign Office to compile 'Bolshevist War Crimes and Crimes Against Humanity' which was passed to the British FO through its Swiss legation.
28 Here is one especially repellent example: http://www.ety.com/HRP/jewishstudies/comcrimes.html
29 BA, Film Archive Berlin/Transit Film, Munich.
30 Quoted in Bruder (2006), p. 231. 'Moskales' is a pejorative characterisation of Russians.

31 Ibid.

32 Oberländer died in 1998. At the time he was deeply involved in far right German political movements and was again under investigation for crimes committed in 1942 when he was leader of the Sonderverband Bergmann, a mixed German and Russian brigade.

33 Quoted in Bruder, op. cit., p. 150.

34 Quoted in Friedländer (2008), p. 214.

35 Quoted in Melnyk, p. 6.

36 Kosyk, op. cit., pp. 510–1.

37 Quoted in Berkhoff & Carynnyk, p. 152. Archive reference: TsDAVOV, 3833/7/6.

38 See Mulligan (1988), p. 34.

39 Bräutigam (1968).

40 Berkhoff (2004), pp. 305ff.

6 Himmler's Shadow War

1 IMT-Blue Series, Vol. 38, pp. 68–94, document L-221 *Der Prozess gegen die Hauptknegsverbrecher vor dem Intennationalen Militargenchtshof.*

2 Himmler, *Dienstkalendar*, p. 195.

3 Quoted in Krausnick & Wilhelm (1981), p. 248.

4 The most important study of the SS brigades is by Cüppers (2005). There is no English translation.

5 The SS Command Staff was established on 7 April as the Einsatzstab (Task Force staff), but was renamed on 6 May as the Kommandostab RFSS. See Cüppers, Wegbereiter (2005); Jürgen Förster, 'Das andere Gesicht desKrieges: Das 'Unternehmen Barbarossa' als Eroberungs- und Vernichtungskrieg', in Roland G. Foerster (ed.), *'Unternehmen Barbarossa': Zum historischen Ort der deutsch-sowjetischen Beziehungenvon 1933 bis Herbst 1941* (Munich, Oldenbourg, 1993), pp. 155–7; Ruth Bettina Birn, 'Zweierlei Wirklichkeit? Fallbeispiele zur Partisanenbekämpfung imOsten', in Bernd Wegner (ed.), *Zwei Wege nach Moskau: Vom Hitler-Stalin-Pakt bis zum 'Unternehmen Barbarossa'* (Piper, Munich, Zurich, 1991), pp. 275–90; Yehoshua Büchler, 'Kommandostab Reichsführer-SS: Himmler's Personal Murder Brigades in 1941', in *Holocaust and Genocide Studies*, Vol. 1, No 1 (1986), pp. 11–25; Yaakov Lozowick, 'Rollbahn Mord: the Early Activities of Einsatzgruppe C', in *Holocaust and Genocide Studies* Vol. 2, No 2 (1987), pp. 221–42; *Unsere Ehre heißt Treue: Kriegstagebuch des Kommandostabes Reichsführer SS. Tätigkeitsberichte der 1. und 2. SS-Inf.-Brigade, der 1. SS-Kav.-Brigade und vonSonderkommandos der SS* (Europa Verlag, Vienna, 1965).

6 Quoted in Cooper (1979), p. 57.

7 Quoted in Krausnick & Wilhelm, *Die Truppe*, p. 248.

8 See P. Blood (2008), an essential source on Hitler's anti-partisan war.

9 Sources: ZStL, B162/19784 contains lengthy interviews with Bach-Zelewski (BZ) conducted in 1960 by lawyers investigating his role in the destruction of Warsaw in 1944. See also: NARA, RG242, A3343-SSO-023, Bach-Zelewski, RG238, T 1270-1, BZ interrogations by IMT, RG238, M 1019-4, BZ interrogations by USMT, BA, SS Bach-Zelewski personal file, BA, R20/45b war diary, NA UK, HW 16/6, MSGP 32, 14/2/42: BZ Diary. Tuviah Friedman, *Bach-Zelewski: Dokumentensammlung* (Haifa, 1996); Machlejd & Wanda (eds), *War Crimes in Poland: Erich von dem Bach* (Warsaw, 1961).

10 Bach-Zelewski's testimony can be accessed in World Jewish Congress Records, C203/Bach-Zelewski.
11 http://www.holocaust-history.org/himmler-poznan/speech-text.shtml
12 Quoted in Büchler, p. 15. A brilliant analysis of the German view of the Pripet can be found in D. Blackburn (2006).
13 See Cüppers (2006), pp. 49–60, 98–107.
14 IMT, Vol. 33, 197, 3839-PS.
15 'Kommandosonderbefehl' Himmler, 28 July 1941, quoted in Norbert Müller (ed.), *Die faschistische Okkupationspolitik in den zeitweilig besetzten Gebieten der Sowjetunion, 1941–1944* (Akademie Verlag, Berlin, 1991), pp. 175–7.
16 Here is a list of names and birth regions according to German prosecution files: Otto Mittelstädt, 4.4.18, Romania; Kurt Zapf, 28.8.20, Łódź; Samuel Grieb, 1.9.20, Neuarzis, Romania; Rudolf Swoboda, 9.7.12, Theresienthal/CSR (Czech Republic); Rudolf Müller, 24.9.1920, Silberbach, CSR; Johann Schütz, 8.2.1918, Grabowa, Poland; Josef Charwat, 30.9.02, Czernowitz, Buchenland?; Alois Chedina, 6.12.09, Cortina, Süd Tirol; Georg Christian, 12.10.03, Kiev, Dersinski (phon?); Otto Deutschmann, 31.7.12, Alsace; Serverin Dörner, 23.10.08, Sudetenland; Werner Dormeyer, 9.1.23, Johannesburg; Oskar Dreger, 24.4.22, Łódź; Dr Med Fritz Eichin, Alsace; Andreas Frank, 29.8.01, Petersburg; Rudolf Fröhlich, 23.3.09, CSR; Adam Fusenecker, 17.7.23, Hungary; Samuel Grieb, 1.9.20, Romania; Karl Kelm, 7.8.18, Marzuki; Franz Klie, 11.2.21, Hungary; Friedrich Maletta, 30.1.07, CSR; Otto Mittelstädt, 4.4.18, Romania; Eduard Moschitz, 2.7.19, Südtirol; Alois Oschowitzer, 9.9.22, CSR; Karl Pangratz, 26.10.12, CSR; Oskar Reinke, 26.8.24, Romania; Anton Scheidenwindt, 1.6.13, Yugoslavia; Johann Serr, 23.2.15, Mangbunar?; Alfred Stegener, Constantinople (d. 1944, Warsaw); Rudolf Swoboda, CSR; Rudold Sucko, 25.1.21, Romania; Helmut Urbainski, 12.3.13, Kattowitz?; Thomas Wolf, 11.12.13, Romanian; Kurt Zapz, ?Łódź.
17 Das Schwarze Korps, 3 September 1942, 3.
18 Büchler, p. 15.
19 BA-MA, RS 4/441: 'Es bleibt kein männlicher Jude leben, keine Restfamilie in den Ortschaften.' SS-Kav. Rgt. 1 Reit. Abt. (Lombard) Abteilungsbefehl Nr. 28.
20 See *Kriegstagebuch des Kommandostabes Reichsfuehrer SS* (Vienna, 1965), pp. 217ff.
21 Reported in the 'Kriegestagbuch', p. 214.
22 Büchler, p. 16; Cüppers, p. 155.
23 Quoted in Burleigh (2010), p. 399.
24 See R. Breitmann, 'Friedrich Jeckeln: Spezialist für die "Endlösung" in Osten', in Smelser & Syring (eds), *Die SS. Elite unter dem Totenkopf.* Cited by Burleigh (2010), p. 400.
25 See Pohl, pp. 27ff., in Bandon & Lower (eds) (2008).
26 Büchler, p. 17.
27 See D. Pohl, in Brandon & Lower (2008), pp. 29ff.
28 Figures from Pohl, op. cit.

7 The Blue Buses

1 Ezergailis (1996), pp. 242–3. The 'Landgericht Hamburg' holds records of a number of war crimes trials of German and Latvian SD recruits who served in Latvia, including testimony by Hemicker. StA Hamburg 141 Js 534/60 Anklageschrift

vom 10.5.1976 gegen Viktors Arājs. I have made significant use of Ezergailis' detailed account of the Latvian Holocaust throughout this chapter. I have, however, drawn different conclusions from his evidence.

2 'Der Dienstkalender Heinrich Himmlers, 1941/42' (Hamburg, 1999), pp. 178ff.
3 See Breitman (1991), p. 170; and *Diesntkalendar*, pp. 180–1.
4 Ibid., p. 195.
5 Longerich (2008), pp. 552–3.
6 See Boog et al., op. cit., p. 589.
7 Kershaw (2000), pp. 400ff.
8 NCA, IMT L-221, Bormann memorandum.
9 BA-MA, RW4/14: Abschrift, RFSS, 25 July 1941.
10 Nuremberg Document PS-1997; see also Dallin, p. 99.
11 See Dallin, p. 183. Lohse's manifesto, 'Ostland baut auf', was published in *Nazionalsozialistische Monatshefte* in January 1942.
12 To be more precise, according to Ezergailis, p. 146, SD powers passed to the Kommandant der Sicherheitspolizei und des SD (KdS), which was subordinated to the HSSPF, who received orders from Berlin.
13 Quoted in Ezergailis (1996), p. 119.
14 Ibid., p. 120.
15 See Lumans, *Latvia in WW2*, pp. 161ff.
16 Quoted in Ezergailis, p. 133.
17 Alfreds Bilmanis, *Latvia under German Occupation* (Press Bureau of the Latvian Legation, Washington DC, 1943).
18 Arad (1989), pp. 27ff.
19 Longer representative extracts can be found in Ezergailis, Chapter 3.
20 Quoted in Ezergailis, p. 159.
21 ZStL B 162/2976-3074, 3076, 3078-3117, 14230, 14486, 14532, 14607, 21481: Schwurgericht, Koln, 1968. Arajs trial: 'wegen gemeinschaftlichen Mordes an mindestens 13.000 Menschen zu lebenslanger Freiheitsstrafe verurteilt' (Fundstelle: Bundesarchiv, B 162/14607).
22 Ezergailis, pp. 226ff.
23 Ibid., p. 161.
24 Arad (1989), pp. 27ff.
25 Quoted in Lumans (2006), p. 237.
26 Ibid., p. 240.
27 Figures cited in Lumans, p. 242.
28 Yad Vashem, 02/870, cited in Ezergailis, p. 167.
29 BA, R 58/215, 'Der Dienstkalender Heinrich Himmlers 189, Ereignismeldungen UdSSR, No. 48', 10 August 1941.
30 See K. Kangeris, '"Closed" Units of Latvian Police …', in *The Hidden and Forbidden History of Latvia under Soviet and Nazi Occupations 1940–1991* (2007), pp. 104–21. Christopher Browning's study 'Ordinary Men' also examined a German closed brigade: the 101st Reserve Police Battalion.
31 See Curilla (2006), pp. 908ff.
32 http://www.time.com/time/magazine/article/0,9171,833564-1,00.html. See also G. Walters (2010).
33 According to this propagandist Cukurs website: http://herberts-cukurs.blogspot.com/
34 Zuroff (2010), pp. 117–20. Testimony by Rafael Shub, Abraham Shapiro and Max Tukacier.

35 Evidence cited by R. Viksne, 'Members of the Arajs Commando in Soviet Court Files; Social Position, Education, Reasons for Volunteering, Penalty', in *The Hidden and Forbidden History of Latvia* ... (2007), pp. 188–208.
36 See H. Krausnick & H-H. Wilhelm (1981), p. 567.
37 Friedländer (2008), p. 281.
38 Kershaw, op. cit., pp. 465–6.
39 'The Topography of Genocide', in Stone (ed.) *The Historiography of the Holocaust*', pp. 221–2. At Lvow, the killing grounds next to the Janowska Camp are called 'The Sands'. The deep sand dunes that fronted the Baltic Sea at Liepāja in Latvia provided another killing site. Babi Yar was a ravine in deep sand deposits that were easy to dynamite. At Treblinka 1, Poles were worked to death in an enormous sand and gravel pit.
40 Ezergailis, p. 244.
41 F. Michelson, *I Survived Rumbula* (New York, 1979).
42 Quoted in Friedländer (2008), p. 262.

8 Western Crusaders

1 Quoted in *The Danish Volunteers in the* Waffen *SS and their Contribution to the Holocaust and the Nazi War of Extermination* by Claus Bundgård Christensen, Niels Bo Poulsen & Peter Scharff Smith. I would like to thank the authors for their assistance with this chapter.
2 NARA, T175, 110/2634766, Berger to Himmler, '20,000 Mann-Aktion'.
3 Boog et al., p. 1049.
4 Quoted in Förster, p. 1051. PA Handakten Ritter No 55.
5 BA-MA Wi/I: OKW/WFSST/Abt L (II Org.), No 001331/41, 6 July 1941.
6 BA-MA RH 22/30, 'Report on the operations of the French Legion and lessons learnt'.
7 Figures from Boog, op. cit.
8 Boog, op. cit., p. 1056.
9 BA MA RH 53-23/49, War diary of training staff, 19 October 1941.
10 BA-MA RH 26-7/19, Combat Report on the French Legion, 23 December 1941.
11 IMT, L-221, xxxviii.
12 NARA T175, 111/2635480, Berger memorandum, 'Vermerk v. Staf de Clerq'.
13 Himmler, *Geheinreden*, p. 157.
14 The definite study is in German: I. Heinemann, *'Rasse, Siedlung, deutsches Blut': Das Rasse- & Siedlungshauptamt der SS und die rassenpolitische Neuordnung Europas* (Göttingen, Wallstein Verlag, 2003).
15 The Staatsarchiv, Nürnberg holds Heydrich's correspondence about Steding's book.
16 'Jacob Burckhardt: Cold War Liberal?' in *The Journal of Modern History*, Vol. 74, No 3 (September 2002), pp. 538–57.
17 Himmler, *Die Schutzstaffel* ... (1937), p. 31.
18 See Klemperer, *The Language of the Third Reich* (London, 2000), pp. 162–3.
19 Herzstein (1982), p. 74. Heinze served under SS leader Franz Alfred Six in the Lebensgebietsmäßige Auswertung (Life Space Evaluation Office).
20 Heinrich Himmler, *Geheimreden* ..., p. 234.
21 Goldhagen (1997), p. 408
22 See Christensen et al. (1999), p. 34.
23 Quoted in Christensen et al. (1999), p. 34.

24 Harald was arrested on 5 May 1945 and his diary confiscated by his captors. It ended up at the editorial office of *Land og folk*, a Danish communist paper, which suggests that it was somehow acquired by communist resistance fighters. Here the diary gathered dust for close to half a century. When *Land og folk* collapsed in 1990, the diary was offered for sale and was bought by the Danish Museum of Freedom.
25 Diary, p. 27.
26 Estes (2005), p. 5.
27 See Boog et al., pp. 1076ff.
28 See Christensen et al. (1998).
29 See Friedlander (1995); Burleigh (1994).
30 Per Sørensen, letter, 6 February 1942, quoted in Christensen et al., p. 33.
31 Ibid., p. 334.
32 Ibid., p. 8.
33 Interview with *Kaj, 2 May 2007, arranged by Christian Barse.
34 *De Jeugd die wij vreisden* (Utrecht, 1948).
35 Estes (2005), Chapter 2, p. 6.
36 For example, Dr Armand Langermann (veterinarian, Auschwitz), Dr Carl Værnet (SS-Sturmbannführer, SS doctor in Buchenwald) who invented an artificial gland, as a treatment for homosexuality.
37 Danish National Archives, Rigspolitiet, Centralkartoteket, Bovruparkivet, B.269, cited by Christensen. In March 1947 a British military court tried and sentenced to death a 39-year-old Danish Waffen-SS officer from southern Jutland. This man had commanded the guards at the Wilhelmshaven-Banterweg camp in north-west Germany. In April 1945 he was ordered to evacuate 200 Jewish prisoners to Bergen-Belsen. Transfers of concentration camp prisoners at the end of the war frequently turned into death marches. In this particular case, the majority of prisoners died on the road from exhaustion, malnutrition or gross mistreatment. On arrival at Lüneburg camp, near Bergen-Belsen, the Danish SS man executed the surviving prisoners 'to avoid spreading typhus'. He personally shot six prisoners. At Wilhelmshaven-Banterweg camp, this particular Danish volunteer had frequently tortured and mistreated inmates.
38 Christensen, letter collection, No 62, November 1941.
39 Widely quoted, see Evans (2008), p. 202.
40 http://www.aftenposten.no/meninger/debatt/article2109354.ece; http://en.wikipedia.org/wiki/Jewish_deportees_from_Norway_during_World_War_II#List_of_Jewish_individuals_deported_from_Norway
41 See Kott (2009), pp. 141ff.; and Embling (2009). The influential German race scientist Hans F.K. Günther idealised Norway and Norwegians. He married a Norwegian and spent some time in Skien conducting 'research'. He argued that Norwegians had conserved their pure Nordic blood because they had been isolated from the rest of Europe. He claimed that Norwegian peasants physically resembled the old German nobility: 'give them a people to conquer and they soon will show their inborn capacity for domination.'
42 See R. Klein, 'Das Polizei-Gebirgsjäger-Regiment 18: Massaker, Deportation, Traditionsplege', in *Zeitschrift für Geschichtswissenschaft* 55 (2007).

9 The Führer's Son

1 http://www.gutenberg-e.org/esk01/frames/feskvid.html.

2 See Kershaw, *Nemesis*, p. 395ff.
3 Conway (1993), p. 261. I would like to thank Professor Martin Conway for his advice while I was researching this chapter. Nigel Jones generously wrote an account of his visit to Degrelle in 1999.
4 See www.radioislam.org.
5 *Guardian*, Ian Traynor, 17 September 2007; and by the same journalist http://www.guardian.co.uk/world/2010/may/09/belgium-flanders-wallonia-french-dutch
6 *Der Spiegel*, 19 September 2007.
7 http://www.newadvent.org/library/docs_pi11qp.htm
8 See Kramer (2007) for a detailed discussion of the Holocaust of Louvain, pp. 6ff.
9 See Degrelle's own self-pitying account (1941).
10 Letter from Hitler to Mussolini, in *Documents on German Policy Series D (1937–1945)*, Vol. 9, p. 439.
11 See Halder, *War Diaries*, 17 May 1940.
12 See Boog et al., *Germany and the Second World War*, Vol. 5, pp. 86–8.
13 Falkenausen had served in Imperial Japan and Ottoman Turkey. He spent most of the 1930s in China, where he was appointed military advisor to Kuomintang leader Chiang Kai-Shek.
14 Gasten (1993), p. 57.
15 See Mazower (2008), pp. 232–8.
16 See *Die Verlorene Legion*, p. 8 and another post-war apologia, *La Cohue*, pp. 517ff.
17 NARA, T-501 Reel 94, fr. 541 Report, September 1941.
18 Derks (2001), pp. 45ff.
19 Ibid., p. 233.
20 See Browning, *The Final Solution and the German Foreign Office: a Study of Referat D3 of Abteilung Deutschland, 1940–1943* (New York, 1978).
21 Goebbels, *Die tagebücher*, Part 2, iv.178.
22 See J.H. Brinks, 'Beyond Anne Frank; Dutch (pre)wartime Collaboration with Nazi Germany and its aftermath', in Alan Stephens & Raphael Walden (eds), *For the Sake of Humanity: Essays in Honour of Clemens Nathan* (Martinus Nijhoff, Leiden, Boston, 2006), pp. 47–62. 'Among the Dutch authorities, especially among the senior staff of police, there were quite a few who already during the interwar years offered their services to the Nazis. They saw Hitler et al. as the most reliable defence against the "Red peril". The police commissioner of Amsterdam, Broekhoff, for example, personally reported in 1935 to the *Gestapo* in Berlin that the Dutch Minister of Defence would co-operate in the mutual fight against *"kommunistische und marxistische Umtriebe"* (communist and Marxist machinations). Under the pen-name of "David" Broekhoff took care of the exchange of information through which 250 German "illegals" who had fled to the Netherlands immediately after the occupation in May 1940 were arrested by the *Sicherheitspolizei*. Rotterdam's then chief commissioner of police, Mr. L. Einthoven, too, figured, together with 17 other Dutch police officers considered to be "*deutschfreundlich*" (pro-German) in a list of names of the *Gestapo*.'
23 Quoted in Conway, p. 293.
24 Degrelle, *Discours prononcé à Liége*, pp. 9–11. Quoted in Conway, p. 302.
25 See H. Möller, *Das NS-Erbe des Auswärtigen Amtes.*
26 In 'a Legion Wallonne sur le front russe', in Robert Aron (ed.), *Histoire de Notre Temps* (Librairie Plon, Paris, 1968). Charles d'Ydewalle argued that Fernand

Rouleau was the real founder of the legion. Amateur historian Eddy de Bruyne proved that it was Rouleau who first approached the Militärverwaltung to propose the formation of a Walloon Legion while Degrelle hobnobbed with Abetz in Paris. Degrelle engineered the dismissal of Rouleau to aggrandise his own power.

27 Degrelle (1985), p. 10.

28 BA-MA, RH 24-4/54, 4th Army Corps, War Diary, 20 December 1941.

29 Quoted in Estes (2007).

10 The First Eastern SS Legions

1 K. Berkhoff, *Holocaust and Genocide Studies* (spring 2001), pp. 1–32.

2 Tooze (2007), pp. 479ff.

3 Solzhenitsen, *The Gulag Archipelago*, p. 218.

4 Berkhoff (2004), Chapter 4; C. Streit, 'Soviet Prisoners of War', in *Encyclopedia of the Holocaust*, Vol. 3 (1990); S.P. Mackenzie, 'The Treatment of Soviet prisoners of War', in *Journal of Modern History*, 66, No 3 (1994).

5 The 'Abel mission' is discussed in letters from Wolfram Sievers, the head of the SS-Ahnenerbe to Himmler, 22.5.43; Josef Grohmann, *Zu den anthropologischen Untersuchungen in russischen Gefangenenlagerns* (1943); Heinemann, pp. 532ff.; and Kater (1974), pp. 208–11.

6 Abel's work in the hellish German POW camps inspired other SS anthropological projects. In 1942, an unsigned letter proposed that the SS-Ahnenerbe begin collecting the skulls of 'Jewish-Bolshevik Commissars'. The following year, Dr Bruno Beger, with the approval of Adolf Eichmann, travelled to Auschwitz where he carried out measurements of a group of Jewish and 'Asiatic' prisoners. The SS transferred the prisoners to another camp near Strasbourg where they were gassed and then dismembered. Their remains ended up in the anatomy department of the University of Strasbourg.

7 http://www.spiegel.de/spiegel/spiegelspecial/d-39863530.html

8 http://www.antisemitism.org.il/

9 http://www.telegraph.co.uk/news/worldnews/europe/estonia/3965268/Russians-protest-at-Estonia-SS-calendar.html

10 http://www.independent.co.uk/news/europe/estonia-accused-of-antisemitism-after-memorial-is-erected-to-ss-executioner-564715.html

11 Information released by the Ministry of Foreign Affairs of the Russian Federation (2004): 'The 658th Eastern battalion under the command of A. Rebane conducted punitive operations against civilians near the town of Kingisepp and the village of Kerstovo (the Leningrad region), committed brutal murders and burnt down the whole villages (Babino, Habalovo, Cigirinka, etc.) to intimidate the partisans. As evidenced by the witnesses and participants in these punitive operations, A. Rebane's unit caught five or six Soviet partisans in the village of Cigirinka in November 1942. In the course of this operation the village was burnt to the ground and three villagers died (РГВА. Ф.451п. Оп.5 Д.149. Л.144-145).'

12 Ezergailis (1996), pp. 194–5.

13 Birn (2001), pp. 182ff.

14 Trials of War Criminals, NMT, Green Series, v. 4. Sandberger was mounting a legal case that he was 'following orders'.

15 A. Weiss-Wendt (2010), p. xvii.

16 Ibid., p. 342.

17 Quoted in Birn, p. 185.
18 See Gurin-Loov (1994).
19 Arad (1989), p. 347.
20 Under German rule, the Omakaitse targeted many different 'enemies of Estonia'. The list included a rural underclass of farm hands and unskilled workers who had benefitted from Soviet land reform and free schooling and were accused of holding socialist views. They had become, according to popular opinion, '*hochmuetig und frech*' (arrogant and uppish); Omakaitse squads shot hundreds. Alcoholics, gossips and troublemakers – they too were targeted. Estonian police arrested women who had done the laundry or other domestic chores for the Soviet occupiers and anyone who was heard to speak Russian. Women who had had Russian boyfriends were all dragged in front of 'Strafprojektierungskommisionen', Estonian commissions appointed by the Germans, and frequently sentenced to death. Ruth Bettina Birn examined the recommendations for execution held in the archives of the occupation Estonian security police: recommended for execution are petty thieves of 'no value to the community'; an alcoholic thief who is judged 'a disadvantage to the community'; a prostitute 'completely useless to the community'; another prostitute whose 'descendants will surely be inferior human beings, detrimental to the interests of society, the same as she is.' Sandberger himself believed that 35 per cent of people arrested by his Estonian colleague were later sentenced to death. Estonian society, dominated as it was by old baronial families, was conservative and reactionary. Under German occupation, Jews had been the first victims; many hundreds of other Estonians judged to be 'useless' or 'detrimental' followed them to the gallows or execution chamber. Sandberger, who had such good relations with his Estonian colleagues, was preoccupied by pressing '*rassepolitischen Grundsaetzen*' (racial political considerations). His colleague Heinrich Bergmann, a National Socialist fanatic, advised action 'devoid of all restraint' to deal with Estonia's 'Roma'. In February 1942, Sandberger ordered his Estonian police to begin treating Roma 'as if they were Jews': this *Zigeuneraktion* was completed by the autumn. In the first phase, the Estonian police arrested gypsies, locked them up in camps and put them to work. By the end of October, many of them had been executed.
21 BA-MA, RH 19 III/492, Kant to Oberste Heeresleitung, 12.8.41.
22 BA NS, 19/382, Berger to Rosenberg, 10.10.1941.
23 BA NS, 19/3522, Jost to Himmler, 7.8.1942 (Estniche SS Legion).
24 BA NS, 19/3522, Berger to Rosenberg.
25 Nuremberg Document NO 3301, USMT IV, Case 11, Himmler, 13 January 1943.
26 NARA, T175, 140/2668141-355.
27 Himmler, *Dienstkalender*, pp. 542–4.
28 Blood (2008), pp. 89–90.
29 After the war, Valdmanis immigrated to Canada where his career ended in ignominy after a conviction for fraud. He was killed in a highway accident in 1970.
30 Bassler (2000), pp. 142ff.
31 BA NS, 19, C.19.2.8, Germanische, fremvolkische und sonstige nicht-deutsche (FW) Verbande: 1506; C.19.2.5.
32 Facsimile in Silgailis (1986), Appendix 1.
33 Quoted in Lacis (2006), p. 36.
34 Bassler, p. 150.
35 Ezergailis (2005), pp. 60–1.

36 Silgailis, Appendix 5, p. 217.
37 Quoted in Bassler, p. 151.
38 Ibid., p. 153.
39 See for example Lacis (2006).
40 This account uses information from R.B. Birn, 'Zaunkönig an Uhrmacher. Grosse Partisanenaktionen 1942/3 an Beispiel des "Unternehmens Winterzauber"', in *Militärgesschichtliche Zeitschrift 60* (2001), pp. 110ff.
41 See ibid., pp. 99–118.
42 Himmler, *Geheimreden* (ed. Smith/Peterson).
43 Sutton (2008), pp. 185ff.
44 BA NS, 19/1446. Minutes on Himmler's conversation with Hitler, 7 September 1943.

11 Nazi Jihad

1 See *Jihad and Jew-Hatred: Islamism, Nazism and the Roots of 9/11* by Matthias Kuntzel (2007) for an extreme version of this argument.
2 NA UK, WO 208/3781, 'Investigation into the Last Days of Adolf Hitler'.
3 See Avi Jorisch, 'Al-Manar: Hizbollah TV', in *Middle East Quarterly* (winter 2004).
4 Julius (2009), pp. 95ff.
5 Matthias Küntzel, *From Zeesen to Beirut National Socialism and Islamic antisemitism*; Seth Arsenian, 'Wartime Propaganda in the Middle East', in *The Middle East Journal*, Vol. 2 (October 1948); Robert Melka, *The Axis and the Arab Middle East 1930–1945* (University of Minnesota, 1966); Heinz Tillmann, *Deutschlands Araberpolitik im Zweiten Weltkrieg* (East Berlin, 1965).
6 Quoted in Gensicke (1988), p. 121.
7 For a clear-sighted and up-to-date analysis of this period and the role of the Grand Mufti of Jerusalem see E. Karsh (2010), Chapter 1.
8 There is a substantial literature on the British mandate and the emergence of Arab nationalism. See, for example, J. Marlowe, *The Seat of Pilate. An Account of the Palestine Mandate* (Cresset Press, London, 1959); . Hyamson, *Palestine under the Mandate 1920–1948* (Methuen, London, 1950); Y. Porath, *The Emergence of the Palestinian-Arab National Movement. Vol. 1: 1918–1929* (Frank Cass, London, 1974) and *The Palestinian Arab National Movement. Vol. 2: 1929–1939. From Riots to Rebellion* (Frank Cass, London, 1977); B. Wasserstein, *The British in Palestine. The Mandatory Government and the Arab-Jewish Conflict 1917–1929* (Royal Historical Society, London, 1978); M. Cohen, *The Origins and Evolution of the Arab-Zionist Conflict* (University of California Press, 1987); Ann Mosely Lesch, *Arab Politics in Palestine, 1917–1939.*
9 H. Cohen, 'The Anti Jewish Farhud in Baghdad 1941', in *Middle Eastern Studies*, 3 (1966), pp. 2–17.
10 See Hans-Jürgen Döscher, *Das Auswärtige Amt in Dritten Reich: Diplomatie im Schatten der Endlösung* (Berlin, 1987), p. 168, cited by Herf (2010).
11 Details from Speer, *Inside the Third Reich*, pp. 158ff.
12 Quoted in Gensicke (2007), p. 124. See also DGFP, Set D, Vol. 12, 'Record of the Conversation between the Führer and the Grand Mufti of Jerusalem on November 28, 1941 in the presence of the Reich Foreign Minister and Minister Grobba in Berlin', pp. 881–5.
13 See Matthias Kuntzel, 'National Socialism and Anti-Semitism in the Arab World', in *Jewish Political Studies Review* (spring 2005).

14 Quoted in Cüppers & Mallmann, 'Rede Mufti zur Eröffnung des Islamischen Zentralinstituts' v. 18.12.1942, PAAA, R 27327; see Matthias Kuntzel, 'Von Zeesen bis Beirut. Nationalsozialismus und Antisemitismus in der arabischen Welt', in Doron Rabinovici, Ulrich Speck & Natan Sznaider (eds), *Neuer Antisemitismus? Eine globale Debatte* (Frankfurt am Main, Suhrkamp, 2004), pp. 271–93.

15 In a debate in the House of Commons on 2 July 1942, Churchill said: 'The military misfortunes of the last fortnight in Cyrenaica and Egypt have completely transformed the situation, not only in that theatre, but throughout the Mediterranean. We are at this moment in the presence of a recession of our hopes and prospects in the Middle East and in the Mediterranean unequalled since the fall of France.'

16 See Klaus-Michael Mallman & Martin Cüppers, '"Elimination of the Jewish National Home in Palestine": The Einsatzkommando of the Panzer Army Africa, 1942', in *Yad Vashem Studies* XXV (available online at http://www1.yadvashem.org/about_holocaust/studies/vol35/Mallmann-Cuppers2.pdf).

17 BA-MA, RW 5/690. Quoted in Cüppers & Mallmann (2007).

18 Quoted op. cit. Niederschrift dess. v. 29.8.1942; ibid., R 27325; see Kriegstagebuch Amt Ausland/Abwehr II v. 13.7.1942, BA-MA, RW 5/498.

19 Details from Satloff (2006), pp. 44ff.

20 BA B, NS 19/1775OKW/WFSt/Qu.IV an RFSS v. 8.12.1942; BA-MA, RH 2/600; Oberkommando Heeresgruppe(HGr) Afrika/Ic an OKH/Gen.St.d.H./Op.Abt. v. 19.4.1943, Ordensvorschlag HoSSPF Italien v. 25.2.1945, BAB, R 70 Italien/19. The experience of the Tunisian Jews is also documented in Daniel Carpi, *Between Mussolini and Hitler. The Jews and the Italian Authorities in France and Tunisia* (Brandeis University Press, Hanover & London, 1994); and Raul Hilberg, *The Destruction of the European Jews* (Harper & Row, New York, 1961), pp. 411–3.

21 *Völkischer Beobachter*, 20 March 1943: '*Aufruf des Großmufti gegen die Todfeinde des Islams, Araber werden für ihre Freiheit an der Seite der Achse kämpfen*'.

22 Gensicke, op. cit., pp. 134–9.

23 NARA T-77, Roll 895, and T-501, Roll 264 and 268. For the Croatian legion see Tomasevich, *The Chetniks*, p. 395.

24 NARA T-120, Roll 5793, and Hoover Institution, Heinrich Himmler Collection, Box 5.

25 Redzic, p. 178.

26 Stein (1996), pp. 179–80.

27 NARA T-175, 94/2614801, see also the Kersten Memoirs, pp. 258ff.

28 Hoover Institution, Himmler Collection, Box 5, File 281. Vrančić's letter was forwarded to Phleps who passed it on, with comments, to Hans Jüttner in the SS Head Office in Berlin.

29 Quoted in Gensicke (2007), pp. 116–9 for an account of el-Husseini's journey.

30 NARA T-120, Roll 5782 and 5783.

31 NARA: T-120, Roll 2908, E464782.

32 Hoover Institution Archives, Himmler Collection, Box 5, File 281.

33 See Trigg (2009), p. 83.

34 Diary of Erich Braun, quoted in Lepre, pp. 41–2.

35 Report by the German police commander in Sarajevo to Einsatzgruppe E in Zagreb, 15 July 1943, National Archives Microcopy T-175, Records of the Reich Leader and Chief of the German Police, roll 140, frame 952; Glaise-Horstenau, Die Erinnerungen, 254; Letter from SS-Obergruppenfuhrer Artur Phleps to Himmler,

7 July 1943, T-175/140/955; Memorandum on the Meeting of Phleps and Himmler, 28 July 1943, T-175/140/952; Letter from Himmler to Phleps, 6 August 1943, T-175/140/949.

36 T-175 roll 70: Dienstanweisung für Imame der 13. SS-Freiwilligen b.h. Geb. Div. (Kroatien).

37 'Rede des RFSS Heinrich Himmler vor den Führen der 13. SS-Frchw. B.h. Goerings-Division (Kroatien) im Fuhrerheim Westlager, Truppenübungsplatz Neuhammer am 11. Januar, 1944', BA NS19/4012, 6–16.

38 See E. Serotta, *Survival in Sarajevo: Jews, Bosnia and the Lessons of the Past* (Vienna, 1994).

39 NARA, T-175, roll 70.

40 *Handzar*, Folge 7, 1943, quoted by Lepre, pp. 77–9.

41 NARA T-175, Roll 126.

42 Report on the Villefranche mutiny, generously shared by George Lepre.

43 Quoted in Lepre, p. 101.

44 NARA T-175, roll 94.

45 NARA T-175, roll 94: Rede des Reichsführers auf der Tagung der RPA-Leiter, am 28 January 1944.

46 Hoare, pp. 279–80.

47 NARA T-175 roll 70 and T-175 roll 21, Berger to Himmler, 26.1.44, Brandt to Berger 31.1.44, and Berger to Himmler 18.2.44.

48 I am grateful to Jonathan Trigg for first-hand descriptions of Bosnia. See *Himmler's Jihadis* (2009), p. 110.

49 Trigg (2009), p. 93.

50 NARA T175, roll 7013 SS-Division Kommandeur, Briefe No 8, 25.2.44.

51 Peter Broucek, *Ein general in Zwielicht: Die Erinnerungen Edmund Glaises von Horstenau* (Vienna, 1988), p. 231.

52 Personal communication, George Lepre, 2008.

53 *Hitlers Lagebesprechungen: Die Protokollfragmente seiner militärischen Konferenzen, 1942–1945*, Heiber, Stuttgart (ed.) (1962), p. 560.

54 See for example *The Berlin-Baghdad Express: The Ottoman Empire and Germany's Bid for World Power, 1898–1918* by Sean McMeekin (2010), which repeats figures from David G. Dalin's biography of the Grand Mufti *Icon of Evil* (2008).

55 R. Birn (1991), pp. 360–1.

12 The Road to Huta Pieniacka

1 Now Гута Пеняцька, Huta Penyats'ka in Ukraine.

2 http://www.ipn.gov.pl/portal/en/19/188/

3 Archiwum Akt Nowich, Warsaw, Signatur 203/XV/14. Source: Hendy papers.

4 Antoni Szczesniak & Wieslaw Szota, *Droga na nikad* (Warsawin, 1973).

5 Hendy papers, interview with Miecelslaw Bernacki, 2004, quoted with permission.

6 The spelling 'Reinhardt' rather than 'Reinhard' suggests another possibility. The new plan may have referred to the German State Secretary of Finance Fritz Reinhardt.

7 I am grateful to Julian Hendy for generously sharing his research materials referred to here as: Hendy, unpublished manuscript (2003). See also: Michael Hanusiak, *Lest We Forget* (New York, 2nd Edn, May 1975); and *Ukrainischer Nationalismus: Theorie und Prazis* (Globus Verlag, Wien, 1979), both cited by Hendy.

8 Trawniki ID file 3191, photograph of SM Fostun in *AUFCGB, Almanac 2 1964–1980* (London, 1980), p. 14; cited in Hendy, unpublished manuscript (2003).

9 Quoted in Black, *Die Trawniki Männer.*
10 http://caselaw.findlaw.com/us-1st-circuit/1531166.html
11 United States District Court, Southern District of Florida, Case 77-2668-Civ NRC, 25 July 1978. USA Plaintiff *v.* Feordor Fedorenko, defendant.
12 Arad (1987), p. 197.
13 Wolfgang Scheffler, Helge Grabitz, *Der Ghettoaufstand Warschau 1943 – aus der Sicht der Taeter und Opfer in Aussagen vor deutschen Gerichten* (München, 1993), pp. 131, 396; fn. 23.
14 See Trawniki personnel files cited in Hendy (2003) for men all entering on 13.02.43: Myron Samyliuk from Tlumacz district, Trawniki ID No 3198; Peter Samiliuk from Tlumacz district, Trawniki ID No 3200; Wasyl Hudyma from Tlumacz district, Trawniki ID No 3202; Iwan Merenda from Tlumacz district, Trawniki ID No 3227; Dmytro Drysliuk from Kolomea district, Trawniki ID No 3245; Jurko Budnyckyj from Kolomea district, Trawniki ID No 3263; Wasyl Bojczuk from Tlumacz district, Trawniki ID No 3293 – OSI documents. It would appear that more than 100 new recruits joined the Trawniki unit on this day. The Germans had a severe shortage of guards at the time.
15 Wolfgang Scheffler, Helge Grabitz, *Der Ghetto Aufstand Warschau 1943: aus der Sicht der Täter und Opfer in Aussagen vor deutschen Gerichten* (München, 1993), pp. 194ff.
16 BA R58/1005: Meeting, 10.6.43. I am grateful to Ray Brandon for drawing my attention to this document.
17 Quoted in Merridale (2005), p. 162.
18 Beevor (1998), p. 398.
19 See for example 'Meldung 51a', December 1942 NARA, T175, 81/260/1524.
20 See Cooper (1979), pp. 78–9.
21 See for example Logusz (1997), Chapter 5.
22 BA NA A3343-SSO-213B, Wächter, Personalakt. For other career details, see Pohl (1996).
23 ZStL, 208 AR 797/66.
24 Malaparte, pp.98ff., 144ff.
25 According to the researcher Robin O'Neil: 'The leading decision-makers in the Nazi hierarchy who argued for total Jewish extermination by gassing were now coming to the fore, and among the leading advocates in the General Government who were in favor of gassing were Dr Wilhelm Dolpheid, SS-Obersturmbannführer, Dr Ludwig Losacker, SS-Obersturmbannführer Helmut Tanzmann, SS-Gruppenführer and Governor, Otto Wächter.' http://www.jewishgen.org/Yizkor/belzec1/bel050.html
26 See Pohl (1997) for a detailed account of infighting between Hitler's potentates.
27 NARA, T175/70/2586239-60.
28 Quoted in Melnyk, p. 12.
29 Tgb. N. 104\43g,28/4.1943.
30 Feldkommandostelle 28/03/1943, Tgb.Nr 47/82/43 Rf/Bn. HFD, H/10/6.
31 Berger, 6/05/1943, cdSSHA/BE/Dr/VS-Tgb.Nr2239/43g. HFD, H/10/12.
32 See Jurgen Thorwald, *The Illusion: Soviet Soldiers in Hitler's Armies* (New York, 1975), pp. 135ff.
33 The minutes are recorded in HFD, H/10/14.
34 For details see Pohl (1996), pp. 182ff., 280ff.
35 A caveat is in order here. In July 1942, the SS recruited two Ukrainian police auxiliary units, the 102nd and 110th. Once training had been completed the units

were assigned to duties in Belarus. Two years later, the SS redesigned the two units as Schuma battalions, which were later absorbed by the 30th Waffen-SS Grenadier Division (Weissruthenische No 2). After the Allied landings in Normandy, the Germans decided to use this bodged together division to combat French Maquis units (now known as the Forces Frangaises de l'lnterieur, or FFI) in the 'Belfort Gap' in the Vosges. This completely unexpected turn of events dismayed the Ukrainians who had expected to fight the Soviet army. In August 1944, the Ukrainians shot their German officers and joined the FFI. After the war, some of the Ukrainians were deported to the USSR and ended up in Soviet camps. Others joined the French Foreign Legion and fought in Indochina and North Africa. See R. Sorebey, *Forum for the Study of the UPA* (2005).

36 Tgb.N10\43g.28/4/1943.
37 Quoted in Melnyk, p. 26.
38 HFD, H/10/26, 10 May 1943.
39 Berger to Himmler, 3 June 1943. HFD, H/10/34.
40 Quoted in Melnyk (2005), p. 30.
41 Cüppers (2006), p. 420.
42 Brandon, personal communication.
43 Quoted in Housden (2003), pp. 206ff.
44 Zvi Weigler, *Yad Vashem Bulletin*, No 1 (1957).
45 See for example Logusz (1997), a partisan account of the SS division's formation and history, pp. 492ff.
46 Minutes of the Meetings of the Military Board 'Galicia': *Brotherhood of Veterans*, (Toronto, 1993), pp. 90–5.
47 See also in Polish S. Wronski & M. Zwolakowa, *Polacy Zydzi, 1939–1945* (Warsaw, 1971); and in English T. Piotrowski (1998), pp. 229ff.
48 NARA, RG242, frames 0689–1843. Quoted in Blood, p. 76.
49 See Förster, *Die Sicherung des 'Lebensraumes'*, p. 1240.
50 Melnyk includes Himmler's complete speech in 'To Battle', Appendix VIII.

13 'We Shall Finish Them Off'

1 Noakes & Pridham (1988), p. 996.
2 Evans (2008), pp. 618ff.
3 See *The Kersten Memoirs* (New York, 1957).
4 Longerich (2008), p. 728.
5 Pridham & Noakes (1998), p. 928.
6 This account of the Polish uprising is based on Davies (2004), Ciechanowski (1974) and Hanson (1982). I am grateful to the Warsaw Rising Museum for their assistance in Warsaw.
7 Frank, Diary, p. 310.
8 Kershaw, *Nemesis*, p. 725.
9 These details are provided in the 'Bach Zelewski files' held at the Zentrale Stelle der Landesjustizverwaltung Ludwigsburg:

Wehrmachtkommandatur Warschau
Korpsgruppe von dem Bach
Kampfgruppe Reinefarth

Waffen-SS:
SS-Pz.Gren.Ausb.u.Ers.Btl (SS-Kampfgruppe)
SS-Kav.Ausb.u.Ers.Abt.8
SS-Flak-Abt.5
SS-Sonder-Rgt. Dirlewanger

Polizei (17)
Inc, Schutzmannschaftsbataillon Kos. Batl.209
Wesiruth.Btl. 13 (Sipo)
Kampfgruppe

Korpsgruppe BZ:
29.Waffen SS (RONA)
SS-Brigade Dirlewanger

Kampfgruppe Reinefarth
Inf.Btl. 1/111 Aserbeidschan
II Btl. Bergmann Aserbeidschan
Kosaken-Regt.3
IV.Kosaken-Sicherungs-Rgt. 57
Kosaken-Abt. 69
Kosaken-Abt. 572

10 ZStL: BA 162/19784, Bach-Zelewski interrogation.
11 NARA, RG 242, T175/R 109/2632683-85.
12 NARA RG 153, interrogation of Dr Wilhelm Gustav Schüppe, April 1945, cited by Breitman (1991).
13 Gerhard von Mende's handwritten curriculum vitae can be accessed at Humboldt University, Berlin: UAHUB, M138 Bd.1, 16/11/1939. His daughter married Bjorn Østring, the Norwegian SS volunteer I interviewed in Oslo, and his son Erling still lives in Berlin.
14 See Johnson (2010), pp. 19–21.
15 Hauner (1981), Chapter Two.
16 NARA RG 238, M-1019/R 33/222, interrogation of Prince Veli Kajum Khan, 27 February 1947.
17 Littlejohn (1987).
18 Heiber & Glatz (2005), p. 73.
19 Bräutigam (1968).
20 Johnson (2010), pp. 26–7.
21 *The Wannsee Conference and the Genocide of the European Jews* (Berlin, 2002), p. 89.
22 Gerhard von Mende and others claimed that the Germans spared the 'Mountain Jews' of the North Caucasus. But a new study by Kiril Feferman has shown that many 'Mountain Jews' were also killed. A few *Bergjuden* communities escaped because German race experts were divided about the ethnic identity of these mysterious peoples. Some believed they were 'Tats', descended from Iranians. HGS, V, No 1 (spring 2007).

23 Trials of the War Criminals before the Nürnberg Military Tribunals under Control Council Law No 10 (Washington DC) Document No 1818: Leibbrandt and Berger: 'Agreement between the Reich Leader SS and Chief of the German Police and the Reich Minister for the Occupied Eastern Territories', 15 March 1943.

24 See P. Klein, 'Curt von Gottberg', in K. Mallmann & G. Paul, *Karrieren der Gewalt: NS Täterbiographien* (Darmstadt, 2004).

25 Quoted in Maclean, pp. 182ff.

26 See Schulte, pp. 172ff.

27 Ibid., p. 175.

28 Teske, *Die Silbernen Spiegel: Generalstabdients unter der Lupe* (Heidelberg, 1952).

29 For more details see: Leonid Rein, 'Untermenschen in SS Uniforms: 30th Waffen-Grenadier-Division of Waffen SS', in *The Journal of Slavic Military Studies*, 1556–3006, Vol. 20, Issue 2 (2007), pp. 329–45.

30 Erickson (1983), pp. 272–3.

31 Ibid., p. 329.

32 Noakes & Pridham, op. cit., VIII, p. 952.

33 Reported by Davies, Blood & Hanson, op. cit.

34 The pretext may have been an incident when RONA men raped and killed two German women of the Kraft durch Freude (Strength through Joy) organisation.

14 Bonfire of the Collaborators

1 Author interview, November 2007.

2 H. Heiber (ed.), *Hitler and his Generals: Military Conferences 1942–1945* (London, 2002), pp. 711–4.

3 Longerich (2007), pp. 754ff.

4 De la Mazière (1922–2006) was interviewed in Marcel Ophuls' documentary about occupied France, *Le Chagrin et la Pitié* (1969), and wrote the autobiographical *La Rêveur Casqué* (Paris, 1973) and *La Rêveur Blessé* (Paris, 2003). His notoriety has obscured his importance.

5 See Susan Zucotti, *The Holocaust, the French and the Jews* (Basic Books, 1993), pp. 13ff.

6 Quoted in E. Weber, *Action Francaise* (Stanford, 1962), p. 199.

7 *Le Pays Réel*, 30 June 1943, quoted in Conway, p. 342.

8 Müller (2008), p. 135.

9 De Bruyne/Rikmenspoel (2004), pp. 120ff.

10 Conway (1993), p. 261.

11 Degrelle (1985), pp. 198ff.

12 Ibid., p. 201.

13 Quoted in Conway, p. 249.

14 As it turned out, the new submarines, delivered in June 1944, immediately developed serious teething problems. They would have been almost useless without a fleet of reconnaissance aircraft – and the production of the Ju290 was halted in the summer after Allied raids damaged production centres. See Evans (2008), p. 672.

15 See Lumans (2006), pp. 353ff.

16 Melnyk, p. 162.

17 Lange report, quoted in Melnyk, op. cit., p. 174.

18 BA, NS 19/323.

19 I am grateful to Mr Hendy for allowing me to make use of his detailed research

notes concerning the events in Slovakia. All quotations from 'Hendy Documents', op. cit.

20 Evening situation reports of the German Commander in Slovakia (*Abendmeldungen des Deutschen Befehlshabers Slowakei*), Czech State Archives, Prague, Fond 110, pp. 117–212, quoted by Hendy.

21 Hendy interview with Dr Jan Stanislav, Director of the Museum of the Slovak National Uprising, Banska Bystrica, 4 December 1998.

22 Quoted in Heike, op. cit., p. 200.

23 Bellamy, *Total War*, pp. 631ff.

24 BA-KO, NS 19/544.

25 Quoted in Evans (2009), p. 680.

26 Interview, 2006.

27 For the unit history of the SS 'Charlemagne' see Trigg (2006), Forbes (2006), Mabire (1975) and R. Soulat, unpublished manuscripts. I would like to thank M. Soulat for his assistance. Strictly speaking the 33rd SS Waffen-Grenadier Division 'Charlemagne' did not fight in Berlin in April 1945, but an SS Sturmbataillon comprising French volunteers who had survived Soviet attack in Pomerania and had chosen not to accept release from their oaths of allegiance.

28 Quoted in Forbes (2006), p. 354.

29 Ibid., p. 383.

30 Also referred to as the 'Overkommando Oberrhein'.

31 Guderian, *Panzer General* (1953), p. 412.

32 Historical Division: United States Army, Europe, 30 June 1959: D-4O8 Personal Notes of Operations Officer Oberst Hans Georg Eismann; 280 pp; 1955. See also *Under Himmler's Command: The Personal Recollections of Oberst Hans-Georg Eismann, Operations Officer, Army Group Vistula, Eastern Front 1945* (2010).

33 Erickson (1983), p. 559.

34 Bernadotte's negotiations remain controversial; he was later assassinated by the Zionist Stern Gang. For more details see: Sune Persson, 'Folke Bernadotte and the White Buses', in *Journal of Holocaust Education*, Vol. 9, Issue 2–3 (2000), pp. 237–68; David Cesarani & Paul A. Levine (eds), *Bystanders to the Holocaust: A Re-evaluation* (Routledge, 2002); A. Ilan, *Bernadotte in Palestine* (Macmillan, 1989), p. 37.

35 Goebbels, *Final entries*, Hugh Trevor-Roper (ed.) (2007), p. 162.

36 Bellamy (2007), pp. 652ff.

37 de Bruyne/Rikmenspoel, pp. 170ff.

38 Forbes (2006), p. 402.

39 Kershaw (2000), pp. 808–9.

40 See Maser (2002), pp. 360–1.

41 Forbes, p. 450.

42 Quoted in Forbes, p. 456.

43 Author interview, op. cit.

15 The Failure of Retribution

1 Degrelle (1985), pp. 326ff.

2 Kersten, p. 292.

3 National Archives, UK: ADM 1/30080. The circumstances surrounding el-Husseini's return to the Middle East remain clouded. The Saudis may have pressurised the French to release their guest, and may have flown the el-Husseini part of the way.

In any case, the Mufti was back.

4 Report by OSS, Quoted by Herf (2010), pp. 243–4.

5 Ibid., p. 234.

6 National Archives, UK CO 537/1317.

7 Arafat's birth name was Mohammed Abdel Rahman Abdel Raouf Arafat al-Qudwa al-Hussein; his clan was only distantly related to the Grand Mufti's.

8 See Ronald Nettler, *Past Trials and Present Tribulations: A Muslim Fundamentalist's View of the Jews* (Oxford, 1987), pp. 72–89, cited in Herf, p. 309.

9 See I. Altman & J. Rubenstein (2009), pp. xixff.

10 See R. Florian, 'Resurging anti-semitism in Romania', in *Society*, Vol. 35, No 1 (November, 1997). http://webcache.googleusercontent.com/search?q=cache:NPtGFyDg_PoJ:www.antisemitism.ro/uploads/251/2003-report-on-anti-semitism-in-romania-eng.pdf+modern+romania+anti-semitism&cd=10&hl=de&ct=clnk&gl=de

11 Judt (2005), pp. 41ff.

12 Earl (2009), pp. 25ff.

13 Quoted by Bower (1981, 1995).

14 Main sources: Margolian (2000), pp. 131ff.; and Cesarani (1992, 2001), pp. 74ff.

15 Cesarani (1992), pp. 79–81.

16 Tom Bower, interview for BBC *Newsnight*, 11 December 1989.

17 NA UK, FO371/66605.

18 Bower (1981, 1995), pp. 258ff.

19 UK NA, FO 371/65754.

Sources

Archives (with abbreviations used in Notes)

Bundesarchiv Berlin-Lichterfelde (BA)
NS 7
SS-und Polizeigerichtsbarkeit
NS 19
Persönlicher Stab ReichsFührer-SS
NS 31
SS-Hauptamt
NS 34
SS-Personalhauptamt
R59
Volksdeutsche Mittelstelle
R 70 Belgien
Polizeidienstellen im Berich des Militäbefehlshabers Belgien und Nordfrankreich
RS
Rasse-und Siedlungshauptamt

Politisches Archiv des Auswärtigen Amtes, Berlin-Mitte (AA/PA)
Inland II – D
Angelegenheiten der Waffen-SS
Inland 11 g
Verbindung zum Reichsführer-SS
Handakten Ettel
Handakten des Staatssekretärs Keppler
Büro des Staatssekretärs im Auswärtigen Amt

Bundesarchiv-Militärarchiv Freiburg im Breisgau (BA-MA)
RS 3 1 1. SS-Panzer-Division Leibstandarte SS Adolf Hitler
RS 3 11 11. SS-Freiwilligen-Panzer-Grenadier-Division 'Nordland'
RS 3 13 13. Waffen-Gebirgs-Division der SS Handschar
RS 3 14 14. Waffen-Grenadier-Division der SS (galizische Nr. 1)

RS 3 15 15. Waffen-Grenadier-Division der SS (lettische Nr. 1)
RS 3 19 19. Waffen-Grenadier-Division der SS (lettische Nr. 2)
RS 3 2 2. SS-Panzer-Division Das Reich
RS 3 20 20. Waffen-Grenadier-Division der SS (estnische Nr. 1)
RS 3 21 21. Waffen-Gebirgs-Division der SS Skanderberg
RS 3 23 23. SS-Freiwilligen-Panzergrenadier-Division Nederland
RS 3 25 25. Waffen-Grenadier-Division der SS Hunyadi (ungarische Nr. 1)
RS 3 3 3. SS-Panzer-Division Totenkopf
RS 3 5 5. SS-Panzer-Division 'Wiking'
RS 3 6 6. SS-Gebirgs-Division Nord
RS 3 7 RS 3-7. SS-Freiwilligen-Gebirgs-Division 'Prinz Eugen'
RS 3 8 8. SS-Kavallerie-Division 'Florian Geyer'
RS 6. Befehlshaber der Waffen-SS
RS 4 Brigaden, Legionen, Standarten sowie Kampfgruppen und Einheiten der Waffen-SS
RH 6 Reichminsiterium für die besetzten Ostgebiete
RS 4 Indische Legion, (Indisches) Infanterie-Regiment 950

Bundesarchib-Koblenz (BAK)
Allg.Proz.21
Prozesse gegen Deutsche in euopäischen Ausland

Bundesarchiv Ludwigsburg: Zentrale Stelle der Landesjustizverwaltung (ZStL)
Records of criminal prosecutions:
Erich von dem Bach-Zelewski and Warsaw Uprising
Viktors Arājs
SS Cavalry Brigades
Schutzmannschaftsbataillon

Vojensky Historical Archive, Prague (VHA)
Btl.d.W-SS z.b.V
Bataillon der Waffen-SS zur besonderen Verfügung

Centre for Historical Research on War and Contemporary Society (CEGES-SOMA)
Archives et Documents de formations de collaboration militaire
Archives et documents du gouvernement militaire allemand, d'institutions allemande civile et concernant la Wehrmacht

National Archives and Records Administration, College Park, Maryland (NARA) Collection of Captured German Records
T-77 OKW
T-78 OKH
T-120 Auswärtiges Amt
T-454 Reich Ministry for the Occupied Eastern Territories
T-175 Himmler materials
T 501 Reeder Tätigkeitsberichten, Légion Wallonie

National Archives of the United Kingdom (NAUK)
Riga Ghetto Case, FO 1060/598
German foreign Ministry archives, the Grand Mufti, GFM 33/437
Disposal of Ukranians, FO 371/57880

Imperial War Museum, Enemy Document Section, HFD
HFD. Papers of the Reichsführer-SS: H/10/1 - H/10/81 (Himmler File Document).

The Hoover Institution Archives (HI)
East European Collection
Germany Collection

Nazi Conspiracy and Aggression: Office of the United States Chief of Counsel for Prosecution of Axis Criminality (NCA)
Green Series in 4 volumes, including Einsatzgruppen trial.
Red Series in 6 volumes available online at: http://www.loc.gov/rr/frd/Military_Law/NT_Nazi-conspiracy.html

United States Holocaust Memorial Museum (USHMM)
Military-Historical Institute (Prague) records, 1941-1944: manuscript RG-48.004M
RG-11.001M.85. Historical Commission of the Reichsführer-SS, Berlin (Fond 701)
RG-11.001M.85. Historical Commission of the Reichsführer-SS, Berlin (Fond 701)
RG-06.027. 'Latvian State Archives KGB Records from Fond 1986 Relating to War Crime Investigations and Trials in Latvia, 1941–1995'
RG-11.001M.90. Alfred Rosenberg, Nazi ideologue (Fond 641)
RG-06. War Crimes Investigations and Prosecutions

Bibliography

Primary Sources

Arad, Y., Israel Gutmann & Abraham Margaliot (eds), *Documents on the Holocaust: Selected Sources on the Destruction of the Jews of Germany and Austria, Poland and the Soviet Union* (Yad Vashem, 1987).

Arad, Y., Shmuel Krakowski, Schmuel Spector & Stella Schossberger, *The Einsatzgruppen Reports: Selections from the Dispatches of the Nazi Death Squad's Campaign against the Jews in the Occupied Territories of the Soviet Union, July 1941–January 1943* (York, 1989).

Auswartiges Amt, *Akten zur deutschen Auswärtigen Politik 1918–1945, Serie D, E* (Göttingen, 1974–79).

Baade, F. (ed.), *Unsere Ehre heißt Treue:Kriegstagebuch des Kommandostabes Reichsführer SS, Tätigkeitsberichte d. 1. u. 2. SS-Inf.-Brigade, d. 1. SS-Kav.-Brigade u. von Sonderkommandos d. SS Erschienen*, (Europa Verlag, Wien, Frankfurt, 1965).

Berger, G., *Auf dem Wege zum Germanischen Reich: Drei Aufsätze* (Berlin, 1943).

———, 'Zum Ausbau der Waffen-SS', *Nation Europa*, Vol. 4 (1953).

Brautigam, O., *So Hat Es Sich Zugetragen* (Verlag Holzner, Wurzburg, 1968).

Ciano, G., *Diary: 1939–1943*, trans. R.L. Miller (Phoenix Press, London, 1947).

Degrelle, L., *La Guerre en Prison* (Les Editions Ignis, Brussels, 1941).

———, *Hitler: Born at Versailles* (Institute for Historical Review, 1969).

———, *Letter to the Pope on his Visit to Auschwitz* (Historical Review Press, 1979).

———, *Campaign in Russia: the Waffen SS on the Eastern Front* (Crecy, Bristol, 1985).

———, *Lettres à mon Cardinal: message aux Belge par Otto Skorzeny* (1975).

———, *Meine Abenteuer in Mexiko* (Literarisches Institut P. Haas & Cie, Augsburg, 1937).

———, *J'accuse M. Segers. J'accuse le minister Segres d'être un cumlard, un bankster, un pillard d'épargne et un lâche* (Kessel, 1935).

Dobroszycki, L. & Jeffrey S. Gurock, *The Holocaust in the Soviet Union: studies and resources on the destruction of the Jews in the Nazi occupied territories of the USSR, 1941–1945* (New York, 1993).

Domarus, M. (ed.), Hitler, A., *Speeches and Proclamations, 1932–1945* (Wauconda III, Bolchazy-Carducci, 1990).

Friedlander, H. & Sybil Milton, *Archives of the Holocaust: an international collection of selected documents* (New York, 1995).
Gigliotti, S. & B. Lang (eds), *The Holocaust: a Reader* (Blackwell Publishing Ltd, London, 2005).
Halder, F., *The Halder War Diary 1939–1942*, ed. Charles Burdick & Hans-Adolf Jacobsen (Greenhill, London, 1988).
Himmler, H., *Der Dienstkalender Heinrich Himmlers: 1941–1942* (Christians, Hamburg, 1999).
———, *Geheimreden 1933 bis 1945 und andere Ansprachen*, ed. Bradley F. Smith & Agnes Peterson (Propyläen Verlag, 1974).
———, *Die Schutzstaffel als antibolshewistische Kampforganisation* (Zentral Verlag der NSDAP, München, 1936).
———, *Festgabe für Heinrich Himmler* (Wittich Verlag, Darmstadt, 1941).
Hioo, T., *Estonia 1940–1945: Reports of the Estonian International Commission for the Investigation of Crimes against Humanity* (2006).
Hitler, A., *Hitler: Reden 1932 bis 1945 Kommentiert von einem Deutschen Zeitgenossen*, Vol. Band 2: Untergang 1941–1945 (1963).
———, *Hitler's War Directives, 1939–1945*, ed. H. Trevor-Roper (Sidgwick & Jackson, London, 1964).
———, *Monologue im Führerhauptquartier 1941–1944: Die Aufzeichnungen Heinrich Heims*, ed. W. Jochmann (Hamburg, 1980).
———, *Hitler's Table Talk, 1941–1944*, trans. Norman Cameron (Phoenix Press, London, 2000).
———, *Hitlers Briefe und Notizen. Sein Weltbild in handschriftlichen Dokumenten*, ed. Stocker Maser (Graz, 2002).
Hopp, G., *Mufti-Papiere. Briefe, Memoranden, Reden und Aufrufe Amin al Husainis aus dem Exil, 1940–1945*, ed. G. Hopp (Klaus Schwarz Verlag, Berlin, 2001).
Klee, E., W. Dressen & Volker Riess, *Those were the Days: the Holocaust as Seen by the Perpetrators and Bystanders*, trans. Deborah Burnstone (London, 1988).
Klein, P. (ed.), *Die Einsatzgruppen in der besetzten Sowjetunion, 1941/1942: Die Tatigskeits und Lageberichte des Chefs der Sicherheitspolizei und des SD* (Hentrich, Berlin, 1997).
Mendelsohn, J. & Donald Detwiler, *The Holocaust: Selected Documents in Eighteen Volumes*, Vol. 10: 'The Einsatzgruppen or Murder Commandos' (New York, 1982).
Neitzel, Sönke, *Tapping Hitler's Generals: Transcripts of Secret Conversations, 1942-45* (Barneley, St Paul, Ms., 2007).
Nollendorfs, V. & Erwin Oberländer, *The Hidden and Forbidden History of Latvia under Soviet and Nazi Occupation 1940–1991* (Symposium of the Commission of the Historians of Latvia Vol. 13, 2005).
Rubenstein, Joshua & Ilya Altman (eds), *The Unknown Black Book: The Holocaust in Soviet occupied Territories* (Indiana University Press, Bloomington IN, 2008).
Sauer, K., *Die Verbrchen der Waffen-SS, Eine Dokumentation der VVN-Bund der Antifaschisten* (Röderberg-Verlag, Frankfurt am Main, 1997).
Schellenberg, W., *The Memoirs of Walther Schellenberg* (Harper & Bros, New York, 1956).
Silgailis, A., *The Latvian Legion* (R. James Bender Publishing, 1986).
Steding, C., *Das Reich und die Krankheit der europäischen Kultur* (Hanseatische Verlagsanstalt, Hamburg, 1942).

Selected Secondary Sources

Abshagen, K., *Canaris* (Hutchinson, London, 1956).
Ailsby, C., *Hitler's Renegades: Foreign Nationals in the Service of the Third Reich* (Spellmount, Staplehurst, Kent, 2004).

Alexander, S., *The Triple Myth: A Life of the Archbishop Alojzije Stepinac* (Columbia University Press, East European Monographs, Boulder, New York, 1987).

Almond, M., *Europe's Backyard War* (London, 1994).

Altschuler, M., *Soviet Jewry on the Eve of the Holocaust: A Social and Demographic Profile* (Yad Vashem, Jerusalem, 1998).

Aly, G., *Final Solution: Nazi Population Policy and the Murder of the European Jews*, trans. A.B. Belinda Cooper (Oxford University Press, New York, 1999).

Andreyev, C., *Vlasov and the Russian Liberation Movement: Soviet Reality and Emigre Theories* (Cambridge University Press, Cambridge, 1987).

Arad, Y., *Bełżec, Sobibór, Treblinka: the Operation Reinhardt Death Camps* (Indiana University Press, 1999).

———, *Ghetto in Flames: the Struggle and Destruction of the Jews in Vilna in the Holocaust* (Yad Vashem Jerusalem, 1980).

Armstrong, J., *Ukrainian Nationalism* (Columbia University Press, New York, London, 1955).

Aron, I., *Fallen Leaves: Stories of the Holocaust and the Partisans* (Shengold, New York, 1981).

Ashton, H.S., *The Netherlands at War* (Routledge, London, 1941).

Bankier, David & Israel Gutman, *Nazi Europe and the Final Solution* (Yad Vashem, Jerusalem, 2003).

Barros, J. & R. Gregor, *Double Deception: Stalin, Hitler and the Invasion of Russia* (Northern Illinois University Press, DeKalb, 1995).

Bartov, O., *The Eastern Front: 1941–1945: German Troops and the Barbarisation of Warfare* (New York, 1985).

Bartov, O., Atina Grossmann & Mary Nolan (eds), *Crimes of War: Guilt and Denial in the 20th Century* (The New Press, New York, 2002).

Bellamy, C., *Absolute War: Soviet Russia in the Second World War* (Macmillan, London, 2007).

Berkhoff, K., *Harvest of Despair: Life and Death in the Ukraine under Nazi Rule* (Harvard University Press, Cambridge MA, 2004).

Berkhoff, K. & M. Carynnyk, 'The Organization of Ukrainian Nationalists and Its Attitude towards Germany and Jews: Iaroslav Stets'ko's 1941 Zhyttiepys', in *Harvard Ukrainian Studies*, Vol. XXIII, No 3/4 (1999), pp. 149–84.

Biber, D., 'The Yugoslav Coup d'Etat, 27th March 1941', in J. Erickson & D. Dilks (eds), *Barbarossa: The Axis and the Allies* (Edinburgh University Press, Edinburgh, 1994).

Biddiscombe, P., *The SS Hunter Battalions: the Hidden History of the Nazi Resistance Movement 1944–45* (Tempus, Stroud, 2006).

Birn, R.B., 'Collaboration with Nazi Germany in Eastern Europe: the Case of the Estonian Security Police', in *Contemporary European History*, Vol. 10, No 2 (2001).

Blackbourn, D., *The Conquest of Nature: Water, Landscape and the Making of Modern Germany* (Cambridge, 2006).

Blau, G.E., *The German Campaign in Russia: Planning and Operations 1940-42* (Department of the Army, Washington, 1955).

Bloch, M, *Ribbentrop* (Abacus, 1994).

Blood, Philip W., *Hitler's Bandit Hunters: the SS and the Nazi occupation of Europe* (Potomac Books Inc., Washington DC, 2006).

Bloxham, D., *Genocide on Trial: War Crimes Trials and the Formation of Holocaust History and Memory* (Oxford, 2001).

Boog, H., Gerhard Krebs & Detlef Vogel, *Germany and the Second World War, Vol 4: The Attack on the Soviet Union* (Clarendon Press, Oxford, 1990) (Militärgeschichtliches Forschungsamt).

Bower, T., *Blind Eye to Murder: Britain, America and the Purging of Nazi Germans – A Pledge Betrayed* (London 1981, 1995).
Braham, R., *The Politics of Genocide: The Holocaust in Hungary* (New York, 1993).
Brandon, R. & Wendy Lower (eds), *The Shoah in Ukraine: History, Testimony, Memorialization* (Indiana University Press and USHMM, 2008).
Brebeck, W.E. & K. Huser, *Wewelsburg 1933–1945: A Cult and Terror Centre of the SS*, trans. R. Benson (Landesbildstelle Westfalen, Münster, 2000).
Breitman, R., *The Architect of Genocide: Himmler and the Final Solution* (Knopf, New York, 1991).
Browder, George C., *Foundations of the Nazi Police State: the Formation of Sipo and SD* (Lexington, 1990).
———, *Hitler's Enforcers: The Gestapo and the SS Security Service in the Nazi Revolution* (New York, Oxford, 1996).
Browning, C.R., *The Origins of the Final Solution: The Evolution of Nazi Jewish Policy, September 1939–March, 1942* (University of Nebraska Press, Lincoln, 2004).
———, *Collected Memories: Holocaust History and Post-war Testimony* (University of Wisconsin Press, Madison, Wisconsin, 2003).
———, *The Final Solution and the German Foreign Office: a Study of Referat DIII of Abteilung Deutschland* (New York, 1978).
———, 'The Wehrmacht in Serbia Revisited', in Bartov et al. (2002).
Bruyne, E. & M. Rikmenspoel, *For Rex and for Belgium: Leon Degrelle and the Walloon Political and Military Collaboration 1940–1945* (Helion & Co., London, 2003).
Brozsat, M., *Anatomy of the SS State* (New York, 1968).
Burgers, N., *Holland under the Nazi Heel* (Stockwell, 1958).
Burleigh, M., *Germany Turns Eastward: A study of Ostforschung in the Third Reich* (Cambridge University Press, Cambridge, 1988).
———, *The Third Reich* (London, 2002).
Burleigh, M. & W. Wippermann, *The Racial State: Germany 1933–1945* (Cambridge, 1991).
Buss, P. & A. Mollo, *Hitler's Germanic Legions: an Illustrated History of the Western European Legions with the SS, 1941–1943* (Macdonald & Jane's, London, 1978).
Butnaru, I.C., *The Silent Holocaust: Romania and its Jews* (Greenwood Press, New York, 1992).
Carp, M., *Holocaust in Romania: Facts and Documents on the Annihilation of Romania's Jews 1940–1944*, trans. S. Murphy (Simon Publications, Safety Harbor, Fl., 2000).
Carrell, P., *Hitler Moves East, 1941–1944* (Little, Brown, Boston, 1964).
Cecil, R., *The Myth of the Master Race: Alfred Rosenberg and Nazi Ideology* (B.T. Batsford Ltd, London, 1972).
Cesarani, D., *Justice Delayed: How Britain became a Refuge for Nazi War Criminals* (Mandarin, London, 1992).
——— (ed.), *The Final Solution: Origins and Implementation* (London, 1994).
Clark, A., *Barbarossa: The Russian German Conflict 1941–1945* (Cassell, London, 1965).
Conway, M., *Collaboration in Belgium: Leon Degrelle and the Rexist Movement 1940–1944* (Yale University Press, Newhaven, London, 1993).
Cookridge, E., *Gehlen: Spy of the Century* (London, 1972).
Cooper, M., *The Phantom War: the German Struggle against Soviet Partisans* (London, 1979).
———, *The German Army 1933–1945: Its Political and Military Failure* (Scarborough House, Lanham MD, 1990).
Cowgill, A., *The Repatriations from Austria: the Report of an Enquiry* (London, 1990).

Cüppers, M., *Wegbereiter der Shoah: Die Waffen-SS, der Kommandostab Reichsführer-SS und die Judenvernichtung 1939-1945* (Wissenschaftliche Buchgesellschaft, Darmstadt, 2005).
Czerniakov, A., *The Warsaw Diary of Adam Czerniakov* (London, 1999).
Dahl, H., *Quisling: a Study in Treachery* (Cambridge, 1999).
Dallin, A., *The German Occupation of the USSR in World War ll: A Bibliography* (Columbia University Press, New York, 1955).
———, *The Kaminsky Brigade 1941–1944: A Case Study of German Exploitation of Soviet Disaffection* (Russian Research Center, Harvard University, Cambridge MA, 1956).
———, *German Rule in Russia 1941–1945: A Study of Occupation Policies* (Macmillan, New York, 1957).
Davies, N., *Rising '44: the Battle for Warsaw* (Macmillan, 2003).
Davies, P., *Dangerous Liaisons: Collaboration and World War Two* (Harlow, UK, 2004).
Debois, Fr.Patrick & J. Fredj, *The Mass Shooting of Jews in the Ukraine 1941-1944: the Holocaust by Bullets* (Memorial de la Shoah, 2009).
De Luca, A., 'Der "Großmufti" in Berlin: The Politics of Collaboration', in *International Journal of Middle East Studies*, Vol. 10, No 1 (1979), pp. 125–38.
Dean, M., *Collaboration in the Holocaust: Crimes of the Local Police in Belorussia and Ukraine, 1941–1944* (St Martin's Press, New York, 2000).
———, 'Local Collaboration in the Holocaust in Eastern Europe' in D. Stone (ed.) (London, 2004).
Deák, I., J. Gross & T. Judt (eds), *The Politics of Retribution in Europe: World War II and its Aftermath* (Princeton, 2000).
Deichmann, U., *Biologists under Hitler*, trans. Thomas Dunlop (Cambridge MA, 1996).
Deletant, D., *Hitler's Forgotten Ally: Ion Antonescu and His Regime, Romania 1940–44* (London, 2006).
Derks, H., *Deutsche Westforshung: Ideologie und Praxis im 20. Jahrhundert* (Akademische Verlagsanstalt, Leipzig, 2001).
———, *Jew, Nomad or Pariah: studies on Hannah Arendt's Choice* (Aksant, Amsterdam, 2004).
Douglas, L., *The Memory of Judgement: Making Law and History in the Trials of the Holocaust* (New Haven, London, 2001).
Dudgeon, A., *The War that Never Was* (Airlife, Shrewsbury, England, 1991).
Duffler, J., *Nazi Germany 1933–1945: Faith and Annihilation* (St Martin's, New York, 1996).
Dunn, Walther S., *Heroes or Traitors: the German Replacement Army, the July Plot, and Adolf Hitler* (Praeger, Westport, 2005).
Earl, H., *The Nuremberg SS-Einsatzgruppen Trial, 1945–1958* (Cambridge, 2009).
Edwards, J., *Berlin Calling* (Praeger, New York, 1991).
Ehrenburg, I. & Vasily Grossman, *The Black Book: The Ruthless Murder of Jews by German Fascist Invaders throughout the Temporarily Occupied Regions of the Soviet Union and in the Death Camps of Poland during the War of 1941–1945* (Holocaust Library, New York, 1980).
Eidintas, A., *Jews, Lithuanians and the Holocaust* (Versus Aureus, Vilnius, 2002).
Erikson, J., *The Road to Stalingrad* (Weidenfeld & Nicolson, London, 1975).
Ezergailis, A., *The Holocaust in Latvia, 1941–1944: the Missing Centre* (The Historical Institute in Latvia & United States Holocaust Memorial Museum, Riga, 1996).
———, *Nazi Soviet Disinformation about the Holocaust in Nazi-Occupied Latvia* (Riga Occupation Museum, Riga, 2005).
———, *The Latvian Legion: heroes, Nazis or victims: a collection of documents from OSS war-crimes investigation files* (Riga, 1997).

Feldman, G. & Wolfgang Seibel, *Networks of Nazi Persecution: bureaucracy, business and the organisation of the Holocaust* (New York, 2005).

Fenyo, M., *Hitler, Hórthy and Hungary: German-Hungarian Relations 1941–1944* (Yale University Press, New Haven, 1972).

Friedman, P., *Roads to Extinction: Essays on the Holocaust* (Jewish Publication Society of America, New York, 1980).

Friedman, S., *Pogromchick: the Assassination of Simon Petiura* (New York, 1976).

Frommer, B., *National Cleansing: Retribution against Nazi Collaborators in Postwar Czechoslovakia* (Cambridge, 2005).

Garrard, J. & C. Garrard, *The Bones of Berdichev: the Life and Fate of Vasily Grossman*) Free Press, New York, 1996).

Gaunt, D., Paul Levine & Laura Palosuo, *Collaboration and Resistance during the Holocaust: Belarus, Estonia, Latvia, Lithuania* (New York, 2002).

Geller, J., 'The Role of Military Administration in German-Occupied Belgium, 1940–1944', in *The Journal of Military History*, Vol. 63, No 1 (1999), pp. 99–125.

Gensicke, K., *Der Mufti von Jerusalem, Amin el-Husseini, und die Nationalsozialisten* (Peter Lang, Frankfurt am Main, Bern, New York, Paris, 1988).

Gerlach, C., 'The Wannsee Conference, the fate of German Jews and Hitler's Decision in Principle to exterminate all European Jews', in *The Journal of Modern History*, 70 No 4 (1998).

Gesin, M., *The Destruction of Ukrainian Jewry during World War II* (The Edwin Mellen Press, Lampeter, Wales, 2006).

Grenkevich, L., *The Soviet Partisan Movement 1941–1944* (Frank Cass, London, Portland OR, 1999).

Gurin-Loov, E., *Suur Having. Eesti juutide katastroof 1941 (Holocaust of the Estonian Jews)* (Horisont, Tallinn, 1994).

Haberer, E., 'History and Justice: Paradigms of the Prosecution of Nazi Crimes', in *Holocaust Genocide Studies*, Vol. 19, No 3 (2005), pp. 487–519.

Hanson, J., *The Civilian Population and the Warsaw Uprising of 1944* (Cambridge University Press, Cambridge, 1982).

Harvey, E., *Women and the Nazi East: Agents and Witnesses of Germanization* (Yale University Press, University of Liverpool, New Haven & London, 2003).

Hathaway, Jay, *In Perfect Formation: SS Ideology and the SS-Junkerschule-Tölz* (Schiffer Military History, Atglen PA, 1999).

Headland, R., *Messages of Murder: A Study of the Reports of the Einsatzgruppen of the Security Police and the Security Service, 1941–1943* (Associated University Presses, Cranbury NJ, 1992).

Heer, H. & K. Naumann, *War of Extermination: the German Military in World War 2* (Berghahn, New York, 2000).

Heiber, H., & D. Glantz, *Hitler and his Generals: Military Conferences 1942–1945* (London, 2002).

Heifetz, E., *The Slaughter of the Jews in the Ukraine in 1919* (Thomas Seltzer, New York, 1921).

Heike, W-D., *Sie wollten die Freiheit: die Geschichte der Ukrainischen Division 1943–1945* (Dorheim Podzun Verlag, 1973).

Heinemann, I., *Rasse, Siedlung, deutsches Blut* (Wallstein Verlag, Frankfurt-am-Main, 2003).

Herb, G., *Under the Map of Germany: Nationalism and Propaganda 1918–1945* (London, New York, 1997).

Herf, J., *The Jewish Enemy: Nazi propaganda during World War II and the Holocaust* (Belknap Press, Harvard University Press, Harvard, 2006).
———, *Nazi Propaganda and the Arab World* (New Haven & London, 2010).
Hilberg, R., *The Destruction of European Jews* (Yale University Press, New Haven, 1985).
———, *Sources of Holocaust Research: an Analysis* (Ivan R. Dee, Chicago, 2001).
Hirschfeld, G., *Nazi Rule and Dutch Collaboration: the Netherlands under German Occupation 1940–1945*, trans. L. Willmot (Berg, Hamburg, 1984).
———, *The Policies of Genocide: Jews and Soviet Prisoners of War in Nazi Germany* (1986).
Hoare, M.A., *Genocide and Resistance in Hitler's Bosnia: The Partisans and the Chetniks, 1941–1943* (British Academy & Oxford University Press, Oxford, 2006).
———, *The History of Bosnia: from the Middle Ages to the Present Day* (Saqi Books, London, 2007).
Hösel, A., *Die Jeugd die wij vreesden* (St Gregorinschuis, Utrecht, 1948).
Hohne, H., *The Order of the Death's Head: the Story of Hitler's SS*, trans. R. Barry (Ballantine, New York, 1970).
Hoidal, O., *Vidkun Quisling: a Study in Treason* (Oslo, 1989).
Housden, M., *Hans Frank: Lebensraum and the Holocaust* (Palgrave, London, 2003).
Hull, I.V., *Absolute Destruction: Military Culture and Practice of War in Imperial Germany* (Cornell University Press, London, 2005).
Hunczak, T., *Symon Petliura and the Jews: a Reappraisal* (University of Toronto Press, Toronto, 1985).
Huneke, D.K., *The Moses of Rovno* (Dodd, Mead & Co., New York, 1985).
Hutchinson, R., *Crimes of War: the Antanas Gecas Affair* (Mainstream Publishing Project Edinburgh, Edinburgh, 1994).
Jelinek, Y., *The Parish Republic: Hlinka's Slovaki People's Party 1939–1945* (East European Quarterly Clumbia University Press, New York, London, 1976).
Johnson, I., *A Mosque in Munich: Nazis, CIA and the Rise of the Muslim Brotherhood in the West* (New York, 2010).
Jones, N., *The Birth of the Nazis: How the Freikorps Blazed a Trail for Hitler* (Carroll & Graf Publishers, New York, 1987).
Jong, L., *The Netherlands and Nazi Germany* (Harvard University Press, Cambridge MA, 1990).
Judt, T., *Post War* (Heinemann, London, 2005).
Kamenetsky, I., *Hitler's Occupation of the Ukraine: A Study of Totalitarian Imperialism* (Marquette University Press, Milwaukee, 1956).
Kater, M., *Das 'Ahnenerbe' der SS: 1935–1945* (DVA, Stuttgart, 1974).
Kemenetsky, I., *Secret Nazi Plans for Eastern Europe: A Study of Lebensraum Politics* (Bookman Associates, New York, 1961).
Kershaw, I., *Hitler, 1889–1936: Hubris* (Norton, New York, 1999).
———, *Hitler, 1936–1945: Nemesis* (Norton, New York, 2000).
Klietmann, K., *Die Waffen-SS: eine Dokumentation* (Munin Verlag, Osnabruck, 1965).
Koehl, R., *The Black Corps* (University of Wisconsin Press, Madison, 1983).
———, *RKFVD: German Resettlement and Population Policy 1939–1945* (Cambridge MA, 1957).
Kohn, N. & H. Roiter, *A Voice from the Forest: Memoirs of a Jewish Partisan* (Holocaust Library, New York, 1980).
Kosyk, Wolodymyr, *The Third Reich and the Ukraine* (Peter Lang, Frankfurt am Main, New York, London, 1993).

———, *The Third Reich and the Ukrainian Question: documents 1934–1945* (Ukrainian Central Information Service, 1991).

Kott, Matthew, *What does the Holocaust in the Baltic States Have to Do with the SS' Plans for Occupied Norway* (Symposium of the Commission of the Historians of Latvia, Vol. 23).

Krausnick, H. & H-H. Wilhelm, *Die Truppen des Weltanschaungskrieges: Die Einsatzgruppen der Sicherheitspolizei und des SD, 1938–1942* (DVA, Stuttgart, 1981).

———, *Anatomy of the SS State* (New York, 1968).

Kuntzel, M., *Djihad and Judenhass: Uber den neuen antijudischen Krieg* (Freiburg, 2003).

Kwiet, K., 'Rehearsing for Murder: The Beginning of the Final Solution in Lithuania in June 1941', in *Holocaust and Genocide Studies*, Vol. 12, No 1 (1998), pp. 3–26.

Lacis, Visvaldis, *The Latvian Legion: According to Independent Observers* (Jumava, 2006).

Langerbein, H., *Hitler's Death Squads* (Texas A&M University Press, 2004).

Lazzaro, R., *Le SS Italiane* (Teti Editore, Milan, 1982, 2002).

Lemkin, R., *Axis Rule in Occupied Europe: laws of occupation, analysis of government, proposals for redress* (Washington, 1991).

Lepre, G., *Himmler's Bosnian Division: The Waffen-SS Handschar Division 1943–1945* (Schiffer Publishing Ltd, Atglen PA, 1997).

Levin, D., 'Disinformation and Antisemitism: Holocaust Denial in the Baltic States, 1945–1999', in J.K. Roth, E. Maxwell & M. Levy (eds), *History* (Palgrave, London, 2001).

———, *The Lesser of two Evils; Eastern European Jewry under Soviet Rule, 1939–1941* (Jewish Publication Society, Jerusalem, Philadelphia, 1995).

Levinson, J. (ed.), *The Shoah in Lithuania* (The Vilna Gaon Jewish State Museum, 2006).

Littlejohn, D., *The Patriotic Traitors: A History of Collaborators in German-Occupied Europe, 1940–1945* (Doubleday, New York, 1972).

———, *Foreign Legions of the Third Reich* (R. James Bender, 1979).

Littman, S., *War Criminal on Trial: Rauca of Kaunas* (Key Porter Books, Toronto, Ontario, 1993).

———, *Pure Soldiers or Sinister Legion: the Ukrainian 14th Waffen-SS Division* (Black Rose Books, Montreal, New York, London, 2003).

Longerich, P., *Heinrich Himmler: Biographie* (München, 2008).

———, *Holocaust: the Nazi Persecution and Murder of the Jews* (Oxford, 2010).

———, 'From Mass Murder to the "Final Solution": the Shooting of Jewish Civilians during the First Months of the Eastern Campaign within the Context of the Nazi Jewish Genocide', in S. Gigliotii & Berel Lang (2005).

Logusz, M., *Galicia Division: the Waffen-SS 14th Grenadier Division 1943–1945* (Schiffer Military History, Atglen PA, 1997).

Lower, W., *Nazi Empire Building and the Holocaust in the Ukraine* (University of North Carolina Press, in association with the United States Holocaust Memorial Museum, 2005).

Lumans, V., *Himmler's Auxiliaries: the Volksdeutsche Mittelstelle and the German National Minorities of Europe, 1933–1945* (University of North Carolina Press, Chapel Hill, 1993).

———, *Latvia in World War II* (Fordham University Press, New York, 2006).

Lyman, R., *Iraq 1941: the Battles for Basra, Habbaniya, Fallujah and Baghdad* (Osprey Publishing Ltd, Oxford, 2006).

Mabire, J., *La Division Charlemagne: les Combats des Français en Pomeranie* (Fayard, Paris, 1974).

———, *Mourir a Berlin: les Francaise Dernier Defenseurs du Bunker d'Adolf Hitler* (Fayard, Paris, 1975).

MacMillan, M., *Peacemakers: The Paris Conference of 1919 and its Attempt to End the War* (John Murray, London, 2001).
MacQueen, M., 'The Context of Mass Destruction: Agents and Prerequisites of the Holocaust in Lithuania', in *Holocaust and Genocide Studies*, Vol. 12, No 1 (1998), pp. 27–48.
Magocsi, P., *A History of Ukraine* (University of Toronto Press, Toronto, 1996).
———, *Galicia: A Historical Survey and Bibliographic Guide* (University of Toronto Press, Toronto, Buffalo, London, 1996).
Mallmann, K-M. & M. Cüppers, *Elimination of the Jewish National home in Palestine: The Einsatzkommando of the Panzer Army Africa, 1942* (Yad Vashem, 2007).
Marrus, M. & R. Paxton, *Vichy France and the Jews* (Stanford University Press, University of Toronto Stanford CA, 1981).
Maslovski, V., *The Tragedy of Galician Jewry* (National Library of Canada, 1997).
Matthäus, J., 'Controlled Escalation: Himmler's Men in the Summer of 1941 and the Holocaust in the Occupied Soviet Territories', in *Holocaust and Genocide Studies*, Vol. 21, No 2 (2007), pp. 218–42.
Mayer, A.J., *Why Did the Heavens not Darken: the Final Solution in History* (Verso, London, 1990).
Mazower, M., *Hitler's Empire* (London 2008).
———, *Inside Hitler's Greece: The Experience of Occupation 1941–1944* (New Haven, 1993).
———, *The Balkans* (Weidenfeld & Nicolson, London, 2000).
McCarthy, J., *The Ottoman Peoples and the End of Empire* (Arnold, London, 2001).
Megargee, G., *War of Annihilation: Combat and Genocide on the Eastern Front, 1941* (Bowman & Littlefield Publishers, Inc., Maryland, 2006).
Van der Meij, L., *The SS in the Netherlands, 1940–1945* (Oxford University Press, Oxford, 1996).
Melnyk, M., *To Battle: the Formation and History of the 14th Galician Waffen-SS Division* (Helion & Company, Solihull, 2002).
Mineau, A., *Operation Barbarossa: Ideology and Ethics Against Human Dignity* (Editions Rodopi, Amsterdam, New York, 2004).
Mosse, G., *The Crisis of German Ideology* (Schocken Books, New York, 1981).
Motyl, A., *The Turn to the Right: the Ideological Origins and Development of Ukrainian Nationalism 1919–1929* (Columbia University Press, New York, 1980).
Müller, K-J., 'The Brutalisation of Warfare, Nazi Crimes and the Wehrmacht', in J. Erickson & D. Dilks (eds), *Barbarossa: the Axis and the Allies* (Edinburgh University Press, Edinburgh, 1994).
Müller, R-D., *Hitlers Ostkrieg und die deutsche Siedlungspolitik: Die Zusammenarbeit von Wehrmacht, Wirtschaft und SS*, Fischer (Frankfurt am Main, 1991).
———, *An der Seite der Wehrmacht: Hitlers ausländische Helfer beim Kreuzzug gegen den Bolschewismus* (Berlin, 2007).
Mulligan, T.P., *The Politics of Illusion and Empire: German Occupation Policy in the Soviet Union, 1942–1943* (Praeger, New York, 1988).
Murphy, D.T., *The Heroic Earth: Geopolitical Thought in Weimar Germany* (Kent State Press, Kent, Ohio, 1997).
Neulen, H., *Eurofaschismus und der Zweite Weltkrieg: Europas verratene Söhne* (Universitas, München, 1980).
———, *An deutscher Seite: internationale Freiwillige vom Wehrmacht und Waffen SS* (Universitas, München, 1985).

———, *Europa und das 3 Reich: Einigungsbestrebungen im deutschen Machtbereich* (Universitas, München, 1987).
Nicosia, F., *The Third Reich and the Palestine Question* (Transaction Publishers, New Brunswick, London, 2000).
Niewyk, D. & F. Nicosia, *The Columbia Guide to the Holocaust* (Columbia University Press, New York, 2000).
Niven, W., *Facing the Nazi Past: United Germany and the Legacy of the Third Reich* (Routledge, London, 2001).
Nollendorfs, V., *Catalogue of the Museum of the Occupation of Latvia 1940–1941* (Museum of the Occupation of Latvia, Riga, 2005).
Overy, R., *Russia's War* (Penguin Books, London, 1997).
———, *The Dictators: Hitler's Germany and Stalin's Russia* (Allen Lane, London, 2004).
Padfield, P. *Himmler: Reichsführer-SS* (Cassell, London, 1990).
Paulsson, G., *Secret City: the Hidden Jews of Warsaw 1940–1945* (Newhaven & London, 2002).
Pohl, D., *Nationalsozialistische Judenverfolgung in Ostgalizien 1941–1944: Organisation und Durchführung eines staatlichen Massenverbrechens* (Oldenbourg, München, 1996).
———, *Die Trawniki-Manner in Vernichtungslager Belzac 1941–1943* (N-S Gewaltherrschaft: Beitrage zur historischen Forschung und Juristischen, Aufarbeitung, 2006).
Poliakov, L., *The Aryan Myth: A History of Racist and Nationalist Ideas in Europe* (New American Library, New York, 1977).
Poprzeczny, J., *Odilo Globocnik: Hitler's Man in the East* (McFarland & Co., Inc., Jefferson NC, London, 2004).
Radu, I., *The Holocaust in Romania* (Dee, Chicago, 2000).
———, 'Revisionism in Post Communist Romanian Political Culture: Attempts to Rehabilitate the Perpetrators of the Holocaust', in D. Roth et al. (London, 2001).
Reitlinger, G., *The House Built on Sand: The Conflicts of Germany Policy in Russia, 1939–1945* (Macmillan, London, 1960).
Remeikis, T., *Lithuania under German Occupation: 1941–1945: despatches from US Legation in Stockholm* (Vilnius, 2005).
Rempel, G., 'Gottlob Berger and Waffen-SS Recruitment 1939–1945', in *Militärgeschichtliche Mitteilungen*, No 27 (1980), pp. 107–22.
Rein, L., 'Local Collaboration in the Execution of the "Final Solution" in Nazi-Occupied Belorussia', in *Holocaust and Genocide Studies*, Vol. 20, No 3 (2006), pp. 381–409.
Rich, D., 'Reinhard's Foor Soldiers: Soviet-era Trials as Sources', in J. Roth, E. Maxwell & M. Levy (eds), *History* (Palgrave, London, 2001).
Rich, N., *Hitler's War Aims: the Establishment of the New Order* (W.W. Norton, New York, 1974).
Ripley, T., *The Waffen-SS at War: Hitler's Praetorians 1925–1945* (Zenith Press, St Paul MN, 2004).
Rossini, Alexander B., *Hitler Strikes Poland: Blitzkrieg, ideology and Atrocity* (Lawrence, Kansas, 2003).
———, 'Nazi Anti-Jewish Policy during the Polish Campaign: the Case of the Einsatzgruppe von Woyrsch', in *German Studies Review*, Vol. 24, No 1 (2001), pp. 35–53.
Rubenstein, J. & Ilya Altman, *The Unknown Black Book: the Holocaust in the German Occupied Soviet Territories* (Indiana University Press, in association with the USHMM, Bloomington and Indianapolis, 2008).
Schmaltz, E. & S. Sinner, 'The Nazi Ethnographic Research of Georg Leibbrandt and Karl Stumpp in the Ukraine, and its North American Legacy', in *Holocaust and Genocide Studies*, Vol. 14, no 1 (2000), pp. 28–64.

Schmitt, H., *European Union: from Hitler to de Gaulle* (Van Nostrand, New York, 1969).
Schulte, T., *The German Army and Nazi Policies in Occupied Russia* (Berg, Oxford, New York, Munich).
Sebastian, M., *Journal 1935–1944*, trans. P. Camiller (Pimlico, London, 2001).
Segev, T., *One Palestine Complete: Jews and Arabs under the British Mandate*, trans. H. Watzman (Henry Holt & Co., New York, 1999).
Shepherd, B., *War in the Wild East: the German Army and Soviet Partisans* (Harvard University Press, Cambridge MA, 2004).
Smith, B., *Heinrich Himmler: A Nazi in the Making, 1900–1926* (Hoover Institution Press, Stanford CA, 1971).
Snyder, T., *The Reconstruction of Nations: Poland, Ukraine, Lithuania, Belarus, 1569–1999* (Yale University Press, New Haven, 2003).
Soucy, R., *French Fascism: the Second Wave 1933–1939* (Yale University Press, New Haven CT, 1995).
Spector, S., *The Holocaust of Volhynian Jews, 1941–1944* (Achva Press, Jerusalem, 1990).
Speer, A., *Inside the Third Reich: Memoirs* (New York, 1970).
———, *Infiltration*, trans. J. Neugroschel (Macmillan, New York, 1981).
Stein, G., *The Waffen-SS: Hitler's Elite Guard at War* (Cornell University Press, Ithaca NY, 1966).
Steinberg, J., *All or Nothing: The Axis and the Holocaust 1941–1943* (Routledge, London, 1990).
———, 'Types of Genocide? Croatians, Serbs and Jews, 1941–5', in D. Cesarani (ed.) (1994).
Stepan, J., *The Russian Fascists* (Harpers, 1978).
Stevenson, D., *1914–1918: The History of the First World War* (Penguin, London, 2004).
Stone, N., *World War One: A Short History* (Allen Lane, London, 2007).
———, *The Eastern Front: 1914–1917* (Penguin Books, London, 1998).
Sutton, K., *The Massacre of the Jews of Lithuania* (Geffen Publishing House, Jerusalem, 2008).
Sydnor, C., *Soldiers of Destruction: the SS Death's Head Division 1933–1945* (Princeton University Press, Princeton NJ, 1977).
Thal, U., *Religion, Politics and Ideology in the Third Reich; Selected Essays* (Routledge, London & New York, 2004).
Thorwald, J., *The Illusion: Soviet Soldiers in Hitler's Armies* (New York, 1975).
Tomasevich, J., *Occupation and Collaboration: War and Revolution in Yugoslavia, 1941–1945* (Stanford University Press, Stanford CA, 2001).
Tooze, A., *The Wages of Destruction: the Making and Breaking of the Nazi Economy* (2006).
Trigg, J., *Hitler's Gauls: the History of the 33rd Waffen SS Division Charlemagne* (Spellmount, Stroud, 2006).
———, *Hitler's Jihadis* (Spellmount, Stroud, 2008).
Troper, H. & M. Weinfeld, *Old Wounds: Jews, Ukrainians, and the Hunt for Nazi War Criminals in Canada* (University of North Carolina Press, Chapel Hill, 1989).
Tsimhoni, D., 'The Pogrom (Farhud) against the Jews of Baghdad in 1941: Jewish and Arab Approaches', in J. Roth et al., *Holocaust: History* (Palgrave, London, 2001).
Tutorow, N., *War Crimes, War Criminals, and War Crimes Trials: an Annotated Bibliography and Source Book* (New York, London, 1986).
Van Creveld, M., *The Balkan Clue: Hitler's Strategy 1940–1941* (Cambridge University Press, London, 1973).

Wegner, B., *The Waffen-SS: Organization, Ideology and Function*, trans. R. Webster (Basil Blackwell, Oxford, 1990).

——— (ed.), *From Peace to War: Germany, Soviet Russia and the World, 1939–1941* (1997).

Weigler, Z., *Two Polish Villages razed for Extending Help to Jews*, Vol. 1 (1957).

Weinberg, G., *A World at Arms: A Global History of World War II* (Cambridge University Press, New York, 1994).

———, *The Foreign Policy of Hitler's Germany: Starting World War II 1937–1939* (University of Chicago Press, Chicago & London, 1980).

Weiss-Wendt, A., *Murder without Hatred: Estonians and the Holocaust* (Syracuse University Press, 2010).

Weitz, E., 'In the Age of Genocide: Race and Nation under Nazi and Soviet Power', in J. Roth et al., *History* (Palgrave, London, 2001).

Westerman, E., *Hitler's Police Battalions: Enforcing Racial War in the East* (University Press of Kansas, Lawrence, Kansas, 2005).

Wette, W., *The Wehrmacht: History, Myth, Reality*, trans. D.L. Schneider (Harvard University Press, Cambridge MA, 2006).

Winckler, H.A, *Germany: the Long Road West, 1933–1990* (Oxford University Press, Oxford, New York, 2007).

———, *The Long Shadow of the Reich: Weighing up German History (The 2001 German Historical Institute Annual Lecture)* (German Historical Institute, London, 2002).

Yeomans, R., 'Of "Yugoslav Barbarians" and Croatian Gentleman Scholars: Nationalist Ideology and Racial Anthropology in Interwar Yugoslavia', in M. Turda & P.J. Weindling (eds), *Blood and Homeland: Eugenics and Racial Nationalism in Central and Southeast Europe 1900–1940* (CEU Press, Budapest, New York, 2007).

Yerger, Mark C., *Allgemeine-SS: the Commands, Units and Leaders of the General SS* (Schiffer Military History, Atglen PA, 1997).

Zuroff, E., *Operation Last Chance: One Man's Quest to Bring Nazi Criminals to Justice* (Palgrave, Macmillan, London, 2009).

Acknowledgements

The very broad scope of this book was challenging. I must thank first of all Simon Young who took time out from his busy schedule to provide invaluable advice and hands on assistance with a recalcitrant manuscript. Professor Michael Burleigh, William Niven and Nigel Jones read parts of the manuscript. Judith Lanio assisted with a mass of German language documents and texts with great efficiency. Christian Barse assisted with Danish materials. A number of historians generously responded to my many questions: Marko Attilla Hoare, Milan Hauner, Clemens Heni, David Cesarani, Andrew Ezergailis, Martin Dean, Wendy Lower, Martin Conway, Saul David, Adam Sisman, Timothy Snyder and Giles MacDonough all provided expert advice. Matthew Kott offered valuable insights into the German occupation of Norway, Latvia and the Baltic. At a critical stage, Ephraim Zuroff and Dovid Katz made valuable contributions. Detlef Siebert provided vital leads. I am grateful to Julian Hendy and Ray Brandon for generously sharing their insights and hard-won information about Ukrainian nationalism and the formation of the SS Division 'Galizien'. George Lepre and Michael Melnyk, who have written accounts of the Bosnian 'Handschar' and the Ukrainian 'Galizien' SS divisions respectively, sent unique documentary materials. From these, I have drawn my own and no doubt different conclusions.

I spent many hours in some excellent libraries, above all the British Library and the Weiner Library in London. The National Archives in Kew was another important resource. I must also thank three German libraries: the German National Library in Leipzig, the State Library and the excellent Library of the 'Topography of Terror' in Berlin. I am especially indebted to the Bundesarchiv/Zentralle Stelle in Ludwigsburg. I discussed the problems of collaboration with Ojārs Ēriks Kalniņš at the Latvian Institute and with historians at the Museum of the Occupation, Riga, the Riga Ghetto and Latvian Holocaust Museum and the Central State Archives in Kiev. Needless to say, I alone am responsible for any errors of fact and judgement. I welcome corrections, comments and further research proposals through the website listed at the end of the Acknowledgements.

I thank Pimlico Books for permission to quote from Mihail Sebastian's *Journal: 1935–1944* (2003) and the United States Holocaust Memorial Museum for permission to use extracts from Frida Michelson's remarkable memoir *I Survived Rumbuli* (1979).

Richard Johnson, Patrick Janson-Smith and Neil Blair backed the project at the beginning of a long haul. Simon Hamlet, Christine McMorris and Lindsey Smith at The History Press bravely took on a long manuscript. I must also thank some good friends: Laurence Peters, who read parts of the book and made valuable suggestions; Gerda Sousa; Karin Kaschner-Sousa; and David Robson, Sarah Dewis (and family), who provided varieties of nourishment and accommodation. My wife Diana Böhmer and our son Jacob put up with my periodic disappearances with fortitude: my love to them.

www.hitlersforeignexecutioners.com

Index